# DESIGN OF SOLID STATE POWER SUPPLIES

# DESIGN OF SOLID STATE POWER SUPPLIES

## Third Edition

## Eugene R. Hnatek

VAN NOSTRAND REINHOLD
New York

Copyright © 1989 by Van Nostrand Reinhold

Library of Congress Catalog Card Number 88-39698

ISBN 0-442-20768-9

Printed in the United States of America

Van Nostrand Reinhold
115 Fifth Avenue
New York, New York 10003

Van Nostrand Reinhold International Company Limited
11 New Fetter Lane
London EC4P 4EE, England

Van Nostrand Reinhold
480 La Trobe Street
Melbourne, Victoria 3000, Australia

Nelson Canada
1120 Birchmount Road
Scarborough, Ontario M1K 5G4, Canada

16 15 14 13 12 11 10 9 8 7 6 5 4 3 2 1

Library of Congress Cataloging-in-Publication Data

Hňatek, Eugene R.
    Design of solid state power supplies / Eugene R. Hňatek.—3rd
  ed.
        p. cm.
    Includes bibliographies and index.
    ISBN 0-442-20768-9
    1. Electronic apparatus and appliances—Power supply.
2. Semiconductors. I. Title.
TK7868.P6H46  1989
621.381'044—dc19                                        88-39698
                                                            CIP

TO
Susan, Stephen, Jeffrey and David

# Preface

Since the publication of the pioneer first edition of *Design of Solid State Power Supplies* in 1971, the importance of power supplies, especially switching supplies, for powering all electronic equipment is evidenced by such phenomena as these:

- Four major annual conferences dedicated to the design and application of power supplies;
- The number of books appearing on the subject (of which the first edition of this book was the forerunner);
- The number of professional seminars being offered on the subject.

Even though the switching power supply, or switcher, or switched-mode power supply (SMPS) has become the dominant form of providing regulated power, the title of this 3rd edition has remained intact. However, that is about all that has remained unchanged. The book has been totally restructured, refocused, and reorganized to emphasize switching power supplies in order to better serve your needs as a timely reference book. Some of the new topics detailing important developments that have been added include the following:

- High-frequency operating conditions
- The resonant converter
- The Cuk converter
- Surface-Mount Technology
- MOSFETs
- State-of-the-art ICs focused for power supply applications
- Rectifier selection criteria
- Distributed power

The book still offers a practical approach to the design of solid state power supplies. Chapter 1 provides an introduction to the subject at hand—the design of switching power supplies. Chapters 2 and 3 divide the topic of switching power supplies into two parts. Chapter 2 presents the basic switching topologies, while Chapter 3 provides design hints and design examples. Chapters 4 through 7 deal with the details of some of the constituent building blocks and/or components of the switcher: transformer and inductor design; power switch considerations; IC voltage regulators; and magnetic amplifiers, respectively. Chapter 8 covers the important topic of electromagnetic compatability, and Chapter 9, the concluding chapter, presents detailed converter design considerations and examples, proceeding from the electrical performance specification requirement and culminating in compact, lightweight, and high performance power supplies.

I have tried to make this book both practical and timely by addressing key design and useage issues. I hope that you find it useful for your needs.

I want to express my deep appreciation for the editorial assistance and untiring efforts of Nomi Vallejo in preparing the manuscript for this 3rd edition for publication.

EUGENE R. HNATEK
San Jose, California

# Contents

# DESIGN OF SOLID STATE
# POWER SUPPLIES

# 1

# Introduction

To meet the need for constant, well-regulated power outputs, modern electronic equipment and energy storage devices require transformer-coupled power conditioning equipment, specifically DC/DC converters. The varying voltage and current characteristics of the power sources, the transients and noise, load isolation needs, and varied supply voltages dictate this need.

The design of solid-state power supplies presents a multitude of challenging problems. A number of these problems arise from the normal sequence of total electronic equipment design. Overall systems are planned first; then on the basis of this preliminary system design, space in the equipment envelope is allocated and performance requirements are established for the various subsystems. Therefore, the design of a power supply frequently follows the design of the equipment it serves, forcing upon the power supply design envelopes and performance levels that are difficult to accommodate.

Other factors with which the designer must cope include, but are by no means limited to, the following:

- Excessive performance requirements.
- Incomplete information concerning load characteristics and operating times.
- Tradeoffs among incompatible parameters, such as ripple vs time or power vs small volume and low weight.
- Detrimental effects of peripheral equipment.
- Other engineering requirements, such as computer-aided design, component analysis, mechanical and thermal considerations, packaging, and electromagnetic interference.

The primary function of any power supply is to hold the voltage in its output circuit at a predetermined value over the expected range of load currents. Working against the supply are variations in load current, input voltage, and temperature. The degree to which a power supply can maintain a constant voltage in the face of these variations is the basic figure of merit. While there is a degree of interaction between these performance-degrading factors, especially between output current and temperature, it is most convenient to consider their effects separately.

The extent to which the output voltage is affected by output or load current is usually called load regulation and expressed as a percentage of the output voltage. Specifically, the general formula for load regulation is:

$$\text{Load Regulation } (\%) = \left(\frac{E_{ml} - E_{fl}}{E_o}\right) 100 \qquad (1\text{-}1)$$

where: $E_{ml}$ = output voltage with minimum rated load,
$E_{fl}$ = output voltage with maximum rated load,
$E_o$ = nominal or reference output voltage (usually $E_{fl}$ to minimize the numerical value of load regulation).

Input regulation is a measure of the effect of changes in the input voltage on the output voltage. Comparable traditional specifications are line regulation and ripple rejection. Both of these terms suggest more limited meanings than input regulation and are not as descriptive. Input regulation is given by

$$\text{Input Regulation } (\%/E_{\text{in}}) = \left(\frac{\Delta E_o}{\Delta E_{\text{in}} \times E_o}\right) 100 \qquad (1\text{-}2)$$

where: $\Delta E_o$ = change in output voltage, $E_o$, for $\Delta E_{\text{in}}$,
$\Delta E_{\text{in}}$ = change in input voltage,
$E_o$ = nominal output voltage.

The $\Delta E_o$ and $\Delta E_{\text{in}}$ terms are frequently replaced with rms AC voltages since sinusoidal input variations are common (power supply ripple). A low value of input regulation is desirable.

Since the predominant form of line power in the modern world is alternating current, a method must be used to transform this AC power to rough DC. It is the function of AC to DC transformer-rectifiers to rectify AC current, providing a pulsating DC which is then filtered to produce a constant level of voltage. Sometimes a DC/DC converter then takes this rough DC voltage and either regulates it more closely at the same level; or reduces (bucks) it or increases (boosts) it to the desired level.

The DC/DC converter function can be either linear or switching in operation. The linear power supply from AC input to DC output consists of the following functional blocks: 50 to 400 Hz transformers; rectifiers; series pass transistors and regulators in feedback loops. Isolation is obtained with transformers the sizes of which are inversely related to frequency. For example, transformers used with 400 Hz power are considerably smaller than those used for 60 Hz. Rectifiers, usually in the form of a bridge, deliver unregulated DC to the pass transistor that is in series with the load. The current through the transistor depends on line voltage changes or circuit loading. The pass transistor acts as a variable resistor and dissipates power continuously. Under the worst case conditions of high-input line voltage and low-output DC voltage, power dissipation becomes excessive and efficiency may drop to 15%. However, even under favorable conditions, efficiency rarely exceeds 35%.

The linear supply offers excellent voltage regulation and is relatively insensitive to shifts in line frequency. Initial cost is moderate, but due to low efficiency, it is expensive to operate. This type of supply is widely used in TV sets, audio amplifiers and stereos, as well as a wide range of commercial and industrial electronics equipment. Most laboratory power supplies are also of the linear variety.

The switching power supply (also called a switcher or switch-mode power supply) has become the most widely used means of providing regulated DC voltage. As such the majority of this book concerns itself with switching power supplies. When compared with the linear supply, the switcher occupies as little as one-third of the volume, with considerably less weight. The switching supply is more efficient than the linear supply (by at least a factor of two) because its pass transistors do not draw current continuously. They are switched "full-on" and "full-off," reducing wasteful power dissipation. However, this generates output noise which must be filtered. The result of the switching action of the power transistors is a square waveform that must be filtered, normally by an LC filter, to provide the desired output voltage. This output is sensed and compared to the reference voltage in the amplifier. It is used to modulate either the pulse width or frequency, which in turn sets the DC value of the output square wave.

Figure 1-1 shows the block diagram of the generic switching power supply. The heart of the supply is the high frequency inverter that chops the rectified line voltage at a high frequency (20 kHz − 1 MHz). The inverter also transforms the line voltages to the correct output level for use by logic or other electronic circuits. The remaining blocks support this basic function. The 60 Hz input line is rectified and filtered by one block and after the inverter steps this voltage down, the output is again rectified and filtered by another. The output voltage is regulated by the control circuit which closes the loop from the output to the inverter. Most control circuits generate a fixed frequency internally and utilize pulse width modulation techniques to implement the desired regulation. The on time of the square wave drive to the inverter is controlled by the output voltage. As load is removed or input voltage increases, the slight rise in output voltage will signal the control circuit to deliver shorter pulses to the inverter and conversely as the load is increased or input voltage decreases, wider pulses will be fed to the inverter.

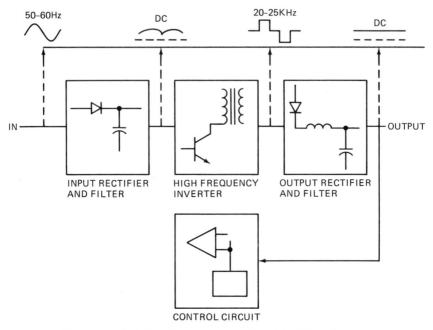

Figure 1-1   Switching Power Supply—Functional Block Diagram.

Table 1-1 compares the salient features of the linear dissipative and switching power supplies.

## DC/DC SWITCHING POWER SUPPLIES

Transistors lend themselves to efficient, reliable conversion of low-voltage direct current to high-voltage direct or alternating current, with output powers from milliwatts to kilowatts, output voltages from tens of volts to tens of kilovolts, and frequencies from tens of hertz to one megahertz.

Conversion is efficient and the power supply is small because it contains a high frequency switching power stage built with one or more semiconductor switches and operated cyclically at cut-off and saturation, which applies high voltage pulses to a transformer. Many units regulate the filtered (averaged) output by varying the duration of the pulse as load and line changes occur.

The power stage commonly takes one of several forms: a full bridge, a half bridge, push-pull, or a flyback. Coupled with this is a modulation method, of which four forms are widely used: pulse-width modulation, pulse-time modulation, frequency modulation, and pulse-frequency modulation. Modulation is applied to the time relationships—both duration and frequency—of the voltage pulses that are applied to the transformer. What power stage configuration and

## Table 1-1 Power Supply Comparison

| *Linear Dissipative Regulators* | *Switching Regulators* |
| --- | --- |
| Constant current source | Constant power source |
| Low noise (EMI) | High noise (EMI) generator requires heavy output filters |
| No active switching elements | Active switching elements |
| Fast response | Fast response |
| Lowest ripple voltage | High ripple voltage |
| Precise regulation | Good regulation |
| Low efficiency | Highest efficiency (depends on switching frequency) |
| Medium-high weight | Low weight (depends on switching frequency) |
| Series element must dissipate excess power; high thermal dissipation | Increase switching frequency, and decrease weight |
| | Low thermal dissipation because of switching mode operation of series pass transistor |
| Simple circuitry | Complex circuitry |
| Low piece parts count | High piece parts count |
| Medium output impedance | Low output impedance |
| Complex circuitry required for current limiting | |
| Should be used for applications having widely varying input voltages | Most widely used power supply |

modulation technique are used depends on the intended application and thus resultant specifications: output voltage, power required, input line range, efficiency, ripple, weight and size.

## TYPES OF SWITCHING POWER SUPPLIES

The term switcher can apply to any number of different designs in which the pass transistors are switched at frequencies from 20 KHz to about 1 MHz, adjusting the output to line and load conditions. Some of the widely used converter topologies are:

- Flyback
- Forward
- Push-pull
- Half bridge and full bridge

Both flyback and forward converters store energy in an inductor, and only while their switches are conducting. But a flyback, or ''parallel,'' converter's load is in parallel with the inductor, so its stored energy passes to the load during the off (flyback) period. On the other hand, a forward, or "series" converter's inductor is in series with the load, so its energy goes into load and inductor simultaneously during the on-period as well as into the load from the inductor during the off-period.

A push-pull converter reduces output-voltage ripple by doubling ripple-current frequency to the output filter. Moreover, it uses small transformer cores because, unlike forward or flyback converters, it excites the transformer core alternately in both directions. But push-pull converter transformers are subject to DC imbalance that can lead to core saturation.

Push-pull converter transformers can be single ended, push-pull, half wave or full wave bridge.

Table 1-2 summarizes the salient features of each of these four DC/DC converter tyes. A detailed discussion of each type can be found by referring to reference (1).

Figure 1-2 depicts several of these categories:

- Ringing-choke single-transistor type for low output powers in the range of milliwatts to watts and mainly for DC outputs.
- Self-oscillating push-pull type for medium output (AC or DC) in the range of hundreds of watts.
- Driven push-pull type for large outputs (AC or DC) from hundreds to thousands of watts.
- Driven half bridge for large outputs from hundreds to thousands of watts.

### Ringing-Choke Converters

The ringing-choke converter circuit (see Figure 1-2(a)) is a simple transformer-coupled relaxation oscillator which switches the transistor on and off periodically. When the transistor is on, power is transferred from the DC supply to the circuit transformer; when the transistor is off, the power stored in the transformer is transferred to the load at a high voltage. Ringing-choke converters are capable of delivering up to 2 W of power with efficiencies of 75%. The operation of the oscillator requires that a rectifier be included in the output circuit to ensure that output power is drawn during one-half of the cycle only. As a result, the output is not ordinarily suited to provide symmetrical alternating current and is used most often for providing a rectified DC output.

The advantages of the ringing-choke circuit are its high efficiency and its peculiar suitability to compact, economical, very-high-voltage, small-power DC supplies. Converters required to yield three to six times the rated maximum dissipation of the transistor are fairly easy to achieve in practice. It is essentially a constant-power output device so that, by a suitable selection of load resistance,

## Table 1-2   DC/DC Converter Comparison

| Type | Advantages | Disadvantages |
|---|---|---|
| Flyback | • Isolation from supply to line<br>• Output doesn't need to be filtered by choke<br>• Simple circuitry<br>• Low component count<br>• Easily provides multiple outputs<br>• Simple power transistor base drive circuitry<br>• Doesn't require output transformer<br>• Useful in high voltage multiple output applications | • High output ripple<br>• Requires large core inductors<br>• Low power output levels |
| Forward (single ended or ringing choke) | • Low output voltage ripple<br>• Easily provides multiple outputs<br>• Needs only single switching transistor<br>• Simple power transistor base drive circuitry<br>• Useful in low power applications | • Requires addition of transformer for line isolation<br>• Extra winding and components required for multiple outputs |
| Push-pull | • Low output voltage ripple<br>• Good regulation<br>• Uses small transformer component cores<br>• Provides automatic line isolation<br>• High power output capability<br>• Useful in high power high performance single output applications | • High component count<br>• Complex circuitry<br>• Complex power transistor base drive circuitry<br>• Subject to dc imbalance which can cause core saturation<br>• Extra winding and components required for multiple outputs |
| Half bridge | • Transistors see supply voltage (approx.)<br>• Primary utilization 100%<br>• Leakage inductance spikes conducted back to DC bus<br>• Low interwinding capacitance allows fast switching<br>• Capacitor C eliminates DC core flux: collector currents are balanced | • Isolated drivers required<br>• Driver circuits must have low capacitance<br>• Capacitor must pass full primary current |

high voltages can be achieved without a prohibitively high number of secondary turns in the transformer.

Fundamentally, this circuit implies a high-impedance output and poor load regulation. Other drawbacks are the limited power available from the single transistor (some 10 W out of each watt internal dissipation in the transistor), the unsymmetrical nature of the output waveform (which tends to restrict it to DC output use), and the complexity of design compared with that of symmetrical circuits.

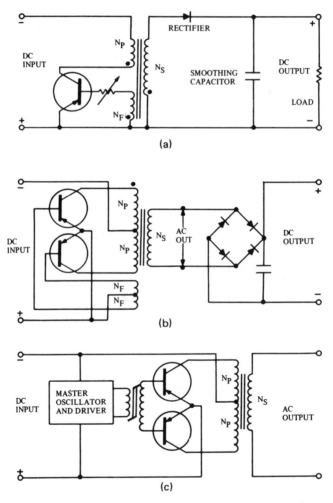

Figure 1-2 Basic types of converter power supplies. (a) Ringing-choke single-ended converter. (b) Self-oscillating push-pull converter. (c) Driven push-pull converter.

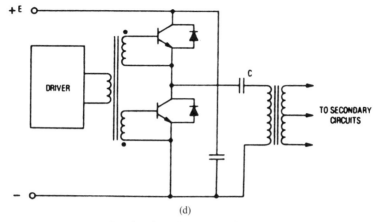

Figure 1-2 (*Continued*) (d) Half-bridge converter.

## Self-Oscillating Push-Pull Converters

For higher power outputs, push-pull circuits are more suitable. In the medium power range (from 10 to 100 W), the self-oscillating push-pull converter along the lines of Figures 1-2(b) is most commonly used.

Basically, the circuit uses two power transistors in a symmetrical square-wave oscillator. It can perhaps be compared best with a mechanical vibrator without contacts. The transistors play the part of switches that are constrained to be alternately on and off to connect the unidirectional input voltage alternately to the separate primary windings of an output transformer. This produces an alternating square-wave output across the transformer secondary.

Circuits using present-day transistors are usually of this type. However, the circuit does have certain limitations. For powers above 100 W, the self-driven converter becomes difficult to design; furthermore, the frequency of the oscillation is inherently of low stability.

## Driven Push-Pull Converters

For very high power outputs, driven converters (Figure 1-2c) are more common than self-oscillating ones, although self-oscillating units of quite high power can be found. For example, 250 W from one pair of power transistors has been achieved. Another circumstance calling for driven converters is that in which a multiphase AC output is required. But one of the most common reasons for preferring driven circuits is the frequency stability that can be achieved by the use of a separate master oscillator.

The development of this type of converter circuit is active at present and much of the conventional circuit design of magnetic and servo amplifiers is applicable to it. Outputs in the kilowatt range are being achieved. This book is primarily concerned with self-oscillating and driven push-pull (parallel) converters.

## Bridge-Type Converters

Many of the disadvantages of the push-pull circuit can be alleviated by the use of half-bridge converters as shown in Figure 1-2d. In this circuit, the transistors switch at the applied DC voltage $E$ instead of $2E$ as in the push-pull case, but there are some problems that must be overcome especially in the drive circuitry. Because the transistors see only half the voltage of the push-pull circuit, it is possible to use this circuit from rectified 230 V AC lines while still specifying only 400 volt transistors. It is also possible to design a dual input converter if the 117 volt input is used with a voltage doubler rectifier.

The leakage inductance problem associated with the push-pull circuit causes large spikes of voltage to appear at the switching transistors. The half-bridge converter does not have this problem because the energy stored in the leakage inductance of the power transformer is conducted back into the DC bus through the commutation diodes.

Unequal turn-off times in the power transistors result in core saturation with the push-pull converter. By using the half-bridge converter, the addition of a series capacitor feeding the primary of the power transformer eliminates the DC core flux problem. The unbalance of turn-off times merely produces a shift in the mean DC level of the waveform so that the transformer now sees balanced volt-seconds. A full-bridge converter doubles the number of power transistors over the half-bridge converter, thus doubling the current carrying capability (and output power), but it requires four isolated drive signals.

## SCR Power Supplies

Silicon Controlled Rectifiers (SCR's) are sometimes used in place of transistors in parallel (push-pull) converters to increase the power capability of the power supply. This book is limited to low and medium power output ranges, that is, up to 2 kW. Thus, the use of SCR's in these power supplies has been restricted to use in starting circuits, voltage regulators, and on-off magnetic firing circuits. The use of SCR's in place of transistors for power levels of less than 2 kW has been limited for several reasons:

■ SCR's are slow-speed switching devices and thus tend to dissipate more power than transistors. Typically, the turnoff time for an SCR may be several microseconds, whereas the turnoff time for a comparable transistor is tens to hundreds of nanoseconds.

- SCR's are capable of supporting much higher load currents than transistors. Therefore, as load requirements increase, SCR's become more desirable than transistors. However, as the current increases, the heat sink and thus the weight requirements increase also.
- SCR inverters are heavier than transistor inverters.
- SCR inverter circuits are complex, requiring extra circuits to turn the SCR's off. This is not so with transistor inverters, which are simpler in nature.
- SCR circuits result in lower reliability because of the need for commutation and the addition of complex logic circuitry to prevent false triggering and provide proper commutation timing.
- For power levels up to 2 kW transistor inverters offer a substantial increase in reliable operation and efficiency as compared to SCR's.

## DESIGNING THE POWER SUPPLY

The trend for present-day power supplies is toward increased complexity; Stemming primarily from overspecification of the performance requirements by systems designers and power supply users. In order to satisfy extremely tight voltage regulation, frequency regulation, low harmonic distortion, and high efficiency requirements, the power supply designer must use many extra circuits in his design. These circuits might not be necessary if a more realistic approach were taken in specifying the performance requirements; most loads do not require such tight tolerances for proper operation. In certain instances the lack of necessary information about the load characteristics also tends to make the power supply design more complex.

### Packaging the Power Supply

Another problem in designing power supplies is the unusual configurations required by systems designers. For some as yet unexplained fundamental law of electronic systems planning, the power supply must be the last item assigned. Naturally, whatever space is left over when all the other equipment is in place goes to the power supply.

This means that no two successive designs are alike. The regulators must be redesigned, different components may have to be used, and completely new heat dissipation schemes may have to be designed, all the while keeping in mind the requirements that have been highlighted in addition to the basic needs of clean design.

In many cases, the system designer has no alternative except to utilize odd-shaped areas for the power supply packages. Wasted or unused spaces in designs are luxuries which cannot be accepted. Therefore, strange configurations are forced on power supplies, posing added tradeoff decisions and increasing design and fabrication costs.

## Specifying the Requirements

The power supply designer must consider a multitude of design factors that affect the successful development and performance of his power supply. These factors are subject to compromise or "tradeoff," but all factors must be given due consideration. The following paragraphs discuss these factors and offer solutions and guidance for their resolution. The principal design factors are as follows: output waveshape, types of oscillators, output power, input voltage, ambient temperature, transistor power ratings, number of transistors used, transistor voltage ratings, transistor current ratings, operating frequency, transformer specifications, heat sink arrangements, starting circuits, transient (spike) suppression, load regulation, line regulation, extreme load conditions, frequency stability, output voltage, transistor current gains, transistor circuit configurations, output rectification, efficiency, interference suppression, and packaging (mechanical design).

Power supplies can be made less complex, more reliable, and less costly if the system designer is careful not to overspecify his performance requirements. Also, the more information supplied to the power supply designer about the load that must be regulated, the simpler and more economical his design will become.

For example, the maximum continuous values specified may actually be peak or surge requirements. Therefore, if this is the case, the duty cycle should be specified to eliminate the possibility of selecting components having safety factors ($V_{CE}$ or $I_C$) far in excess of actual circuit needs. The power supply designer already considers reasonable safety factors in his design ($V_{CE} \simeq 150\%$ and $I_C \simeq 150\%$) of the maximum continuous circuit requirements. Lack of definition of the actual loads has been one of the main causes for increasing power supply volume, weight, complexity and cost.

Since certain circuits require tight voltage regulation (0.05%) and low output impedance, it may benefit the system designer to consider that the regulator be placed directly into his own equipment while the other outputs can be contained in a separate power supply package. This type of tradeoff can only be made by a power supply designer knowledgeable about the load requirements of the system designer.

Another power supply requirement is the need for low output impedance (0.015 ohm from DC to 100 Hz). After carefully selecting his circuit gain and components to provide minimum impedance, the power supply designer finds that in actual system operation the interconnection between his power supply and the load contributes three times or greater the output impedance he was required to hold. To top it off, system operation often is not affected at all by this extra impedance in the power supply lines. This overspecification certainly has not benefited the system designer but it has increased the price, size, and weight of the power supply.

Certain power supply requirements dictate the use of nondissipative (switching regulators) which are much more efficient than the dissipative (pass) regulators. The use of a switching regulator requires the use of more components, and results in added circuit complexity, increased output impedance at higher frequency (see Figure 1-3), and RFI generation. Thus RFI filters must be added in each of the input and output leads, which tends to increase power supply weight and size.

The use of a switching regulator increases the power supply's output impedance substantially at high frequencies. This increase in output impedance must be balanced against the improved efficiency.

Methods for cooling must be considered early in the design stages of power supplies, whether forced air is used or the conduction capabilities of heat sinks are employed. As new functions are added with resulting additional circuitry and components and the continual demand remains to obtain more power from less volume, the elimination of heat becomes a more serious problem.

## Output Waveshape

Transistor efficiency is highest with a symmetrical square wave, since high voltage is then associated with low current and high current with low voltage; therefore, in both the on and off conditions, the internal dissipation of the transistor (product of current and voltage) is lowest with a square wave. The load power that can be controlled by transistors operating substantially at cutoff is many times their rated dissipation. For this reason, the square wave is used in high-power converters. There may, however, be certain applications in which the load requirements call for the less efficient sine wave.

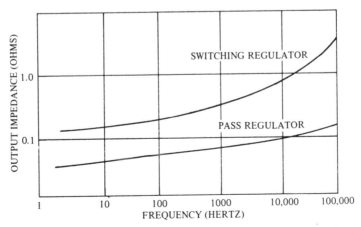

Figure 1-3  Typical switching regulator and pass regulator frequency versus impedance curves.

## Output Power

The output power required is normally prescribed. Because of the transformer coupling to the output, the output voltage is not as important as the output power. Assuming that the voltage requirements are not in the kilovolt range (where transformer winding difficulties arise), the designer first wants to know the power required.

## Input Voltage

The supply voltage for a converter is also a requirement that is prescribed for the designer.

## Ambient Temperature

Since semiconductors (diodes, rectifiers, transistors, and integrated circuits) are temperature-sensitive devices, the designer must know the ambient temperatures in which the equipment is expected to operate. In the initial stages of design, the most important effect of temperature to consider is the derating of the maximum permissible internal dissipation in the transistor that must be made to ensure that the maximum junction temperature recommended by the manufacturer is not exceeded. Each manufacturer publishes information on the derating of $P_{max}$ with elevated ambient temperatures. The highest ambient temperature to be expected in the equipment should be estimated and the $P_{max}$ of the transistors reduced accordingly. In closed "unblown" equipment, temperatures on the order of 45°C to 50°C and deratings of some 33% are fairly normal.

The engineer must, however, also take into account the expected lower limit of ambient temperature, because of a decrease in current gain and an increase in input impedance. When a converter is designed to operate at temperatures of 25°C, commonly quoted by manufacturers, it may have insufficient loop gain at lower temperatures and cease to oscillate or fail to start.

## Load Regulation

Output frequency and voltage are to a large degree independent of output loading in push-pull converters so that load regulation is basically good over a wide range of power outputs. In the case of ringing-choke circuits, which are constant-power-output devices, voltage regulation is poor, but circuits have been developed to stabilize output voltage in this case.

## Line Regulation

Variations in the output voltage and frequency regulation as a function of input voltage are small for the driven converter in which the master oscillator is usually stabilized from this point of view. In the ringing-choke and self-oscillating push-pull circuits, both the output voltage and frequency are directly proportional to the input direct voltage. Voltage regulation circuits to achieve output voltage stability against input line variations are widely used.

## Extreme Load Conditions

In most conventional self-oscillating feedback converters, heavy overload can be dangerous only if the converter is so designed that it continues to oscillate under such conditions. Usually the circuit is self-protecting, because it goes out of oscillation before the transistors exceed their maximum permissible power dissipation. The gross overload reduces oscillator loop gain to a point at which the oscillation ceases. Normally, no precautions against load short-circuit are necessary with self-oscillating circuits. In driven converters, precautions against overloads are much more necessary, as an overload does not drive the converter out of oscillation. Also, in self-oscillating circuits with separate feedback transformers, the feedback is less dependent on the load, and therefore overload protection is not automatic. In such circuits proper fusing of both input and output circuits is recommended.

One other point should be noted. Under peak drive conditions, the transistor may operate in avalanche conditions on the negative-resistance part of its characteristic curve for part of the cycle and the transistor can then act as a self-destructive dynatron oscillator. This occurs when low load resistance is combined with a low-impedance feedback circuit. Usually, a limiting resistor in the feedback circuit will prevent it.

## Efficiency

Converter efficiency is mostly affected by transistor and transformer losses. The transformer losses are relatively constant over a wide range of frequency. The transistor and diode losses, on the other hand, are sensitive to frequency.

Transistor efficiency is commonly high, since the transistor makes a good switch with a very low cutoff leakage current and saturation voltage. In most transistors currently used in power switching, the major internal losses are due to high currents across the saturation resistance when the transistor is on. Considerable losses also occur during switching between on and off, particularly if the operational frequency is so high that the switching times (which remain relatively constant for different frequencies) are more than a few percent of the total periodic cycle. Good transistors have relatively low leakage currents, and losses in the off condition of the transistor are relatively unimportant, provided high junction temperatures are not used. These leakage currents rise exponentially with junction temperature doubling approximately with every 9°C, and at high dissipation or ambient temperature levels, this may make the leakage dissipation losses comparable with the other losses. Transistor efficiency of over 90% is common.

In an ideal converter with perfect square waveshape and loss-free transistors and associated circuits, an overall converter efficiency of 100% would be obtained. In an actual converter, losses fall into six groups:

- Eddy-current, hysteresis, and copper losses in the transformers.
- Spike suppression-circuit losses.
- Feedback-resistor-network losses.
- Starting-circuit losses.
- Rectifier losses.
- Transistor losses.

The magnitude of these individual losses is of course dependent upon the precise circuit, but as a rough generalization, the following losses might be met.

| Type of Loss | Percentage Loss |
|---|---|
| Transformer | 5 |
| Spike suppression, feedback network and bias network | 7 |
| Rectifier | 5 |
| Transistor | 8 |
| Total | 25% |

This implies an overall efficiency of 75% and a transistor efficiency of 92%. The total losses and their distribution over the various sections of the circuit vary with the level of operation, ambient temperature, etc., so the figures given above must be taken as very rough, intended only to convey some idea of the various contributions to the total.

These figures are based on the assumption that the converter is working toward the maximum output. When power much less than the designed maximum is being drawn, lower efficiencies are obtained.

*Transient Response*

Behavior of the power supply when its load changes suddenly is determined by the output filter impedance and the supply's feedback response. The key here is not only how long the voltage deviates from its static value, but how far it deviates. For example, during a load application, a supply that deviates 50 mV on a 5-V output for 5 ms might be much more acceptable for a transistor-transistor logic (TTL) system load than one that deviates 750 mV for 0.5 ms. While the latter's duration is considerably shorter, its extreme voltage dip would be harmful to the system.

Transient deviation is determined primarily by passive components, i.e., the output filter capacitors. Their equivalent series resistance (ESR) and equivalent series inductance (ESL) can determine the extent of supply output changes during a load application or removal. Voltage deviation can be expressed as change in current times ESR, plus rate of change of current times ESL; or

$$\Delta V = \Delta I \times \text{ESR} + \Delta I / \Delta T \times \text{ESL}$$

However, the deviation is modified by the closed-loop response, which is a function of the output choke size (inductance) and output capacitance as well as pulse-width variation.

In addition, lead length between the supply and load introduces a resistive and inductive drop, which must be included when determining voltage deviation and load capacity.

### Switching Noise

Although common-mode noise is often ignored in switching specifications, it is, nevertheless, a problem, especially in large computer systems. This noise, which is common to both input or both output leads, is generated through stray capacitances within the supply and is difficult to eliminate after other design aspects are complete.

Some manufacturers try to eliminate common-mode noise with relatively large capacitors between line and chassis and output and chassis; however, this often makes system grounding more difficult. Suppose that the output of a switching power supply is connected to chassis ground through a capacitor to shunt common-mode-noise (Figure 1-4). Suppose also that all the loads share a common ground connection, established to eliminate or at least to minimize problems associated with "ground-loop" currents (Figure 1-5). A data line connects two of these loads, A and B, and the return for the data line is necessarily through the common ground line. However, another path is present through the chassis to the power supply and back along the voltage line to the originating circuit. If these paths have different impedances, they can cause errors in power supply load B (here assumed to be also the signal load for A). To avoid such problems, either the common-mode shunt should not be used on the power supply, or it should be made with the smallest possible capacitance.

Differential noise, like transient response, is a function of passive components: the output filter capacitors and filter chokes. The filter choke determines capacitor ripple current which, in turn, determines ripple voltage generated by the capacitor ESR, ESL, and, to a lesser degree, capacitance.

Switching noise should be defined over a wide frequency range, beginning at 47 Hz, and going up to 10 or 50 MHz, depending on system sensitivity. It is reasonable to ask for ripple and noise of less than 50 mV peak-to-peak for any

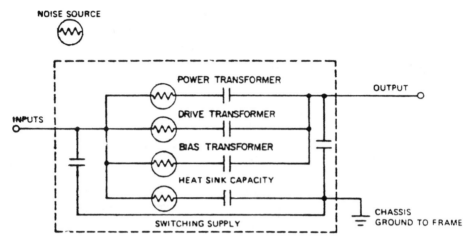

Figure 1-4 Common-mode noise sources. Shunting common-mode noise to chassis will reduce input and output common-mode noise but can lead to other problems.

voltage from 2.0 to 60.0 V. While this may seem excessive compared with a linear (non-switching) supply, which generates less than 5 mV ripple and noise, the noise actually is attenuated with distance from the supply and is greatly reduced at the load. Noise spikes usually have a very high peak-to-average ratio, and are of fairly short duration. A fixed frequency of operation helps avoid potential interaction with noise sources in the load.

## SWITCHING POWER SUPPLY TRENDS

Switching power supplies, being used in virtually every type of electronic equipment, have seen widespread acceptance and use, especially over the past five

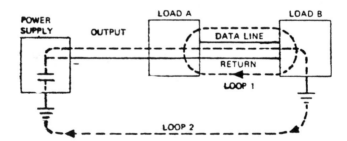

Figure 1-5 Common ground for loads. If all loads are connected to single ground point, there can be two paths for signal returns. If paths have different impedances, their existence can cause errors in one or more of the loads.

years in such diverse applications as computers, telecommunications, automobiles, and toys and games, in addition to military/aerospace systems.

Switching power supplies are using new technology to meet system demands for more power. They are moving to megahertz switching frequencies, shrinking in size, and becoming more efficient. As a result, switching supplies are changing to deliver the most power in the smallest package with the best efficiency.

To deliver the most power with the best efficiency, switching supplies are also undergoing fundamental changes in their components and circuit configurations: new topologies and design techniques such as resonant-mode supplies and current mode control, higher power density, higher operating frequency, component miniaturization, development of new and improved components, use of distributed power supplies, and a shift to low voltage requirements. Each of these areas of change is now discussed.

## Higher Power Density and Operating Frequency

Switching supply power density continues to improve. System designers want smaller supplies in order to have more room for the electronics (typically, the power supply occupies only 25% of the total system volume), and the supply must put out more power from that limited volume.

Furthermore, electronic equipment designers keep increasing the number of functions their systems perform. Even though individual circuit elements may consume less power, there are so many more elements for implementing special features that total system power demand is higher than ever.

But systems are getting smaller because of the increased use of larger scale ICs, hybrid circuits, and other smaller components. Moreover, switching supplies have decreased in size because switching frequencies have increased and components technology has improved. Early switching supplies operated from 1 kHz to 10 kHz and produced objectionable, audible noise. Later versions moved up in the 20 kHz to 50 kHz range where they ran silently. The current trend is for switching frequencies in the 100 kHz to 1 MHz region. Virtually all new switching power supply designs operate at 100 kHz or better. Higher frequencies allow smaller component values to be used, but parts also have gotten smaller over the years. Capacitors, inductors, and transformers have also benefited from improved materials and manufacturing processes. In addition, specialized ICs and sophisticated packaging techniques such as surface mount technology have helped shrink switching supplies even further.

So, as switching frequencies climb to 1 MHz and beyond, power supplies must produce an increasing amount of power while taking up less space, causing their power density to increase dramatically. For power supply technology, such a situation only intensifies the crucial need for improved efficiency.

## Component Improvement and Miniaturization

Improvement and miniaturization is accelerating across all component types, the resultant of which is a smaller and higher performance power supply than was available with earlier technologies and components.

### MOSFETs

A major component trend has been the development and increased use of power MOSFETs in switching power supply designs.

Power MOSFETs are becoming the preferred switching device over bipolar transistors because they enable power supply designers to achieve megahertz switching operation due to their majority carrier nature and can be driven directly by the new control ICs. MOSFETs simplify dealing with magnetic saturation because they do not store charge, thereby decreasing their switching transition times by an order of magnitude or more over bipolar transistors. In addition, since MOSFETs are voltage controlled and require very little current drive, simplified drive circuitry results. Moreover, the absence of destructive secondary breakdown, which is inherent with bipolar transistors, reduces or even eliminates the need for speed limiting snubber circuitry. Other advantages of MOSFETs include ease of application, space savings on PC boards due to less complicated turn-on/turn-off drive circuitry than for bipolar, and higher voltage operation. Also, they do not current hog when paralleled, and they exhibit excellent gain and switching time temperature stability. However, they suffer from high on resistance (4-5 $\Omega$ versus bipolar's 0.1 $\Omega$), sensitivity to reverse voltage spikes and large die size.

Currently, the growing use of MOSFETs is taking place primarily in the low voltage area. Bipolar transistors are still used in high voltage applications primarily because of cost.

However, new approaches and devices have also appeared in the bipolar area—especially in terms of switching speed. A variety of devices is now available that can switch amperes of current in 2 $\mu$s and less and exhibit voltage ratings of over 1000 V. Improved parametric specifications, power capability, packaging, reliability, and cost have been achieved because of the emphasis of bipolar power transistor use in power supply applications. (Details of both bipolar power transistors and MOSFETs are presented in Chapter 5.)

### Integrated Circuits

The capabilities of power IC technology are advancing. The earliest switching supplies used such ICs as comparators, op amps, and voltage regulators. Later, pulse-width modulation (PWM) control ICs became available and eliminated many discrete components. Further improved control ICs use current-mode control in a double loop technique that optimizes the performance of switching regulators. The latest generation of control ICs were developed for high switching frequencies. (Chapter 6 discusses the development of integrated circuits for use in switching power supplies.)

In the past bipolar transistors limited the IC voltage regulator (and thus power supply) operating frequency to 80 kHz or less. But with the role of the transistor switch destined to go to power MOSFETs, and with the realization of DMOS power devices on a chip with bipolar control circuitry, IC regulators have been designed for higher frequency operation (1 MHz and greater).

Built into the chips are error amplifiers, protection circuits, (overvoltage, current limiting, and thermal shutdown for example) and bipolar transistors that drive external bipolar or MOS transistors. In a few instances (as with some IC DC/DC converters such as MAXIM's MAX631-632 family, Maxim's MAX 641 and Siliconix's Si9100), a power MOSFET resides on the regulator IC itself, effectively turning the chip into a stand-alone power system. However, it should be noted that some outboard components, including diode and reactive elements are still required to complete the subsystem. The missing link in the power IC technologies is, so far, the integration of good power diodes. Good recovery characteristic *P-N* diodes with gold doping or good Schottky barrier diodes have not been produced.

### Rectifiers

Output rectifiers represent the largest single source of generated heat in a power supply. As such, higher speed lower dissipative rectifiers are in the design and development stage at many semiconductor manufacturers. However, in the range of 200 A and greater, there are no available synchronous rectifiers. Products need to be developed to address this need. (Selection considerations for rectifiers used in switching supplies is discussed in Chapter 3.)

### Capacitors

Capacitors have been designed to minimize equivalent series resistance and series inductance for more effective operation at high frequencies. Specifically, large-value multilayer ceramic capacitors for filtering and improved electrolytic capacitors with lower equivalent series resistance (ESR) and higher capacitance per-unit-volume have been developed for power supply use. Megahertz operation allows capacitance values to be lower and, therefore, capacitor size to be smaller.

### Magnetics

For switching power supplies, toroidal transformers have enjoyed widespread use in medium to high power levels, where they are cost effective. At lower power levels, ferrite E-cores are commonly used because of their low cost. However, the difference in efficiency and size between modern ferrite E-cores and toroids is not nearly as great as it is at power line frequencies, so there is litte reason to change.

The trend toward higher operating (switching) frequencies in the 100 kHz to 1 MHz range presents some problems to power transformer designers that do not exist at lower frequencies. In fact, a large concern of computer manufacturers when operating in the 300 kHz–1 MHz frequency range is the problem with magnetic material characteristics at these frequencies. Flux densities in magnetic materials become loss-limited instead of saturation-limited at high frequencies. Thus, more attention must be given to the core and core materials used. For example, many ferrite materials work well at 100 kHz, but not at 250 kHz, where the losses start to increase rapidly with frequency. As a result, new low loss ferrites have been developed for the inductors and transformers used in megahertz frequency range power supply applications. And such high frequency magnetics are substantially smaller than their low frequency counterparts.

Advances in magnetic materials are at the heart of almost all developments in magnetics. A new generation of high frequency magnetic structures is expected to result from the development of advanced materials, such as Type F ferrite, from TDK Corporation. At 25°C, Type F material has about one-third the losses of conventional H7C4 material and about one-half the losses at 100°C. Fuji Electrochemical Company, Ltd., has also announced a new material, H63A, intended for use in high frequency power converters. Similar developments are occurring at other companies: Micro Metals Corporation has introduced its new "dash 52" iron powder mix which has about half the high frequency core losses of previous materials; and Ceramic Magnetics Corporation has developed advanced ferrites for use in high frequency power transformers.

Amorphous alloys are another case where developments in basic materials are leading to new applications. Until recently, amorphous magnetic materials have been used primarily in magnetic amplifier circuits. By selecting the proper alloy and annealing process, amorphous materials can be produced with a wide range of BH loop shapes and permeabilities. Metglas Products has recently developed specific amorphous alloys for use in EMI and output filters in switching power converters. Another relatively recent use of amorphous alloys is in magnetic snubbers.

Also, at high frequencies, proximity and skin effects in magnetic windings become dominant and limit the amount of copper that can be used in a winding without increasing losses. Litz wire, foil, and printed conductors are used to reduce these losses.

High-frequency power chokes also require careful choice of the core and attention to winding technique to minimize stray fields. However, since energy is not being coupled to another circuit, the problems facing the manufacturer of high frequency transformers are not much of a factor in designing inductors.

Additionally, power supply designs dictate the need for low profile transformers and inductors that are easily PC board mountable. Currently, many magnetics manufacturers are offering so-called pancake-style line frequency transformers which are horizontally oriented and have profiles as low as an inch or less.

Planar magnetic structures are another way in which designers are attempting to shrink both the size and profile of high-frequency transformers and inductors. Planar structures have been fabricated with various degrees of complexity, depending mostly on the types of materials and the fabrication equipment available to the designer. In their simplest form, planar structures consist of a spiral pattern (the windings) etched on one or both sides of a printed circuit board with a pair of ferrite cores sandwiched on either side of the board. This type of magnetic structure was demonstrated by AT&T Bell Labs in an experimental 22 MHz 50 W DC/DC converter with a profile of only 0.5 in., including heatsink. Another development in planar structures was a 6 MHz, 40 W experimental power transformer produced at the Virginia Polytechnic Institute. This device was fabricated with windings printed on a ferrite substrate and half of a conventional low profile pot core was used to close the magnetic path. This construction technique can result in power transformers about one half as high as the PC board technique employed by AT&T.

An even further decrease in profile is possible if the pot core is eliminated. At Osaka University in Japan, a novel miniature planar inductor structure has been produced experimentally which has planar coils and magnetic layers in place of conventional cores and wires. The resulting structure consists of a series of copper-foil coils surrounded by silicon oxide insulation and embedded in a ferrite substrate. This approach promises magnetic structures less than 1 mm thick. Even though each of these planar structures is in the early stages of development and offers different cost/performance trade-offs, they serve to point out the critical link between cost-effective high-frequency magnetic structures and advances in materials science and improved fabrication techniques.

The simplest and most reliable regulated power supply is the ferroresonant supply, which consists of a ferroresonant transformer, a resonating capacitor, a rectifier, and an output filter. It contains no electronic regulation circuitry whatsoever. Instead, regulation is achieved through magnetic processes within the transformer core that result in a constant output voltage, even though the input voltage shifts over a fairly wide range. Although ferroresonant power supplies are still popular for a variety of industrial and home appliance applications such as microwave ovens, they are rarely used in mainstream electronic applications. However, high-frequency development work is currently in progress with the aim of producing a power supply combining the simplicity and low generated EMI of ferroresonant technology with the small size and high efficiency of switch-mode technology. (Chapter 4 presents some of the considerations involved in the design of high frequency transformers and inductors.)

## Surface-Mount Technology

Surface-mount technology (SMT) is being used to miniaturize control circuitry and, in some cases, entire power supplies. Because of the importance being placed on SMT, a brief tutorial is presented.

Surface-mount technology is a systems-level assembly method where the leads of a component are connected to the surface of the interconnecting substrate or PCB, resulting in smaller electronic assemblies than those built with DIP (dual-in-line package) or discrete component package through hole-insertion mounting methods. The electronic equipment manufacturers need for reduced component size, lower device weight, increased portability, improved performance, and lower manufacturing costs has challenged the traditional component-piece part packaging techniques. Furthermore, power consumption is reduced and greater integration of IC circuits is achieved by shrinking the size of circuit elements. Shrinking the size of IC elements also reduces on-chip delay times in and between the elements to improve performance and reduce costs per function.

While the DIP has been an industry standard for two decades, other IC packaging techniques have emerged to present a more efficient and better performance approach to the problems of chip density, circuit performance, and cost.

As far back as the early 1960s, the application needs of the military drove manufacturers to reduce the size of IC components and systems. The earliest attempts to satisfy these requirements resulted in packages called flat packs, which used a surface mount technology (SMT) that allowed ICs to be attached to ceramic substrates in missile guidance systems. The earliest flat packs were made in different sizes, with straight or gullwing-shaped leads on two sides. The leads were soldered to the top of a circuit board, as opposed to protruding through the board. This saved valuable space by reducing the height or thickness of the device in relationship to the board.

Shortly after the development of flat packs, small-outline integrated circuit (SOIC) packaging was developed. The SOIC package is a miniature version of a DIP, with gullwing leads. It is about one-third the size and height of a DIP and its use is limited to relatively unsophisticated small- and medium-scale integrated circuits because of the small number of leads it can accommodate.

After 1970, square packages called chip carriers were developed to satisfy the requirements of hybrid circuit manufacturers needing small packages that allowed the pretesting of LSI/VLSI devices. Chip carriers are currently being used for packaging all families of monolithic ICs. A chip carrier is basically the central portion—or die cavity—of a ceramic DIP, with contacts on all four sides. Leadless ceramic chip carriers (LCCCs) were the earliest of these devices. They have a leadless ceramic body with a metal or ceramic cap to cover the die cavity after the IC is die-attached and wire-bonded. They are one third to one sixth the size of the corresponding DIP, with a pin count ranging from 16 to 156.

In 1975, the Joint Electronic Device Engineering Council (JEDEC), a ruling body for the standardization of all package designs, began to work on standards for chip carriers. JEDEC soon approved two other standard chip carriers to go along with LCCCs: the ceramic leaded chip carrier (CLCC) and the plastic leaded chip carrier (PLCC). The CLCC has a lead frame clipped to its edges and folded under the body of the package to provide solder contacts for board mounting. The PLCC, in contrast, uses a plastic body that is injection-molded around a J-bend leadframe that provides solder contacts.

Why the change from the DIP? DIP technology is incapable of meeting the newer demands for state-of-the-art speed, greater interconnection count and density, improved thermal capability, and more efficient use of chip "real estate." For instance, a conventional DIP has 0.100 in. pin spacing, and each additional pair of pins adds a 0.100 in. increment to the length of the DIP. Thus, as the pin count of a DIP increases, its PCB space utilization efficiency drops sharply. For example, an 18-pin DIP with a 0.039 in.$^2$ chip occupies about 0.28 in.$^2$ of board space for a space efficiency rating of about 14% (0.039/0.28). But a 40-pin DIP accepting a maximum die size of 0.05 in.$^2$ requires 1.25 in.$^2$ of board space for a 4% space efficiency rating.

Similarly, increased pin count increases weight. An 18-pin DIP weighs 2.48 g, compared to the 10.92 g of a 40-pin package. This aspect of DIPs is more objectionable in aerospace applications than in commercial and industrial systems.

Critical to all applications, however, is the increase in parasitic resistance, inductance, and capacitance as pin count grows and increases conductor length from the die to the terminals toward the ends of the package. Impedances generated by DIP traces during circuit operation limit digital clock rates to about 500 MHz. Most system designs do not suffer from this speed limitation, but the faster systems may be victim to a related effect. In a 40-pin DIP, the longest trace is about six times longer than the shortest lead, and eight times longer in a 64-pin device. Resultant parasitic impedance difference between the longest and shortest leads could upset timing signals, causing a source of system malfunction difficult to pinpoint. Finally, the high aspect ratio (length/width) of conventional DIPs with over 40 pins makes the package susceptible to breaking during handling.

These problems are being resolved by surface mount technology: small-outline packages (for ICs and transistors) and passive components, chip carriers, and quad flat packs (with leads on all 4 sides). Small outline (SO) packages are used for transistors (SOTs) and low lead count ICs (SOICs). Although smaller in every dimension, these transistor and IC packages are similar in appearance to their full-sized counterparts, but they are specially designed for surface mounting.

The four-sided chip carrier (CC) presents a far more economical solution for packaging high leadcount ICs. "Chip carrier" is the generic term used for any package with interconnect provisions along the periphery of all four sides. The package can be either leaded (with either gullwing or J-leads) or leadless, and does not need to be square. It can be made of ceramic, plastic, epoxy-glass, or metal. The two most common chip carriers are (1) the ceramic leadless chip carrier for high reliability applications and (2) the plastic quad pack for commercial products. The leadless carrier may also be viewed as a flat pack without leads, which is useful since chip carriers are attached to their supporting substrates using reflow soldering, the same basic process used to attach flat packs to PCBs.

Propagation delay is more nearly constant among the leads of the chip carrier because of less variation in the length of the wire traces within the package. Like the SOIC, however, the chip carrier is more difficult to handle than the DIP. But advantages far outweigh disadvantages. The chip carrier's four-sided configuration consumes less substrate area, typically at least 70% less, than its most popular competitor, the widely used DIP. In addition, the resultant shorter lead lengths improve circuit performance. For example, a 40-pin LCC is 0.48 in.$^2$ and accepts a 0.25 in.$^2$ die for a space efficiency rating of 26% compared to 4% for a standard DIP. A 64-lead chip carrier has dimensions of .720 in. × .720 in., while a comparable 64-lead DIP is .900 in. × 3.200 in. as a direct result of a lead accommodation.

For weight sensitive applications, the comparison between chip carriers and standard DIPs is equally impressive (see Table 1-3). An 18-lead DIP weighs about eight times more than an equivalent LCC; a 40-lead DIP weighs about 16 times more than its LCC counterpart.

The chief performance advantage of CCs over DIPs derives from the shortened as well as the more uniform and more closely matched signal path lengths made possible by the package configuration. By reducing inductance and capacitance of internal conductors, higher frequencies, generally up to about 4 GHz, can be handled by the circuits packaged in CCs.

### Table 1-3   Weight Comparison

| No. Leads | Chip Carrier vs. DIP (grams) Chip Carrier | DIP |
|:---:|:---:|:---:|
| 24 | — | 1.37 |
| 28 | 0.13 | — |
| 40 | — | 4.02 |
| 44 | 0.33 | — |
| 48 | — | 9.0 |
| 52 | 0.60 | — |
| 64 | — | 12.11 |
| 68 | 1.63 | — |

### Table 1-4 Propagation Delay Comparison (ns)

| Package | Propagation Delay |
|---|---|
| 64 pin DIP | 0.3 ns |
| 64 pin FP (50 mil center) | 0.1 ns |
| 64 pin Leadless Chip Carrier (40 mil center) | 0.05 ns |
| Basic chip with wire bond | 0.018 ns |

In general, the chip carrier performs better than a comparable DIP particularly as lead count increases (Table 1-4), in large part because of differences in lead and conductor lengths. The longest trace on a 64-lead DIP is almost eight times that of the corresponding trace on a 64-lead chip carrier. The ratio of the longest to shortest trace is about 12.5:1, while the corresponding ratio for the chip carrier is 1.4:1. Long leads mean increased lead resistance and inductance. Unequal trace lengths affect system and device performance by restricting power and ground capabilities. Long side-to-side conductor traces result in significant lead-to-lead capacitance that can affect some devices.

A comparison of small outline packages and plastic leaded chip carriers with dual-in-line packages shows that DIPs have larger parasitic impedances because of longer leadframe interconnections. Because of these reduced parasitic impedances, the SO and PLCC packages provide the following:

1. Reduced self-inductance and pin-to-pin capacitance (Table 1-5). For example, the self-inductance for a 16-pin DIP varies from about 3.5 nH to 11 nH. End pins exhibit the maximum values. Since both the DIP and the SO

### Table 1-5 Self Inductance and Pin Capacitance Comparisons for DIP, SO and PLCC Packages

| Package | Self Inductance (nH) | | Pin Capacitance (pF) | |
|---|---|---|---|---|
| | Maximum | Minimum | Maximum | Minimum |
| 14 pin DIP | 10.2 | 3.2 | 1.13 | 0.38 |
| 14 pin SO | 3.8 | 2.6 | 0.54 | 0.22 |
| 20 pin DIP | 13.7 | 3.4 | 1.49 | 0.53 |
| 20 pin SO | 8.5 | 4.9 | 0.85 | 0.45 |
| 20 pin PLCC | 5.0 | 4.2 | 0.61 | 0.61 |

packages have leads on only two sides, both have similar self inductance profiles. Size differences make the SO package values proportionally smaller than those of the DIPs. The PLCC package has a flatter, more uniform, lead self-inductance distribution because it has leads on four sides. Interestingly, 14- and 16-pin SO packages have less lead inductance than 20-pin equivalent PLCCs. However, as pin count increases, 28-pin PLCC packages present less lead self-inductance than 24-pin SO packages. This occurs because the SO package leads become longer, similar to those of a DIP.

2. Reduced ground movement.
3. Improved high frequency switching.
4. 50% reduction in pin-to-pin crosstalk.
5. Better high frequency noise margin (because leads are short and have small cross-sectional areas, reducing stray capacitive coupling).
6. Less undershoot for high speed switching, providing a clearer signal.

As SMT packages begin to replace DIPs and other standard packages, they present new problems that must be resolved. For example, first it is difficult to test SMT devices after they have been installed on boards. Second, there is a lack of standards among vendors producing devices in a variety of widths, heights, and lengths. This complicates board layout. In some cases, lead finishes and the thickness of metals used in the finishes vary. Third, the coplanarity of SMT leads (the flatness of leads with respect to their board mounting plane) is often outside acceptable tolerances, making it difficult to achieve strong soldered contacts and attain a high level of manufacturing efficiency. Fourth, with the absence of leads in some leadless chip carriers, mismatches can result in the thermal coefficient of expansion (TCE) between the SMT component and the board material to which it is mounted. Fifth, the soldered leads cannot be readily inspected. And lastly, manufacturing standards and infrastructure must be developed for surface mount packages.

In summary, more and more components (ICs, transistors, diodes, resistors, and capacitors—and to some extent relays, inductors and transformers) are appearing in surface mount packages, thus making products that use surface mount technology more common. Figure 1-6 depicts the physical configurations of the aforementioned SMT packages. In switching power supplies, the low power control circuitry can be fabricated with surface mount technology, considerably reducing the board area of the supply.

## Low-Voltage Output Requirement

Historically, the lowest output voltage found in switching supplies was 5 V DC. This has become the most popular voltage because it is the standard logic voltage used with TTL and CMOS. Five volts is also used for other non-logic appli-

- LEAD TYPE: GULL-WING
- PACKAGING: TAPE AND REEL, TUBE AND BULK
- PIN COUNT: 3 (SOT 23 AND 89), 4 (SOT 143)

**b. SMALL OUTLINE TRANSISTOR (SOT)
PACKAGE FOR TRANSISTORS,
DIODES, AND LEDs**

S.O.T. CONFIGURATIONS:

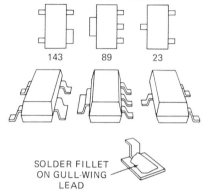

143     89     23

SOLDER FILLET
ON GULL-WING
LEAD

| LOW | MEDIUM | HIGH |
|---|---|---|
| .0004-.004 IN. | .003-.005 IN. | .004-.010 IN |
| 0.01-0.10 mm | 0.08-0.13 mm | 0.10-0.25 mm |

PROFILE CHARACTERISTICS SHOWING
BEARING CLEARANCES (NOT TO SCALE)

END
TERMINATION

SOLDER
FILLET

- LEAD TYPE: MEATALLIZED END
  TERMINALS
- PACKAGING: TAPE AND REEL, BULK
- PIN COUNT: 2

**a. RECTANGULAR PACKAGE FOR RESISTORS
CAPACITORS, INDUCTORS AND CRYSTALS**

14-PIN
GULL-WING
PACKAGE

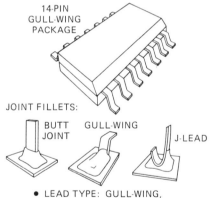

JOINT FILLETS:

BUTT   GULL-WING
JOINT         J-LEAD

- LEAD TYPE: GULL-WING,
  LESS COMMONLY,
  J-LEADS (SOJ PACKAGE)
- PACKAGING: TUBE,
  LESS OFTEN TAPE & REEL
- PIN COUNT: 8-28

**c. SMALL OUTLINE INTEGRATED CIRCUIT
(SOIC) PACKAGE FOR LOW PIN COUNT ICs**

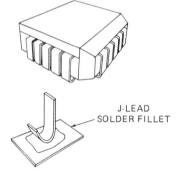

J-LEAD
SOLDER FILLET

- LEAD TYPE: J-LEADS MOST COMMON;
  GULL-WING ON MINIATURE PACKAGE
- PACKAGING: PREDOMINANTLY TUBE,
  SOME TAPE AND REEL
- PIN COUNT: 16-156

**d. PLASTIC LEADED CHIP CARRIER (PLCC)
PACKAGE FOR HIGH PIN COUNT ICs**

Figure 1-6   SMT Component Configurations.[1]

[1] Courtesy of *Electronics Purchasing*, February 1988.

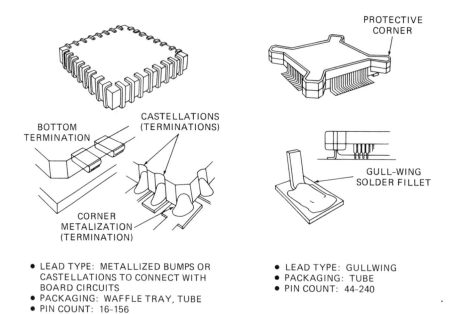

PROTECTIVE
CORNER

CASTELLATIONS
(TERMINATIONS)

BOTTOM
TERMINATION

GULL-WING
SOLDER FILLET

CORNER
METALIZATION
(TERMINATION)

- LEAD TYPE: METALLIZED BUMPS OR
  CASTELLATIONS TO CONNECT WITH
  BOARD CIRCUITS
- PACKAGING: WAFFLE TRAY, TUBE
- PIN COUNT: 16-156

- LEAD TYPE: GULLWING
- PACKAGING: TUBE
- PIN COUNT: 44-240

e. LEADLESS CHIP CARRIER (LCC) FOR
   HIGH PIN COUNT ICs

f. QUAD FLAT PACK FOR HIGH PIN COUNT ICs

Figure 1-6  (*Continued*)

cations and as a general DC system bus voltage from which other voltages such as $\pm 15$ V DC are derived using a DC/DC converter.

The advent of VLSI (Very Large Scale Integrated) circuits and ULSI (Ultra Large Scale Integrated) circuits, and the continual advances in computer technology, will affect power supply output requirements of the future: high current (hundreds of amps) at low voltage (2–3.5 V). Such voltages as 2.0, 2.5, 3.3, and 3.5 V DC are among those likely to become standard logic voltages. These lower voltages will challenge power supply designers to produce such outputs at relatively high efficiency.

## Power Supply Design Techniques

### Current Mode Control

The availability of high-speed MOSFETs has caused power supply designers to reevaluate the traditionally used switching practices. With high frequencies and light loads, for example, maintaining the narrow pulses dictated by traditional pulse-width modulation (PWM) can affect duty cycle and degrade power supply efficiency. To remedy the situation, many IC regulator and power supply designs now incorporate a form of frequency modulation called "pulse-interval modulation." This modulation method varies the interval between pulses, yet keeps the width of the pulses steady. The lighter the load, the longer the interval between pulses and the lower the duty cycle.

As for regulation, current-mode control is supplementing or even replacing voltage mode control, which is the customary technique for switching power supplies. Voltage-mode control uses one loop to sense the voltage at the power supply's output, feed back the value, and correct any deviations. Current-mode control, on the other hand, adds an inner feed-forward loop that regulates the peak conductor current and thereby compensates for deviations in input supply voltage. In this way, the switch duty cycle is controlled not by the output voltage but by the switch current. Higher frequencies, greater voltage and current handling, and lower power dissipation are among the results. Current mode control has several significant advantages, mostly due to its tight regulation: it is inherently self-limiting; given the dual feedback loops, voltage levels can be adjusted immediately—before an improper voltage appears at the output; good loop stability can simplify compensation; and finally, current mode supplies can be connected in parallel, easing the use of modular power systems.

## Resonant Converters

Pulse-width modulated (PWM) supplies are inadequate for megahertz frequencies. Their losses are excessive because they are called "hard switching circuits." With hard switching, circuit action causes a transistor switch still carrying current (and thus still turned on) to turn-off, or a transistor switch still blocking voltage (and thus still turned off) to turn-on. And the more often the switch turns on and off, the greater its power losses. Moreover, the transistor's transition time (the time it takes to turn-on or to turn-off) should be as short as possible.

Ideally, for minimal losses, a transistor switch should be turned off when the current through it is zero (known as zero current switching), and it should be turned on when the voltage across it is zero (known as zero voltage switching). With such switching conditions, the transistor's transition time becomes unimportant.

In addition, as switching frequency increases, the topology of the power supply becomes extremely complex. At high frequencies, parasitic inductances and capacitances are no longer negligible. The switching transistor, be it bipolar or MOSFET, must drive not only the winding of a transformer but also an array of reactive components largely created by parasitics. And the energy stored in the magnetics and reactive components must be dissipated during the transistor's switching interval.

Currently, the best topologies for high frequency switching supplies appear to be resonant circuits, which are also called resonant converters. Unlike PWM supplies, resonant circuits soften the transition interval so as to minimize switching losses. Thus, even though they are more complex, they are more efficient when operating at the same frequency.

In a resonant electrical circuit, energy "bounces" between being stored as a current in an inductor and being stored as a voltage across a capacitor. The voltage and the current oscillate between extremes, passing through zero periodically. Ideally, if turn-off happens during a naturally occurring zero of current or if turn-on happens during a naturally occurring zero of voltage, then switching efficiency should be maximized.

## Multiple Output Supplies

Switching supplies, with multiple outputs, are becoming more prevalent. Early varieties of switching supplies had regulation on the main output, but not on the auxiliary outputs—they were quasi-regulated (good line regulation but poor load regulation). Present day applications require not only regulated 5 V DC but also other well regulated voltages like 12, 15, 24, 36 and 48 V. Many analog circuits such as operational amplifiers, instrumentation amplifiers, A/D and D/A converters, and sample-and-hold circuits require precision $\pm 12$ V DC or $\pm 15$ V DC. Industrial control loops require regulated 24 V DC and telecommunication circuits need regulated 48 V DC. Also, many disk drives require faster transient response times than unregulated switching supplies provide.

Well regulated auxiliary outputs on a switching supply demand additional regulation circuits, which can be established by adding a linear regulator to the auxiliary output or a separate switching regulator or magnetic amplifier regulator. Another technique is to add an "adder regulator," a simple switching circuit added at the output. More switching power supplies in the future will have regulated auxiliary outputs.

## Distributed Power

The usual way to obtain multiple DC outputs is to generate them centrally, in a system's main power supply, and then route them throughout the system to the points where they are needed. But, due to the proliferation of low power ICs, the use of high switching frequencies, and the availability of MOSFETs and small magnetics and capacitors, this approach is beginning to yield to one in which only the supply's main output is distributed to PC boards throughout the system. In this case, local power supplies residing close to or right with their respective loads provide the required voltage and/or power conditions, as shown in Figure 1-7. Typically, the distributed power supplies operate from a common input voltage produced by a single bulk power supply, which also

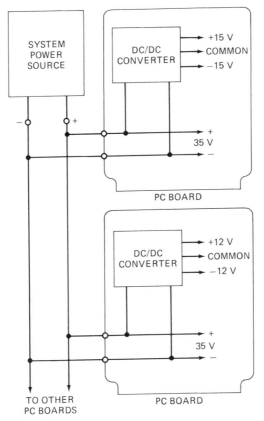

Figure 1-7  The local distribution of power as a major function of DC/DC converters that mounts on PC boards.

provides isolation, transformation, regulation, and supervision for the distributed, or intermediate, voltage. Because this intermediate voltage is usually DC, the distributed power units are generally DC/DC converters.

Distributed power offers a number of significant system advantages: it cuts power losses and thus allows the use of small power supplies; it provides enhanced system performance; it reduces noise pickup and minimizes ground loop and decoupling problems; it improves regulation; it makes fault tolerance simpler; it eases thermal management and conserves energy; and it helps provide system design flexibility by making it easier to (1) power a new board or modify an existing board, and (2) to provide a new operating voltage not found elsewhere in the system.

With distributed power, the intermediate voltage level is high enough to transport sufficient power at nominal currents using moderately sized wire gauges, so power losses are low. Also, because the short, manageable wiring does not act like an antenna, interference problems such as noise and crosstalk are minimal. Moreover, the unavoidable drop of the intermediate voltage through the wiring is immaterial because the local supplies are regulated; and the regulating circuit of each local supply buffers its load from changes on other supplies, thereby eliminating cross-regulation. Besides on-board regulation, each local supply or DC/DC converter also has short-circuit protection and current-limiting to protect itself from overload and the load it is powering from over dissipation.

Distributed power is also fault tolerant. Modular circuit blocks are powered separately, so a faulty block can be isolated and only its power shut down. For critical circuits, redundant DC/DC converters connected in parallel can maintain system operation during a failure.

Power supplies are inherent generators of heat. With distributed supplies, much of the heat is distributed, too. The convection or low-velocity circulating air that cools the electronics is often sufficient to cool the power modules as well.

Distributed supplies also provide the means to power auxiliary monitoring circuits independently of other parts of the system. Conversely, distributed supplies make it possible to conserve energy by powering down just those portions of the electronics not in use.

Distributed power even enhances system design flexibility. Since the power modules come in a wide range of voltages, designers can select the optimum parts for their system, independent of their required operating voltages. The design can be modified up until the time the system is released to production, and the correct standard power modules added to accommodate any changes. Furthermore, system upgrades, maintenance, and repair are also easy, because modules can serve multiple purposes.

# 2

## Power Supply Topologies

### INTRODUCTION

All electronic equipment needs regulated voltage(s) by which to operate. Normally, AC main supply voltage is taken, rectified into a pulsating DC voltage, and filtered to further reduce ripple by an AC/DC converter to provide rough-regulated DC. This DC voltage is then taken and regulated by a DC/DC converter (or regulator) to the required operating voltage specifications. The DC/DC converter chops the DC input voltage into a time varying voltage and then filters it to obtain the desired average or DC value. In some instances, a DC system voltage already exists, but must be converted by a DC/DC converter (as above) to the desired level or levels. It should be noted that the words converter, switcher, switching power supply, switchmode power supply, and regulator are used interchangeably in this book. This chapter focuses on the types of DC/DC converters available to satisfy user voltage requirements. First, however, a brief discussion of AC/DC conversion is presented.

### AC/DC CONVERSION[1]

The nature of power supplies has changed. Older designs needed 60 Hz transformers to isolate the DC load from the AC line, and to step the AC voltage down to a level suitable for a low voltage rectifier, capacitor input filter, and linear regulator. As shown in Figure 2-1a, switchmode power supplies do not have a 60 Hz transformer. Instead, the AC line voltage is applied directly to the rectifier, and a capacitor input filter develops a DC voltage that approaches the peak voltage of the AC line, as shown in Figure 2-1b. This DC voltage is converted to high frequency AC, passed through a small, efficient transformer, and rectified.

The transformer coupled approach of Figure 2-1a steps down the AC input voltage in proportion to the transformer primary-to-secondary winding ratio

---

[1]Portions of this section used with permission from "Switcher Architecture-key to selecting the right power supply," by L. Illingworth, *Electronic Products,* Sept. 30, 1983.

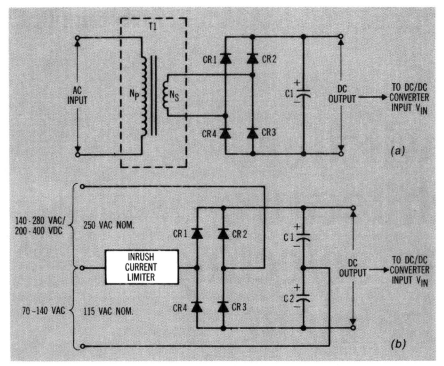

Figure 2-1  Both the transformer coupled (a) and the off-line (b) rectifier/filter derive DC/DC input from the AC power line. (Reprinted with permission from "Switching Architectures—Key to Selecting the Right Power Supply," *Electronic Products*, September, 1983, Hearst Business Communications Inc.)

$N_p/N_s$. The stepped-down voltage is rectified by diodes CR1-CR4, then filtered by capacitor C1 for presentation to a DC/DC converter.

The advantages of this approach, which is particularly suitable for switcher topologies requiring low DC operating voltages, include simplicity, isolation, and inherent filtering of high frequency electromagnetic interference. However, the one drawback, transformer size, is compounded by the need of European countries for 50 Hz operation. While the increased resistance of the longer winding at first appears to involve a simple 20% increase over the number of turns for 60 Hz operation, it actually requires an increase in wire gauge to prevent an increase of copper *IR* loss. This results in a 50 Hz transformer that is 40 to 50% larger than the 60 Hz prototype. The transformer coupled approach, therefore, is most suitable for 20 W (and under) power supplies, with higher power requirements better served by the other design techniques.

The off-line rectifier/filter of Figure 2-1b takes up less space than the transformer coupled approach and affords several operational advantages. For a nominal 115 V AC input, this circuit functions as a voltage doubler, providing a 200 to 400 V DC operating voltage over the AC input range. For a 250 V AC nominal input, the circuit is configured as a bridge rectifier, delivering the same 200 or 400 V DC for converter power over the same 2 : 1 input range. In each case, the converter switch must withstand a maximum of 400 V and tolerate a minimum of 200 V across the high and low line conditions. These limits are generally transient in nature, but either line condition can last for an uncomfortably long period. Input surge protection limits the in-rush current to turn-on.

Both of the AC/DC conversion circuits of Figure 2-1 use a full-wave rectifier and a filter capacitor. The capacitor discharges some of its stored energy into the load resistance while the rectifier is not conducting. When the instantaneous AC voltage exceeds the voltage on the capacitor, the rectifiers conduct and recharge the capacitor. The AC line current flows in pulses, once each half cycle, as shown in Figure 2-2.

In general, higher peak currents and shorter rectifier on-times provide a more economical design. Unfortunately, the pulsating current has a higher rms value than a sine wave current for the same power. As the current pulse peak increases and its duration decreases, the rms current and VA input increase. Although the voltage and power remain constant, the load's power factor decreases.

The pulsing AC current drawn by the capacitor input filter is equivalent to a combination of a line frequency component and a line frequency harmonic component. Only the fundamental component draws power from the AC line. The harmonic component contributes to the total rms current value, but not to the power consumption.

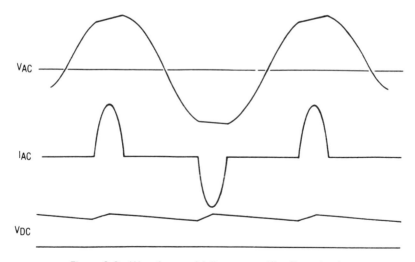

Figure 2-2   Waveforms of full-wave rectifier-filter circuit.

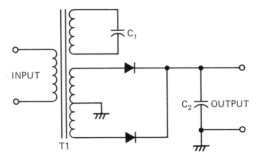

Figure 2-3 Schematic diagram of a ferroresonant supply.

For the same input AC voltage and load power, decreasing current on time has these effects: (1) the line frequency component remains the same; (2) the harmonic component increases; (3) the rms value of the current increases; (4) the total input VA increases; and (5) the power factor, watts per volt-ampere, decreases.

Another AC/DC converter (which is a variation of the transformer-coupled approach) is the ferroresonant converter. A simplified schematic diagram of the ferroresonant converter with surge protected input and overvoltage and short-circuit protected output and using as few as five parts is shown in Figure 2-3. The key components in the circuit are a ferroresonant transformer ($T1$) and its associated capacitor $C1$.

More than one magnetic flux path exists in the transformer core, allowing the secondary circuit to saturate while the primary circuit operates in the linear region of the core material characteristic. As a result, the output (load) circuit is both voltage and current limited while the primary (line) circuit, not completely coupled to the load, can operate normally. The effect is similar to the operation of a series resistor-zener diode DC shunt regulator, except that the excess flux is not dissipated as heat.

The isolating and stabilizing action of the $T1$–$C1$ combination yields the following benefits:

■ The secondary AC waveform is a quasi-square wave which is much easier to filter than the usual sine wave obtained from a linear transformer or rectifier.
■ The input power factor is high (0.90 to 0.95), compared to the values typical of other methods (0.60 to 0.70).
■ The input and output are free of wideband noise.

Since the magnetic circuits that yield these advantages must operate at the power line frequency, size (watts per cubic inch) and weight (watts per pound) of the ferroresonant converter are similar to those of the linear regulated power supply. Similarly, the "tuned" secondary circuit ($C1$ and the winding reactance of the transformer) must be properly balanced at the operating frequency. This makes the regulation sensitive to changes in power line frequency. Typically, a 1 Hz shift in frequency (say 60 to 59 Hz) will cause the output voltage to shift by about 2.5 percent. This shift can be a limitation where the line frequency may vary. Further, the absence of high gain feedback in ferroresonant supplies makes their load regulation and ripple relatively poor (several percent compared with less than one percent of the output voltage in electronically-stabilized linear and switching supplies).

Standard ferroresonant products have output power ratings from 60 to 20,000 W, possibly giving ferroresonant units the broadest spectrum of any regulated power supply type. Because their inherent simplicity yields exceptionally high field reliability, ferroresonant AC/DC converters are widely used in the telecommunications industry, making 48 V DC the most popular standard output.

## GENERAL CLASSIFICATIONS OF DC/DC CONVERTERS

DC/DC converters can be classified as being either linear (dissipative) or switching. Figure 2-4 shows the block diagrams of each of these converters.

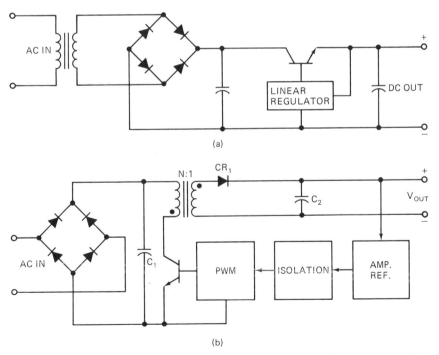

Figure 2-4   The linear supply (a) requires bulkier components, while the switcher (b) is more complex.

Switching power supplies can be further classified as being either self-oscillating or driven. The design issues of these types are presented in the section on push-pull converters. Today, the overwhelming majority of switching converters are driven rather than self-oscillating.

The linear converter requires a bulky power transformer since it must operate off the line frequency of 50 or 60 Hz. Both input and output filter capacitors must be large because they must have large capacitance to filter the 100 or 120 Hz ripple that results from full wave rectification of the line voltage. The linear converter employs a series-pass power transistor which dissipates a significant amount of power and, therefore, must have a large heat sink.

In the linear converter, the line voltage is converted to a lower voltage by a power transformer, then rectified and filtered. The switching converter, also called an off-line switcher, rectifies and filters the line voltage, then converts the resulting DC voltage into a high frequency square wave which drives the power transformer. The use of a high frequency, generally in the range of 20 kHz to 200 kHz (and now to 1 MHz) allows the use of a smaller power transformer, smaller input filter capacitor $C1$, and smaller output capacitor, since the higher the frequency the lower the value of capacitance required to filter it. Even the input filter capacitor, $C1$, of the switcher is smaller than the input filter capacitor of the linear supply. The input capacitor in the switcher is charged to a much higher voltage, the rectified line voltage, rather than a stepped-down voltage. Since energy is proportional to CV, this capacitor can be smaller than the one in the linear supply and still store the same energy.

The largest loss is in the linear regulator circuit. The linear regulator has transformer core and winding losses, losses in the rectifying diodes, and a large loss in the series-pass power transistor, which continuously conducts the power supply's output current. The transistor always has a voltage drop across it, the minimum of which is determined by the minimum line voltage at which the supply is designed to operate. This voltage drop is higher for nominal line voltage and still higher at high line voltage. The voltage drop multiplied by the current through the transistor results in a power loss, which is significant.

The switching converter is more efficient. Most linear converters operate with typical efficiencies of only 30%. However, some linear power supplies operate with up to 50% efficiency, but these are areas where line variations are minimal.

A typical switching supply also has transformer and diode losses, though the transformer losses are somewhat smaller than those of the linear supply. The losses in the regulation circuitry are extremely small, giving the switching supply its high efficiency.

In the switcher, regulation is accomplished by pulse width modulation of the switching transistor. Since this transistor is either on or off, its losses are very small, though there are additional small losses in the feedback circuitry ahead of the switching transistor. As a result, switcher efficiencies run from 70–80%

but occasionally fall to 60–65% when linear post regulators are used for the auxiliary outputs.

One can gain an appreciation of the significant efficiency improvement of the switching converter by looking at a specific example. A medium power linear converter with a 100 W output rating typically requires 200 W input due to its 50% efficiency. 100 W are dissipated within the supply and must be disposed of in some way, generally by convection cooling. A 100 W switcher, on the other hand, typically has 75% efficiency and requires only 133 W of input power. Now, only 33 W of internal dissipation must be disposed of. From a thermal point of view, this means that a switcher can be one-third the size of a linear supply with the same output power.

Table 2-1 summarizes some of the key features of linear and switching power supplies.

The switcher is more complex than the linear supply because of its complex feedback loop and pulse width modulation. Yet its power density, its output power divided by its volume, is significantly higher. In addition, the smaller magnetic and capacitive components in switchers result in power densities anywhere from two to ten times as great as those in linears. In most applications, the switcher is three to five times more compact than the linear supply.

Linear converters have the ability to achieve low leakage currents from primary to secondary and primary to chassis. This results from the use of either a split-bobbin transformer winding, which physically separates the primary and secondary, or from the use of a Faraday shield between primary and secondary. In either case, the capacitive coupling from primary to secondary can be made extremely low so that AC leakage currents can be kept very low. This coupling is much more difficult with switching supplies since, at high frequencies, the

### Table 2-1  Comparison of Power Supply Characteristics

| Switcher | Linear |
|---|---|
| Complex design | Simple design |
| High efficiency | Moderate efficiency |
| High power density | Low power density |
| High power/weight | Low power/weight |
| High leakage | Low leakage |
| High EMI | Low EMI |
| Requires input line filter | No input line filter |
| Loose line and load regulation | Tight line and load regulation |
| High output ripple | Very low output ripple |
| Slow transient recovery | Fast transient recovery |

primary and secondary windings must be closely coupled, so capacitive feed-through of AC current is much higher.

Linear converters have better regulation, particularly load regulation. The switcher usually has good regulation of 0.1 to 1% on its main output, which is part of the regulation loop. The auxiliary outputs can have somewhat worse load regulation, up to about 5%. In many cases, a linear regulator or a mag-amp regulator is added to one or more auxiliary outputs for improved regulation.

Output ripple and noise is generally higher for switchers. While additional output filtering can reduce ripple and noise, it adds lag time in control loop response.

## The Linear or Series Dissipative Regulator

The series regulator is well suited for medium current applications with nominal voltage differential requirements. Modulation of a series pass control element to maintain a well-regulated prescribed output voltage is a straightforward design technique. Safe-operating-area protection circuits such as overvoltage, fold-back current limiting, and short-circuit protection are easily adapted. The primary drawback of the series regulator is its consumption of power. The series regulator (Figure 2-5) will consume power according to the load, proportional to the differential-voltage to output-voltage ratio. This becomes considerable with increasing load or differential voltage requirements. This power represents a loss to the system, and limits the amount of power deliverable to the load since the power dissipation of the series regulator is limited.

$$P_{reg} = E_{in}I_{in} - E_{out}I_{load} \tag{2-1}$$

$$I_{in} = I_{reg} + I_{load} \tag{2-2}$$

Since $I_{load} \gg I_{reg}$

$$I_{in} \approx I_{load}$$

$$\therefore P_{reg} \approx I_{load}(E_{in} - E_{out}) \tag{2-3}$$

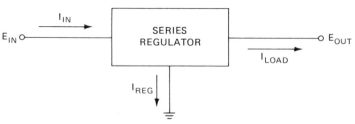

Figure 2-5  Series regulator.

The efficiency of a series regulator is given by

$$\text{Max. efficiency} = \eta = \frac{P_{\text{out}}}{P_{\text{in}}} = \frac{E_{\text{out}} I}{E_{\text{in}} I} = \frac{E_{\text{out}}}{E_{\text{in}}}$$

where: $E_{\text{in}} I$ = input power to the regulator,
$I$ = load current,
$E_{\text{out}} I$ = output power of the regulator.

Thus, for the greatest efficiency, the output voltage should be as close as possible to the input voltage. However, the regulator's output can never be greater than the least value of the input. The power lost in the regulator $(E_{\text{in}} - E_{\text{out}})I$ is all dissipated in the series control element which must then be adequately chosen and heat-sinked. The voltage drop across the control transistor $Q_1$ in Figure 2-7 was found to be 1.5 V. This voltage drop has been verified in many designs as being typical; thus, the series dissipative regulator is not very efficient.

The series regulator provides good dynamic regulation and stability, low output impedance over a wide frequency range, and low output ripple. The low output impedance of the series regulator results in a low EMI susceptibility threshold (a factor of 10 to 20 better than a switching regulator). A curve of the output impedance of a typical series regulator as a function of frequency is shown in Figure 2-6. This curve was obtained from the regulator of Figure 2-9.

Since the basic series regulator needs no magnetics, it can be made quite small. The series regulator can operate over wide input voltage, output load, and temperature ranges. The output ripple of a series regulator contains the frequencies of the oscillator and its harmonics.

## Examples of Series Dissipative Regulator Designs

Figures 2-7 and 2-9 depict the schematic diagrams of two linear regulators. The design criteria for the regulator of Figure 2-7 is as follows:

| | |
|---|---|
| Input voltage: | 22 VDC to 29.5 VDC |
| Output voltage: | $-28.3$ VDC $\pm$ 1% |
| Efficiency: | 85% minimum |
| Load: | 20 W |
| Temperature extremes: | $-30°$F to $+160°$F |
| Output ripple: | 100 mV peak-to-peak |
| Output impedance: | 0.1 ohm DC |
| | 0.4 ohm maximum 4 Hz to 200 kHz |

The design criteria for the series regulator in Figure 2-9 are:

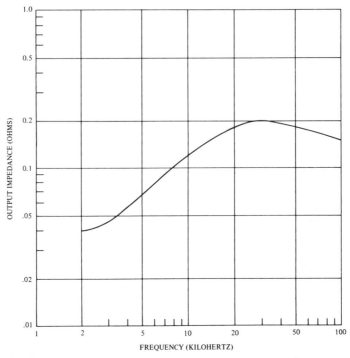

Figure 2-6 Typical output impedance characteristics for a series regulated power supply.

| | |
|---|---|
| Input voltage: | 22.0 VDC to 29.5 VDC |
| Output voltage: | 28.3 VDC ± 2% |
| Output load current: | 20 mA maximum |
| Output ripple: | 50 mV peak-to-peak |
| Temperature extremes: | −30°F to +160°F |
| Output impedance: | 50 ohms maximum DC to 150 kHz |

Figure 2-8 depicts the output voltage variation of the regulator in Figure 2-7 with variations of the input voltages of the converter as based on tests conducted for three load conditions: no load, half load, and full load. It should be noted that Figure 2-9 contains an overload protection circuit composed of components $R_{29}$ and $Q_{20}$. Figure 2-10 shows the effects of a series regulator on the output ripple characteristics of a DC/DC converter: it decreased the ripple of the regulated output by more than a magnitude of two.

A brief description of the circuit operation of the regulator shown in Figure 2-9 is as follows.

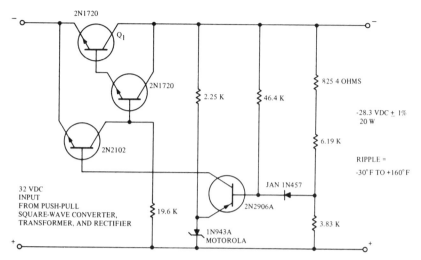

Figure 2-7  DC/DC power converter negative output series voltage regulator.

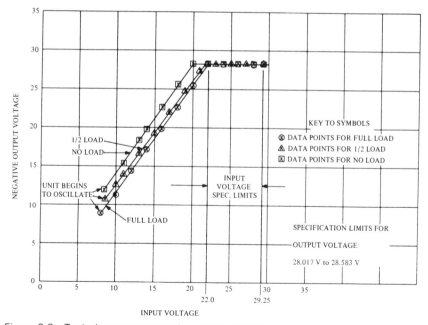

Figure 2-8  Typical converter negative output voltage versus input voltage for various loads.

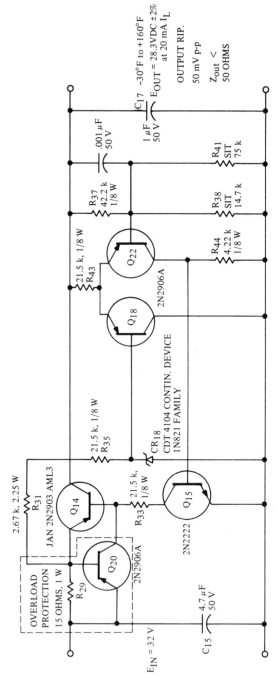

Figure 2-9   20 mA load series voltage regulator. $E_{in}$ produces a voltage across capacitor $C_{17}$ developed by current flowing through resistor $R_{31}$. Voltage across $C_{17}$ is applied to diff. amp. ($Q_{18}$, $Q_{22}$). When diff. amp. begins to operate it allows $Q_{15}$ (driver transistor) to conduct, thereby providing drive power to negative transistor $Q_{14}$.

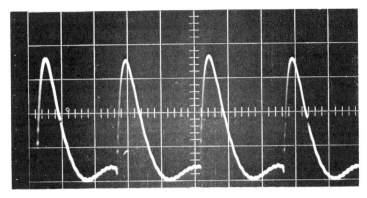

Ripple Waveform Without Series Regulator
Vertical:     .005 V/cm
Horizontal: 500 μsec/cm

Input:        20 V
Input to +28.3 V Reg.
Full Load

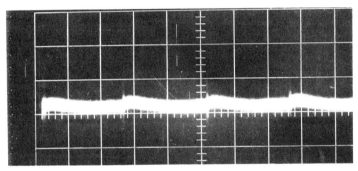

Ripple Waveform With Series Regulator
Vertical:     .005 V/cm
Horizontal: 500 μsec/cm

Input:        20 V
Output:      28.3 V
Full Load

Figure 2-10   DC/DC converter output ripple waveforms.

The series regulator consists of regulator transistor $Q_{14}$, driver transistor $Q_{15}$, differential amplifier transistors $Q_{18}$ and $Q_{22}$, and reference Zener diode $CR_{18}$. Transistor $Q_{20}$ and resistor $R_{29}$ form the overload protection portion of the circuit.

The AC voltage appearing at the secondary of the oscillator power transformer is rectified and applied to the input of the series regulator. This voltage produces a voltage across capacitor $C_{17}$ which was developed by the current flowing through resistor $R_{31}$. The voltage across $C_{17}$ is applied to the differential amplifier. When the differential amplifier begins to operate, it allows driver transistor $Q_{15}$ to conduct, thereby providing drive power to regulator transistor $Q_{14}$.

As the output load current increases, the voltage across resistor $R_{29}$ increases until transistor $Q_{20}$ is turned on. This decreases the drive power to regulator transistor $Q_{14}$, thereby limiting the output current to a constant value. When the output voltage drops to a value less than 5 V, the differential amplifier is unable to supply sufficient drive power to drive transistor $Q_{15}$ to allow it to conduct. With no drive power supplied to its base, regulator transistor $Q_{14}$ stops conducting.

## Switching Power Supplies

Since their introduction over two decades ago, switching power supplies have offered improved efficiency, weight, and size when compared with linear supplies, and they have now found their way into most new electronic equipment designs.

Switching supplies were first developed in the early 1960s for military and aerospace applications where high efficiency and high power density were paramount considerations. While the solid state technology of that decade made switching supplies feasible, the supplies were expensive. In recent years, however, switching technology has improved, and the cost has dropped dramatically due to better quality components at lower costs.

A switching power supply is a high frequency power conversion circuit whose primary advantages are high efficiency, small size, and light weight as compared with the linear supply. But the switching supply is a more complex design, with a higher piece-parts count, and it can not meet some of the performance capabilities of linear supplies. The switching supply also generates a considerable amount of electrical noise. Early switching supplies operated at relatively low switching frequencies, about 4 to 10 kHz. At these frequencies, the supplies produced an annoying audible sound. More recent designs have been in the 20 to 50 kHz range—above audibility. Moreover, many new designs are routinely in the 100 to 500 kHz range, with some going to 1 MHz primarily due to the introduction of MOS power transistors, along with advances in other components.

Higher switching frequencies allow the use of smaller magnetic and capacitive components, so power supplies can be shrunk further. However, size does not decrease in direct proportion to frequency. Thermal considerations also play an increasing role as power densities increase.

Switchers have become cost effective because designers have been able to simplify the control circuits, and they have found even lower cost alternatives in the passive component area. In addition, high frequency MOSFETs have become available. Because of this, and the demanding requirements of microprocessor based systems—especially personal computers and minicomputers—switching power supply usage has increased rapidly.

## The Basic Switching Power Supply

The switching regulator uses the same series transistor as the simple series or linear regulator, but turns it either full on or full off depending on how the output voltage compares with a reference signal.

The transistor switches between cutoff and saturation at high rates, with pulse-width modulation of the resulting square wave. The pulse-width modulated wave is then filtered through a low-pass filter, and the resulting direct current is controlled in amplitude. The efficiency of such a regulator can be as high as 95%.

Basically, the switching regulator consists of a power source $E_{DC}$, available duty cycle switch $S_1$, an LC filter assumed to provide a constant output $E_{out}$, and the load $R_L$ as shown in Figure 2-11(a). Figure 2-11 shows the typical waveforms of the switching regulator. The input filter in this regulator design serves three purposes: (1) to smooth out spikes and high-frequency transients with large peak values and small volt-second integrals; (2) to eliminate input ripple having frequency components at or near the modulating frequency of the switching transistor which would produce low-frequency components by heterodyning; and (3) to attenuate AC components produced by transistor switching. The switching transistor chops the DC output of the input filter in such a way as to deliver constant volt-second energy pulses to the integrator.

The integrator portion of the circuit serves to smooth the pulsating DC. When the transistor is off, the diode conducts, permitting continuous energy flow to

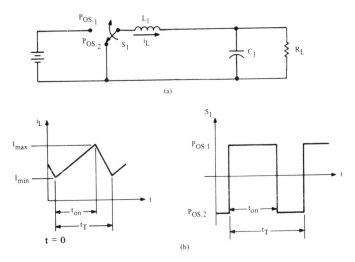

Figure 2-11   Switching regulator.

the load from the integrator circuit. The inductor, along with its capacitor, becomes a means for storing sufficient electrical energy during the transistor on period for delivery to the load during the off period to provide a regulated output voltage.

As indicated in Figure 2-11(a), switch $S_1$ is periodically maintained in position 1 for a time $t_{on}$ and in position 2 for a time $t_T - t_{on}$. When $S_1$ is in position 1, a constant voltage $E_L = E_{DC} - E_{out}$ is impressed across $L_1$, and when $S_1$ is in position 2, a constant voltage $E_L = E_{out}$ is impressed across $L_1$. Since

$$E_L = L_1 \frac{di_L}{dt},$$

$$iL(t) - i_L(t = 0) = \frac{E_L}{L_1} \times t \quad (E_L \text{ is constant})$$

Therefore,

$$i_1 = \frac{E_{DC} - E_{out}}{L_1} t_{on}$$

$$i_2 = \frac{-E_{out}}{L_1} (t_T - t_{on})$$

(2-4)

where $i_1$ and $i_2$ are the increases in current through $L_1$ when $S_1$ is in positions 1 and 2, respectively. For the regulator operation to be steady-state, $i_1$ must equal $-i_2$, and this leads to the requirement that

$$\frac{t_{on}}{t_T} = \frac{E_{out}}{E_{DC}}$$

(2-5)

Ideally speaking, equation (2-5) shows that the duty cycle of switch 1 is determined only by the input voltage $E_{DC}$ and the desired output voltage $E_{out}$, and is independent of the load current $i_{out}$. Since equation (2-4) determines only the changes in current through $L_1$, the actual currents must be determined by equating the power flowing from the input source $E_{DC}$ to the output power $E_{out}/R_L$. Thus $E_{out}^2/R_L = E_{DC} \times I_{AVE_2} \times t_{ON}/t_T$ where $I_{AVE_2}$ is the average current through $L_1$.

Therefore $I_{AVE_2} = E_{out}^2/R_L$. Using equations (2-4) and (2-5)

$$I_{min} = \frac{E_{out}}{R_L} - \frac{t_T}{2L_1} \times E_{out}\left(1 - \frac{E_{out}}{E_{DC}}\right)$$

$$I_{max} = \frac{E_{out}}{R_L} + \frac{t_T E_{out}}{2L_1} + \left(1 - \frac{E_{out}}{E_{DC}}\right)$$

(2-6)

where $I_{min}$ and $I_{max}$ are the minimum and maximum values of the current $i_L$ through the inductor. As shown in Figure 2-11b, the current through $L_1$ is a sawtooth waveform. The ideal switching regulator has 100% efficiency. In actual practice, the efficiency of the regulator is limited by losses in the switch and inductor, but these losses are in principle considerably less than those found in most other regulating approaches.

The switch is voltage-actuated and is on when the control voltage is below a threshold and off when it is above the threshold. A pulse network provides an exponentially decaying waveform which is amplitude and frequency dependent upon the input voltage. The exponential waveform is superimposed upon a DC voltage output of a regulator stage. The on time is thus dependent upon the amplitude of the pulse and the output of the regulator. A variation in the amplitude of the pulse due to a variation in input, in essence, provides preregulation or line regulation. A variation in regulator output voltage provides load regulation and compensates for losses in the voltage-activated switch and filter. Ideal waveforms are shown in Figure 2-12. The regulator DC voltage output is designated as $E_{reg}$, the fixed switch threshold is designated as $E_{sw}$, and the pulse amplitude is designated as $E_o$. The fixed time constant of the decaying waveform is designated as $\tau$. Figure 2-12a shows the switch input for low input voltage. The dotted lines indicate a change in regulator voltage to compensate for losses due to an increased load. Figure 2-12b is the corresponding switch operation. Figure 2-12(c) shows the switch input for a high input voltage. Both the pulse amplitude $E_{out}$ and the frequency $1/t_\tau$ have increased, and as shown in Figure 2-12d a decreased $t_{on}$ ratio results. This is in accord with the principle indicated by equation (2-5). Again, in Figures 2-12(c) and 2-12(d) the dotted lines indicate a regulator change due to an increased load.

Even in the ideal case the preregulator does not provide 100% line regulation, and some changes in the regulator voltage are required (thus indicating some change in the output). The variation of pulse frequency with pulse voltage is given by $1/t_\tau = f = (f_{nom}/E_{nom})E_{out}$, where $f_{nom}$ is the frequency at the nominal pulse amplitude $E_{nom}$. The pulse amplitude $E_{out}$ is directly proportional to the input voltage. Hence both frequency and pulse amplitude increase directly with frequency. Since

$$E = E_{reg} + E_{out}e^{-t/\tau}, \tag{2-7}$$

where $E$ is the switch control voltage, the switch changes position when

$$E = E_{sw} = E_{reg} + E_{out}e^{-(t_\tau - t_{on})/\tau}$$

Hence, $t_{on} = t_T - \tau \ln(E_{out}/E_{sw} - E_{reg})$, or

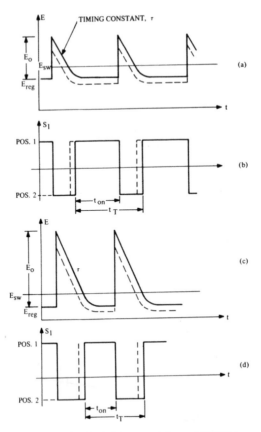

Figure 2-12   Switching regulator waveforms.

$$\frac{t_{on}}{t_T} = 1 - \frac{f_{nom}}{E_o \text{ nom}} + E_o \tau \ln \frac{E_{out}}{E_{sw} - E_{reg}} \tag{2-8}$$

where the relationship between $t_T$ and $E_{out}$ has been used. Note that this deviates considerably from equation (2-5) which can be written as

$$\frac{t_{on}}{t_T} = \frac{E_{out}}{E_{DC \, nom}} \times \frac{E_{nom}}{E_{out}} \tag{2-9}$$

where $E_{DC \, nom}$ is the nominal input voltage corresponding to the nominal pulse amplitude $E_{nom}$. However, by proper choice of parameters, equations (2-8) and (2-9) can be made to approximately coincide. As indicated previously, the regulator stage can compensate for any differences. The decrease in equation (2-8) below equation (2-9) is advantageous because switch losses decrease at high $E_{DC}$; hence, less change in $E_{reg}$ is required for line regulation.

This brings us to $L_1$ and $C_1$.

The inductance required of $L_1$ depends on the choice of $C$. Ordinary reactance charts do not work at a 20 kHz switching frequency for most aluminum and tantalum capacitors. If special low ESR capacitors are to be used, the manufacturer will supply total impedance/ESR data in his specification. If not, data must be gained by testing. Generally the most effective capacitors are the dry slug (CSR style) tantalum type. Wet tantalum (CL67 style) have approximately twice the impedance of the dry type but suffer reliability problems. For aluminum capacitors, figures vary widely depending on manufacturer and construction. The lowest ESR in aluminum construction is in what is known as the "stacked" foil type; with all other styles it is typically an order of magnitude higher. Representative values would be 0.5 milliohm for a stacked foil device and 5 to 20 milliohms for other styles. Due to this limitation of capacitor impedance, $L_1$ must be greater than critical in most cases. Typically, if the capacitor impedance is one milliohm and the maximum p-p ripple is 10 millivolts, $\Delta I$ for $L_1$ would be 10 amperes. If $E_{in}$ equals 20 volts (5 volt output) and $f = 20$ kHz, $L$ minimum would be calculated as follows:

$$L = (E_{in} - E_{out}) \cdot \frac{\Delta T}{\Delta I}$$

$$= \frac{(20 - 5) \cdot 25 \cdot 10^{-6}}{10}$$

$$= 42.5 \ \mu H \tag{2-10}$$

If the supply is a 100 ampere supply, $\Delta I$ represents a small increment of the flux since most of it is obtained from the average DC. The magnetic circuit is then determined by the large gap required, and core losses are not as significant as might be expected assuming a small gauge grain-oriented silicon is used. Ferrite and toroid inductors could be used if $\Delta I$ represents a large portion of $I_{out}$.

It is important that the core material used for the inductor have a "soft" saturation characteristic. Cores that saturate abruptly produce excessive peak currents in the switch transistor if the output current becomes high enough to run the core close to saturation.

Powdered molybdenum-permalloy cores are recommended. They exhibit a gradual reduction in permeability with excessive current so that output currents above the design value cause only a gradual increase in switching frequency.

Another precaution, frequently overlooked in the design of switching circuits, is the proper ripple rating of the filter capacitors. High-frequency ripple can cause capacitors to fail—an especially important consideration for capacitors used on the unregulated input, because the ripple current through them can be higher than the DC load current. The situation is eased somewhat for the filter capacitor on the output of the regulator, where ripple current is only a fraction of the load current. Nonetheless, proper design usually requires that the voltage rating of this capacitor be higher than that dictated by the DC voltage across it for reliable operation.

$L_1 C_1$ have an effect on the response time of the regulator, since the loop gain must be less than zero when $X_L = X_C$ or stability problems will result. Typically, when an $LC$ combination is chosen, the frequency of this corner is between 100 and 500 Hz. Response times of 0.5 to 5 milliseconds are representative for these values.

The pulse-width modulated switching regulator circuit shown in Figure 2-13 provides more detail than the circuit of Figure 2-11. In this circuit a DC voltage $E_{in}$ is developed across $C_1$ and chopped by transistor $Q_1$. Filter $L_1 C_2$ restores the chopped DC to a constant voltage, $E_{out}$. Diode $CR_1$ provides a path for the current through $L_1$ when $Q_1$ is off. Regulation is achieved by varying the on time of $Q_1$, with respect to the off time. The output voltage $E_2$ is equal to the average DC voltage.

$$E_{out} = \frac{E_{in} \cdot t_{on}}{t_{on} + t_{off}} \qquad (2\text{-}11)$$

The frequency of operation is the first parameter to be determined in designing a PWM regulator. The lower the frequency the smaller the switching losses but the larger the components. The switching frequency should be high enough to keep the values of inductor $L_2$ and capacitor $C_2$ small, but not so high that $Q_1$ and $CR_1$ become prohibitively expensive. Typical operating frequencies range from 20 kHz to 50 kHz. Since the faster the switching, the greater the noise, this must be traded off against EMI requirements.

The minimum voltage possible would occur when $E_{in}$ minus the drop in $Q_1 L_1$

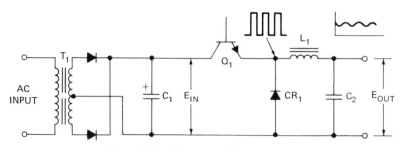

Figure 2-13   Basic switching regulator.

equals $E_{OUT}$. Proper selection of each component of Figure 2-13 is extremely important.

$C_1$ because of its capacitance must usually be an aluminum electrolytic. It is good practice to add a paper capacitor or small tantalum in parallel with $C_1$. This capacitor will serve to provide a low impedance at the switching frequency. The aluminum capacitor may then be placed where convenient if the smaller capacitor is wired as close by as possible between the collector of $Q_1$ and the anode of $CR_1$.

The choice of diode $CR_1$ involves, first, the capability of handling the full output current with a low forward drop, and, second, having a recovery time consistent with the switching speed.

The diode's recovery time is important because of its influence on output noise. After the switching transistor shuts off, the diode conducts, charging capacitor $C_2$. When $Q_1$ turns on again, $CR_1$ is still in its conducting state and shorts $Q_1$ to ground for a short time. This double conduction dissipates power in both $Q_1$ and $CR_1$ and is a prime source of noise.

It will also be necessary to investigate other characteristics usually not specified. When $Q_1$ turns on, the output load current is through $CR_1$, causing $Q_1$ to be looking into a dead short for the diode's recovery time. The maximum current capability of $Q_1$ (which may be many times the actual load current) will pass through $CR_1$ before it turns off. Different diodes present different characteristics under this condition. Observing the voltage and current waveforms at $CR_1$ and measuring the input power for different diode types and manufacturers will allow the designer to select the optimum diode for a given circuit. Maximum efficiency means minimum diode and transistor power.

When $Q_1$ is turned off, $L_1$ attempts to continue the current flow through $CR_1$. A commonly overlooked fact is that the forward recovery time (the time required to start full conduction) is sometimes an order of magnitude longer than the reverse recovery time. If $L_1$ is not damped, large high frequency spikes can occur during this time. Elimination of these spikes is accomplished by adding an $RC$ across $CR_1$. Care must be used in choosing the value so as to obtain maximum spike reduction while dissipating minimum power.

Typically, the time constant of the $RC$ would be 0.1 to 0.5 $\mu$s. The value for $R$ would be determined by the formula:

$$R = \frac{2E_{in}}{I_o} \tag{2-12}$$

where

$E_{in}$ = Unregulated DC
$I_o$ = Output current

Switching transistor $Q_1$ must be selected not only for maximum voltage and current requirements but also for switching speed and second breakdown capability. Care must be taken not to either overdrive or underdrive this transistor. Too little drive will not allow the device to saturate completely or turn off $CR_1$ fast enough, and too much drive will increase storage times, causing storage and turn off problems. In switching supplies that must operate over a great dynamic load range, it is sometimes necessary to vary the drive with load to compensate for this condition. Drive current is chosen not by the $h_{FE}$ specification but by the $V_{ce}$(sat) specification. This specification gives the drive necessary for saturation as a forced beta (typically 5 to 10).

When more than one transistor is used in parallel in a switching regulator, it is not necessary to insert emitter resistors for current sharing if care is used in choosing devices whose gains fall quickly after their rating and if sharing resistors or some other means are used to keep base currents equal.

Turning off $Q_1$ presents other problems. The faster it turns off, the more efficient the supply, but this is generally achieved through reducing the stored base charge very rapidly by returning it to a negative voltage. The second breakdown rating of transistors is greatly affected by how the device is turned off. Maximum reliability will be achieved in turning the transistor off by using the minimum "pull off" current consistent with required storage and turn off times. With current transistors, it is often possible to use a simple base emitter resistor. Storage times do not impair efficiency, only turn-off time does.

A second $LC$ filter can be added to a circuit as shown in Figure 2-14. This circuit will allow capacitors with higher ESR to be used. Typically, a 10 to 1 reduction would be used at $L_2C_2$ to allow for greater ripple reduction and a good response time. Stability integrity is achieved by $R_1$ and $C_3$, allowing DC information of the output without the second phase lag information.

Although this circuit generally involves lower component costs, it has the disadvantage of allowing large voltage excursions during load transients.

The regulator design above has the main advantage of high efficiency over

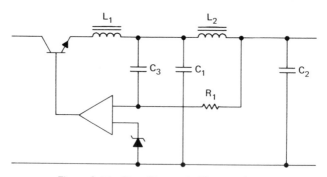

Figure 2-14  Two filter switching regulator.

great input ranges. There generally is a small advantage in size over linear regulators due to a small input transformer and fewer sinking requirements. Cost is slightly higher, because of the addition of the switching inductor and diode and the requirement for faster transistors and lower ESR capacitors. When used in systems, this design makes an excellent post regulator. One unregulated DC voltage can be made available, and local switching regulators may be used. The necessity for a low frequency transformer, however, keeps the size and weight of the design high.

*Switching Supply Variations*

The duty cycle $D = t/T$ of a switching regulator can be altered in a number of different ways.

One technique is to constantly maintain a fixed or predetermined ''on'' time ($t$, the time the input voltage is being applied to the $LC$ filter) and vary the duty cycle by varying the frequency ($1/T$). This method provides ease of design in voltage conversion applications (step-up, step-down, or invert) since the charge developed in the inductor of the $LC$ filter during the on-time (which is fixed) determines the amount of power deliverable to the load. Thus, calculation of the inductor is fairly straightforward.

$$L = \frac{E}{I}\,t \qquad\qquad (2\text{-}13)$$

where

$L$ = value of inductance in microhenrys
$E$ = differential voltage in volts
$I$ = required inductor current defined by the load in amps
$t$ = on-time microseconds

The fixed on-time approach is also advantageous from the standpoint that a consistent amount of charge is developed in the inductor during the fixed on-time. This eases the design of the inductor by defining the operation area to which the inductor is subjected.

The operating characteristic of a fixed on-time switching voltage regulator is a varying frequency, which changes directly with changes in the load. This can be seen in Figure 2-15.

A physical implementation of the fixed on-time variable frequency circuit is shown in Figure 2-16.

Referring to Figure 2-16 both switching transitions are controlled by the sensed voltage, and the regulator cycles between an upper and a lower threshold

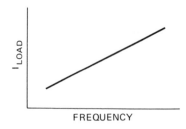

Figure 2-15   Frequency versus load current for fixed on-time switching regulator.

of the output voltage; both the duty cycle and frequency are free to vary. Frequency is primarily a function of $L_2$, $C_2$ and threshold range, but it also varies with $E_{in}$ and, to some extent, with the load current.

Since the free-running regulator's bandwidth is limited only by the rise and fall times of the various transistors in the circuit, response times run from 2 to 4 $\mu$s. Now, in any switching regulator, the output ripple depends on the frequency (higher frequency gives lower ripple but also lower efficiency) and its input voltage. Since the free-running regulator operates from its own ripple—which remains fairly constant—the frequency adjusts to the minimum necessary for the specified output ripple. The free-running circuit also operates well over a wide range of output load currents.

In the fixed off-time switching voltage regulator, the average DC voltage is varied by changing the on-time ($t$) of the switch while maintaining a fixed off-time ($t_{off}$). The fixed off-time switching voltage regulator behaves opposite to

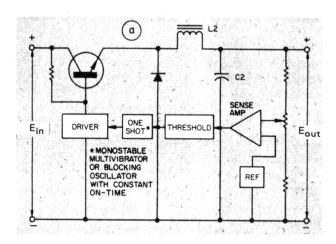

Figure 2-16   Free running switching regulator in which the transistor operates at constant pulse width and variable frequency.

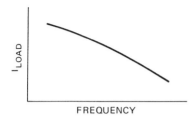

Figure 2-17 Frequency versus load current for fixed off-time switching regulator.

that of the fixed on-time regulator in that as the load current increases, the on-time is made to increase, thus decreasing the operating frequency; this can be seen in Figure 2-17. This approach provides for the design of a switching voltage regulator that will operate at a well-defined frequency under full-load conditions. The fixed off-time approach also allows a DC current to be established in the inductor under increased load conditions, thus reducing the ripple current while maintaining the same average current. The maximum current flowing through the inductor under transient load conditions is not as well defined as that above. Thus additional precautions should be taken to ensure that the saturation characteristics of the inductor are not exceeded.

The fixed-frequency (pulse width modulated, PWM) switching regulator varies the duty cycle of the pulse train to change the average power. The fixed-frequency concept is particularly advantageous for systems employing transformer-coupled output stages because it enables efficient design of the associated magnetics. Taking advantage of its compatibility in transformer-coupled circuits and the advantages of the transformer in single and multiple voltage-conversion applications, the fixed-frequency switching voltage regulator is used in many computer power supply control circuits. As with the fixed off-time switching regulator, in single-ended applications the fixed-frequency regulator establishes a DC current through the inductor for increased load conditions to maintain the required current transferred with minimal ripple current.

In the fixed-frequency regulator, Figure 2-18, one switch transition is always performed by the external source; the other transition occurs when the output voltage reaches a predetermined threshold level. The sampling period is fixed by the external frequency, and the duty cycle is free to vary.

The PWM regulator is often called a "ripple regulator" because its operation is based on regulation through the output ripple. Regulation is accomplished by using an error, or difference voltage, to control the output and minimize the error. The voltage difference (shown in Figure 2-19) between a reference level ($CR_1$) and a sample of the regulator output ($R_1$), is amplified by $Q_5$ and $Q_6$ (differential amplifier) and fed to a Schmitt trigger ($Q_3$ and $Q_4$). When the error voltage exceeds a preset voltage, the Schmitt trigger "fires," and the change-

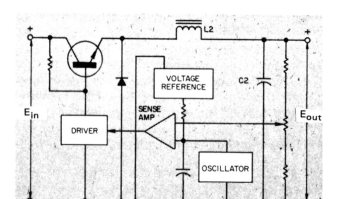

Figure 2-18   Fixed frequency switching regulator in which the transistor on time varies.

of-state condition is applied to switching transistor $Q_1$ to correct the output voltage. Therefore, as the output varies, switching transistor $Q_1$ is turned on or off as necessary to restore the regulator voltage to the correct level.

Circuit losses of the ripple regulator are a function of the transistor saturation resistance, which is very small. Transistor switching losses are minimized by using high-speed transistors. Thus, keeping the losses at a minimum maximizes

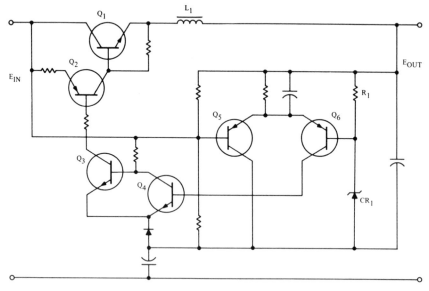

Figure 2-19   Ripple-regulator.

the circuit efficiency. With the ripple regulator, input transients are eliminated and circuit response is limited only by the switching transistors and capacitor $C_1$, which offers a low impedance path for the regeneration of the Schmitt trigger. Output voltage changes due to input transients are not detectable when high-speed transistors and the right value for capacitor $C_1$ are used. Transients due to load changes are determined by (1) the regulator filter impedance, (2) the input voltage, and (3) the regulator response time. The high switching frequency of the regulator requires less inductance in the filter section. Low filter inductance $(L_1)$ keeps the filter impedance low enough to give the circuit the required response time. Thus, load transients do not affect the regulator performance.

Increasing the gain of the differential amplifier ($Q_5$ and $Q_6$) and matching these transistors of the amplifier preceding the Schmitt trigger greatly reduces the output ripple.

The ripple output of any switching-mode regulator is a function of the frequency (higher frequency gives lower ripple but also lower efficiency) and the input voltage. Since the free-running regulator operates from its own ripple, that ripple remains fairly constant. The frequency adjusts itself to the minimum necessary for the required output ripple at a given input voltage. A typical plot of these parameters is shown in Figure 2-20. A constant-frequency-type regulator would have to run at the highest frequency in order to meet the ripple specification at high input voltage, but would far exceed the specification (at the cost of efficiency) at low input voltages. The free-running switching-mode regulator also has an advantage in that it operates well over the wide range of load currents required and, indeed, even at no load.

*Improved Ripple Rejection in a PWM.*    By substituting an exponentially-increasing ramp voltage for the conventional linear ramp, the line-ripple rejection of a pulse-width modulator is more than doubled, yet the circuit is simplified.

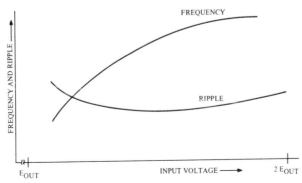

Figure 2-20  Frequency and ripple of free-running switching-mode regulator.

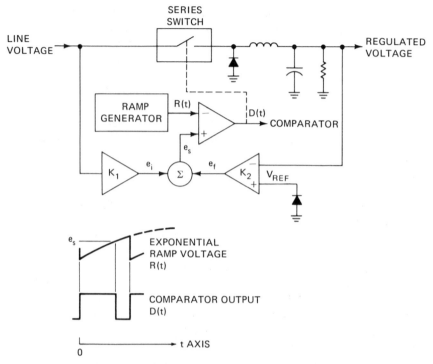

**Figure 2-21** An exponentially increasing voltage $R(t) = A_1 (1 - e^{-t/\tau})$ from a ramp generator determines the pulse width in this PWM. When the amplifier gain constants are optimized for this ramp characteristic, line-ripple rejection is substantially improved.

An exponentially-increasing ramp voltage $R(t)$ controls the duration of pulses at the output of the pulse-width-modulated buck regulator shown in Figure 2-21. The start of a ramp ($t = 0$) begins an output pulse. When the ramp voltage equals the sum of two signals $ej$ and $ef$ ($ej$ is derived from the input line voltage, and $ef$ is an error signal derived from the output voltage), the output pulse is terminated.

The gain constants $K_1$ and $K_2$ are optimized to cancel ripple on the output voltage. If this is done assuming an exponential ramp (and using conventional circuit analysis), ripple rejection is substantially improved over circuits that use a linear ramp. Moreover, the exponential ramp can be implemented by a simple RC circuit, whereas a linear ramp requires a constant source to charge a capacitor.

With the exponential ramp voltage in a nominally 28 V input, 21 V output, pulse-width modulator, ripple rejection is greater than 20 dB as the input is varied from 25 to 30 V, and it is greater than 15 dB for input variations over a 23.5 to 32 V range.

This large ripple rejection is particularly valuable when applying the modulator to power converters for which size and weight restrictions prevent the use of large capacitive filters. The high ripple rejection may also be helpful in computers, scientific and biomedical instruments, and other equipment demanding stable power sources.

*Digital Control Signal Processing for Switching Power Supplies.*   One of the main problems with existing switching-regulator technology is that of reliability. One means to increased reliability is to employ circuit techniques to limit, instantaneously, the electrical stresses in all power processor components. Existing regulators often omit this limiting function and, instead, rely on generous derating of all power components for achieving reliable operation. This practice places the regulator at the mercy of uncontrolled transient stresses, and becomes impractical in the prevailing trend to higher power.

A digital signal processor (DSP) for DC-to-DC converters has been developed that processes all incoming signals and transmits the correct output signal to operate the power switch, thus limiting stress and providing predictable transient operation.

A DC/DC converter must be oscillatory in nature due to the finite flux capability of its inductive elements. The converter control system, therefore, must provide the proper discrete time intervals in controlling the on/off of the power switch. In terms of timing implementation, the various means of duty-cycle control include the following:

1. Constant "on" time $T_n$, variable "off" time $T_f$.
2. Constant $T_f$, variable $T_n$.
3. Constant $(T_n + Tf)$, variable $T_n$ and $T_f$.
4. Variable $T_n$, $T_f$, and $(T_n + T_f)$.

Basic building blocks of the DSP consist of the following: isolator, time delay, memories, oscillator, control gates, and power/control interface, as shown in Figure 2-22.

The function of the isolator is to provide the control circuit input/output isolation. The time delay is used to effect the proper duty-cycle control during either steady-state or transient operations. The memories are used to effect logical state changes as a result of various input control signals. The oscillator produces signals to set timing constraints for various duty-cycle operations. The control gates are used to gate logic signals. The power/control interface maintains electrical compatibility between the control circuit and the base drive of the power switch.

The processing of digital signals is explained by reference to key points identified as points 1 through 12 in Figure 2-22. Signal 9 is logical 1 if and

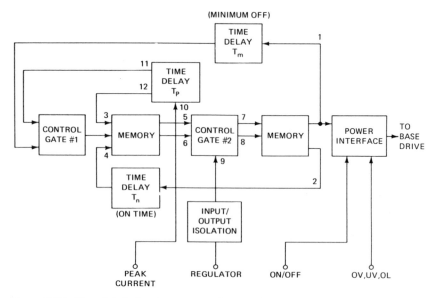

Figure 2-22   The digital signal processor shown utilizes a constant $T_n$ (or $E_iT_n$). It improves switching-regulator control in two major ways: (1) the ability to perform different modes of duty-cycle control to achieve flexibility and standardization and (2) the ability to instantaneously limit power-component stress to enhance regulator reliability. The DSP is capable of working in unison with all types of power circuits and analog controllers, thus making it useful with many switching-regulator applications.

only if the DC converter output voltage is less than the regulator reference. This condition is equivalent to no regulator action. When signal 9 is logical 1, the blocks in the figure (except for the power-interface blocks) combine to form a free-running oscillator. The oscillating frequency depends upon whether signal 10 is logical 1 or logical 0. Signal 10 is defined to be the peak-current protection signal and is logical 1 when the inductor current is less than the predetermined peak protection current.

When the DSP receives the converter "on" command, a few cycles of current buildup must occur before the peak current in the energy-storage inductor reaches the predetermined protection level. During this time of buildup, signal 10 is logical 1, and the time delay $T_p$ is not actuated by the peak-current sensor. The oscillating frequency of this DSP is $1/(T_n + T_m)$ where $T_n$ is "on" time and $T_m$ is minimum "off" time of the converter. These time intervals are coupled to the base drive circuit to control the power switch through the power-interface block. Also during this time of (peak current) buildup, the output voltage is less than the regulator reference so that signal 9 remains logical 1 throughout.

The time required for peak current buildup is less than the time required for output voltage buildup. Thus, when the time of peak current buildup has come to an end, signal 10 becomes logical 0 in each cycle before $T_n$ is timed out while signal 9 remains at logical 1. Time delay $T_p$ is activated. $T_n$ is shortened to $T'_n$, and $T_m$ is replaced by $T_p \geqslant T_m$. The oscillating frequency during this interval of output voltage buildup is $1/(T'_n + T_p)$.

Eventually, the converter reaches its intended regulation level. Signal 9 becomes logical 0 in each cycle as the output voltage intersects the voltage reference. The circuit design insures that the inductor current remains below the peak-current protection level at all times, so that signal 10 is always logical 1. Minimum "off" time $T_m$ is disabled, "on" time becomes $T_n$, while "off" time is increased to $T_f$. The oscillating frequency in steady state is thus $1/(T_n + T_f)$.

The converter on/off command signal and the converter shutoff signal are processed by the power-interface block. These signals respond to either external command or internal protection for the initiation or the termination of the converter operation. A logical 0 input always terminates the operation, and reset is required before restart.

Due to the latching function associated with the overvoltage (OV), undervoltage (UV), and overload (OL) shutoff, these signals are normally processed by another memory block before entering the power interface. This is different from the peak-current signal and the regulator signal, which are required to exhibit both logical 0 and logical 1 cyclically. Consequently, the regulator signal 9 and the peak-current signal 10 are generally obtained from a threshold detector.

A constant-frequency DSP can also be developed by using some of the basic building blocks of the constant-"on" DSP by processing the same signals. By utilizing an oscillator for defining a constant "on" time plus "off" time, the regulator signal determines both $T_n$ and $T_f$. The time-delay block $T_n$ is thus eliminated. Furthermore, due to the availability of essentially the entire period as the time delay following an excessive peak-current detection, time delay $T_p$ can also be eliminated. Excluding these blocks along with their control gates results in a simple constant-frequency DSP.

## BASIC CONVERTER TOPOLOGIES

The basic converters from which all other converter topologies evolved are the buck (or step-down) converter, the boost (or step-up converter) and the buck-boost (step-down/step-up) converter. These are now discussed.

### The Buck (Step-Down) Converter

Figure 2-23 shows the basic buck-converter topology. The buck circuit interrupts the line and provides a variable-pulse-width square wave to a simple averaging filter such that the voltage applied to $L1$ is either $V_{in}$ or 0. Notice that when $S1$

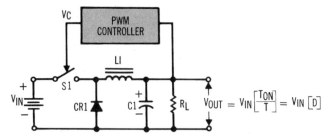

Figure 2-23  Buck (Step-down) Converter: $V_{out} < V_{in}$. (Reprinted with permission from "Switching Architectures—Key to Selecting the Right Power Supply," *Electronic Products*, September, 1983, Hearst Business Communications, Inc.)

is closed, $CR1$ is reverse biased (off) and that when $S1$ opens, the current flow through $L1$ forces the diode to become forward biased (on). DC output voltage is then the average voltage applied to $L1$. If $t_{ON}$ is the time $S1$ is closed, and $T_{off}$ is the time it is open, $V_{out}$ is equal to

$$V_{out} = V_{in} \frac{t_{on}}{t_{on} + t_{off}} \tag{2-14}$$

where $t_{on} + t_{off} = T = $ period

$$\therefore V_0 = V_{in} \frac{t_{on}}{T} = V_{in} D \tag{2-15}$$

where $D = $ duty cycle $= \dfrac{t_{on}}{T} = \dfrac{t_{on}}{t_{on} + t_{off}}$

This formula shows

1. That the basic property of buck converters is that the DC output is *always* less than the DC input.
2. That the output voltage of a switching converter (not only a buck converter) depends *only* on the duty cycle of the switching network and the input voltage, not on $L1$, $C1$, frequency or load current (in the continuous mode of operation).

The pulse-width modulator (PWM) controller of Figure 2-23 manipulates duty cycle $D$—the fraction of switching period $T$ when switch $S1$ is on—to produce an output $V_{out} = V_{in} \times D$. The controller compares $V_{out}$ to an internal reference and adjusts $D$ to obtain the value of $V_{out}$ that will balance the control loop. Diode $CR1$ catches, or clamps, the negative swing of inductor $L1$ when $S1$

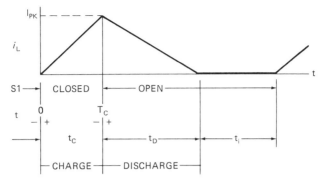

Figure 2-24   Inductor current waveform.

opens. The $L1$–$C1$ combination forms a low-pass filter network. The current through the inductor $(i_L)$ at any given time $(t)$ is

$$I = \frac{V_{in} - V_{out}}{L1} t \tag{2-16}$$

For a constant $V_{in}$, $V_{out}$, and $L1$, $I$ varies linearly with $t$.

The current increases while $S1$ is closed according to the waveform shown in Figure 2-24. The peak current in the inductor, therefore, is dependent on the period of time $S1$ is closed, which is the on time of the switch $(t_{on})$ and is given by

$$I_{pk} = \frac{V_{in} - V_{out}}{L1} t_{on} \tag{2-17}$$

When $S1$ opens $(t = t_{C+})$, the current through the inductor is $I_{pk}$ since the current cannot change instantaneously, the voltage across the inductor inverts, and the blocking diode $(CR1)$ is forward biased to provide a current path for the discharge of the inductor into the load and filter capacitor. The inductor current then discharges linearly, as illustrated in Figure 2-24. The capacitor current follows the inductor current. The relationship between these current waveforms is shown in Figure 2-25.

For the output voltage to remain constant, the net charge delivered to the filter capacitor must be zero. This means that the charge delivered to the capacitor from the inductor must be dissipated in the load. Since the charge developed in the inductor is fixed (constant on time), the time required for the load to dissipate that charge will vary with the load requirements. The actual operating frequency is, therefore, dependent on the load requirements.

The previous discussion has assumed that the converter is operating in the

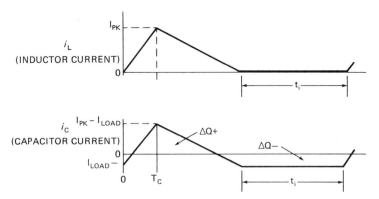

Figure 2-25  Inductor and capacitor current waveforms.

discontinuous mode. This means the inductor current is discontinuous ($I_L = 0$) some time during $S1$ off time if the load current is low enough.

The buck converter will operate in the discontinuous mode for any load current less than

$$I_{out} \le \frac{V_{out}\left(1 - \dfrac{V_{out}}{V_{in}}\right)}{2fL1} \qquad (2\text{-}18)$$

where $f$ = switching frequency

Discontinuous mode operation alters the original statement that output voltage depends only on input voltage and switch duty cycle. When the load is continually increased, the idle time ($t_i$ in Figures 2-24 and 2-25) decreases to the point where the converter initiates a charge cycle at or before the complete discharge of the inductor. This condition is called the "continuous mode of operation" ($I_L$ never equals 0, $t_i = 0$). In this mode, a DC idle current is passed through the inductor.

Filter Capacitor $C1$ has an equivalent series resistance (ESR) which establishes the minimum ripple voltage achievable.

$$V_{ripple}(min) = I_{pk}(ESR) \qquad (2\text{-}19)$$

When the filter capacitor size has been increased such that the voltage change experienced by the filter capacitor equals the minimum ripple voltage, an additional increase in capacitance value will not reduce the ripple voltage. It is important, therefore, to employ a filter capacitor with minimal ESR. Note, however, that due to its architecture, some ripple voltage is required for proper operation of the regulation circuit.

Waveforms for voltage and current of $S1$, $CR1$, $L1$, $C1$, and the input source are shown in Figure 2-26 for both continuous and discontinuous modes of operation.

A general property of "ideal" switching regulators (converters) is that they do not dissipate power in the process of converting from one voltage or current to another; i.e., they are 100% efficient and the following formula applies:

$$P_{out} = P_{in} \text{ or, } (I_{out})(V_{out}) = (I_{in})(V_{in})$$

and

$$I_{in} = I_{out}\left(\frac{V_{out}}{V_{in}}\right) \tag{2-20}$$

Equation 2-20 shows that the average current drawn by the input of a switching regulator can be much higher or lower than the load current, depending on the ratio of output to input voltage. If this fact is ignored by the designer, problems can arise: a low-voltage to high-voltage converter will draw more current from the low voltage supply than it is capable of handling.

The design equations for the buck converter are summarized as follows:

$$I_{pk} \geq 2I_{load} \text{ (for discontinuous operation)} \tag{2-21}$$

$$L1 = \frac{V_{in} - V_{out}}{I_{pk}} t_{on} \tag{2-22}$$

$$f_0 = \frac{2I_{load}}{I_{pk}} \times \frac{V_{out}}{t_{on}V_{in}} \tag{2-23}$$

$$\text{where: } t_D = \frac{I_{pk}}{V_{out}} L_1 \tag{2-24}$$

$$t_i = \frac{I_{pk} - 2I_{load}}{2I_{load}} \times \frac{t_{on}V_{in}}{V_{out}} \tag{2-25}$$

$$C = \frac{(I_{pk} - 2I_{load})^2}{V_{ripple} \times 2I_{pk}} \times \frac{t_{on}V_{in}}{V_{out}} \tag{2-26}$$

The buck converter forms the basis for many types of transformer-coupled DC/DC converters.

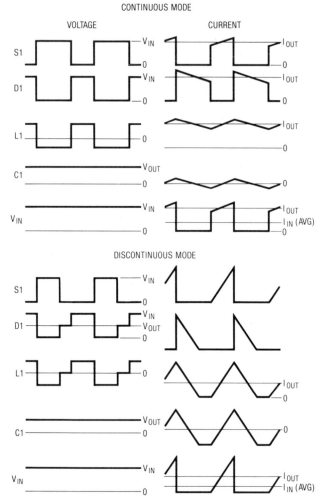

Figure 2-26  Buck converter waveforms.

## An Example of a Buck Regulator Design

The design procedure for the buck switching regulator is best illustrated by a practical example. The specifications for a desired regulator are as follows:

Output: $E_{\text{out}} = 20$ V

$I_L = 1$ A

Ripple $= 60$ mV p-p

Input:                         $E_{in} = 28(+3, -4)$ VDC

Efficiency:                    85%

Operating frequency:           15 kHz

Here is how all the required parameters can be determined:

- Step 1:
Pick $I_{max} = 1.125$ A.
- Step 2:
Determine choke inductance by:

$$L = (E_{in} - E_{out})E_{out}/2fE_{in}(I_{max} - I_C)$$
$$= (28 - 20)20/2(15 \times 10^3) \times 28(1.125 - 1)$$
$$= 1.524 \text{ mH} \tag{2-27}$$

- Step 3:
With $E_{pp} = 60$ mV and $L = 1.53$ mH, calculate the value of $C$ where $\Delta H = 0$:

$$C = (E_{in} - E_{out})E_{out}/f^2(E_{pp} - \Delta H)8LE_{in}$$
$$= (28 - 20)20/[(15^2) \times (10^6) \times (60) \times (10^{-3})$$
$$\times (8) \times (1.53) \times (10^{-3}) \times (28)]$$
$$= 34 \ \mu\text{F} \tag{2-28}$$

Also compute the choke losses. The overall choke losses include AC and DC losses. The DC loss is simply $I_L^2 R_{ch}$, so where $R_{ch} = 1$ ohm and $I_L = 1$ A it is 1 W. The expression for the AC loss in watts is:

$$(f^2L^2/6Q)E_{in}(I_{max}^3 - I_o^3)/E_{out}(E_{in} - E_{out}) \tag{2-29}$$

With the value of $L$ calculated in Step 2 and $Q = 20$ at 15 kHz:

$$\text{AC loss} = 575 \text{ mW}$$

where:   $E_{in} =$ DC input voltage,
$E_{out} =$ DC output voltage,
$E_B =$ bias voltage.
$E_d =$ diode forward voltage drop,
$E_{pp} =$ peak-to-peak ripple voltage,

$f$ = operating frequency in hertz,
$H$ = hysteresis of voltage comparator, in volts,
$I_B$ = transistor base current,
$I_C$ = transistor collector current,
$I_c$ = capacitor current,
$I_{ch}$ = choke current,
$I_i$ = input current,
$I_L$ = DC load current,
$I_{max}$ = maximum value of $I_i$,
$I_o$ = minimum nonzero value of $I_i$,
$I_p$ = peak transistor current,
$R_{ch}$ = choke DC resistance,
$V_{CEO}$ = transistor collector-to-emitter breakdown voltage,
$V_{CES}$ = transistor collector-to-emitter saturation voltage,
$t_T(\text{off})$ = transistor turn-off time,
$t_T(\text{on})$ = transistor turn-on time,
$t_D(\text{off})$ = diode turn-off time,
$t_D(\text{on})$ = diode turn-on time,
$\eta$ = efficiency.

■ Step 4:

Choose the transistor. A 2N3879 silicon npn transistor was selected. Its parameters are:

$$V_{CES} = 200 \text{ mV with } I_C = 1\text{A}, I_B = 0.1 \text{ A}$$

$$t_{T(\text{on})} = t_{T(\text{off})} = 110 \text{ nsec},$$

$$V_{CEO} = 75 \text{ V}$$

■ Step 5:

Choose the diode. A Unitrode fast-recovery UTR02 diode was used. It has a reverse recovery time of 1 $\mu$s, turn-on time of 100 ns, and a forward voltage drop of 600 mV at 1 A.

Once the transistor and diode have been selected, their losses can be calculated. The transistor conduction losses are:

$$V_{CES} E_{\text{out}} I_L / E_{\text{in}} = (0.2)(20)(1)/28$$

$$= 143 \text{ mW} \tag{2-30}$$

The diode conduction loss is:

$$E_d I_L (E_{in} - E_{out})/E_{in} = (0.6)(1)(28 - 20)/28$$

$$= 171 \text{ mW} \tag{2-31}$$

The switching losses are as follows:

$$f I_p E_{in} t_{D(off)} + f E_{in} I_{max}(t_{T(off)^2} + t_D(off)^2)/6t_T(off) \tag{2-32}$$

$$= (3)(28)(1 \times 10^{-6})(15 \times 10^3)$$

$$+ (15 \times 10^3)(28)(1.125)[0.11^2 \times 10^{-12} + 0.1^2 \times 10^{-12}]/$$

$$6(0.11 \times 10^{-6})$$

$$= 1260 \text{ mW} + 15.8 \text{ mW} = 1276 \text{ mW} \tag{2-33}$$

Drive loss with a 50-V bias supply is as follows:

$$E_B I_L E_{out}/h_{FE} E_{in} = (5)(1)(20)/(10)(28) = 358 \text{ mW} \tag{2-34}$$

The total losses are therefore:

| | |
|---|---|
| Choke | 1.000 |
| | 0.575 |
| Transistor | 0.143 |
| | 1.260 |
| | 0.0158 |
| Diode | 0.171 |
| Drive | 0.358 |
| Total | $\overline{3.4728} \simeq 3.5$ W |

Thus the efficiency becomes the following:

$$\eta = 20/(20 + 3.5)$$

$$\eta = 85.1\% \tag{2-35}$$

This example is an actual design. The overall efficiency of the operating supply was measured at 85%. From the breakdown of the losses, the designer may see exactly where he may trade watts for dollars. It may be noticed that losses due to leakage currents of off semiconductors have been neglected. This is a valid approximation, particularly in the case of silicon semiconductors, which have very low leakage currents.

The complete schematic of this switching regulator is shown in Figure 2-27.

### The Boost (Step-Up) Converter

Figure 2-28 shows a schematic diagram of the basic boost or step-up converter. The boost converter generates a DC output voltage that is *always* greater than the input voltage and is given by

$$V_{out} = \frac{V_{in}}{1 - D} \tag{2-36}$$

Where $D$, the duty cycle of switch $S1$ is less than 1 and is the ratio of switch on time to off time. In the boost converter, the $L1$–$C1$ network does not form an inherent low-pass filter, making the ripple and noise much higher than for the buck regulator topology.

In operation, during the charging cycle ($S1$ closed) inductor $L1$ is charged directly by the input potential.

$$i_L = \frac{V_{in}}{L1} t_c \tag{2-37}$$

$$\text{thus } I_{pk} = \frac{V_{in}}{L1} t_{on} \tag{2-38}$$

In the step-up converter, the peak current is not related to the load current as in the buck converter. The reason is that during the inductor charge-cycle, blocking diode $CR1$ is reversed-biased, and no charge is delivered to the load. The circuit in Figure 2-28 delivers power to the load only during the discharge cycle of the inductor (when $S1$ is open); diode $CR1$ is forward-biased, and the inductor discharges into the load capacitor. The potential across the inductor during this phase of the charge/discharge cycle is $V_{out} - V_{in}$. The discharge time of the inductor then becomes:

$$t_D = \frac{I_{pk}}{V_{out} - V_{in}} L1 \tag{2-39}$$

Thus, $L1$ first stores energy and then delivers this energy plus energy from the input line to the load.

The following design equations for the boost converter are derived from the inductor and capacitor current waveforms, as shown in Figure 2-29.

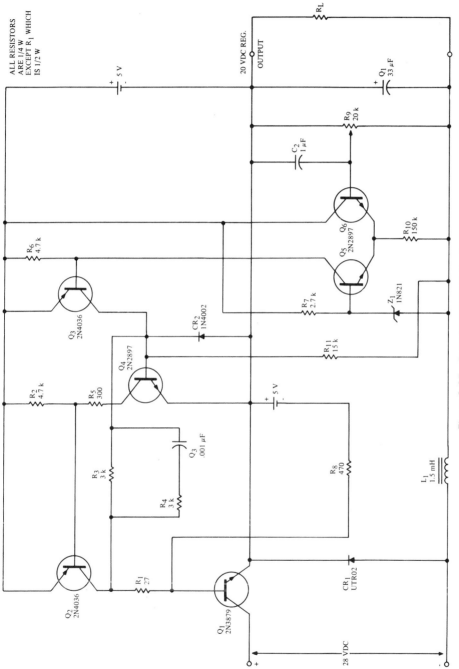

Figure 2-27 Buck Regulator Example.

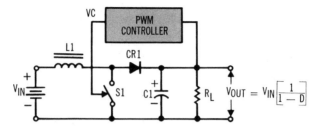

Figure 2-28 Boost (Step-up) Converter: $V_{out} > V_{in}$. (Reprinted with permission from "Switching Architectures—Key to Selecting the Right Power Supply," *Electronic Products*, September, 1983, Hearst Business Communications Inc.)

$$I_{pk} = 2I_{load} \left( \frac{V_{out}}{V_{in}} \right) \tag{2-40}$$

$$L1 = \frac{V_{in}}{I_{pk}} t_{on} \tag{2-41}$$

$$f_0 = \frac{2I_{load}}{I_{pk} - t_D} \tag{2-42}$$

$$C_1 = \frac{(I_{pk} - I_{load})^2}{V_{ripple} \times 2I_{pk}} t_D \tag{2-43}$$

$$t_D = t_{on} \left( \frac{V_{in}}{V_{out} - V_{in}} \right) \tag{2-44}$$

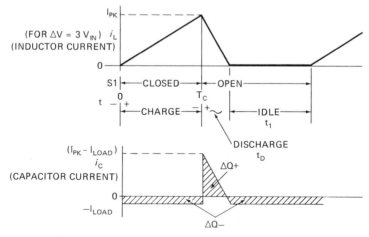

Figure 2-29 Boost converter inductor and capacitor current waveforms.

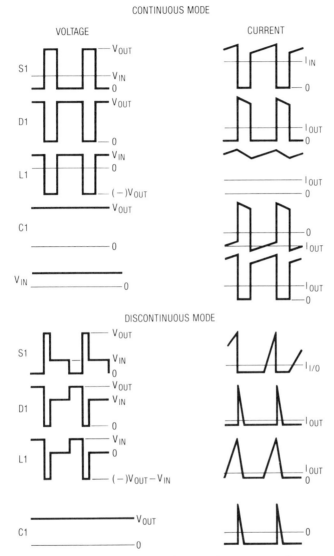

Figure 2-30   Boost regulator voltage and current waveforms.

Figure 2-30 shows the voltage and current waveforms for all components of Figure 2-28, both for continuous and discontinuous mode operation. Note that the current drawn from the input and delivered in pulses to the load is significantly higher than the output load current. The amplitude of input current and peak switch and diode current is equal to:

$$I_{pk} = I_{out} \left( \frac{V_{out}}{V_{in}} \right) \quad \text{(continuous mode)} \qquad (2\text{-}45)$$

Average diode current is equal to $I_{out}$ and average switch current is $I_{out} \times (V_{out} - V_{in})/V_{in}$, both of which are significantly less than peak current. The switch, diode, and output capacitor must be specified to handle the peak currents as well as average currents. Discontinuous mode operation requires even higher ratios of switch current to output current.

One drawback of the boost converter is that it cannot be current-limited for output shorts because the current steering diode, $CR1$, makes a direct connection between input and output.

Because it delivers a fixed amount of power to the load regardless of load impedance (except for short circuits), the boost regulator is widely used in photoflash and capacitive-discharge (CD) automotive ignition circuits to recharge the capacitive load. It also makes a good battery charger. The boost converter sees little use in high power switcher designs. For an electronic circuit load, however, the load resistance must be known in order to determine the output voltage:

$$V_{out} = \sqrt{P_0 R_L} = \sqrt{L1 f_0 R_L/2} \qquad (2\text{-}46)$$

where

$R_L$ = the load resistance

In this case, the choke current is proportional to the on time or duty cycle of the switch, and regulation for fixed loads simply involves varying the duty cycle as before. However, the output also depends on the load, which was not the case with the buck regulator, and results in a variation of loop gain with load.

The transient response, or responses, to step changes in load are very difficult to analyze. They lead to what is termed a "load dump" problem, which requires that energy already stored in the choke or filter be provided with a place to go when the load is abruptly removed. Practical solutions to this problem include limiting the minimum load and using the right amount of filter capacitance to give the regulator time to respond to this change.

### A Boost Regulator Example[2]

The design requirements for a boost regulator are as follows:

|   |   |
|---|---|
| Input Voltage (Battery) | 16 V to 26.1 V DC |
| Regulated Output Voltage | +27.9 ± 0.2 V |

---

[2]"Switching Power Supplies For Satellite Radiation Environments," by C. A. Berard, *Solid State Conversion*, September/October 1977, used with permission.

Load Range             0 to 450 W; 530 W pk.
Conversion Efficiency  $>89\%$ (110 $-$ 280 W load)
                       $>87\%$ (450 W load)

A practical boost regulator is shown in block diagram form in Figure 2-31.

Output voltage is sensed and compared to a voltage reference by a comparison amplifier. The error signal is amplified and loop stability compensation performed before pulse-width modulation occurs. A duty-cycle limited, constant frequency pulse-width modulator circuit converts voltage error into a time ratio error which controls transistor power switches. When the transistor switch is on the inductor current increases storing energy in the magnetic field. When the transistor is off, energy is transferred via the flyback diodes to the loads and energy storage capacitors. Operation of the transistor switches is at a high frequency relative to the resonance of the inductor-capacitor network.

In simplified form the transfer function can be easily derived by equating the net change of inductor current to zero over a period of the transistor switch. Assuming the input and output voltage are essentially constant,

$$\Delta I_{L_1} = \frac{E_s \alpha}{L_f} \text{ when the switch is on, and} \tag{2-47}$$

$$\Delta I_{L_2} = \frac{(E_o - E_s)(1 - \alpha)}{L_f} \text{ when the switch is off} \tag{2-48}$$

where

$E_s$  is the input source voltage
$E_o$  is the output voltage
$\alpha$   is defined as the proportion of the period $1/f$ when the transistor switch is on
$L$   is the choke inductance value, and
$f$   is the operating frequency

Equating (2-47) and (2-48) to get the steady state solution gives

$$E_o = \frac{E_s}{1 - \alpha} \tag{2-49}$$

Dealing with real switching regulators requires the above derivation to be expanded to include (at least) the first order non-ideal parameters. The applicable equation is

$$E_o = \frac{1}{1 - \alpha} (E_s - E_{sw} - I_s R_L) - (E_d + I_s R_w) \tag{2-50}$$

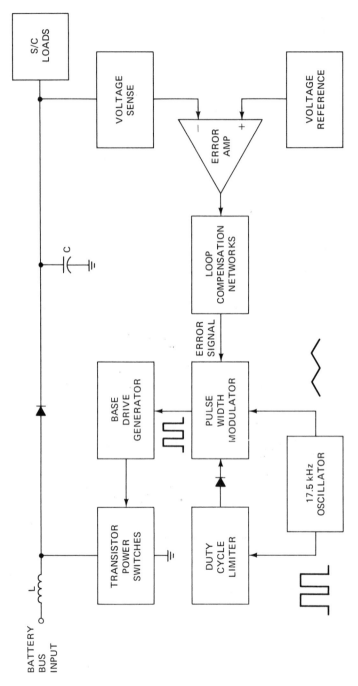

Figure 2-31   Boost regulator functional diagram.

where

$E_d$    is the flyback diode forward voltage drop,

$E_{sw}$    is the transistor switch-on voltage drop

$I_s$    is the source current

$R_L$    is the resistance of the inductor, and

$R_w$    is the equivalent resistance in the flyback circuit

Use of this equation permits accurate determination of the maximum required duty cycle and is helpful in calculating power transfer efficiency.

Applying switching regulators to high-current applications usually presents serious problems with component selection. Paralleling transistors and diodes is undesirable in practical terms due to the difficulty and high cost of achieving (and maintaining over life and temperature) suitable matching. Very high current transistors exhibit relatively slow switching times which are incompatible with high frequency and high efficiency.

Parallel-path power sections provide a way around some component problems while having beneficial effects on weight, efficiency and reliability. A three-section approach (Figure 2-32) provided maximum advantage in this application. Component weight is reduced from a single-section design and a more advantageous configuration, which has no single-point failures, results. Components in each section are rated to carry one-half the load to further enhance reliability and ensure continued operation under failure mode conditions. All are driven simultaneously from a common PWM error signal, are connected to common input and output points, and with normal component tolerances share current with rather close tolerance (measured data confirms +4% tolerance with full 450-W load and 23 amperes total input current).

Identical base drive circuits for each section enhance high-efficiency performance. Each section (Figure 2-33) provides one ampere constant current base-drive to the switching transistor and the resonant flyback network generates a reverse current of about 400 mA for rapid turn-off. Efficiency and ground isolation are enhanced by the transformer coupling technique; 3 A total base drive is provided using less than 165 mA from the output bus. Operation is (by design) controlled by the passive components; thus performance is not influenced significantly by semiconductor parameter variations.

Continuous application of base-drive to the switching transistors in the boost regulator will cause over-current failure of those devices. Thus, a positive means to restrict their operation to a maximum on-time is needed. The technique illustrated in Figure 2-34 provides such protection using signals available within the sawtooth oscillator, and a diode-resistor network. During a portion of the period, a saturating level is diode-gated to hold the PWM output low (no base drive to switches) while the sawtooth provides normal modulator action during

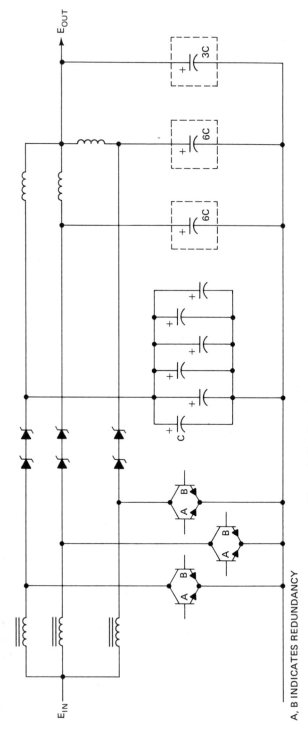

Figure 2-32 Parallel power sections of boost regulator.

A, B INDICATES REDUNDANCY

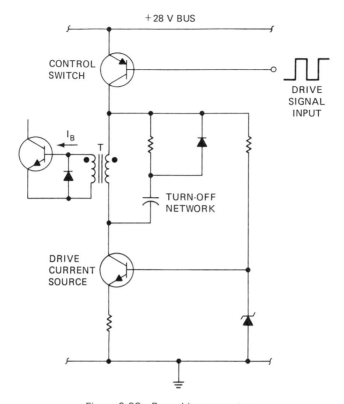

Figure 2-33 Base drive generator.

the remainder of the period. By using an asymmetrical oscillator, any duty cycle limit may be generated; this design incorporates a 65% limit.

Component selection for a switching regulator is a complex process requiring trade-offs of performance, weight, cost, and (for space applications) reliability and previous applications history. Semiconductors are among the most difficult to select.

Switching transistors were the most difficult part to optimize. Fast switching times, high current gain, low saturation voltages, broad safe-operating-area limits, high voltage and current ratings, and isolated case construction were all desired. A 100-V, 30-A device in a TO-61/I package was chosen for its performance characteristics. It is a double-diffused epitaxial type (similar to a 2N5330) whose second breakdown resistance is acceptably high though not as rugged as a single-diffused chip.

Flyback diode selection was driven by several requirements, all indicating an almost ideal diode. Low forward drop was needed for efficiency especially considering the desirability of using series-connected, redundant diodes. Fast

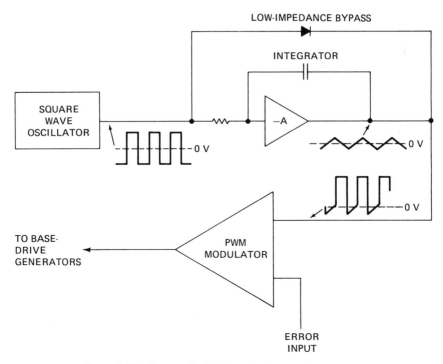

Figure 2-34   Duty cycle limiting of pulse-width modulator.

switching and low recovery current spiking are necessary to meet the low ripple and noise specifications without excessively complex filtering. Hot-carrier rectifiers were the only parts capable of satisfying these stringent requirements; a screened version of the TRW type SD-51 was ultimately selected and used.

Inductors in the power switching network are required to withstand high DC bias and be of low-loss design. A gapped C-core of 3% silicon-steel wound with a 1.4 in. by 0.004 in. copper strip and Kapton insulation provides suitable characteristics. Each 500-$\mu$h, 15-A choke weighs 10 ounces and has 20 milliohms DC resistance. The core is rugged and does not need special mounting provisions to achieve resistance to vibration and shock.

Predicted performance goals were achieved on all units produced, with an adequate margin from specification limits.

Efficiency behavior is generally 90% or greater and exceeds the specification requirements as shown in Figure 2-35. Full load is applied only for 5-10 minute durations so the slightly lower full-load efficiency, due primarily to resistive losses, is acceptable and allows a small weight savings. At light loads, efficiency is reduced by the bias losses of the base drive and control circuits.

Ripple and noise measurements are taken under worst-case conditions of min-

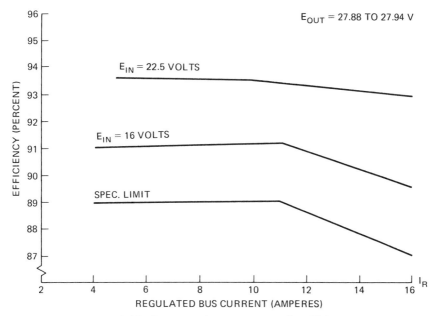

Figure 2-35   Boost regulator power transfer efficiency.

imum input voltage and maximum load. Actual performance of 25 mV peak-to-peak (0.09%) reflects the conservative worst-case design philosophy.

## The Buck-Boost (Step-down/Step-up or Inverting) Converter

The buck-boost converter, depicted schematically in Figure 2-36, is similar to the boost converter except that the load is referred to the inductor side of the input instead of the switch side. The output voltage of the buck-boost converter

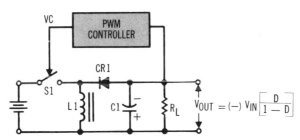

Figure 2-36   Buck-boost converter schematic diagram. (Reprinted with permission from "Switching Architectures—Key to Selecting the Right Power Supply," *Electronic Products*, September, 1983, Hearst Business Communications, Inc.)

can be greater than, equal to, or less than the input voltage by varying the duty cycle ($D$) of switch $S1$, and with the opposite polarity of the input voltage:

$$V_{out} = -V_{in}\left(\frac{D}{1 - D}\right) \tag{2-51}$$

Again, the inductor is used for energy storage and does not act as a part of a filter.

During the charging cycle ($S1$ closed), inductor $L1$ is charged only by the input potential and the load is isolated from the input, similar to the boost converter.

$$I_{pk} = \frac{V_{in}}{L1}\, t_{on} \tag{2-52}$$

Like the boost converter, the input provides no contribution to the load current during the charging cycle, and thus the maximum load current for discontinuous operation will be limited by the peak current, in accordance with that observed in the boost converter.

$$I_{L\,max}(\text{discontinuous}) = \frac{I_{pk} t_D}{2(t_D + t_c)} \tag{2-53}$$

The discharge rate ($t_D$), however, differs due to the difference in the potential across the inductor during its discharge which is $V_{out}$.

$$\therefore t_D = \frac{I_{pk}}{|V_{out}|}\, L1 \tag{2-54}$$

The current waveforms for the buck-boost converter are similar to those shown for the boost converter and, like for the boost converter, the peak switch, diode, and output capacitor currents can be significantly higher than the output current. Thus, these components must be sized accordingly.

$$I_{pk} = \frac{I_{out}}{1 - D} = I_{out}\left(\frac{V_{out} + V_{in}}{V_{in}}\right) \quad \text{(continuous mode)} \tag{2-55}$$

The maximum switch voltage is equal to the sum of the input voltage plus the output voltage. The forward turn-on time of $CR1$ is, therefore, very important in higher voltage applications to prevent additional switch stress.

The following design equations apply for the buck-boost converter

$$I_{pk} \geq 2I_{load}\left(1 + \frac{|V_{out}|}{V_{in}}\right) \qquad (2\text{-}56)$$

$$L1 = \left(\frac{V_{in}}{I_{pk}}\right)t_{on} \qquad (2\text{-}57)$$

$$f_0 = \frac{2I_{load}}{I_{pk}t_D} \qquad (2\text{-}58)$$

$$C = \frac{(I_{pk} - I_{load})^2}{V_{ripple} \times 2I_{pk}} \times t_D \qquad (2\text{-}59)$$

$$t_D = t_{on}\left(\frac{V_{in}}{|V_{out}|}\right) \qquad (2\text{-}60)$$

The buck-boost converter is the prototype for the transformer-coupled flyback converter.

Currently used switcher topologies have evolved from the buck, boost, and buck-boost circuits:

- The flyback (or ringing choke) converter
- The forward converter
- The Cuk converter
- The half-bridge converter
- The full-bridge converter
- The push-pull converter, both self oscillating and driven

These topologies are now presented.

## LOW-POWER CONVERTER TOPOLOGIES

### The Flyback Converter[3]

The flyback, or ringing choke converter, is a constant-output power rather than a constant-output current source. The flyback converter works by cyclically

---

[3]Portions of this section used with permission from:
  a. Design of Single-Ended DC/DC Converters by W. F. Slack, *Proceedings of Powercon I*, March 1975.
  b. "Use A Single Ended Switching Regulator to Convert DC to DC with Simple Circuitry," W. P. Steele. Excerpted with permission from *Electronic Design* 1, January 6, 1972, copyright Hayden Publishing Co. Inc. (1972).
  c. V. Brunstein, "Design Flyback Converter for Best Performance." Excerpted with permission from *Electronic Design* 26, December 29, 1976, copyright Hayden Publishing Co., Inc. (1976).
  d. N. Kepple, "High Power Flyback Switching Regulators," *Solid State Power Conversion*, January/February 1978. Excerpted with permission.

storing energy in a magnetic field and then dumping this stored energy into a load. By varying the amount of energy stored and dumped per cycle, the output power can be controlled and regulated. Needless to say, the heart of the flyback system is the magnetic circuitry which governs the amount of power achievable, the efficiency of the converter, and the operating range of the converter.

The flyback converter (see Figure 2-37) is, circuit-wise, the simplest of the low-power topologies. This circuit uses a transformer to transfer energy from input to output, providing DC I/O isolation.

During $S1$ on time (switch $S1$ closed), energy builds up in the core due to increasing current in the primary winding. At this time, the polarity of the secondary winding is such that $CR1$ is reverse-biased, blocking secondary winding load current flow and making capacitor $C1$ the sole source of load current during this phase. When $S1$ opens at the start of the off time, the primary winding inductive swing transfers the total stored energy to the secondary winding in the proper polarity for $CR1$ to conduct current to the load. The output voltage is given by:

$$V_{\text{out}} = V_{\text{in}} \left( \frac{D}{1 - D} \right) \left( \frac{N_s}{N_p} \right) \tag{2-61}$$

The turns ratio $N_s/N_p$ of the transformer can be adjusted for optimum power transfer from input to output. The flyback converter can have an output voltage which is higher or lower than the input voltage.

The principle of operation of the flyback converter is based on the energy storage in a choke during time period $t_1$, and the discharge of the energy to a load during a second period $t_2$. Normally, due to isolation requirements, the primary winding of a transformer is the storage choke in addition to its normal

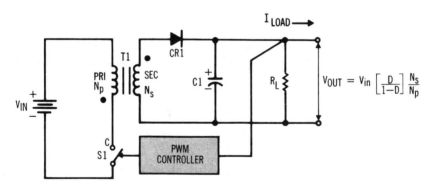

Figure 2-37  The flyback converter.

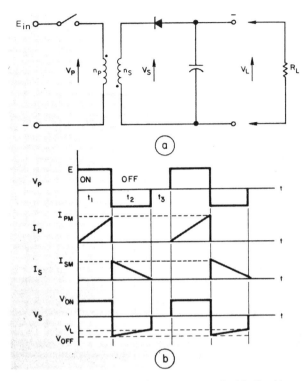

Figure 2-38 The power stage of a flyback converter can be idealized into an ideal switch and a linear transformer for analysis (a). The primary and secondary voltages and currents of the ideal converter (mode A operation) are shown in (b). Reprinted with permission from *Electronic Design 26*, December 20, 1976, copyright Hayden Publishing Co. Inc. (1976).

role. Figures 2-38 and 2-39 will be used to discuss the operating details of the flyback converter in both operating modes:

- Mode *A*, or discontinuous mode of operation, in which energy is dumped from the choke before the transistor turns on again.
- Mode B, or continuous mode of operation, in which the transistor is turned on while energy is being dumped into the load.

In mode A, (Figure 2-38b) during $t_1$, the input current increases linearly until it reaches a maximum value, $I_{pM}$, given by:

$$I_{pM} = \frac{E}{L_p} t_1, \qquad (2\text{-}62)$$

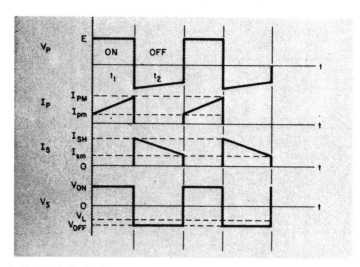

Figure 2-39  Mode B. If all energy isn't discharged before the next cycle begins, then the flyback currents and voltages are somewhat different from those for complete discharge. Energy is stored during interval $t_1$. Reprinted with permission from *Electronic Design 26*, December 20, 1976, copyright Hayden Publishing Co. Inc. (1976).

where $E$ is the battery voltage, and $L_p$ the primary inductance. During $t_2$, the current in the secondary decreases from a maximum of $I_{sM}$ given by:

$$I_{sM} = \frac{V_s(\text{off})}{L_s} t_2, \qquad (2\text{-}63)$$

where $V_s$ is the voltage across the secondary, and $L_s$ the secondary inductance.

Note that the primary current and secondary current do not flow at the same time. Thus the input and output are not coupled at the same time. Energy does not flow through the transformer as in the fashion of a conventional DC/DC converter. It is transferred in a bucket brigade fashion, first being drawn from the input source to the core, then released from the core to the output.

This fact implies significantly different transformer design considerations which will be discussed later. In addition, it indicates that noise of fundamental frequency higher than the switching frequency will not magnetically couple from input to output or vice versa.

The energy stored in the primary during $t_1$, expressed by $1/2 L_p I_{pM}^2$, is transferred to the secondary with efficiency $\eta_{tr}$, a number that indicates the quality of the transformer. From the relationship, $1/2 L_s I_{sM}^2 = \eta_{tr} 1/2 L_p I_{pM}^2$, comes:

$$I_{sM} = \sqrt{\eta_{tr}} \frac{n_p}{n_s} I_{pM}, \qquad (2\text{-}64)$$

where $n_p$ is the primary turns and $n_s$ the secondary turns. The load current is obtained by integrating the secondary current for the period $T$ and getting this result:

$$I_L = \frac{1}{2} I_{sM} \frac{t_2}{T}.$$

(2-65)

Since the rectifier loses power, a rectifying efficiency, $\eta_R$ (significant for low output voltages) must be introduced. Thus:

$$\eta_R = \frac{V_L}{V_s(\text{off})}.$$

(2-66)

Using the notation, $\tau = t_1/T$, and introducing the overall efficiency, $\eta = \eta_{tr}\eta_R$, the power delivered to the load can be deduced from Equations 2-62 through 2-66:

$$P_L = \left(\frac{1}{2} \eta \tau^2\right) \frac{E_{in}^2}{f L_p}.$$

(2-67)

The value of $t_2$ as a function of the load can also be found:

$$t_2 = \sqrt{\frac{2\eta_R L_s}{f R_L}}.$$

(2-68)

Expressions 2-65 and 2-66 can be reshaped for easier use in design. Thus:

$$I_L = \frac{1}{2}\left(\frac{n_p}{n_s} \sqrt{\eta_{tr}}\right) \frac{\tau(1 - \tau) E_{in}}{f L_p}$$

(2-69)

$$V_L = \eta_R \sqrt{\eta_{tr}} \left(\frac{n_s}{n_p}\right) \left(\frac{\tau}{1 - \tau}\right) E_{in}$$

(2-70)

Analyzing the equations reveals a number of things about converters working in mode A:

1. The maximum input current $I_{pM}$ does not depend on load variations.
2. The power delivered to the load does not depend on the load value $R_L$. The converter provides a constant power, and the battery is isolated from the load for any output conditions, including no-load. For the no-load condition, the output voltage increases until the dissipation of the leakage resistances equals the constant power of the converter. At that point, $t_2$ is very short.

3. The power delivered by the converter depends on the product, $fL_p$, for a given duty cycle and battery voltage, $E$. In other words, for a given power, an increase in the frequency decreases the inductance, $L_p$, and, hence, the size of the ferrite core.
4. The values of $\tau$ or $f$, or both can be modified through a feedback loop (usually the frequency is kept constant and only $\tau$ is varied) to stabilize values $I_L$ or $V_L$ with load variations.

It may be worthwhile to mention one big advantage of the flyback converter at this time, that is that the power transistor (switch) has some impedance in series with it at all times. The rise time of the collector current is well controlled and has a relatively slow rise time. With this controlled rise time it is possible to turn off the transistor before the core $T1$ saturates.

Another feature is that during the charge cycle the secondary or load is completely disconnected from the primary, so that faults which occur on the secondary are not directly reflected to the primary, although there is an indirect reflection from the secondary to the primary.

The dump cycle (mode B) is more complex than the storage cycle. In this mode of operation, the primary and secondary currents and voltages are similar to mode A (Figure 2-39). During $t_1$, energy is stored by a linear growth of the current in the winding from a minimum value, $I_{pm}$ (residual from the previous cycle), to a maximum, $I_{pM}$, where:

$$I_{pM} - I_{pm} = \frac{E_{in}}{L_p} t_1. \tag{2-71}$$

During $t_2$, current appears only in the secondary and decreases linearly from a maximum of $I_{sM}$ to a minimum of $I_{sm}$ as follows:

$$I_{sM} - I_{sm} = \frac{V_s(\text{off})}{L_S} t_2. \tag{2-72}$$

Since switching occurs before the secondary current drops to zero, energy remains stored in the choke and is transferred back to the primary through the current $I_{pm}$. The relationship between the currents—similar to the one in mode A—is given by:

$$I_{sM} - I_{sm} = \frac{n_p}{n_s} (I_{pM} - I_{pm}), \tag{2-73}$$

and the load current is:

$$I_L = \frac{I_{sM} + I_{sm}}{2} \left( \frac{t_2}{T} \right).$$  (2-74)

Using the same notations as in mode A results in the following:

$$I_L = \frac{1}{2} \left( \sqrt{\eta_{tr}} \frac{n_p}{n_s} \right) \tau (1 - \tau) \frac{E_{in}}{fL_p} + \sqrt{\eta_{tr}} \frac{n_p}{n_s} (1 - \tau) I_{pm}.$$  (2-75)

$$V_L = \eta_R \sqrt{\eta_{tr}} \frac{n_s}{n_p} \left( E_{in} \frac{\tau}{1 - \tau} \right).$$  (2-76)

$$P_L = \frac{1}{2} \eta \tau^2 \frac{E_{in}^2}{fL_p} + \eta \tau E_{in} I_{pm}.$$  (2-77)

The values of $I_{pm}$ and $I_{pM}$ are given by:

$$I_{pm} = I_{pM} - \frac{1}{2} \frac{E_{in} \tau}{fL_p};$$  (2-78)

$$I_{pM} = \frac{\eta_R}{\tau} \left( \frac{\tau}{1 - \tau} \right)^2 \left( \frac{n_s}{n_p} \right)^2 \frac{E_{in}}{R_L}.$$  (2-79)

In mode B, therefore:

1. The output voltage is not dependent on the load or the frequency. However, a weak dependency exists between $V_L$ and the load through the efficiency, which depends on $R_L$.
2. The power delivered by the converter is a function of $R_L$, so protection against overload is necessary.
3. Voltage $V_L$ can be stabilized against variations in $E_{in}$ only by varying $\tau$.
4. Equation 2-78 shows that $I_{pm}$ varies with the load and, at a certain moment, can be zeroed. At that moment, a change in the working mode occurs since the increase of resistance $R_L$ leads to mode A operation.
5. For the condition, $I_{pm} = 0$, the equations for mode B become the same as those for A, where $t_3 = 0$, and $t_1 + t_2 = 1/f$. This case corresponds to a blocking-oscillator mode (self-oscillating flyback). The limiting value for $R_L$ can be obtained from any of the cited conditions.

For $R_L$ in the blocking-oscillator mode, the frequency and the duty cycle vary according to:

$$R_L = \eta_R \frac{n_s^2}{n_p} \left[ \frac{2fL_p}{(1 - \tau)^2} \right]. \tag{2-80}$$

An important feature of the flyback regulator is its ability to operate at duty cycles less than the clock period (Figure 2-38b) $V_p$. This gives the flyback regulator a very large dynamic range.

Comparing the two modes of operation, mode A is advisable for those converters with constant output current (for instance, remotely supplied amplifiers in cable-communication equipment), because the current variation to be stabilized for a given $R_L$ is small. On the other hand, mode B is better for a converter of constant output voltage, because $V_L$ is only mildly dependent on load variations.

To obtain the same output power in mode B, as in mode A, the maximum current should be smaller and approach a limit that is half the value required in mode A.

Diminishing the number of turns in mode A decreases copper losses and the ripple induced in the supply is greater in mode A because of the greater variation in mode A's switching current.

The major difference between a flyback transformer's design and a feed-through converter's is a DC-current component in the flyback windings that requires a properly chosen air gap.

Transistor waveforms for both mode A (discontinuous) and mode B (continuous) operation are shown in Fig. 2-40. Mode B operation generally provides an advantage for the switching power transistor in that it needs to switch only half as much peak current in order to deliver the same power to the load. In many instances, the same transformer may be used with only the gap reduced to provide more inductance. Sometimes the core size will need to be increased to support the higher $LI$ product (2 to 4 times) now required because the inductance must increase by almost 10 times to effectively reduce the peak current

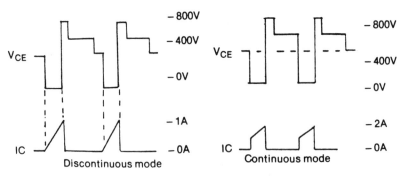

Figure 2-40 Flyback transistor waveforms.

by two. In dealing with the continuous mode, the transistor has from 500 to 600 V connected across the collector and emitter rather than 400 V, because there no longer is any dead time to allow the flyback voltage to settle back down to the input voltage level. Generally, it is advisable to have $V_{CEO(SUS)}$ ratings comparable to the turn on requirements.

Both input ripple current and output ripple current are high in a flyback converter because output filter chokes are not required. The output capacitors feed from an energy source rather than a voltage source. This disadvantage is somewhat offset by the ability to achieve current or voltage gain and the inherent isolation afforded by the transformer.

The switch input current waveform is triangular. It begins at the start of the switch on cycle ($S1$ closed) and rises linearly to the end of the cycle to a peak value approximately twice that of the forward converter. The rms input ripple current is about 70 to 80% greater than that of an equivalent forward converter. This requires higher power switches and diodes to maintain low switch saturation and diode forward voltages. These voltages are dissipative losses that reduce efficiency and generate internal heat.

Peak switch current in a flyback converter is equal to:

$$I_{peak}(S1) = I_{out}\left(\frac{V_{in} + V_{out}}{V_{in}}\right)\left(\frac{N_s}{N_p}\right) \qquad (2\text{-}81)$$

Notice that peak switch current can be reduced to a minimum by using a very small value for $N_s/N_p$. This has two negative consequences, however; the switch voltage and diode current become very large during switch off time, and for a given maximum switch voltage, optimum power transfer occurs at $V_{in} = V_{max}/2$.

Also triangular, the current waveform to the output capacitor(s) has a maximum value at the start of the switch off cycle ($S1$ open) and ramps down linearly to zero during the off cycle. The peak value depends on circuit and load conditions, but it can be estimated as four times the average output current. The rms ripple output current will again exceed that of an equivalent forward converter by 70 to 80%. This imposes stringent requirements on the output capacitor(s) which must handle higher input and output ripple currents that generate heat. A disadvantage of the flyback converter is the high energy which must be stored in the transformer windings in the form of DC current requiring a low-inductance, high-current primary, which in turn requires larger cores than would be necessary with pure AC in the windings. Conventional E and pot core ferrites are difficult to work with because their permeability is too high even with relatively large gaps (.50 to .100 in.).

Because switch $S1$ of the flyback converter is open when load current flows (due to the transformer phasing and $CR1$), overload protection is easier to achieve in a flyback converter than in a forward converter. Provided the transfer is allowed to fully reset at the end of each cycle, the flyback switch never sees the sudden high transient load currents that are seen by the forward converter switch.

To exploit this, the flyback-switch-current limit is held proportional to the output voltage, thereby maintaining a constant output current. This is important because the flyback output current—the load current—depends on the output voltage and the energy stored in the transformer primary. Because this is a constant power output source, a fall in output voltage can produce an increase of output current that can endanger the supply and its load. Many flyback converters reduce the switching frequency under startup and overload conditions to limit output current to safe levels, while minimizing internal supply dissipation.

Figure 2-41 depicts the schematic diagram of the basic single ended DC/DC converter which results from the previous discussion. Note that a square waveform is used to drive the base of switching transistor $Q_1$. What is now needed is a circuit to drive the base and provide the required waveform.

A clamp winding is often used (Figure 2-42) to allow energy stored in the leakage reactance to return safely to the line instead of avalanching the switching transistor.

A 120/220 V AC flyback design requires transistors that block twice the peak line plus transients or about 1 kV. Motorola Semiconductor's MJE 13000 and 16000A series transistors with ratings of 75 V to 1000 V are used here. These bipolar devices are reasonably fast (100 ns) and are typically used in the 20 kHz to 50 kHz operating frequency range. The availability of 900 V and 1000 V

BASIC CIRCUIT

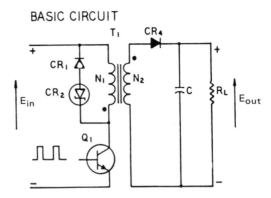

Figure 2-41   Basic single-ended DC/DC converter circuit. Reprinted with permission from *Electronic Design 1*, January 6, 1972, copyright Hayden Publishing Co. Inc. (1972).

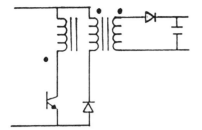

Figure 2-42 Flyback converter with clamp winding.

MOSFETs allows operation in the 50–100 kHz range, with some square wave designs achieving 300 kHz operation.

The two transistor variation of this circuit (Figure 2-43) eliminates the clamp winding and adds a transistor and diode to effectively clamp peak transistor voltages to the line. With this circuit, a designer can safely use the faster 400 V to 500 V bipolar or FET transistors and push operating frequencies considerably higher. But, there is the added cost over the single transistor circuit due to the extra transistor, diode, and base drive circuitry.

Figure 2-44 represents a base drive circuit in which the base drive is taken from a tertiary winding of $T_1$ much in the fashion of a blocking oscillator. This circuit is only one of several possibilities, but for low power applications, offers several features for a minimal parts cost. These features are: (1) self starting is provided, (2) $i_p$ is held constant as $E_{in}$ varies, and (3) overcurrent and short circuit protection are inherent.

Circuit operation is as follows. $R_3$ provides a small base current to bring the collector current of $Q_1$ into a region where $\beta$ is large enough to cause sufficient regenerative feedback. This drives $Q_1$ into saturation via the tertiary winding. The $R_3$ current is minimal as $CR_4$ is reversed biased during $Q_1$ saturation thereby resulting in maximum loop gain. As a result, start up is reliable even with a shorted output.

$CR_3$ and $R_2$ are chosen to yield an $i_p$ consistent with previous calculations. When $i_p$ is reached, $Q_1$ becomes a current source, due to $CR_3$ and $R_2$, thereby

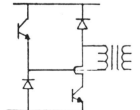

Figure 2-43 Two-transistor forward or flyback converter circuit (clamp winding is not needed).

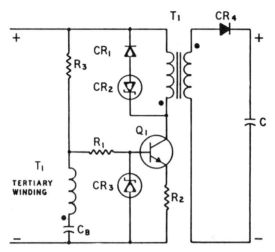

Figure 2-44   Base drive circuit where the base drive is taken from a tertiary winding of the transformer. Reprinted with permission from *Electronic Design 1*, January 6, 1972, copyright Hayden Publishing Co. Inc. (1972).

dropping the voltage across $T_1$ primary and the base drive winding. This action results in $Q_1$ being switched off. $C_B$ is chosen large enough such that the base charge delivered during one cycle of $Q_1$ saturation does not significantly change the voltage across $C_B$. $C_B$ recharges primarily through forward biased $CR_3$ during the $t_2$ flyback interval. Recharge of $C_B$ through $R_3$ is generally impractical as it would dictate a significantly lower $R_3$ value resulting in excessive dissipation.

Overload protection is inherent as $i_p$ is limited.

Some helpful comments regarding practicalities of this circuit follow:

1. Do not run the output unloaded as this will cause all of the power to be delivered to $CR_2$ with possible failure due to excessive dissipation.
2. Try to achieve tight coupling in the transformer. This will minimize the energy stored in the primary leakage reactance that must be absorbed by $CR_2$.
3. $CR_1$ need not be a fast switching diode as it has almost the full $t_2$ time to recover.
4. $CR_4$ need not be a fast switching diode although performance is degraded somewhat if one is not used. Long reverse recovery time of $CR_4$ prevents $Q_1$ from switching on to initiate the next cycle because the secondary winding is held positive during the recovery interval by the output capacitor.

    This lengthens the $t_2$ time and dissipates some of the energy just delivered to the output. Both of these effects reduce delivered output power. However, because $Q_1$ cannot be driven on until $CR_4$ is recovered, large noise-producing current spikes normally associated with slow reverse recovery are avoided.

5. Multiple outputs, of the same polarity, may be obtained by tapping the secondary and adding a rectifier and filter capacitor for each output. All outputs will track in ratio to each other as they are locked together by the transformer turns ratio. To insure tight tracking, the secondary copper current density may be lowered.

6. Multiple outputs of opposite polarity may be obtained by use of a second secondary winding, oppositely phased.

The circuit of Fig. 2-44 represents one type of base drive circuit, but there are others that are more widely used. Two base drive control methods found to be popular among the many existing converters using flyback circuits are (1) the proportional voltage-control method and (2) the bistable control method. Figure 2-45 shows a general block diagram of a converter employing a two-winding energy-storage converter. For a proportional voltage-controlled converter, the comparator often is a differential amplifier, which detects the error voltage between the output and the given reference, amplifies it, and provides a proportional DC voltage. This voltage then controls the ratio $t_{on}/t_{off}$ of the power transistor through some sort of voltage to duty-cycle encoder. A practical implementation of this method of control is shown in Figure 2-46.

This circuit (Figure 2-46) operates in the following manner, assuming the circuit is in regulation.

The error amplifier generates a voltage proportional to the difference of the $E_{out}$ and the $V_{reference}$. This voltage is divided down by $R_3$ and $R_4$ and is applied to the level detector inverting input.

When the 20 kHz clock pulse goes high, the $Q$ output goes high and turns

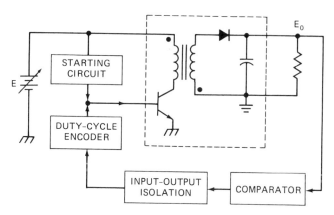

Figure 2-45  Converter block diagram using a two-winding energy-storage circuit as an example. Reprinted with permission from *Solid State Power Conversion*, January/February 1978.

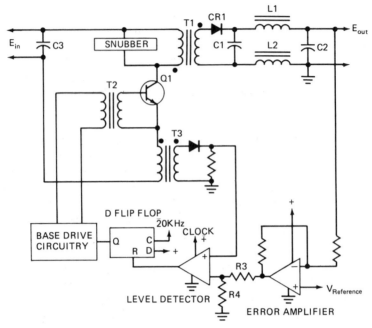

Figure 2-46   Basic circuit of an isolated single stage, single output flyback regulator. Reprinted with permission from *Solid State Power Conversion*, January/February 1978.

on transistor $Q_1$ through the base drive circuitry and transformer $T_2$. Current builds up as was shown in Figure 2-38b and allows energy to be stored in $T_1$.

This primary current also flows through transformer $T_3$ which serves as a current transformer; therefore, the voltage applied to the noninverting input of the level detector is a ramp directly proportional to the collector current of $Q_1$.

When the current ramp voltage equals the inverting input voltage from the error amplifier, the $D$ flip-flop is reset and $Q_1$ is turned off. The energy now stored in $T_1$ is dumped into the load until the clock sets $Q$ high which restarts the whole cycle again.

It can be seen that when the output voltage is low, the output of the error amplifier increases and causes more energy to be stored and dumped, each cycle thereby increasing the amount of power being delivered to the load. Of course, the inverse of this is also true. If the output voltage decreases, the amount of energy stored and dumped each cycle will decrease. This action gives the required regulation by regulating the amount of power delivered to the load.

The output filter, made up of $L_1$, $L_2$, and $C_2$, is used for filtering out the 20 kHz ripple caused by the ESR of the capacitor and the ripple current. Generally, this filter can be very small.

Also shown in Figure 2-46 is a snubber circuit around transformer $T_1$. Needless to say, there is leakage inductance associated with $T_1$. When $Q_1$ is turned off, this leakage inductance can cause a large voltage spike that may destroy $Q_1$ by exceeding the safe operating voltage. So, a snubber circuit is placed around the transformer to absorb this energy.

Using the current ramp of $Q_1$ to generate the regulating ramp has some unique advantages that should be discussed. One of the most common causes of failures in a switching regulator is that the current in the power transistor gets too high. The current transformer allows continuous monitoring of this current. If the voltage applied to the level detector inverting input is limited, the emitter collector current is limited, which therefore protects $Q_1$.

Another advantage of limiting the current of $Q_1$ is that the amount of energy stored is limited which means the amount of power delivered to the load is limited. Since the output is a fixed voltage, then the output current is limited without the use of a series resistor.

However, the current will not fold back but will follow a constant power curve. Foldback can be achieved by feeding $E_{out}$ to the junction of $R_3$ and $R_4$ through a diode and resistor network. This will reduce the amount of current delivered to the load at a given voltage.

Another advantage of using the current ramp for regulation is that high ripple rejection from the input line is achievable. The transistor $Q_1$ will stay on long enough to store the required amount of energy each cycle. If the input voltage drops, $Q_1$ will conduct longer and vice versa. Therefore, the amount of energy delivered to the load is a constant for a constant output of the error amplifier.

A bistable controller converter, on the other hand, generally uses a bistable comparator such as a Schmitt trigger circuit to generate a high-voltage or a low-voltage signal which in turn either directly controls the $t_{on}$ and $t_{off}$ of the power transistor or indirectly controls the ratio of the two times as some modified function of the two-level output from the comparator.

Figure 2-47 shows an example of a bistable controlled DC/DC converter. The output voltage is 28 V regulated to within $\pm 0.26\%$ for an input voltage range of 24 to 32 V and an output of zero to 8 W. The conversion frequency remains in a narrow range from 3.7 to 3.85 MHz. Efficiency is in the low 40% range at an output power of 4 W and increases only to the mid 50% range at an output of 7 W. At first glance, this efficiency may appear surprisingly low. However, after consideration of a breakdown of the power losses in the circuit, the low efficiency becomes convincingly reasonable. For example, this converter has a measured efficiency of 57.1% at an input voltage of 24 V and an output power of 7.25 W. The measured input power was 12.7 W. The difference 12.7 − 7.25 = 5.45 W is the regulated-converter power dissipation.

With 0.2 W saturation loss, the power switch alone contributes about 2.9 W. The remaining 2.55 W is made up of 1.44 W from the duty-cycle encoder, 0.51 W from the bistable comparator, 0.15 W from the diode, and 0.45 W from other losses such as the energy-storage core, copper losses, etc. It may be noted

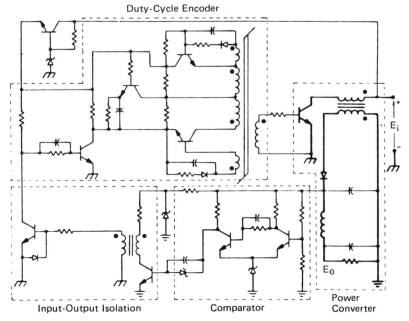

Figure 2-47   Regulated DC-to-DC converter with bistable control. Reprinted with permission from *Solid State Power Conversion*, January/February 1978.

that the switching loss in this circuit is about two and a half times larger than that of a comparably designed proportional voltage controlled DC/DC converter. This is caused by the shorter period $T$ and the fact that both the current and the voltage of the power transistor are higher in this configuration than in the previous converter.

Multiple regulated outputs are achievable using the flyback regulator because the flyback converter provides better cross-regulation than other converter topologies (i.e., changing the load on one winding has little effect on the output voltages of the other windings). A transformer with multiple secondary windings is shown in Figure 2-48. The energy is stored in the core, as described before, by turning $Q_1$ on. However, when $Q_1$ is turned off energy is dumped into the secondary windings and during this cycle the equation

$$\frac{N_1}{V_1} = \frac{N_2}{V_2} = \frac{N_3}{V_3} = \frac{N_4}{V_4}$$

(2-82)

will hold for the generated voltage.

The easiest way to see this is if one considers that all of the secondary windings, $N_1$, $N_2$, and $N_3$ have an equal number of turns. If, for instance, the

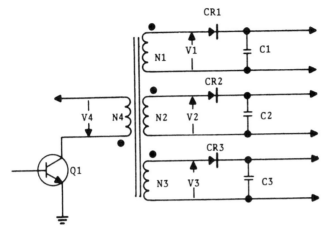

Figure 2-48  Multiwinding flyback transformer. Reprinted with permission from *Solid State Power Conversion*, January/February 1978.

voltage on $C_3$ was less than that on $C_1$ and $C_2$, the winding $N_3$ would clamp to $C_3$ plus one diode drop. But since the other capacitors have a higher voltage, their respective diodes would not become forward biased. Therefore, all of the energy stored would begin to dump into $C_3$. If the voltage $V_3$ was raised high enough to allow the other diodes to become forward biased, then some of the energy stored in the core would be dumped into $C_1$ and $C_2$.

Regulation therefore can be obtained on two of the windings by regulation of the third. The regulation of the transformer is not as good as a conventional DC to DC converter since the core material gives poorer coupling, however, the regulation can be tightened by using multifilar and parallel winding techniques.

Increasing the output of a flyback regulator can be accomplished by simply paralleling the number of stages for the desired power. A schematic of a parallel stage regulator is shown in Figure 2-49. As indicated, the number of stages is unlimited, so the output power deliverable by a flyback regulator is unlimited. The sharing of these parallel stages is inherent since the transformers are essentially current sources rather than voltage sources.

Paralleling stages also can decrease the amount of ripple current applied to $C_1$. By timing the parallel stages so that the transformers alternately dump their stored energy into the load, the current flow will be spread over more of the cycle, thereby decreasing the ripple current. In fact, it is possible to time the energy dump periods so that the secondaries essentially deliver constant current to the capacitor and load. This decreases the size requirements of $C_1$ since the transformers can be used as a storage means.

Reliability can also be increased by using parallel stages. If the number of stages can be increased to the point where failure of one or more stages will

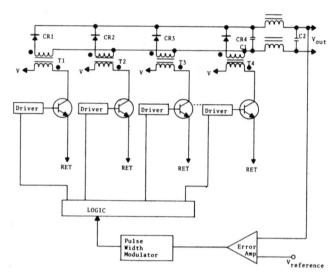

Figure 2-49  Flyback regulator with paralleled stages to obtain higher output power. Reprinted with permission from *Solid State Power Conversion*, January/February 1978.

not affect the load, then the reliability can be improved. This can be accomplished by fusing the transistor on the primary side and fusing the secondary diode. The failure of these components will cause that stage to be removed from the primary and/or load, allowing the rest of the stages to furnish the required power.

In summary, the flyback converter offers some unique advantages over other switching converter topologies. These are as follows:

1. Reduced number of components are required.
2. Inherent short circuit protection is afforded.
3. Inherent power transistor protection is provided.
4. The regulator has a large dynamic range.
5. Many stages can easily be paralleled to provide higher output power.
6. Multiple regulated outputs can be achieved from a single stage.
7. Reduced conducted susceptibility.

Efficiencies of up to 90% are obtainable with flyback switching regulators, while power losses are similar to those in other types of high efficiency regulators. The flyback converter is probably the most widely used converter for output levels below 150 W.

### The Forward Converter

The single-transistor forward converter is shown in Figure 2-50. Although it initially appears very similar to the flyback converter, it is not. Yet like the

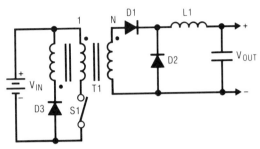

Figure 2-50   Forward converter schematic diagram. (Reprinted with permission from "LT1070 Design Manual," Application Note 19. Courtesy of Linear Technology Corporation)

flyback converter, it is a very popular low power converter and is practically immune from transformer saturation problems.

The operating model for this circuit is actually the buck regulator discussed earlier. But instead of storing energy in the transformer and then delivering it to the load, this circuit uses the transformer in the active, or forward mode, and delivers power to the load while the transistor is on. It does this, however, at the expense of an extra winding on the transformer, two more diodes, and an additional output filter inductor. Power is transferred from the input to the load through $D1$ during switch on time. When the switch turns off, $D1$ reverse biases and inductor ($L1$) current flows through $D2$. The output voltage is equal to:

$$V_{\text{out}} = V_{\text{in}} \left( \frac{N_{\text{sec}}}{N_{\text{pri}}} \right) D \qquad (2\text{-}83)$$

The additional winding and $D3$ are required to define switch voltage during switch off time. Without this clamp, switch voltage would jump all the way to breakdown at the moment the switch is opened due to the magnetizing current flowing in the primary winding. This reset winding generally has the same number of turns as the primary, is usually bifilar wound, and clamps the reset voltage to twice the line voltage. However, its main function is to return energy stored in the magnetizing inductance to the line and thereby reset the core after each cycle of operation. Because it takes the same time to set and reset the core, the duty cycle of this circuit cannot exceed 50%.

The output voltage ripple of the forward converter is low because of $L1$, but input ripple current is high due to the low duty cycles normally used. A smaller core can be used for $T1$ compared to the flyback converter, because there is no net DC current to saturate the core. Standard ferrite cores work well here.

In the forward converter, the transformer generally has no gap because the high permeability core material (3000 gauss) minimizes the magnetizing current ($I_M$). Thus, the core should be chosen large enough so that the resulting $LI$

product insures that $I_M$ is less than $I_{sat}$ at the operating voltage. For flyback designs, which require lower permeability cores than other switcher topologies, a large gap is necessary.

$$L_g \gg L_m/\mu$$
$$L_g = \text{gap length}$$
$$L_m = \text{magnetic path length} \qquad (2\text{-}84)$$
$$\mu = \text{permeability}$$

The gap directly controls the $LI$ parameters, and doubling it will decrease $L$ by two and increase $I_{sat}$ by two. Again, the anticipated switching currents must be less than $I_{sat}$ when the core is gaped for the correct inductance.

The transistor waveforms shown in Figure 2-51 illustrate that the voltage requirements for the forward converter are identical to those for the flyback converter. For the single transistor versions, 400 V turn on and 1 kV blocking devices like the MJE13000 and MJE16000 deflection transistors are required.

Again, the two-transistor circuit variation shown in Figure 2-43 adds a cost penalty but allows a designer to use the faster 400 to 500 V devices. With this circuit, operation in the discontinuous mode refers to the time when the load is reduced to a point where the filter choke runs "dry." This means that choke current starts at, and returns to, zero during each cycle of operation. Most designers prefer to avoid this mode of operation because of higher ripple and noise, even though there are no adverse effects on the components themselves.

Few high-speed switching transistors can withstand these stresses. The voltage and current capabilities of available bipolar transistor switches limit the maximum output power of the single-ended forward converter to about 150 W. The switching speed limitations of these devices restrict the maximum efficient operating frequency to about 40 kHz.

A multiple-output forward converter is shown schematically in Fig. 2-52. This higher power circuit typically derives a 200 to 400 V DC operating voltage

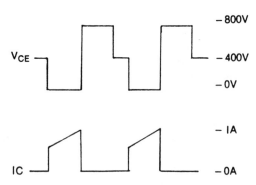

Figure 2-51   Forward converter transistor waveforms.

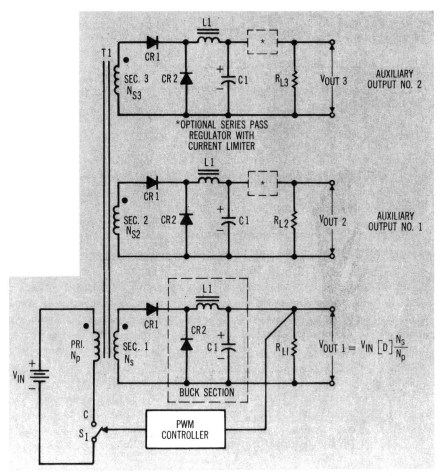

Figure 2-52 Multiple output forward converter with DC Input/Output isolation. (Reprinted with permission from "Switching Architectures—Key to Selecting the Right Power Supply," *Electronic Products*, September, 1983, Hearst Business Communications, Inc.)

from an off-line rectifier/filter section. In operation, the collector of the transistor switch (switch S1, position *C*) ideally swings positive during the reset cycle by an amount equal to the 200 to 400 V DC operating voltage. In practice, the overswing can reach 400 to 800 V, with 1000 V possible, due to leakage inductances and stray capacitance.

## The Cuk Converter

The Cuk converter, which is basically a boost converter preceeding a buck converter, is the result of a deliberate effort by Professor Slobodan Cuk of Cal

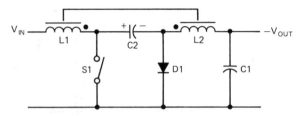

Figure 2-53   The Cuk converter.

Tech to synthesize a switching power supply with as many desirable properties as possible. Like a buck-boost converter, the Cuk converter's output voltage is the opposite polarity of the input voltage, but with the advantage of low ripple current at both input and output.

From the basic Cuk converter schematic diagram of Figure 2-53, a single switch and a single commutator diode are required in conjunction with two inductors. $C1$ serves as the main energy transfer link between input and output.

The Cuk converter, like the buck-boost converter, can provide an output voltage magnitude above, equal to, or below the value of its input voltage, depending on the duty cycle of $S1$. Ideally, the relationship between input and output voltages is simply $D/(1 - D)$. Switch, diode and capacitor currents are comparable in the Cuk converter to those of their counterparts in the buck-boost converter operating with the same voltage transformation ratio and input power level. When operating in the continuous mode, the input and the output currents of the Cuk converter are nonpulsating, while associated AC ripple currents can be made arbitrarily small by increasing the values of inductors $L1$ and $L2$, often eliminating the need for additional input/output EMI filters in some applications. Because the Cuk converter contains two inductors, it is possible to operate the converter so that one inductor is operating in the continuous current mode and the other in a discontinuous mode. However, it is important to prevent this particular possibility from occurring since it leads to peculiar power conversion transfer characteristics.

The need for two inductors can be eliminated by winding them both on the same core, with exact $1:1$ turns ratio. With slight adjustments to $L1$ or $L2$, either input ripple current or output ripple current can be forced to zero. An improved version even exists which results in both ripple currents going to zero, considerably easing the requirements on size and quality of input and output capacitors and without requiring filters.

The switch must handle the sum of the input and output currents:

$$I_{pk}(S1) = I_{in} + I_{out} = I_{out}\left(1 + \frac{V_{out}}{V_{in}}\right) \qquad (2\text{-}85)$$

The ripple current in $C1$ is equal to $I_{out}$, so this capacitor must be large. It can be electrolytic, however, so physical size is not normally a problem.

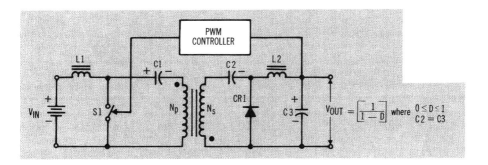

Figure 2-54 The transformer-isolated Cuk converter. (Reprinted with permission from "Switching Architectures—Key to Selecting the Right Power Supply," *Electronic Products*, September, 1983, Hearst Business Communications, Inc.)

DC isolation can be added to the basic Cuk converter by the addition of a transformer, as shown in Figure 2-54. It should be noted that the transformer does not store energy as part of the conversion action. Because of the presence of capacitors in series with each winding, no DC voltage can appear across the transformer winding. This allows full utilization of core flux capability. Thus, the transformer is much easier to design, has better cross-regulation properties between outputs, and is usually smaller in physical size than the transformer of a flyback converter. The transformer-isolated Cuk converter has high operating efficiency and the lowest ripple of all switching power supply topologies.

The Cuk converter exhibits a characterstic that it shares with the boost and buck-boost converters: the transfer function contains a right half-plane zero. This makes stable feedback loops with good transient response difficult to achieve. Additionally, those design implementations of the Cuk converter that utilize integrated magnetics (winding the inductors or the inductors and transformer on the same core) present the following disadvantages: high output voltage ripple due to complex physical construction and required gap "tweaking"; high frequency harmonics; and source noise coupled to the output due to capacitive and inductive coupling. It is both for these reasons and because the Cuk converter is a new and patented topology that it is not widely used.

## HIGH-POWER CONVERTER TOPOLOGIES

The high-power bridge and push-pull converter topologies shown in Figure 2-55 all operate the transformer in the bipolar or push-pull mode and require two or four power transistors. Because the transformers operate in this mode, they tend to be almost half the size of those in the equivalent single-transistor converters, thereby providing a cost advantage over their counterparts at power levels of 200 W to 2 kW.

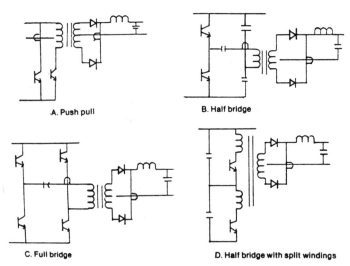

A. Push pull

B. Half bridge

C. Full bridge

D. Half bridge with split windings

Figure 2-55    High-power converter topologies (200W–2kW).

In comparing the single-ended converter with the bridge and push-pull types, the transistor utilization of the single-ended converter is only one half as effective as either the bridge or push-pull converters. Transformer utilization is also lower. Since the single-ended converter drives transformer flux in one direction, it requires twice the core cross-sectional area for the same volts-per-turn. In addition, copper utility is lower.

The rms winding current equals 1.64 times the DC input or output current. This compares to a 1 : 1 ratio of delivered DC to rms winding current in a bridge or totem pole converter with a full-wave output rectifier. Multiplying the flux swing utilization by the copper utilization requires the single-ended converter transformer to have 3.28 times the $A_c A_w$ product of a the bridge and push-pull converter transformers.

Because the transistor and transformer utilizations are significantly lower in the single-ended converter, cost disadvantages become evident at higher power levels where these items become a large fraction of the total parts cost. Yet at low power levels, this disadvantage disappears since the drive circuitry for the single-ended converter is extremely simple and cost effective.

Bridge and push-pull converters suffer from three major problems which are absent in the single-ended converter. First, the transformer primary must be AC-coupled or by other means employed to insure symmetrical primary volt-second balance, otherwise the core will saturate. This is not a problem in the single-ended converter because the flux must reach $B_r$ before the next cycle can be initiated. Second, in a bridge and push-pull converter, means must be provided to insure that overlap (both switching transistors being on during the switching interval) does not occur. Otherwise destructive current spikes will occur. Third, bridge and push-pull DC/DC converters are not inherently overload protected. Because of the necessary circuitry complications attendant with overcoming these problems, the single-ended converter may be cost effective at power levels higher than that dictated by transistor and transformer utilization only. In addition, if fine regulation is required, it is more easily implemented in a single-ended converter than in a bridge or push-pull converter.

## The Half-Bridge Converter

The most widely used topology after the single-ended forward and flyback converters is the buck-derived half-bridge configuration (shown in Figure 2-56). In this circuit, two switches alternately connect one end of transformer $T1$ to $V_{in}$ or ground, with the other end of $T1$ maintained at a voltage set by $V_{in}$ and the $C1$–$C2$ divider. $V_{in}$ can range between 200 and 400 V DC, but in this circuit the maximum voltage across switches $S1$ and $S2$ (switching transistors) is 400 V DC. This lower peak voltage allows the use of MOSFETs to operate at switching frequencies above 100 kHz; approximately 40 kHz is the limit for bipolar transistor designs. The output choke is smaller because it operates at twice the frequency of the single-ended equivalent.

The half-bridge converter provides an output voltage $V_{out} = (D) \times (N_s/N_p) V_{in}$ and requires more complex control circuitry, including isolated drive for switch $S1$: the switch floats because it is not directly connected to ground. Somewhat more efficient in transformer utilization, the half-bridge forward converter has a higher parts count than its single-ended counterparts.

Compared with the push-pull converter, the half-bridge converter has several distinct advantages over the push-pull converter:

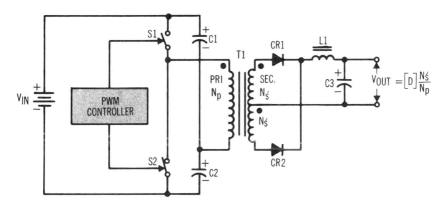

Figure 2-56 The half-bridge forward converter. (Reprinted with permission from "Switching Architectures—Key to Selecting the Right Power Supply," *Electronic Products*, September, 1983, Hearst Business Communications Inc.)

1. The transistors switch at the applied DC voltage $V_{in}$ instead of $2\ V_{in}$ as in the push-pull case and readily available 400 V bipolar transistors may be used.
2. Transformer saturation problems are easily minimized by use of a small coupling capacitor (about 2-5 $\mu$F), as shown in Figure 2-56. Because the primary winding is driven in both directions, a full-wave output filter, rather than half, is now used and the core is actually utilized more effectively.
3. The leakage inductance problem associated with the push-pull circuit causes large spikes of voltage to appear at the switching transistors. The half-bridge converter does not have this problem because the energy stored in the leakage inductance of the power transformer is conducted back into the DC bus through the commutation diodes.
4. Unequal turn-off times in the power transistors results in core saturation with the push-pull converter. By using the half-bridge topology, the addition of a series capacitor feeding the primary of the power transformer eliminates the DC core flux problem. The unbalance of turn-off times merely produces a shift in the mean DC level of the waveform so that the transformer now sees balanced voltseconds, as shown in Figure 2-57.
5. The input filter capacitors are placed in series across the rectified 220 V line which allows them to be used as the voltage doubler elements on a 120 V line. This still allows the converter transformer to operate from a nominal 320 V bus when the circuit is connected to either 120 V or 220 V.
6. This topology allows diode clamps across each transistor to contain destructive switching transients. Transistors that can handle a clamped inductive load line at rated current are now available to aid the power supply designer. However, the older designs in this area still use snubbers to protect the

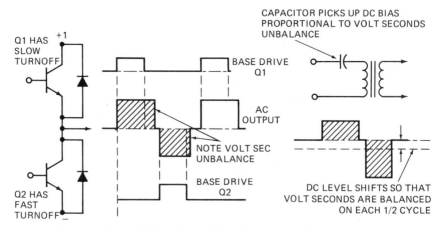

Figure 2-57   AC coupling of power transformer.

transistor which sacrifices both cost and efficiency. Another variation of the half-bridge converter is the split winding circuit (as shown in Figure 2-55 D). A diode clamp can protect the lower transistor, but a snubber or zener clamp must still be used to protect the top transistor from switching transients. Because both emitters are at an AC ground point, expensive drive transformers can now be replaced by lower cost, capacitively-coupled drive circuits.

The half-bridge converter is used in applications where it must supply from 500 W to 2 kW of output power. The effective current limit of TO-3 encased power transistors (300 mil die) is somewhere in the 10 to 20 A range. Once this limit is reached, the designer generally changes to the full-bridge topology (as shown in Figure 2-55 C).

## The Full-Bridge Converter

The buck-derived, full-bridge converter (see Figure 2-58) is generally found in higher-power-output power supplies, 500 W to 2 kW. This topology delivers an output voltage of $V_{out} = (2) \times (D) (N'_s/N_p) V_{in}$ and except for its higher power capability, it has the same advantages and disadvantages as the half-bridge configuration. This design requires four 400V switches and even more complex drive circuitry (four isolated drive signals are required) than the half-bridge converter. Because full line rather than half is applied to the primary winding, the power out is double that of the half-bridge converter with the same switching transistors. The circuit of Figure 2-59 is designed to provide isolated

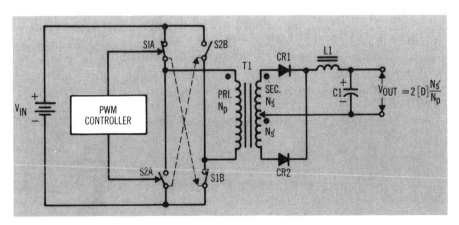

Figure 2-58 The full-bridge forward converter. (Reprinted with permission from "Switching Architectures—Key to Selecting the Right Power Supply," *Electronic Products*, September, 1983, Hearst Business Communications Inc.)

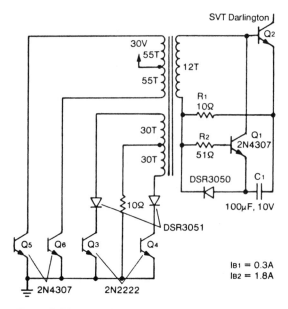

Figure 2-59   Transformer isolated base drive circuit.

base drive in converters configured as half- or full-bridge. The transformer secondary voltage is a quasi-square wave. As the secondary switches positive, a large initial pulse of base current is drawn due to the charging of $C_1$. Following the rapid charging of $C_1$, the base drive current is maintained at a level determined by $V_{BE}$ of $Q_2$ and $R_1$. During the turn-off, the transformer secondary goes to zero due to the shorting of the transformer primary by $Q_3$ and $Q_4$. At this time, the base of $Q_1$ is forward-biased by the capacitor and turns on. This connects $C_1$ across the base-emitter of $Q_2$ resulting in a large reverse $I_{B2}$ through the base of Darlington $Q_2$. $I_{B2}$ is determined by capacitor and circuit resistances and the characteristics of $Q_1$ and $Q_2$. Following the turn-off of $Q_2$, current out of $C_1$ is determined by the base resistances of the Darlington and the base drive current required for $Q_1$. This current can provide reverse bias voltage for $Q_2$ during the turn-off interval for properly selected component values. $Q_1$ should be fast and capable of handling peak collector currents of up to 5 A. A tantalum slug-type capacitor is suitable for $C_1$.

Minimum on time constraints must be observed in this circuit in order to insure that $C_1$ is charged. (In addition, duty cycles of greater than 50% cannot be achieved in this circuit without imposing DC operating conditions on the transformer.)

With the addition of another winding and diode as shown in Figure 2-60, this last limitation is overcome. The primary circuit remains the same but now operates at half the desired output frequency.

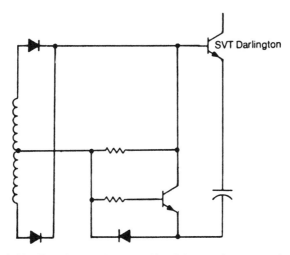

Figure 2-60   Transformer drive capable of duty cycles greater than 50%.

For circuits using the Darlington in the grounded emitter configuration, the circuit shown in Figure 2-61 can be used. The negative bias required for turn-off is created by charging capacitor $C_1$ during the on interval. $I_{B1}$ is set by resistor $R_1$. When $Q_3$ is turned on, capacitor $C_1$ is effectively connected to the base emitter of $Q_4$ causing reverse drive current $I_{B2}$ to flow. $I_{B2}$ is set by the gain of $Q_3$, the charge on $C_1$, and the circuit impedances. Zener diode $DZ_1$ limits the charge on $C_1$, and provides a path for base drive currrent to $Q_4$. Diodes $CR_1$, $CR_2$, and $CR_3$ prevent $Q_3$ from saturating, resulting in faster circuit response to the input signal. Grounding the base of $Q_1$ causes $Q_4$ to conduct, whereas a high level to $Q_1$ initiates turn-off.

With the circuit values indicated, the circuit will generate an $I_{B1}$ of 0.5 amps and $I_{B2}$ of 2.7 amps at 20 kHz with a pulsewidth of 10 $\mu$s. As the on pulse decreases, at some point the voltage on $C_1$ drops below that of the zener clamp. At reduced pulsewidth the turn-off becomes less energetic with resulting longer storage and fall times. At the shorter pulsewidths, $I_{B1}$ increases due to the lower voltage on $C_1$. This circuit is usable down to pulsewidths of about 5 microseconds.

In applications where the transistor frequencies are low or balanced voltsecond content cannot be assured, it becomes difficult to provide a driver transformer with good performance characteristics and low cost. In this situation it is recommended that isolated bias supplies be provided for each transistor. The bias voltage can economically be provided by small 60 Hz transformers followed by bridge rectifiers and light filtering. Control isolation is provided by a high-speed optical coupler which permits the driver to be controlled directly from logic. The forward drive current reaches its final value in about 200 ns from application

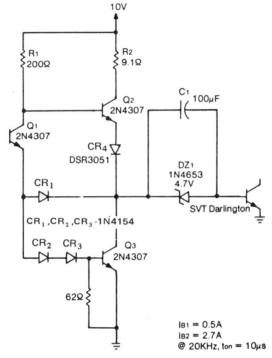

Figure 2-61   Capacitor coupled base drive circuit.

of the input signal. The turn-off drive ($I_{N2}$) responds to the input command in less than 150 ns. (Chapter 3 contains a section on commonly used base drive circuits.)

## Push-Pull Converters

Since the buck-derived, push-pull converter shown in Figure 2-62 is historically one of the oldest converter circuits, its characteristics will be developed at length for illustrative purposes. The push-pull converter delivers an output voltage equal to

$$V_{out} = 2D \frac{N_{s'}}{N_{p'}} V_{in} \tag{2-86}$$

and like the buck derived full-bridge converter, this converter is much more economical at higher power levels than at low power levels.

The push-pull converter topology has the advantage over a single-ended converter in that the voltage across the transformer and hence the peak voltage

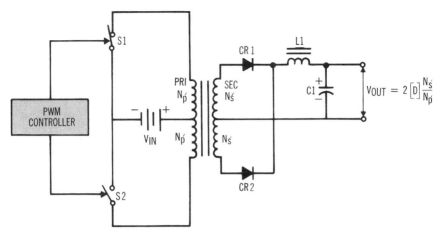

Figure 2-62   The push-pull converter. (Reprinted with permission from "Electronic Architecture – Key to Selecting the Right Power Supply," *Electronic Design*, September, 1988, Hearst Business Communications Inc.)

applied to the transistors is limited to twice the supply voltage. Also, the power supplied to the load is never stored in the transformer, but passes from the conducting transistors through the transformer and into the load. In this way, more power can be handled at greater efficiency and with better regulation than with a single-ended circuit.

For the push-pull circuit to perform at maximum efficiency, the circuit should work into the load on both halves of the operating cycle; that is, the output circuit should be either a full-wave or a voltage-doubler rectifier system. When a half-rectified output is required, only one of the transistors sees the load. Therefore, the other transistor may be a low power type which is required only to pass inductive current for resetting the transformer core.

## Methods of Push-Pull Operation

Push-Pull converters can be constructed from the following configurations.

- *Sinusoidal*: Class C. This method of operation is the most efficient way of providing AC power in sinusoidal form, but in cases where the output is to be rectified, neither the DC/AC inversion nor the final rectification is as efficient as square-wave operation.
- *Square-Wave: Driven Output Stage.* The main disadvantages of a driven system are the low overall efficiency due to the drive stage and the necessity for either matched output transistors or a large output transformer to carry the difference current without saturating. On the other hand, such a system

is advantageous in that the frequency of operation is completely independent both of the applied load and of the supply voltage to the output stage.
- *Square-Wave*: *Self-Oscillating*. With this system of operation, feedback is
  applied to the output transistors from the output transformer. The voltage
  across the transformer is limited by the bottoming of the transistors so that
  the feedback voltage is determined by the supply voltage and the feedback
  turns ratio. The mode of operation of the device depends strongly on
  whether or not the transformer is designed to saturate.

In the 1960s and 1970s, the majority of push-pull converters were of the self-
oscillating variety. However, today this has changed such that driven converters
are the predominant configuration. Table 2-2 compares the main features of the
single-ended, sine-wave, self-oscillating and driven square wave, and transistor
bridge oscillators.

### Using a Nonsaturating Transformer

During one-half of the operating cycle, one transistor is bottomed and a constant
voltage is developed across the transformer primary winding. A certain amount
of drive voltage is therefore applied to the transistor from the output transformer.
   Since the bottomed transistor has a low output resistance, its collector current
is determined by the load, which consists of a reflected load resistance in parallel
with the transformer shunt inductance. The current through the resistive portion
($I_R$) is constant, and the current through the inductance ($I_L$) increases linearly
with time. When the total collector current reaches its maximum value ($I'$),
determined by the drive, the voltage across the transformer falls and feedback
causes the transistor to switch off. The time for the half-cycle ($T$) is therefore
given by:

$$I' = I_R + (V_s T/L) \qquad (2\text{-}87)$$

where $V_s$ is the supply voltage and $L$ the transformer shunt inductance. Due to
the self-balancing action of the transformer, it can be shown that the appropriate
value of $I'$ is the average of the peak currents available from the two transistors.
   It also can be seen that the half-cycle time $T$ depends on $I_R$, which means
that there is considerable variation of operating frequency with load.

### Using a Saturating Transformer

When the transformer is designed to saturate using a square-loop material, the
resistive current $I_R$ is constant as before, but the inductive current is initially
small and increases rapidly as saturation is reached. In this way, the half-cycle

**Table 2-2   Comparison of Primary Types of Oscillators**

| Sine-Wave Oscillators | Single-Ended Oscillators | Parallel Oscillators | | Transistor Bridge Oscillators |
|---|---|---|---|---|
| | | Self-Oscillating | Driven | |
| Low efficiency | Higher efficiency | Highest efficiency | Lower efficiency than self-oscillating type | High efficiency |
| Low-power applications (to about 50 W) | Low-power applications (to about 50 W) | Power levels to 400 W | Power levels to 2 kW | Power levels to 2 kW |
| Poor voltage regulation | Good regulation for varying loads | Inherent regulation | Inherent regulation | About same level of generated EMI as parallel oscillators and same size of filter components |
| Less EMI—no switching transistors | More generated EMI | More generated EMI | More generated EMI | Less complex circuitry than parallel oscillators |
| Less filtering problems with base drive circuits | Large filter components | Smaller filter components / Output regulation independent of load | Smaller filter components / Output regulation independent of load | High weight and large size |
| Less circuit complexity / Good frequency stability | More circuit complexity / Poor frequency stability | Complex circuitry / Poor frequency stability | Very complex circuitry / Good frequency stability achieved by master oscillator | Expensive / Symmetrical square waveform requires symmetrical rectifier |
| Low weight and small size | Low weight and small size | Higher weight and larger size | Higher weight and larger size | Twice as many transistors: four / Higher base drive required / Each transistor sees only the supply voltage / Less switching losses |

*Continued]*

**Table 2-2** (*Continued*)

| Sine-Wave Oscillators | Single-Ended Oscillators | Parallel Oscillators | | Transistor Bridge Oscillators |
|---|---|---|---|---|
| | | Self-Oscillating | Driven | |
| Inexpensive | Inexpensive | Expensive | Expensive | Better transformer utilization |
| Low distortion for reactive loads | High peak voltages | Symmetrical square waveform | Symmetrical square waveform | Small transformer size |
| Used for low-noise, low-power applications | One power transistor Limited output power Unsymmetrical output waveform | Two or more power transistors | Contains master and slave oscillators | Inherent regulation |
| Typical examples: Wein-bridge oscillator Colpitts oscillator | High breakdown voltage required | High transistor stresses High collector current and voltage spikes due to transistor storage time | Two or more power transistors | Used for high-voltage power supplies |
| | No storage loss problems | Partially self-protective against voltage spikes | Reduces transistor stresses | |
| | Long duty cycle | Each transistor sees twice supply voltage when cut off | Prevents storage time losses | |
| | Good utilization of transformer | Higher transistor switching losses | Matched output transistors to carry current without saturation | |
| | Separate transformer winding to reset flux: thus greater power loss | Poor transformer utilization—use winding half of time | Partially self-protective against voltage spikes | |
| | Constant power output drive | Critical transformer design | Each transistor sees twice supply voltage when cut off | |
| | High impedance output | Requires more drive power for feedback to maintain stable oscillations | Less transistor switching losses | |

Utilizes half-wave rectifier for DC output

Output voltage and frequency proportional to input voltage

Used for high-voltage, low-power applications

Typical example: Ringing-choke oscillator

Requires symmetrical (full-wave) rectifier to enable load to draw power on both half-cycles

Output voltage and frequency independent of output loading

Most widely used circuits

Flexible design

Less required drive power for feedback

Good transformer utilization

Small transformer size, less critical transformer design

Requires symmetrical (full-wave) rectifier to enable load to draw power on both half-cycles

Most widely used circuit

Good output voltage and frequency regulation against input voltage variation

Flexible design

Used where extremely tight frequency regulation is required (among other factors)

time is made relatively independent of the load and also of the peak current of the transistor.

## Simple Parallel One-Transformer Oscillators

The basic self-oscillating circuit with a saturating transformer is shown in Figure 2-63. The circuit is a simple push-pull oscillator, with the transistors operated in grounded emitter configuration.

The saturable transformer connected with the push-pull amplifier and the

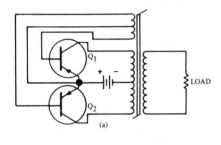

(a)

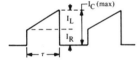
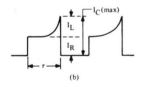

(b)

Figure 2-63  (a) Schematic diagram of the single-transformer push-pull oscillator. (b) Collector current waveforms. Upper: non-saturating transfomer; lower: saturating transformer. (c) Collector voltage waveforms. Upper: normal; middle; insufficient residual inductance; lower: primary not bifilar-wound (insufficient coupling).

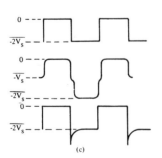

(c)

feedback winding provide positive feedback to the bases of the push-pull pair. The transformer alternately saturates in each direction, which reduces the loop gain to less than unity, causing the transistors to switch "states." The output is a square wave whose frequency is determined by the transformer characteristics. The transformer may be designed to provide a nominal frequency of 400 Hz, but this frequency is directly proportional to input voltage and is somewhat dependent on load. There is, of course, very little voltage regulation.

With one transistor on and bottomed, the collector current has a resistive component $(I_R)$ through the reflected load and an inductive component $(I_L)$ through the transformer shunt inductance.

Because the voltage across the transformer is constant and approximately equal to the supply voltage, the current $I_R$ is constant and the current $I_L$ is initially small (high inductance), and increases rapidly as the transformer nears saturation (Figure 2-63(b), lower). When the transistor can no longer supply the rising current needed to maintain the voltage across the transformer, this voltage falls and feedback causes the conducting transistor to be cut off. The rapid reduction of current in the shunt inductance causes an overshoot voltage to appear across the transformer, and this is applied to the other transistor (which was previously cut off), causing it to conduct. The result is rapid regeneration as the circuit switches over. It is apparent that the switching operation is affected by the value of the shunt inductance in the saturated region, so that too few turns for the area/length ratio of the transformer core cause poor regeneration resulting in an output waveform as shown in Figure 2-63(c), lower. This design parameter is normally covered by other requirements, except at the higher operating frequencies ($> 1$ kHz).

The time $(T)$ for one-half cycle of oscillation may be calculated from the formula:

$$E_{in} = -N_1 \frac{d\phi_t}{dt} 10^{-8} \tag{2-88}$$

where: $E_{in}$ = the supply voltage,
$N_1$ = the number of turns,
$\phi_t$ = the flux in the transformer core.

Integrating equation (2-88) yields

$$E_{in}T = 2N_1\phi_t 10^{-8} \tag{2-89}$$

But $\phi_t = A_c B_m$, where $B_m$ is the saturation flux density in gauss and $A_c$ is the cross-sectional area in square centimeters so that

$$T = \frac{2N_1 A_c B_m 10^{-8}}{E_{in}}$$

and, therefore, the frequency of operation is:

$$f = \frac{1}{2T} = \frac{E_{in} 10^{-8}}{4N_1 A_c B_m} \qquad (2\text{-}90)$$

Another condition that must be satisfied for proper operation is that there should be sufficient current available to saturate the core, having regard to the number of turns on the transformer and the length of the magnetic path. Thus:

$$H = \frac{1.25NI_L}{\ell} > H_o \qquad (2\text{-}91)$$

where $H_0$ has an arbitrary value obtained from the characteristics of the material used, $I_L$ is the inductive current, and $\ell$ is the magnetic path length in centimeters.

Figures 2-64 and 2-65 show actual collector voltage-current traces and collector voltage vs time and collector current vs. time waveforms for the converter of Figure 2-63. From these figures it is seen that this circuit has high current spikes at transistor turnoff which if not suppressed could damage $Q_1$ or $Q_2$ or both. The various techniques available to reduce these spikes are discussed in chapter 3. An improved version of the circuit of Figure 2-63 is discussed at length in a later section (Figure 2-75) where it will be shown that no large collector current spikes are present.

*Efficiency—Power Losses.*   The main losses that occur in the circuit are transistor and transformer losses. These losses are discussed because they must be minimized for high efficiency.

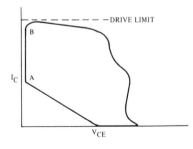

Figure 2-64   Collector voltage-current trace of Figure 2-63. (Redrawn from James Lee Jensen, "An Improved Square Wave Oscillator Circuit," *IRE Trans. on Circuit Theory*, September 1957.)

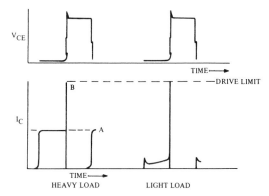

Figure 2-65   Collector voltage and current versus time for Figure 2-63. (Redrawn from James Lee Jensen, "An Improved Square Wave Oscillator Circuit," *IRE Trans. on Circuit Theory*, September 1957.)

- Transistor Losses. Transistor losses are composed of three elements which occur during the "on" and "off" half-cycles and during the transients. The loss during the "on" period depends on the collector current, the bottoming voltage, and the base voltage and current. Loss during the "off" half-cycle depends on the value of $I_{co}$ at twice the supply voltage; at high temperatures this loss may be considerable. In addition, transient losses may be significant, particularly if the switching time is more than a small fraction (1%) of the cycle time.
- Transformer Losses. Transformer losses are composed of two elements: copper loss and hysteresis loss. These losses are determined mainly by the size of the transformer, and are not entirely independent since a large transformer designed for low copper loss has a high hysteresis loss and vice versa. There is, therefore, an optimum size of transformer for a given power handling capacity, although size is not generally a critical factor. In the circuit of Figure 2-63 a toroidal transformer is usually required to obtain optimum performance (low losses).

*Starting Circuits.*   Most converter circuits are self-starting for light loads. However, the basic circuit shown in Figure 2-63 does not necessarily oscillate when a large working load is applied as both transistors are almost cut off and the loop gain is less than unity.

Starting under load is generally achieved by one of three basic circuit techniques:

- Placing a large initial forward bias on the transistors.
- Applying an initial heavy asymmetrical pulse to the circuit.
- Reducing the initial load by a series choke or by a feedback circuit.

When the DC supply is switched on, some current flows in both transistors. As no two transistors are electrically identical, one initially passes more current than the other. This difference or starting current induces a small voltage in the transformer windings. The winding polarities are such that the transistor initially passing the greater current is biased to conduct even more heavily and the other one is cut off. The current in the on transistor increases more and more until the transformer core saturates. Then the induced voltages drop to zero and the base current in the on transistor falls to zero also. The transistors switch over and oscillation starts. When the load is too heavy for the oscillator to start without some auxiliary starting circuit, one of the three arrangements noted above can be used.

The converter transistors may be biased into conduction by means of a voltage divider in the base circuit as shown in Figure 2-66a. Initially, the transistors are both biased on, but as oscillations build up, the base current flowing through $R_2$ causes the mean voltage at the bases to go positive, and the circuit is self-

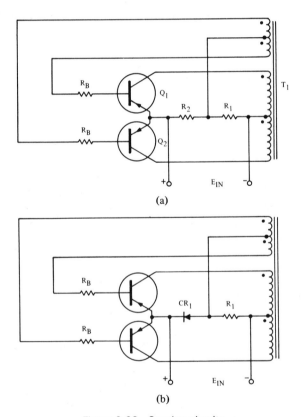

(a)

(b)

Figure 2-66  Starting circuits.

biased into Class C operation. This method of starting is effective, but requires either a large amount of feedback, giving high drive losses or a low value of starting resistor $R_2$, and involving a large current drain through the potentiometer chain. Replacing the resistor $R_2$ with a diode (Figure 2-66b) gives a high resistance when starting and a low resistance, with an almost constant voltage bias, when running, and thereby eliminates the compromise required for the value of $R_2$. The disadvantage of the diode circuit with respect to the resistance circuit is that, in the event of failure to oscillate under overload conditions, the high-resistance diode in the base circuits may cause thermal instability at high ambient temperatures. A resistor may be used to shunt the diode, thus gaining the best of both systems.

The values of the resistors required for starting may be calculated by considering the loop gain of the circuit with a given load. With a total load $R_L$ appearing across the primary of the transformer, the loop gain of the circuit is $g_t R_L/n$ where $g_t$ is the transfer conductance for each transistor stage and $n$ is the feedback turns ratio from the primary winding.

The $g_t$ of each transistor stage is given by:

$$\frac{1}{g_t} = r_e + \frac{R_B}{\alpha_o'} \text{ for } T = 25°C$$

$$= \frac{0.025}{I_E} + \frac{R_B}{\alpha_o'}$$

where $I_E$ is the DC emitter current and $R_B$ is the resistance in series with each base. The value of $r_e$ is both temperature- and current-dependent. For the loop gain to be greater than one:

$$\frac{1}{g_t} < \frac{R_L}{n}$$

$$\therefore \frac{0.025}{I_E} < \frac{R_F}{n} - \frac{R_B}{\alpha_o'}$$

$$I_e > \frac{0.025}{(R_L/n) - (R_B/\alpha_o')} \tag{2-92}$$

But $I_E = \alpha_o' I_B$. Thus

$$I_B > \frac{n}{40(\alpha_o' R_L - n R_B)} \tag{2-93}$$

In the case of the diode circuit the value of $R_1$ can be calculated approximately from

$$R_1 \simeq \frac{1}{2} \times \frac{E_{in}}{I_B} \tag{2-94}$$

When there is a resistor $(R_2)$ between emitter and base, the minimum value of base emitter voltage $V_{BE}$ for the required $I_B$ must be determined, then:

$$R_1 \simeq \frac{E_{in}}{2I_B + [(V_{BE} + I_B R_B)/R_2]} \tag{2-95}$$

The value of $R_2$ is not critical and is usually in tens of ohms.

When the load consists of a rectifier system with a reservoir capacitor, more bias may be required, and this will depend on the value of capacitance used and the leakage inductance of the transformer.

Various starting circuits have been proposed, some of which are illustrated in Figure 2-67. In general, starting devices that introduce high impedances in the feedback circuit tend to introduce undue distortion of the waveform.

Another starting method, allows the converter to operate in a current limiting mode during startup and limits the rate of increase of conduction time with power switches until the filters are charged, without incurring $I_{SB}$ failure of the power switches.

This circuit provides a soft start and operates as follows in Figure 2-68:

Initially the 100 $\mu$F capacitor is discharged and charges through the 470 k$\Omega$ resistor from the +30 V DC source.

Pulses from the 9601 one shot are width controlled by this voltage such that as the 100 $\mu$F capacitor charges the width increases until regulation of the power supply output is achieved.

If an overload is detected, the SCR fires and discharges the 100 $\mu$F capacitor, which inhibits the one-shot pulse output. At this time, the reset circuit, consisting of a 1M$\Omega$ resistor, 2.2 $\mu$F capacitor, and 4-layer diode is energized. After approximately one second, the 4-layer diode fires and resets the SCR by reverse biasing its anode. The circuit then functions in the soft start mode until the regulated output is again established.

Failures at shutdown can be prevented by terminating the drive pulses with some form of crowbar circuit on the waveform generator although this is unnecessary if the DC supply to the waveform generation circuits decays fast enough.

*Practical Design Considerations.*   Using the basic schematic in Figure 2-63, the simplest form of DC converter consists of a square-wave oscillator and an

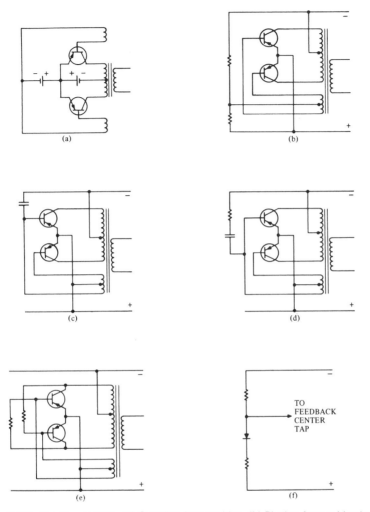

Figure 2-67 Starting circuits. (a) Standing battery bias. (b) Biasing from a bleeder network across the supply. (c) A capacitor between the negative side of the supply and the
base of one transistor. (d) A capacitor and series resistance from the negative side of
the supply to the base of one transistor. (e) Resistances cross-connected between the
collectors and bases. (f) a capacitor between one collector and ground.

output rectifier filter. Schematically, the converter appears as shown in Figure
2-69. When voltage is applied to the input, current flows through $R_1$ to both
bases. Because there is always a slight asymmetry in the circuit, one transistor
turns on. (Assume $Q_1$.) Current flows in winding 2-1, magnetizing the transformer core and all dotted winding ends take negative polarities (with respect

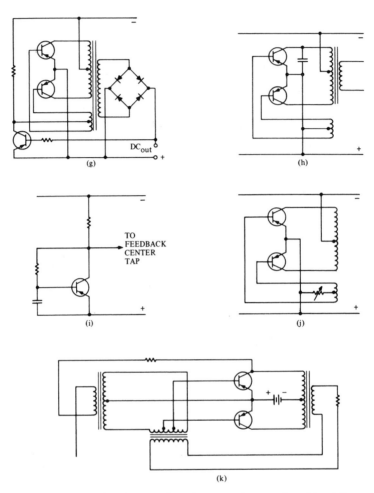

Figure 2-67 (*Continued*) (g) a transistor between the negative side of the supply and the center tap of the feedback windings. (h) Extra feedback turns to increase the feedback more than necessary for normal operation, and a series resistor in the base circuit to adjust the feedback. (i) Combined current and voltage feedback. (j) Capacitance-resistance parallel network from the negative side of the supply to the center tap of feedback windings, with time-constant of the order of the oscillation period. (k) An inductor or saturable reactor in series with the load.

to the undotted ends). Enough voltage must be induced into $N_2$ to make the voltage at winding 5 negative with respect to ground. This will hold $Q_2$ off. $R_1$ must be small in order to supply enough base current to drive the load current that is reflected into the collector windings. After a length of time [$t = (N_1\phi/E_{in})\ 10^{-8}$], $T_1$ will saturate, energy will no longer be coupled to the base wind-

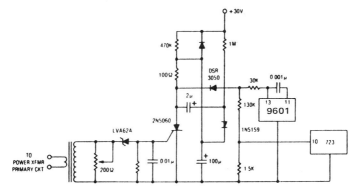

Figure 2-68   Soft starting and overcurrent trip.

ing, and $Q_1$ will shut off. Because of the voltage which had been impressed on $N_2$, there will be a voltage held across $C_2$. The capacitor will keep winding 6-7 at a level below the threshold of either transistor and let the core do the switching. The winding 2-1 will have some inductance, even on the saturated slope of the core. Like any inductor, $N_1$ has stored energy. When $Q_1$ shuts off, winding 2-1 reverses polarity trying to keep current flowing. All the other windings reverse polarity along with 2-1, thus forward-biasing the base of $Q_2$ and starting a new timing interval. After another period of time ($t$), the transformer again saturates and $Q_1$ conducts. Typical waveforms to be expected in this circuit are shown in Figure 2-70. They could only be obtained by having $R_1$ open initially and then closing it after $E_{in}$ is applied.

Two things happen which are undesirable. As the collector voltage rises

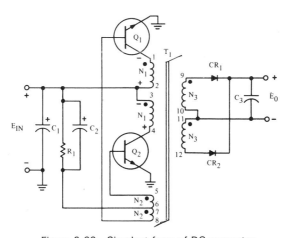

Figure 2-69   Simplest form of DC converter.

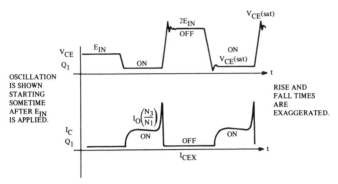

Figure 2-70   $Q_1$ collector current and collector-emitter voltage waveforms. Rise and fall times are exaggerated. Oscillation is shown starting sometime after $E_{in}$ is applied.

because of inductive phase reversal in the winding, it overshoots. This overshoot can be absorbed to some degree with a capacitor from collector to collector. Since the voltage spike is reflected as an undershoot when the transistor is turning on, a fast diode from emitter to collector will clamp the undershoot. Neither technique will eliminate the overshoot, but both will reduce it. The spike can destroy the transistors if it is high enough in amplitude; it also produces noise in the output that is difficult to filter. The second undesirable feature is the sharp increase in collector current as the transistor is switching. As the core saturates and begins to switch, collector current is no longer limited to the reflected load current. It is *only* limited by $E = L(di/dt)$ or $\Delta i = E\Delta t/L$ where $E$ is input voltage. $L$ is the saturated inductance and leakage inductance of the core, and $t$ is the time necessary for the core and transistors to switch. The permeability and thus the inductance of the collector winding falls off very rapidly as the core saturates. Thus, unless the transistors can switch quickly, this current rises to very high values. The use of fast-switching transistors results in an extremely rapid pass through each transistor's active operating region and minimizes transistor power dissipation. Often, if the transistor switching speed is not fast enough, speed-up capacitors are employed in the circuit. However, the drawback of increased transistor switching speed is that it creates transients that must be suppressed (filtered).

The voltage delivered to the output (Figure 2-71) is $E_{out} = E_{in}(N_3/N_1)$ less transistor and diode drops. Without a capacitor (C3) the output voltage would appear as in Figure 2-72.

There is a dead time in the output while one of the transistors is turning off and the other is turning on. (This is often referred to as crossover distortion or commutation). Even with the capacitor, there are spikes at the switching time. Their magnitude depends on the frequency impedance of the capacitors. When switching at 10 kHz, the spike frequencies can be as high as 50 MHz. Tantalum

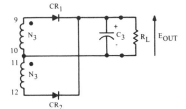

Figure 2-71   Converter output circuit.

capacitors are usually necessary to give enough coulomb storage in a small volume to take care of the dead time. Ceramic capacitors shunting tantalum help reduce high-frequency noise if enough space is available to get in a reasonable amount of capacitance. Filter chokes are necessary to substantially reduce the noise.

A unique advantage of this simple converter is that it can be turned on and off with a very small amount of power. If $R_1$ gets its drive from a separate supply, $E_{in}$ can be permanently applied and $I_{CEO}$ is the only power drain. The converter can be turned on and off by applying power to $R_1$.

Another unique and peculiar feature of this converter is that its frequency increases as $R_1$ increases. Increasing $R_1$ starves the bases. In any other circuit this would cause the converter to stall or to slow down.

One distinct disadvantage of the circuit shown in Figure 2-69 stems from its advantage—all the base current must come through a resistor that must drop the full line voltage. A small modification (see Figure 2-73) can be made to decrease the power that must be wasted in order to drive the bases. The circuit shown in Figure 2-73 operates like the circuit in Figure 2-69 except that the base current now comes through $CR_1$ and $R_2$ instead of $R_1$. $R_1$ can be a much larger resistance; its only function now is starting. Once a transistor turns on and current flow is induced in a base winding, the current takes the easiest path, which is through a diode. $R_2$ serves to limit the base current and $CR_1$ channels the entire starting current into the bases. Once started, this circuit runs without the starting resistor; it stops only upon removal of $E_{in}$. If the load current increases past the capabilities of the base current to keep the transistors saturated, the transistors come out of saturation. This lowers the base drive voltage and, therefore, the base drive current, which further unsaturates the transistors. Thus, the circuit stalls,

Figure 2-72   Appearance of output voltage-waveform without $C_3$.

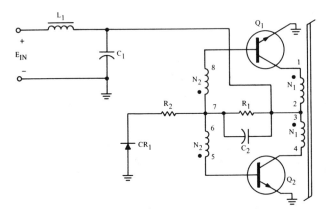

Figure 2-73   Modification of circuit shown in Figure 2-69 to increase efficiency.

providing some degree of overload protection. The circuits of Figures 2-69 and 2-73 belong to a class of converters which could be called *simple one-transformer voltage-driven converters*. These converters have the advantage of simplicity and a fairly good tolerance to asymmetry of component values.

Some care must be taken in selecting the turns for $N_2$. The reverse base voltage applied to the off transistors ($V_{BE} - 2E_{N_2}$) should be kept under $-7$ V.

$R_2$ is selected in the following manner: The maximum base current needed will equal the maximum required collector current divided by the minimum expected beta. The voltage drop across $R_2$ will be $E(R_2) = E(N_2) - V_{BE}(Q_1) - V_f(CR_1) = I_B R_2$. While this type of drive offers no load compensation, it can be temperature-compensated by making $R_2$ a thermistor.

The circuit runs at any load from zero to its stall point. As the load falls under its design maximum, the transistors become more and more overdriven and store charge. This results in longer turnoff times, higher current, and higher voltage spikes at switching.

Still another single-transformer, voltage-driven circuit is shown in Figure 2-74. The mechanics of switching are the same as before: The core supports the voltage until it saturates, the on transistor loses drive and turns off, and the off transistor is kicked on by the inductive reversal in the windings.

The advantage here is the common collector connection that allows both transistors to be mounted to a heat sink without using insulating washers. Diodes $CR_1$ and $CR_2$ give a path around the base limiting resistor for reverse base current to turn the transistor off. This reverse base current (which is instantaneously higher than the forward drive) is necessary to sweep the base region of carriers and to quickly turn the transistor off. A shunt capacitor is sometimes used across the turnoff diodes to further improve turn-on and turnoff; $CR_3$ facilitates turn-on. Since $R_1$ is much greater than $R_2$, the divided voltage would be below

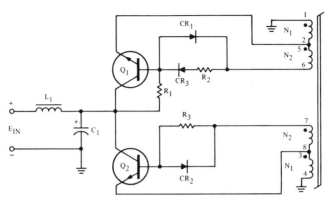

Figure 2-74    Another single-transformer voltage-driven converter.

threshold and the transistor would never turn on the first time. This circuit is less symmetrical than the circuits shown in Figures 2-69 and 2-73. For the latter reason there can be more current spikes. The transformer voltage waveforms and collector current waveforms are quite similar to those shown in Figure 2-70.

## Parallel Multiple-Transformer Oscillators

*Basic Circuit Description.*    The basic oscillating mechanism of the circuit shown in Figure 2-63 places restrictions on the construction and performance of the oscillator. Better performance, higher reliability, decreased size and weight, and improved reproducibility are possible if these restrictions are circumvented by other circuitry. Since a source of the restrictions on this circuit is the method of reducing loop gain by loading, these restrictions are alleviated if the loop gain can be interrupted or reduced by some other means.

The improved oscillator circuit developed by Jensen and described here contains similar mechanisms for positive feedback and means of reversal, that is, transformer coupling having loop gain greater than unity and using energy stored in inductive components to institute reversal. However, a different mechanism is used to reduce loop gain to less than unity. Rather than loading the output circuit until the fixed feedback drive is insufficient to sustain the operating condition, the feedback drive is reduced until it becomes insufficient to supply the existing load and to sustain the operating condition. Figure 2-75 shows one variation of the improved oscillator circuit using a voltage-dependent feedback signal.

A push-pull amplifier supplies the load through output transformer $T_1$, which does not saturate. The positive feedback is taken from a winding on $T_1$, through a feedback resistor $R_{FB}$, and applied to the transistor inputs through a small

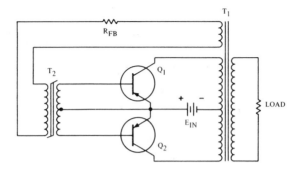

Figure 2-75 (Figures 2-75 through 2-81 redrawn from James Lee Jensen, "An Improved Square Wave Oscillator Circuit," *IRE Trans. on Circuit Theory*, September 1957.)

saturable transformer $T_2$. With sufficient positive feedback in the presence of a disturbance signal, one transistor becomes fully conducting, causing an induced voltage in one-half of the output transformer primary that is nearly equal to the DC supply voltage. The other transistor is back-biased to the nonconducting state. After a certain period of time, as determined by the design of $T_2$ and its associated circuitry, the core of $T_2$ becomes saturated and demands an increased magnetizing current. This increase in magnetizing current results in an increased voltage drop in the feedback resistor $R_{FB}$. The net result is a decrease in the feedback power or a reduction in loop gain. With further increase in magnetizing current, the drive ultimately becomes less than that necessary to sustain the fully conducting condition of the transistor in the presence of the combined collector loads. The loop gain thereby becomes less than unity. The induced voltages decrease, and the energy stored in the inductive components causes induced voltages of the opposite polarities to appear. These cause reversal of the oscillator.

The operation of this circuit is significantly different from the one shown in Figure 2-63, especially with respect to the duty cycle imposed upon the transistors. The total collector load comprises the exciting requirements of $T_1$, the feedback current, and the output load current component. The excitation requirements of $T_1$ are quite flexible since the core does not saturate and can be made small. Figures 2-76 and 2-77 show the collector voltage-current trace and collector voltage and current vs time for this improved converter. The feedback current is limited from increasing appreciably from its low initial value by the large feedback resistor. Consequently, the collector current is determined primarily by the load on the oscillator, and no large current spike is present. The "heavy-load" condition shown in Figure 2-76 represents only about one-half of the available output.

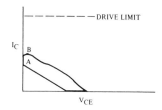

Figure 2-76   Collector voltage-current trace of Figure 2-75.

In turning on and off, the voltage-current trace of the collector circuit approaches the resistance load line with the resistive output load. With a minimum of energy stored in the leakage inductances, the inverse voltage overshoot is minimized, even at the no-load condition. In brief, duty cycle severity on the transistors is much reduced in the new circuit.

The transformers used in this circuit are not too critical, and considerable freedom of design is available. Since only the small feedback transformer need operate at high flux densities, high operating frequencies may be used without incurring large core losses. The proper operation of the oscillator is not dependent upon square hysteresis-loop materials or on the use of toroidal cores. Conventional transformer construction techniques with silicon steel cores and multiple wound coils may be employed while still achieving high efficiency and small size. Figures 2-78 through 2-81 show a few variations of this circuit.

*Circuit Variations.*   In Figure 2-78, the feedback power is taken directly from the primary of the output transformer, eliminating the extra winding and often improving the switching time.

Figure 2-79 shows a scheme for maintaining constant frequency. The frequency of operation is directly proportional to the rate of change of flux in the feedback transformer or to the induced voltage in any winding of that transformer. Variations in the input voltage or in the transistor input impedance normally cause variation in the transformer induced voltage and consequently in the frequency. Through the use of the voltage clipper shown in Figure 2-79,

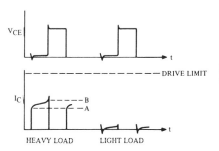

Figure 2-77   Collector voltage and current versus time for Figure 2-75.

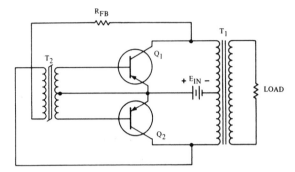

Figure 2-78   Variation of the improved oscillator using direct feedback connection.

however, the induced voltage may be held constant, making the frequency constant and independent of variations in input voltage or transistor input impedance.

Figure 2-80 shows a series addition of a voltage-dependent feedback signal and a current-dependent feedback signal. Either transformer $T_2$ or transformer $T_3$ may saturate, but more consistent results are obtained if the voltage-dependent transformer $T_2$ is saturated. Current feedback facilitates starting the oscillator under heavy-load conditions and usually shortens the switching time. Figure 2-81 illustrates parallel summation of voltage and current feedback.

Another feature of this circuit is the ease with which the frequency may be controlled. Until reversal occurs, it is a bistable inductively coupled circuit. A signal may be inserted in the low-power feedback circuit to cause reversal at any time before the natural reversal occurs. Consequently, the frequency may be easily synchronized with an external signal for frequency or phase control, permitting accurate frequency control from a crystal or tuning fork standard for production of polyphase signals.

These square-wave oscillator circuits offer the following advantages: im-

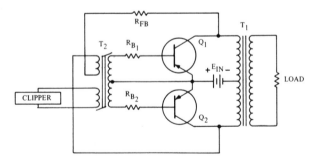

Figure 2-79   Variation of the improved oscillator which maintains constant frequency.

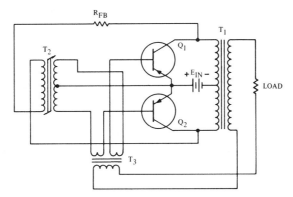

Figure 2-80   Variation of the improved oscillator using series combinations of current and voltage feedback.

proved performance with less severe duty on the transistors; flexibility of transformer design, permitting the use of conventional transformers, rather than toroidal types; adaptability to a variety of operating requirements, such as variable load conditions, frequency regulation, and phase control.

The three converters shown in Figure 2-69, 2-73, and 2-74 all have fairly high current spikes at transistor turnoff. These current spikes, even though of short duration, occur while the collector voltage is rising. The instantaneous power loss during this time can be many watts, thus decreasing the efficiency. If something other than the transformer that supports the collector voltage does the timing, the current spikes and the wasted power decreases. Using a second transformer to drive the bases and to perform the timing improves the switching.

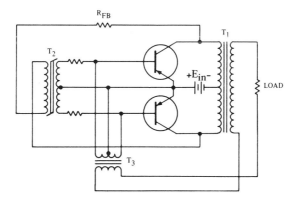

Figure 2-81   Variation of the improved oscillator using parallel combination of voltage and current feedback.

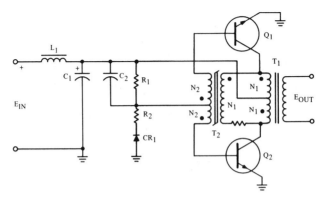

Figure 2-82

There are several varieties of two-transformer voltage-driven converters. Of these, the most straightforward is illustrated in Figure 2-82; circuit operation is similar to that of Figure 2-73. When voltage is applied, current flows through $R_1$ to the bases. One transistor turn on, creating a voltage drop in $N_1$ $(T_1)$. Since $N_1$ $(T_2)$ is shunting $N_1$ $(T_1)$, a voltage is impressed across $N_1$ $(T_2)$ and transformed to $N_2$, driving current into the base which was turned on. The base current is once gain limited by $R_2$. The voltage across $N_1$ $(T_2)$ is twice that of $N_1$ $(T_1)$ less the drop in $R_3$ due to reflected base current; this drop across $R_3$ should be small. The time to saturate $T_2$ is constructed to be less than $T_1$. When $T_2$ saturates, the on transistor once again loses drive, and the off transistor turns on. The transistor switching causes current to reverse in $T_1$ and square waves are produced as before. There is still a current spike when $T_2$ saturates and can no longer support voltage. But the current is now limited by $R_3$ and is much lower than in a one-transformer converter. Because there are two transformers, slight asymmetry in the components causes the length of one half-cycle to be slightly different from that of the next half-cycle. This pushes the power transformer $(T_1)$ closer and closer to saturation on one side. If $T_1$ saturates, $T_2$ loses drive and the transistors will switch anyway. Usually, the unbalance is not enough that $T_1$ actually saturates. A balance is struck when $T_1$ just approaches saturation and the voltage transformed above $E_{in}$ begins to drop. This reduces base drive slightly, which lets the transistors come out of saturation, and switching occurs. There is a higher current spike than with completely matched transistors, but it is still less than with a one-transformer converter.

A possible modification of Figure 2-82, which allows slightly better reverse base current, is shown in Figure 2-83. The circuit in Figure 2-83 can be further modified to a common collector connection (Figure 2-84) which operates similarly to that of Figure 2-74.

The amount of power necessary to drive the bases is rather small. As a result,

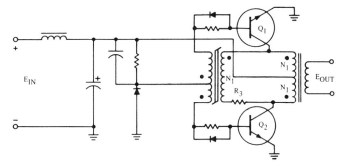

Figure 2-83

the core for the timing transformer can be small. Small cores usually have small cross sections, and the number of turns in the drive winding is high. If the wire size is chosen to be proportional to the reflected base current, the area occupied by the drive winding would be the same as that occupied by the base windings. However, wire that is smaller than No. 33 AWG (finer than No. 33) is quite fragile. Number 33 wire is a safe minimum size unless extra special care and protection can be assured. With this lower limit of wire size, the drive winding can easily take up a disproportionate area. Thus, we drive the second transformer from a lower voltage, tap the collector winding closer to the center tap, tie onto a low-voltage output if there is one, or add a low-voltage drive winding. An adaptation of the last alternative offers some turnoff advantages as well as fewer turns. This is achieved by splitting the collector winding and placing about 10% of the winding between each transistor emitter and ground. Now, when the

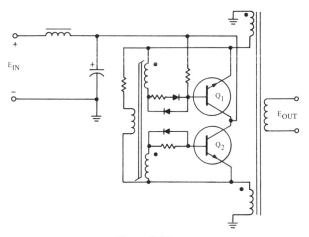

Figure 2-84

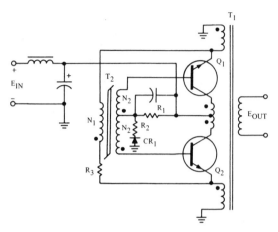

Figure 2-85

timing transformer begins to collapse, there is a larger reverse voltage (than previous circuits) available to shut off the transistor (Figure 2-85).

The voltage rise across $N_2$ is greater in this configuration (Figure 2-85) than in any previous circuit since the on base is elevated by $0.1\ E_{in}$. Therefore, the turnoff voltage is higher when $T_2$ saturates. The hold-off reverse voltage on the off transistor is no higher than it would be without the emitter windings, base current is still limited by $R_2$, and timing is accomplished by the voltage between the emitters.

*Simple Current-Driven Converters.* All of the two-transformer schemes shown so far have been voltage-driven, that is, a voltage is applied to the bases through a resistor to give base drive. The timing has been dependent on the drive voltage to the second transformer. Drive can also be generated with a current transformer. This method has some advantages over voltage drive (short-circuit protection not being one).

A look at current transformers is in order before current drive is explored. Any transformer transforms both voltage and current. Usually the drive is a voltage with low output series impedance. Thus, the output has whatever voltage is transformed at a current determined by the load. If, on the other hand, the driving impedance is high, the source looks like a current source. Thus, the secondary has some transformed current at a voltage determined by the load $(I_R)$.

A slight modification to Figure 2-85 circuit gives it a form of current drive as indicated in Figure 2-86. The modified circuit has a current transformer since the secondary voltage is limited only by the drops around the base loop. The

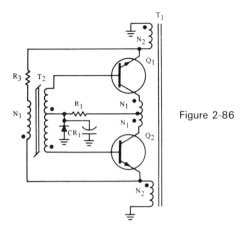

Figure 2-86

only current limiting now is done by $R_3$. The base capacitor is moved to between the center tap and ground. This further stabilizes the center tap and gives a path for reverse base current.

Operation of the circuit is somewhat similar to previous models. When input voltage is applied, current flows through the starting resistor ($R_1$) to the bases, and one transistor turns on. Once it turns on, there is a voltage difference between the emitters and current flows in $N_1$ ($T_2$). This current is transformed into the base windings and keeps the circuit going. The voltage that appears across $N_2$ ($T_2$) is the voltage across $N_2$ ($T_1$) plus the base emitter and diode drops. Since the two diode drops are temperature-, load-, and component-sensitive, the voltage of the emitter winding helps to stabilize the loop voltage drop by swamping out the small changes. The difference between the emitter-to-emitter voltage and the reflected base voltage [in $N_1$ ($T_2$)] appears across $R_3$ and sets the base current. The voltage across $N_2$ ($T_2$) does the timing. When $T_2$ saturates, an inductive kick causes the transformer to switch. Because there is very little series resistance in the base winding, a fairly healthy reverse current can flow and turn the transistors off.

Figure 2-86 is current-driven only by definition of a current transformer. A few more modifications give a true current drive in which base current automatically adjusts for the required collector current (Figure 2-87).

Now, after voltage has been applied and current flows through $R_1$ to the bases, one transistor turns on, causing collector current to flow. The collector current in turn forces a transformed base current to flow, thus keeping the transistor on. A voltage equal to $E(N_1 T_1) + V_{BE} + V_f$ appears across winding $N_2$ ($T_2$) and does the timing. Switching occurs just as before. Large reverse base currents occur at switching and quickly turn off the transistors. This type of circuit is capable of switching faster and more efficiently than the previous circuits. The

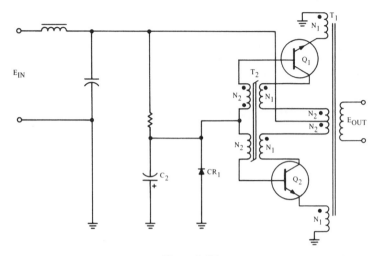

Figure 2-87

emitter windings are necessary to swamp out the minor transistor differences so that symmetrical timing is possible. Even so, if the transistors are not moderately well matched, there will be enough timing asymmetry to push the main transformer toward saturation and allow large collector current switching spikes. As shown here, the circuit will not work under no-load conditions. There has to be enough collector current both to drive the bases and to magnetize the core.

Selecting a current transformer core is much the same as selecting a core for any other square loop transformer. For a given input voltage;

$$E_{in} = 2\phi_t N_1 f \times 10^{-8} \tag{2-96}$$

$$\phi_t = 2 B_m A_c \tag{2-97}$$

$$\text{and} \quad W_a \phi_t = E_{in} A_w / 2kf \times 10^{-8} \tag{2-98}$$

where: $E_{in}$ = supply voltage,
  $N_1$ = primary transformer windings,
  $f$ = operating frequency,
  $\phi_t$ = total core flux in maxwells,
  $A_w$ = wire area in circular mils,
  $W_a$ = case window area,
  $k$ = winding factor.

There are probably at least two or three cores that satisfy the $W_a \phi_t$ requirement. To select the optimum, we check each for magnetizing current $I_{mag} = (H)(ml)/(0.4)(W)(N_1)$. The required $H$ to drive square permalloy at 10 kHz is about 0.14 Oe.

$I_{mag}$ should be less than half the minimum expected base current, that is, $I_{mag} \leq I_C(N_2/N_1)$. Since the current transformer can be small, it usually requires only short lengths of wire. This allows use of lighter wire, smaller windows, shorter mean lengths, and thus less magnetizing current. Magnetizing current can also be reduced by making $E_{in}$ greater or going to lower flux capacity cores. Either of these require more turns on $N_1$.

If a wide load range or a very low load is expected, there simply are not cores available with large windows and small mean lengths.

A hybrid combination of circuits shown in Figures 2-86 and 2-87 allows zero to full load operation; this configuration is shown in Figure 2-88. Enough current is supplied through $N_3$ to keep the core magnetized even if $I_C$ falls to a very low value. $I_C$ never goes to zero since there are always fixed losses to be supplied.

The forced $\beta(N_2/N_1)$ can vary over the range 1:1 to 20:1 or wider depending on the transistor $\beta$ and the environment. As the driven $\beta$ gets higher and higher, the emitter-coupled sustaining winding becomes more necessary.

The tape material in the current transformer should be the same (both in thickness and in type) as in the main transformer for the best switching.

Figure 2-88 should be considered as a typical current-driven square-wave oscillator. In some earlier designs the emitter windings on the main transformer were not used. The result was a lower tolerance to mismatched transistors and a greater difficulty in operating to low load since the current transformer saw less voltage and had fewer turns.

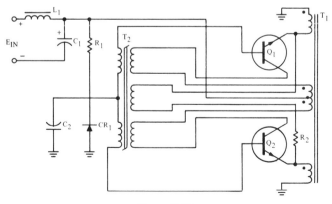

Figure 2-88

## The Resonant Converter[4]

The full-wave resonant converter circuit of Figure 2-89 is suited for use in power supplies with an output load rating of 200 W or greater. This circuit combines zero current switching with full-wave operation.

In this circuit, the frequency controller, which could be an integrated circuit or a set of ICs, produces square-waves that alternately turn transistor switches $Q_1$ and $Q_2$ on and off. The frequency of the controller, and thus the square-waves it produces, is usually variable. A frequency controller is an integral part of every resonant converter, exactly as a PWM circuit is the inherent controller of conventional switching converters.

When transistor switch $Q_1$ turns on, the current in inductor $L_1$ builds gradually. This current equals the sum of the transformer primary current and the charging current of capacitor $C_1$, which is parallel with the primary. Once the voltage across $C_1$ and the primary equals the input voltage, the voltage drop across $L_1$ is zero, and $L_1$ begins to release its stored energy into $C_1$. At a time determined by the natural resonant frequency of $L_1$ and $C_1$, the current in $L_1$ and therefore $Q_1$ reaches 0. The current in $L_1$ now reverses, and $C_1$ begins to discharge its stored energy, maintaining current flow through diode $D_1$.

Ideally, for zero current switching, the changeover from one transistor switch to the other occurs at that point, with $Q_1$ turning off and $Q_2$ turning on. Now, $C_1$ begins to recharge in the opposite polarity, and an identical half-cycle begins.

Because the time before the changeover from one transistor switch to the other is longer than a natural resonant half-cycle, the full-wave circuit of Figure 2-89 operates below its resonant frequency. Turn-off at zero current is very important for bipolar transistors because the majority of losses occur during their turn-off.

The transfer characteristics versus frequency for the full-wave circuit of Figure 2-89 typify the frequency-domain response of a resonant circuit (see Figure 2-90). Varying the frequency of operation controls the output voltage. (In contrast, with PWM supplies, varying duty cycle controls the output voltage.) For a given load factor ($Q$), the output voltage (which equals the ratio of $V_{pri}$ to $V_{in}$) rises to a peak at resonance and then falls again as the operating frequency increases beyond resonance. Such characteristics define two regions of operation: one below the resonant point, and the other above it. In general, operation at frequencies below resonance produces zero current switching, and operation above resonance yields zero voltage switching. But zero current and zero voltage switching cannot be implemented together in the same topology.

The half-wave resonant converter (Figure 2-91) is used for systems requiring

---

[4]Bassett, John "Switching Supplies: Changing with the Times," *Electronics*, January 7, 1988, copyright © the McGraw-Hill Book Company 1988. Used with permission.

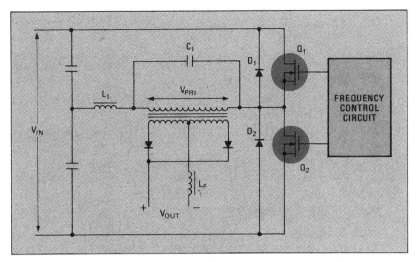

Figure 2-89   Full-wave resonant converter. (Reprinted with permission from *Electronics* magazine, copyright © January 7, 1988, VNU Business Publications)

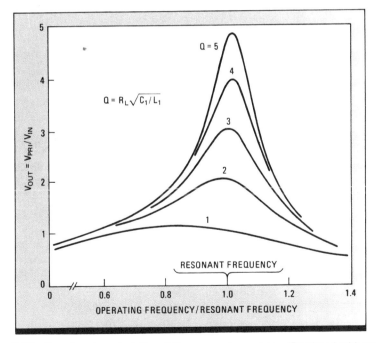

Figure 2-90   Transfer characteristics of the resonant converter. (Reprinted with permis-son from *Electronics* magazine, copyright © January 7, 1988, VNU Business Publications)

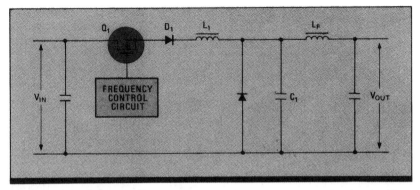

Figure 2-91   The half-wave resonant or quasi-resonant converter. (Reprinted with permission from *Electronics* magazine, copyright © January 7, 1988, VNU Business Publications)

5 to 200 W. Since this circuit allows the resonant capacitor to charge but not to discharge, it is often called a quasi-resonant converter.

Circuit operation is fairly straightforward. When transistor switch $Q_1$ turns on, the combination of the load current and the charging current of capacitor $C_1$ stores energy in resonant inductor $L_1$. When the voltage of $C_1$ equals the input voltage, the voltage of $L_1$ passes through zero and begins to reverse. This allows $L_1$ to release its stored energy to $C_1$. This capacitor continues to charge above the input voltage until the current in $L_1$ and $Q_1$ reaches 0. Then, diode $D_1$ blocks reverse current flow so that $Q_1$ turns off when its current reaches 0.

Although the circuit of Figure 2-91 prevents the resonant discharge of $C_1$, the capacitor can and does discharge into the filter and load circuits, preparing it for the next cycle. The capacitor charges during one half-cycle (when the transistor is on), and then discharges (powers the load) during the other half-cycle (when the transistor is off). The delay time between charging cycles determines the amount of energy transferred to the output and, therefore, the output voltage.

Both the half-wave and full-wave resonant converter topologies are more complex than nonresonant circuits, but they are more efficient—80% vs. 75%— at operating frequencies up to several hundred kilohertz. To reap the reward of smaller size through megahertz operation and yet maintain that excellent 80% efficiency, resonant topologies are changing to take advantage of MOSFET characteristics and zero voltage switching.

Until recently, transistor turn-on was believed to be relatively free of losses, since no current flows before turn-on. Moreover, current should build gradually, just before turn-on, through the circuit inductance only after the voltage at the collector (bipolar) or drain (MOSFET) falls. In reality though, when a transistor is off, substantial capacitance exists at the collector or drain node, be it in the

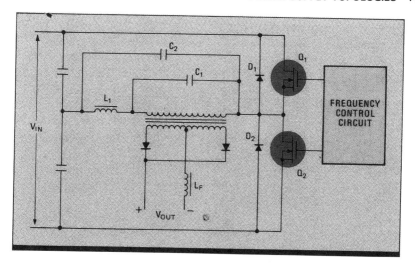

Figure 2-92   Zero voltage switching resonant converter. (Reprinted with permission from *Electronics* magazine, copyright © January 7, 1988, VNU Business Publications)

transistor or from circuit parasitics. When it turns on, the transistor switch totally dissipates the energy stored in that capacitance by rapidly shorting it out each operating cycle. With such hard turn-on, losses can be as high as 25 W for a capacitance of only 2 nF, an operating frequency of 1 MHz, and a transistor voltage in excess of 150 V. Clearly, hard turn-on does not work at high frequencies.

In addition, unlike bipolar devices, which can experience large losses during turn-off, MOSFETs turn-off without any losses even when carrying current. MOSFET switching times are determined primarily by device capacitance and secondarily by the channel transit time of electrons. Because the gate capacitance is large, existing topologies have actually slowed MOSFET turn-off by limiting the drive current to a few hundred milliamperes. In fact, turning the channel on and off effectively requires pulses of several amperes to charge and discharge

## Table 2-3   Switching Topology Output Power Comparison

| Topology | Output Range |
|---|---|
| Resonant | 5 to 300 W |
| Flyback | 20 to 150 W |
| Forward | 150 to 450 W |
| Half-Bridge, full-bridge and push-pull | 200 to 2000 W |

# Table 2-4 Transistor and Diode Requirements for Switching Converters (Reprinted from *Switchmode Power Conversion*, K. Kit Sum, Marcel Dekker, Inc. 1984.)

| | (A) BUCK (STEP DOWN) | (B) BOOST (STEP UP) | (C) BUCK–BOOST |
|---|---|---|---|
| CIRCUIT CONFIGURATION | | | |
| TYPE OF CONVERTER | (A) BUCK (STEP DOWN) | (B) BOOST (STEP UP) | (C) BUCK – BOOST |
| IDEAL TRANSFER FUNCTION | $\frac{V_O}{V_{IN}} = \frac{\tau}{T} = D$ | $\frac{V_O}{V_{IN}} = \frac{T}{T-\tau}$ | $\frac{V_O}{V_{IN}} = \left(\frac{\tau}{T-\tau}\right)(-1)$ |
| COLLECTOR CURRENT ($I_C$) ✕ | $I_{C\,MAX} = I_{RL} + \Delta I_{L1}/2$ | $I_{C\,MAX} = I_{RL}\left(\frac{T}{T-\tau}\right) + \frac{\Delta I_{L1}}{2}$ | $I_{C\,MAX} = I_{RL}\left(\frac{T}{T-\tau}\right) + \frac{\Delta I_{L1}}{2}$ |
| COLLECTOR VOLTAGE RATING ✕ | $V_{CEO} = V_{IN}$ | $V_{CEO} > V_O + I$ | $V_{CEO} > V_{IN} + V_O$ |
| DIODE CURRENTS ✕ | $I_{CR1} = I_{RL}\left(\frac{T-\tau}{T}\right)$ | $I_{CR1} = I_{RL}$ | $I_{CR1} = I_{RL}$ |
| DIODE VOLTAGES ($V_{RM}$) ✕ | $V_{RM} = V_{IN}$ | $V_{RM} = V_O$ | $V_{RM} = V_O + V_{IN}$ |
| VOLTAGE AND CURRENT WAVEFORMS | | | |
| ADVANTAGES | HIGH EFFICIENCY, SIMPLE, NO TRANSFORMER, HIGH FREQUENCY OPERATION. EASY TO STABILIZE REGULATOR LOOP. | HIGH EFFICIENCY, SIMPLE, NO TRANSFORMER, HIGH FREQUENCY OPERATION. | VOLTAGE INVERSION WITHOUT USING A TRANSFORMER, SIMPLE, HIGH FREQUENCY OPERATION. |
| DISADVANTAGES | NO ISOLATION BETWEEN INPUT AND OUTPUT. REQUIRES A CROWBAR IF Q1 SHORTS. C1 HAS HIGH RIPPLE CURRENT. CURRENT LIMIT DIFFICULT. ONLY ONE OUTPUT IS POSSIBLE. | NO ISOLATION BETWEEN INPUT AND OUTPUT. HIGH PEAK COLLECTOR CURRENT. ONLY ONE OUTPUT IS POSSIBLE. POOR TRANSIENT RESPONSE. REGULATOR LOOP HARD TO STABILIZE. | Q1 MUST CARRY HIGH PEAK CURRENT, NO ISOLATION BETWEEN INPUT AND OUTPUT, ONLY ONE OUTPUT IS POSSIBLE, POOR TRANSIENT RESPONSE. |

# Table 2-4 (Continued)

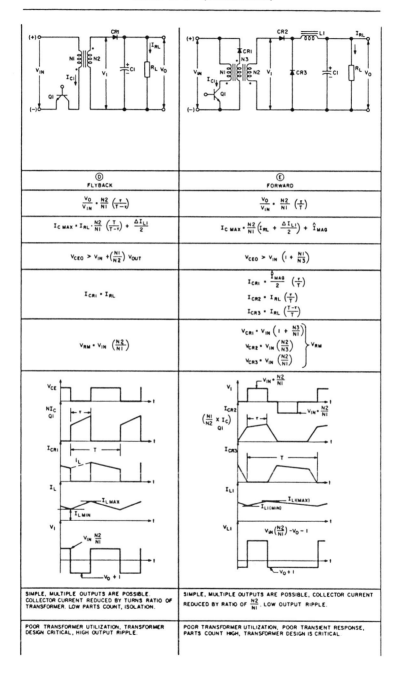

| FLYBACK (D) | FORWARD (E) |
|---|---|

$$\frac{V_O}{V_{IN}} \cdot \frac{N2}{N1}\left(\frac{\tau}{T-\tau}\right)$$ 

$$\frac{V_O}{V_{IN}} \cdot \frac{N2}{N1}\left(\frac{\tau}{T}\right)$$

$$I_{C\,MAX} \cdot I_{RL} \cdot \frac{N2}{N1}\left(\frac{T}{T-\tau}\right) + \frac{\Delta I_{L1}}{2}$$ 

$$I_{C\,MAX} \cdot \frac{N2}{N1}\left(I_{RL} + \frac{\Delta I_{L1}}{2}\right) + \hat{I}_{MAG}$$

$$V_{CEO} > V_{IN} + \left(\frac{N1}{N2}\right)V_{OUT}$$ 

$$V_{CEO} > V_{IN}\left(1 + \frac{N1}{N3}\right)$$

$$I_{CR1} \cdot I_{RL}$$ 

$$I_{CR1} \cdot \frac{\hat{I}_{MAG}}{2}\left(\frac{\tau}{T}\right)$$
$$I_{CR2} \cdot I_{RL}\left(\frac{\tau}{T}\right)$$
$$I_{CR3} \cdot I_{RL}\left(\frac{T-\tau}{T}\right)$$

$$V_{RM} \cdot V_{IN}\left(\frac{N2}{N1}\right)$$ 

$$\left.\begin{array}{l}V_{CR1} \cdot V_{IN}\left(1 + \frac{N3}{N1}\right)\\ V_{CR2} \cdot V_{IN}\left(\frac{N2}{N3}\right)\\ V_{CR3} \cdot V_{IN}\left(\frac{N2}{N1}\right)\end{array}\right\}V_{RM}$$

SIMPLE, MULTIPLE OUTPUTS ARE POSSIBLE. COLLECTOR CURRENT REDUCED BY TURNS RATIO OF TRANSFORMER. LOW PARTS COUNT, ISOLATION.

SIMPLE, MULTIPLE OUTPUTS ARE POSSIBLE, COLLECTOR CURRENT REDUCED BY RATIO OF $\frac{N2}{N1}$. LOW OUTPUT RIPPLE.

POOR TRANSFORMER UTILIZATION, TRANSFORMER DESIGN CRITICAL, HIGH OUTPUT RIPPLE.

POOR TRANSFORMER UTILIZATION, POOR TRANSIENT RESPONSE, PARTS COUNT HIGH, TRANSFORMER DESIGN IS CRITICAL.

# Table 2-4 *(Continued)*

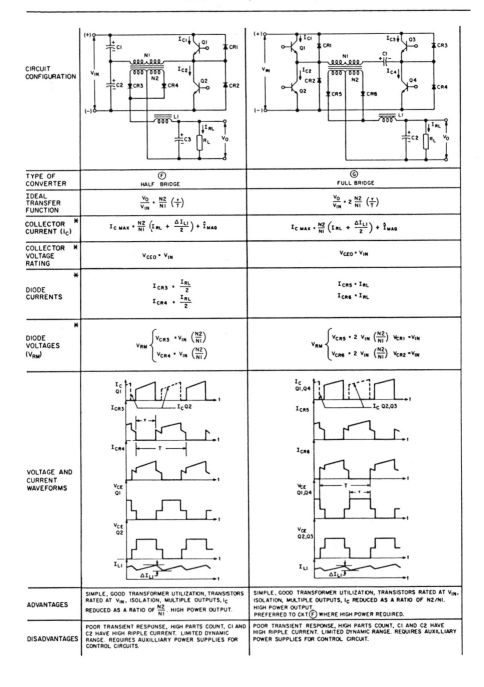

| | | |
|---|---|---|
| CIRCUIT CONFIGURATION | | |
| TYPE OF CONVERTER | (F) HALF BRIDGE | (G) FULL BRIDGE |
| IDEAL TRANSFER FUNCTION | $\frac{V_O}{V_{IN}} = \frac{N2}{N1}\left(\frac{\tau}{T}\right)$ | $\frac{V_O}{V_{IN}} = 2\frac{N2}{N1}\left(\frac{\tau}{T}\right)$ |
| COLLECTOR CURRENT $(I_C)$ ✱ | $I_{C\,MAX} = \frac{N2}{N1}\left(I_{RL} + \frac{\Delta I_{LI}}{2}\right) + \hat{I}_{MAG}$ | $I_{C\,MAX} = \frac{N2}{N1}\left(I_{RL} + \frac{\Delta I_{LI}}{2}\right) + \hat{I}_{MAG}$ |
| COLLECTOR VOLTAGE RATING ✱ | $V_{CEO} = V_{IN}$ | $V_{CEO} = V_{IN}$ |
| DIODE CURRENTS ✱ | $I_{CR3} = \frac{I_{RL}}{2}$ $I_{CR4} = \frac{I_{RL}}{2}$ | $I_{CR5} = I_{RL}$ $I_{CR6} = I_{RL}$ |
| DIODE VOLTAGES $(V_{RM})$ ✱ | $V_{RM}\begin{cases} V_{CR3} = V_{IN}\left(\frac{N2}{N1}\right) \\ V_{CR4} = V_{IN}\left(\frac{N2}{N1}\right) \end{cases}$ | $V_{RM}\begin{cases} V_{CR5} = 2\,V_{IN}\left(\frac{N2}{N1}\right) & V_{CR1} = V_{IN} \\ V_{CR6} = 2\,V_{IN}\left(\frac{N2}{N1}\right) & V_{CR2} = V_{IN} \end{cases}$ |
| VOLTAGE AND CURRENT WAVEFORMS | | |
| ADVANTAGES | SIMPLE, GOOD TRANSFORMER UTILIZATION, TRANSISTORS RATED AT $V_{IN}$, ISOLATION, MULTIPLE OUTPUTS, $I_C$ REDUCED AS A RATIO OF $\frac{N2}{N1}$. HIGH POWER OUTPUT. | SIMPLE, GOOD TRANSFORMER UTILIZATION, TRANSISTORS RATED AT $V_{IN}$, ISOLATION, MULTIPLE OUTPUTS, $I_C$ REDUCED AS A RATIO OF N2/N1. HIGH POWER OUTPUT. PREFERRED TO CKT (F) WHERE HIGH POWER REQUIRED. |
| DISADVANTAGES | POOR TRANSIENT RESPONSE, HIGH PARTS COUNT, C1 AND C2 HAVE HIGH RIPPLE CURRENT. LIMITED DYNAMIC RANGE. REQUIRES AUXILLIARY POWER SUPPLIES FOR CONTROL CIRCUITS. | POOR TRANSIENT RESPONSE, HIGH PARTS COUNT, C1 AND C2 HAVE HIGH RIPPLE CURRENT. LIMITED DYNAMIC RANGE. REQUIRES AUXILLIARY POWER SUPPLIES FOR CONTROL CIRCUIT. |

## Table 2-4   (*Continued*)

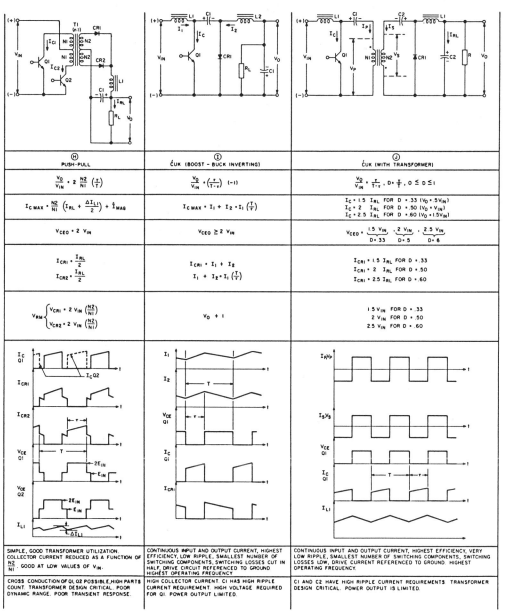

|  | ⓗ<br>PUSH·PULL | ⓘ<br>ĆUK (BOOST – BUCK INVERTING) | ⓙ<br>ĆUK (WITH TRANSFORMER) |
|---|---|---|---|

$$\frac{V_O}{V_{IN}} = 2\frac{N2}{N1}\left(\frac{\tau}{T}\right)$$

$$\frac{V_O}{V_{IN}} = \left(\frac{\tau}{T-\tau}\right)(-1)$$

$$\frac{V_O}{V_{IN}} = \frac{\tau}{T-\tau}, D = \frac{\tau}{T}, 0 \le D \le 1$$

$$I_{C\,MAX} = \frac{N2}{N1}\left(I_{RL} + \frac{\Delta I_{L1}}{2}\right) + I_{MAG}$$

$$I_{C\,MAX} = I_1 + I_2 = I_1\left(\frac{T}{\tau}\right)$$

$$I_C = 1.5\ I_{RL}\ \text{FOR}\ D = .33\ (V_O = .5V_{IN})$$
$$I_C = 2\ I_{RL}\ \text{FOR}\ D = .50\ (V_O = V_{IN})$$
$$I_C = 2.5\ I_{RL}\ \text{FOR}\ D = .60\ (V_O = 1.5V_{IN})$$

$$V_{CEO} = 2\ V_{IN}$$

$$V_{CEO} \ge 2\ V_{IN}$$

$$V_{CEO} = \frac{1.5\ V_{IN}}{D = .33}, \frac{2\ V_{IN}}{D = .5}, \frac{2.5\ V_{IN}}{D = .6}$$

$$I_{CR1} = \frac{I_{RL}}{2}$$
$$I_{CR2} = \frac{I_{RL}}{2}$$

$$I_{CR1} = I_1 + I_2$$
$$I_1 + I_2 = I_1\left(\frac{T}{\tau}\right)$$

$$I_{CR1} = 1.5\ I_{RL}\ \text{FOR}\ D = .33$$
$$I_{CR1} = 2\ I_{RL}\ \text{FOR}\ D = .50$$
$$I_{CR1} = 2.5\ I_{RL}\ \text{FOR}\ D = .60$$

$$V_{RM}\begin{cases}V_{CR1} = 2\ V_{IN}\left(\frac{N2}{N1}\right)\\ V_{CR2} = 2\ V_{IN}\left(\frac{N2}{N1}\right)\end{cases}$$

$$V_O + 1$$

$$1.5\ V_{IN}\ \text{FOR}\ D = .33$$
$$2\ V_{IN}\ \text{FOR}\ D = .50$$
$$2.5\ V_{IN}\ \text{FOR}\ D = .60$$

SIMPLE, GOOD TRANSFORMER UTILIZATION. COLLECTOR CURRENT REDUCED AS A FUNCTION OF $\frac{N2}{N1}$. GOOD AT LOW VALUES OF $V_{IN}$.

CONTINUOUS INPUT AND OUTPUT CURRENT, HIGHEST EFFICIENCY, LOW RIPPLE, SMALLEST NUMBER OF SWITCHING COMPONENTS, SWITCHING LOSSES CUT IN HALF, DRIVE CIRCUIT REFERENCED TO GROUND HIGHEST OPERATING FREQUENCY

CONTINUOUS INPUT AND OUTPUT CURRENT, HIGHEST EFFICIENCY, VERY LOW RIPPLE, SMALLEST NUMBER OF SWITCHING COMPONENTS, SWITCHING LOSSES LOW, DRIVE REFERENCED TO GROUND. HIGHEST OPERATING FREQUENCY.

CROSS CONDUCTION OF Q1, Q2 POSSIBLE, HIGH PARTS COUNT. TRANSFORMER DESIGN CRITICAL. POOR DYNAMIC RANGE. POOR TRANSIENT RESPONSE.

HIGH COLLECTOR CURRENT. C1 HAS HIGH RIPPLE CURRENT REQUIREMENT. HIGH VOLTAGE REQUIRED FOR Q1. POWER OUTPUT LIMITED.

C1 AND C2 HAVE HIGH RIPPLE CURRENT REQUIREMENTS   TRANSFORMER DESIGN CRITICAL. POWER OUTPUT IS LIMITED.

*For reliable operation, it is suggested and recommended that all voltage and current ratings be increased to 125% of the required maximum.

the gate rapidly. Fortunately, the gate voltage is only 10 V and the energy required is small, but the drive components must be rugged.

With the increased use of MOSFETs for high frequency operation, the care that existing topologies take to soften the turn-off interval through zero current switching has become unnecessary. Instead, new topologies are employing zero-voltage switching to soften the turn-on interval.

In most cases, it is relatively easy to modify an existing resonant converter to handle zero voltage switching. For example, the resonant converter of Figure 2-92 uses a parallel snubber capacitor ($C_2$) to modify the full-wave circuit of Figure 2-89 for zero voltage switching. And the circuit's operating frequency is higher than the resonant frequency.

A resonant half-cycle begins when $Q_1$ turns off and before $Q_2$ turns on. While both transistors are off, the energy in $L_1$ transfers to $C_2$, charging this capacitor and causing the voltage in $Q_1$ to increase while that of $Q_2$ to decrease. When the voltage across $Q_2$ reaches 0, it turns on without loss, and diode $D_2$ returns any remaining energy in $L_1$ to the input. The next half-cycle is identical, and begins when $Q_2$ turns off. Now the voltage across $Q_2$ increases and that of $Q_1$ decreases until it is 0, allowing $Q_1$ to turn on without loss. As with other resonant converters, varying the operating frequency changes the output voltage. Capacitor $C_1$ simply functions as it does in the circuit of Figure 2-89.

## SUMMARY

The choice of a converter topology is based on output current and voltage, frequency of operation, component cost, and expected performance levels. Table 2-3 compares the general output capabilities of the popular converter topologies.

Table 2-4[5] summarizes the characteristics of each of the topologies just presented as well as their transistor and diode requirements on a comparative basis. By understanding each of these topologies, the system designer will be able to intelligently select the switcher topology that will best meet his cost/performance needs.

---

[5]Used with permission from *Switchmode Power Conversion*, by K. Kit Sum, Marcel Dekker Inc., 1984.

# 3

---

# Switching Supply Design Hints

## INTRODUCTION

As a continuation of Chapter 2, this chapter focuses on the practical aspects of switching power supplies—design hints and selection criteria for each of the major constituent elements of a switcher and associated topics: input capacitors, rectifiers, transistors, base drive circuits, control circuits, transient protection, current limiting and generated EMI. Also, several design problems and illustrative examples are presented. As will be shown, there are a myriad of issues that come into play and must be addressed—and tradeoffs made—to effect a viable power supply design in accordance with a given set of requirements.

## INPUT CAPACITORS[1]

The best practical procedure for the design of capacitor-input filters is based on the graphical data presented by Schade[1] in 1943, and shown in Figures 3-1 through 3-4. These curves provide all the required design information for half-wave and full-wave rectifier circuits. It should be noted that the rectifier forward-drop often assumes more significance than the dynamic resistance in low voltage supply applications, since the dynamic resistance can generally be neglected when compared with the sum of the transformer secondary-winding resistance plus the reflected primary-winding resistance. The forward-drop may be of considerable importance, however, since it is about 1 V, which clearly cannot be ignored in supplies of 12 V or less.

Referring to the full-wave circuit of Figure 3-2 as an illustrative example, it can be seen that a circuit must operate with $wCR_L \geq 10$ in order to hold the voltage reduction to less than 10% and $wCR_L \geq 40$ to obtain less than 2% reduction. However, it also can be seen that these voltage reduction figures

---

[1]Courtesy of Motorola Semiconductor Products Division, Inc.

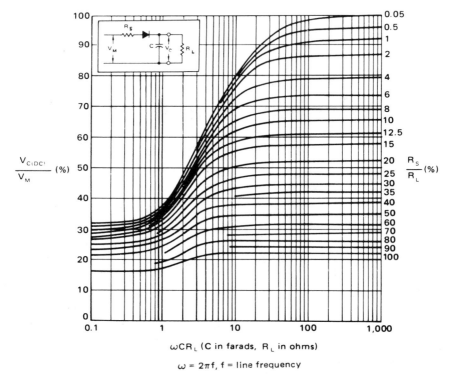

Figure 3-1   Relation of applied alternating peak voltage to direct output voltage in half-wave capacitor-input circuits. (From O. H. Schade, *Proc. IRE*, vol. 31, p. 356, 1943. Copyright of Motorola, Inc. Used by permission)

require $R_S/R_L$, where $R_S$ is now the total series resistance, to be about 0.1%, which, if attainable, causes repetitive peak-to-average current ratios from 10 to 17 respectively, as can be seen from Figure 3-3. These ratios can be satisfied by many diodes; however, they may not be able to tolerate the turn-on surge current generated when the input filter capacitor is discharged and the trans-former primary is energized at the peak of the input waveform. The rectifier is then required to pass a surge current determined by the peak secondary voltage less the rectifier forward-drop and limited only by the series resistance $R_S$. In order to control this turn-on surge, additional resistance must often be provided in series with each rectifier. Thus, a compromise must be made between voltage reduction on the one hand and diode surge rating and average current carrying capacity on the other hand. If small voltage reduction, that is, good voltage regulation, is required, a much larger diode is necessary than that demanded by the average current rating.

The input filter capacitor allows a large surge to develop because the reactance

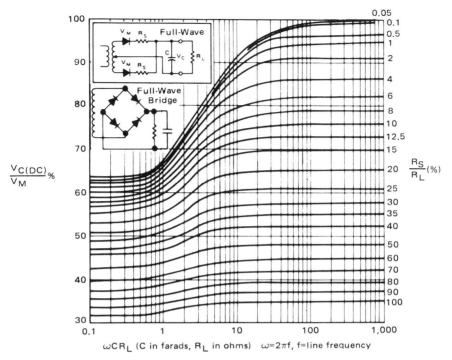

Figure 3-2 Relation of applied alternating peak voltage to direct output voltage in full-wave capacitor-input circuits. (From O. H. Schade, *Proc. IRE*, vol. 31, p. 356, 1943. Copyright of Motorola, Inc. Used by permission)

of the transformer leakage inductance is rather small. The maximum instantaneous surge current is approximately $V_M/R_S$ and the capacitor charges with a time constant $t = R_S C_1$. As a rough check, the surge will not damage the diode if $V_M/R_S$ is less than the diode $I_{FSM}$ rating and $t$ is less than 8.3 ms. $R_S$ should be made as large as possible and tight voltage regulation should not be pursued; thus, reducing the surge and allowing rectifier and transformer ratings to more nearly approach the DC power requirements of the supply.

A typical input filter design procedure, using the curves of Figures 3-1 to 3-4, is now presented. The following parameters are normally known and specified:

$V_{C(DC)}$ = The required full load average DC output voltage of the capacitor input filter

$V_{ripple}(p-p)$ = the maximum full load peak-to-peak ripple voltage

$V_M$ = the maximum no load output voltage

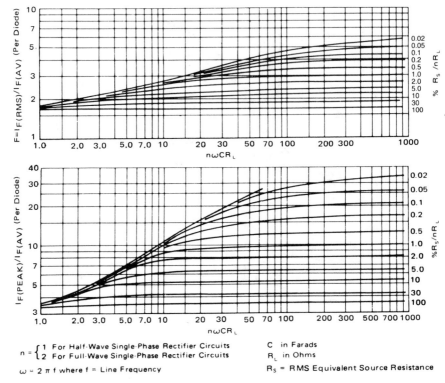

Figure 3-3   Relation of rms and peak to average diode current in capacitor-input circuits. (From O. H. Schade, *Proc. IRE*, vol. 31, p. 356, 1943. Copyright of Motorola, Inc. Used by permission)

$I_o$ = the full-load filter output current

$f$ = the input AC line frequency

From Figure 3-4, one can determine a range of minimum capacitor values to obtain sufficient ripple attenuation. First determine $rf$:

$$rf = \frac{V_{ripple}(p\text{-}p) \times 100\%}{2\sqrt{2}V_{C(DC)}} \tag{3-1}$$

a range for $wcR_L$ can now be found from Figure 3-4.

Next, determine the range of $R_S/R_L$ from Figure 3-1 or 3-2 using $V_{C(DC)}$ and the values for $wCR_L$ just found. If the range of $wCR_L$ values initially determined

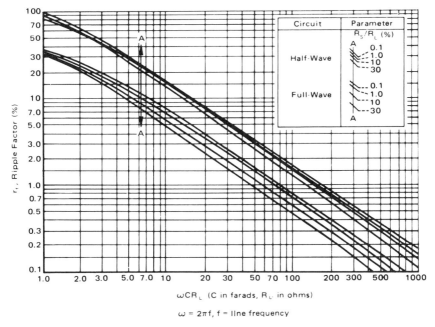

Figure 3-4   Root-mean square ripple voltage for capacitor-input circuits. (From O. H. Schade, *Proc. IRE*, vol. 31, p. 356, 1943. Copyright of Motorola, Inc. Used by permission)

from Figure 3-4 is above $\simeq 10$, $R_S/R_L$ can be found from Figures 3-1 and 3-2 using the lowest $wCR_L$ value. Otherwise, several iterations between Figures 3-1 or 3-2 and 3-4 may be necessary before an exact solution for $R_S/R_L$ and $wCR_L$ for a given $rf$ and $V_{C(DC)}/Vm$ can be found.

Once $wCRl_L$ is found, the value of the filter capacitor, $C$, can be determined from:

$$C = \frac{wCR_L}{2\pi f\left(\dfrac{V_{C(DC)}}{I_o}\right)} \tag{3-2}$$

The rectifier requirements may now be determined:

1. Average Current Per Diode

$$I_{F(avg)} = I_o \text{ for half-wave rectification}$$
$$= I_o/2 \text{ for full-wave rectification} \tag{3-3}$$

2. *Rms* and Peak repetitive rectifier current ratings can be determined from Figure 3-3.
3. The rectifier *PIV* rating is $2V_M$ for the half-wave and full-wave circuits, and $V_M$ for the full-wave bridge circuit. In addition, a minimum safety margin of 20% to 50% is advisable due to the possibility of line transients.
4. Maximum Surge Current

$$I_{surge} = V_M/(R_S + ESR) \qquad (3-4)$$

where  $ESR$ = minimum equivalent series resistance of filter capacitor from its data sheet

Several practical considerations are now presented. Input filter capacitors carry an *rms* ripple current that is 60 to 70% of the DC input current, depending on the converter's duty cycle. The higher the input voltage, the narrower the duty cycle and, consequently, the higher the ripple current. To minimize the internal heating that is caused by ripple current, the designer should choose capacitors that have the lowest equivalent series resistance (*ESR*). Several other points to keep in mind include these:

- Specify the best filter capacitors that the design budget allows and derate them for ripple current.
- Mount capacitors as far from heat generating components (heat sinks, power transistors, and power resistors, for example) as possible.
- Make sure that the PC board traces on which the capacitors are mounted can carry the input current (thus, keeping generated heat to a minimum and not raising the internal temperature of the capacitor).
- Specify capacitors that offer the highest working temperature.

## RECTIFIERS

### Circuit Considerations[2]

Switching power supply output rectifiers and "catch" diodes are subject to unusual stresses due to the fast switching rates and very low impedance seen by the diode during the reverse transient (diode turnoff) and a momentary high impedance during diode turn-on.

Square-wave switching supplies are limited in efficiency and frequency by transistor stress and switching losses, some of which is due to diode switching characteristics. Faster transistors and diodes are helping to increase efficiency

---

[2]Portions of this section were excerpted from Application Note U-73A, "The Importance of Rectifier Characteristics in Switching Power Supply Design." Unitrode Corporation. Used with permission.

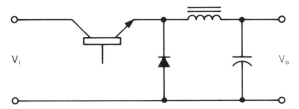

Figure 3-5  Buck converter schematic. (Reprinted with permission from "The Importance of Rectifier Characteristics in Switching Power Supply Design," Application Note U-73A, Unitrode Corp.)

and/or frequency. At low output voltages, and lower frequency, the DC characteristics ($V_{CE(\text{sat})}$ and $V_F$) have the major influence on efficiency. However, as frequency and/or input voltage increase, the switching characteristics become increasingly important.

The subject of rectifier circuit considerations is addressed by examining both the buck and the pulse width modulated (PWM) converters. In the buck-converter of Figure 3-5, the transistor turn-on transient, when the diode is switching from forward conduction to reverse blocking, results in the transistor and diode waveforms shown in Figure 3-6.

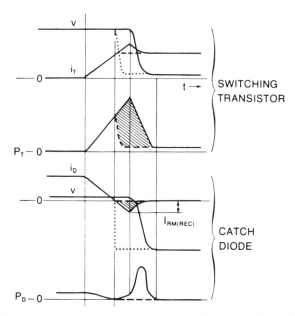

Figure 3-6  Diode and transistor waveforms during switching conditions. (Reprinted with permission from "The Importance of Rectifier Characteristics in Switching Power Supply Design," Application Note U-73A, Unitrode Corp.)

The dashed lines show what the current and power would be if the diode were ideal to the extent of having no reverse recovery time or junction capacitance. (Dotted lines show the voltage for the ideal diode case.) The reverse diode current caused by diode capacitance and recovered charge is shown by the cross-hatched area of the $i_D$ curve. The transistor must conduct this reverse diode current as well as the inductor current. The grey area represents additional transistor dissipation due solely to the diode recovered charge and capacitance.

Faster switching transistors will not necessarily result in reduced switching losses. Unless a diode with a recovery time 2 or 3 times faster than the transistor current rise time is used, a faster transistor will increase the peak recovery current in the diode, and thus increase overall switching losses. Furthermore, a diode with a "soft" recovery characteristic will cause more dissipation than an "abrupt" type with the same peak recovery current. With many currently available switching transistors, a 200 ns fast-recovery rectifier will have a peak recovery current $I_{RM(rec)}$ greater than shown in the $i_D$ waveform of Figure 3-6, where it is about 1/3 of the forward current. This additional collector current (33% above that limited by an ideal diode) can cause increased transistor power dissipation of 100 to 150% during the turn-on period. Other serious problems can occur from high peak currents, such as noise transients in the line, the transistor coming out of saturation, and forward-biased second breakdown.

Rectifiers are now available with recovery characteristics to keep these problems minimal. Their use is required for a switching supply of maximum reliability and efficiency.

When the transistor of Figure 3-5 turns off, the diode turn-on characteristic usually has little effect on power dissipation, but may cause voltage spiking, with resulting noise and the possibility of exceeding the transistor voltage ratings. The voltage spike is due to the forward recovery characteristic and, when present, will occur as shown (dotted) in Figure 3-7. To correct it, a snubber (series $RC$ across the diode, as shown in Figure 3-29) may be needed. However, the choice of an optimum diode will minimize or eliminate this need.

DC losses in the buck-regulator circuit of Figure 3-5 occur alternately when the diode is forward-conducting and when the transistor is turned on. Reduced efficiency due to DC losses is greatest when the output voltage is low, with diode loss being more significant when the input voltage is relatively high and transistor loss dominating when input voltage is close to the output voltage.

Transient (switching) losses in the regulator vary considerably with voltage, being highest at high input voltage. Furthermore, high-voltage transistors and rectifiers generally have longer switching times than low-voltage types. Speed and "recovery characteristic" and consequently, losses, can vary greatly between different device types and manufacturing processes. The other component (turn-off interval) can be similarly developed but it is not significantly affected by diode selection. However, when transistors and/or drive techniques are cho-

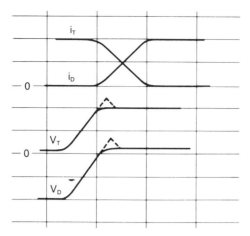

Figure 3-7   Voltage spiking at transistor turn-off. (Reprinted with permission from "The Importance of Rectifier Characteristics in Switching Power Supply Design," Application Note U-73A, Unitrode Corp.)

sen for shorter fall times, overall losses are reduced and the benefits of optimum diode selection become more significant. Proper diode (and transistor) selection is important in all switching supplies, but the higher the voltage (and frequency) the more significant will be the effect of selection on switching losses.

The pulse-width-modulated converter (PWM) of Figure 3-8 has much in common with the buck-regulator; however, in this circuit the output rectifiers also perform the catch diode function. Current waveforms are shown in Figure 3-9 with overshoot due to diode reverse recovery and capacitance. Slow diodes cause additional transistor stress, usually not reduced significantly by transformer impedance. Leakage reactance will often require the use of a snubber to protect the transistor.

One difference between the PWM converter of Figure 3-8 and the buck converter of Figure 3-5 is that in Figure 3-5 the DC diode losses are more significant because $D_1$ and/or $D_2$ are conducting for the full cycle, regardless of the input-voltage-to-output-voltage ratio, and the diode recovery is from half, rather than full, load current.

## Primary Electrical Characteristics[3]

Some of the characteristics of switching rectifier diodes that are important in switching power supplies are as follows:

---

[3]Ibid. 2.

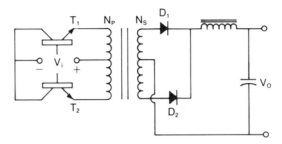

Figure 3-8   PWM converter schematic diagram. (Reprinted with permission from "The Importance of Rectifier Characteristics in Switching Power Supply Design," Application Note U-73A, Unitrode Corp.)

*Peak Inverse Voltage.* The peak inverse voltage, *PIV*, of catch diodes must at least equal the highest input voltage, while the *PIV* of center-tap output rectifiers must be at least twice the maximum output voltage in a square-wave converter and much greater in the pulse-width-modulated converter. More significant are the transient voltages in practical fast-switching circuits due to wiring inductance and a rectifier's own recovery. Unless these are intentionally clipped, damped, or "designed out," it is advisable to use a safety factor of 2 or 3. The *PIV* selected should apply over a range from lowest to the highest expected junction temperature.

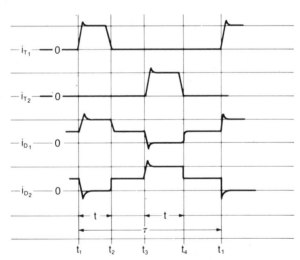

Figure 3-9   PWM switching transistor and rectifier diode waveforms. (Reprinted with permission from "The Importance of Rectifier Characteristics in Switching Power Supply Design," Application Note U-73A, Unitrode Corp.)

*Reverse Recovery Time.* The reverse recovery time, $t_{rr}$, must be much lower than the rise time of the transistor with which it will be used, preferably by at least 3 times when measured at conditions similar to circuit operation. Selection is complicated because rectifiers are normally specified at conditions less severe than in power switching circuits.

Following preliminary selection from available data the devices should be compared in a circuit developing the highest current, junction temperature, and rate of current switching (*di/dt*) expected.

The desired goal is to minimize peak recovery current, $I_{RM(rec)}$, and switching loss. These are the same order of magnitude with Schottky rectifiers (due principally to high capacitance) as with the fastest pn rectifiers.

*Forward Voltage.* The forward voltage, $V_F$, should be as low as possible to optimize efficiency, especially for converter output rectifiers and regulators with high $E_{in}/E_{out}$ ratios. Loss of efficiency due to $V_F$ is most significant at low output voltages.

Schottky rectifiers have the lowest $V_F$ and are, therefore, widely used as output rectifiers for 5 V supplies. Their limitations in *PIV*, transient voltage capability, and temperature must be considered when applying them for use in other voltage applications.

Their selection should be based on conditions where losses are most significant—at rated supply output current and anticipated junction temperature.

*Maximum Average Rectified-Output Current.* The maximum average rectified-output current at maximum expected case or ambient temperature must always be considered. It should be noted that the standard current rating is based on a half-sine waveform. The square-wave switching power supply with an average current equal to this rating will usually dissipate somewhat lower power, and, thus, be used conservatively. However, regulators with $E_{in} \leq 1.5E_{out}$ should use a catch diode with a higher rating than the average current it conducts at full load.

*Peak Voltage During Forward Recovery.* The peak voltage $V_{F(DYN)}$ during forward recovery is significant when using transistors with fast fall times at close to the $V_{CE}$ rating. At lower values of *di/dt* the peak voltages will be lower.

*Surge Current.* Surge current (8.3 ms) is not of great importance because transistor saturation limits fault current. However, if the power supply is designed to provide rapid charging of a large output capacitor, the "overload" requirement for the charge time (perhaps 0.1 to 2 seconds or so) must be considered.

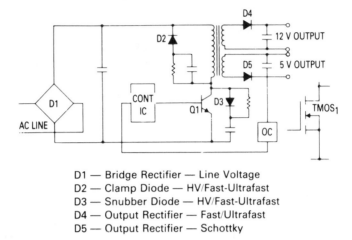

D1 — Bridge Rectifier — Line Voltage
D2 — Clamp Diode — HV/Fast-Ultrafast
D3 — Snubber Diode — HV/Fast-Ultrafast
D4 — Output Rectifier — Fast/Ultrafast
D5 — Output Rectifier — Schottky

Figure 3-10   Switch-mode power supply flyback or boost design. (Copyright of Motorola, Inc. Used by permission)

## Selection Process[4]

The schematic diagram of the switch-mode power supply of Figure 3-10 shows the circuit location of the rectifier diodes that determine the power conversion of this circuit. The input rectifier is generally a standard recovery bridge that operates off the AC line and into a capacitive filter. For the output section, most designers use Schottky rectifiers for efficient rectification of the low voltage, 5 V output windings; and for the higher voltage, 12 to 15 V outputs, fast recovery or ultrafast diodes are used.

### Input Rectifier

When choosing an input rectifier, it is useful to visualize the circuit shown in Figure 3-11 in which small limiting resistors or thermistors and a large bridge are used. The bridge must be able to withstand the surge currents that exist from repetitive starts at peak line voltage. The procedure for selecting the appropriate diode is as follows:

1. Choose a rectifier with 2 to 5 times the average $I_o$ required.
2. Estimate the peak surge current $I_p$ and time ($t$) using:

$$I_p = \frac{1.4 V_{in}}{R_s}; \; t = R_s C \qquad (3\text{-}5)$$

---

[4]Ibid. 1.

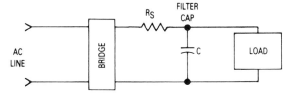

Figure 3-11   Choosing input rectifiers. (Copyright of Motorola, Inc. Used by permission)

where $V_{in}$ is the rms input voltage, $R_s$ is the total series resistance, and $C$ is the filter capacitor size.

3. Compare this current pulse to the subcycle surge current rating ($I_s$) of the diode itself. If the curve of $I_s$ versus time is not given on the data sheet, the approximate value for $I_s$ at a particular pulse width ($t$) may be calculated knowing:

$I_{FSM}$—the single cycle (8.3 ms) surge current rating and using $I^2\sqrt{t} = K$ which applies when the diode temperature rise is controlled by its thermal response as well as power (i.e. $T = K'P\sqrt{t}$ for $t < 8$ ms.).

This gives:

$$I_s^2\sqrt{t} = I_{FSM}^2\sqrt{8.3 \text{ ms}} \text{ or} \tag{3-6}$$

$$I_s = I_{FSM}\left(\frac{8.3 \text{ ms}}{t}\right)^{1/4}, \; t \text{ is in milliseconds}$$

4. If $I_s < I_p$, consider either increasing the limiting resistor ($R_s$) or utilizing a larger diode.

### Output Rectifiers

In the output section, where high-frequency rectifiers are needed, there are several types available to the designer: Schottky rectifiers and a lumped terminology category called "fast-recovery diodes." Schottky rectifiers are based on a metal-to-silicon junction called the Schottky barrier, and are the faster of the two types. Their theoretical recovery times are too short to measure, but in the real world, junction capacitance and external factors result in recovery times as long as 10 ns.

The fast-recovery diodes have recovery times approaching those of Schottky diodes as a result of improved processing techniques. Fast-recovery diodes are subdivided into several speed categories: Fast recovery; superfast; and ultrafast. These vary from one manufacturer to another. Comparative performance for

**Table 3-1    Typical Rectifier Parametric Comparison (Copyright of Motorola, Inc. Used by permission)**

| Parameter | Schottky | Ultrafast | Fast Recovery | Standard Recovery |
|---|---|---|---|---|
| Vғ (VOLTS) Forward Voltage | 0.5–0.6 | 0.9–1 | 1.2–1.4 | 1.2–1.4 |
| tᵣᵣ Reverse Recovery Time | <10 ns | 25–100 ns | 150 ns | 1 μs |
| tᵣᵣ Form | Soft | Soft | Soft | Soft |
| Vʀ (VOLTS) dc Blocking Voltage | 20–60 V | 50–1000 V | 50–1000 V | 50–1000 V |
| Cost Ratio | 3:1 | 3:1 | 2:1 | 1:1 |

devices with similar current ratings is shown in Table 3-1, the point being that lower forward voltage improves efficiency and lower recovery times reduce turn-on losses in the switching transistors. However, the tradeoff is higher cost.

In switching-power-supply applications, recovery time is very important because significant current can flow during the recovery period when the diode is biased off, but conduction continues. This current flow stresses other components and reduces overall efficiency. But recovery time is not the only factor to be considered. Another very important characteristic is the off-state blocking voltage which determines the highest off-state voltage the device can handle.

Depending on the manufacturer, Schottky rectifiers are available with blocking voltage ratings as high as 200 V, while fast recovery types are available with ratings as high as 1000 V or more. Because of their limited blocking voltage, Schottky devices cannot be used in switchers designed to output more than about 15 V DC. Even the few higher voltage devices that are available shouldn't be used in supplies delivering much more than 28 V. So for switching supplies rated at 30 V or more, fast recovery and ultrafast rectifiers are used. But at lower voltages, Schottky rectifiers compete with fast recovery types for use as output rectifiers and catch diodes. Ultimately, the choice of which type to use depends on a variety of factors including price and ratings.

An important factor in determining the external hardware requirements of switching power supplies is the forward voltage drop ($V_F$) of each device. The forward drop for Schottky devices generally runs between 0.4 and 0.6 V; for the other fast recovery types, between 0.9 V and 1.5 V.

The importance of the difference in forward voltage drop is readily apparent when the current passing through the diodes is taken into consideration. For every 0.1 V increase in $V_F$, 0.1 W/A of additional heat must be dissipated. This becomes significant when high current is involved. If a fast-recovery diode with a forward drop of 1.5 V is used in a 100 A supply, for example, 100 W more heat would have to be dissipated than if a Schottky device with a 0.5 drop were used instead. This difference might be the difference between using a small metal tab and a large heat sink, or even between natural convection cooling and forced air cooling.

The Schottky rectifier, though, does have a problem: relatively high off-state leakage current. If a fast-recovery device has a leakage of 5 μA, for example, an equivalent Schottky rectifier's leakage current under the same conditions might be 5 mA or more. This higher leakage current places additional stress on other components, increases heat dissipation, and reduces overall circuit efficiency. Another drawback is the Schottky diode's much higher junction capacitance—often 20 to 30 times that of fast recovery types. Although modern semiconductor designs incorporate a guard ring to reduce the effect of the junction capacitance, older designs still in production require external snubbing networks with larger, more expensive components than those required for fast recovery types.

EMI, as produced by the output rectifiers, is still another issue. The amplitude of the EMI contributed by the rectifiers depends to a large extent on the so-called "softness" of the diode's on-to-off state recovery. In this context, the term refers to the ratio of the fall time to the rise time of the recovery current waveform. A device with a low ratio has a nearly symmetrical, pulse-like waveform and is said to have a snappy recovery. On the other hand, if the ratio is high and the reverse current decays slowly, the device is said to have a soft recovery. The softer the recovery, the lower the EMI and, as a rule, fast recovery rectifiers recover more softly than do Schottky rectifiers.

Table 3-2 provides a guideline summary for determining the rectifier rating requirements of buck, push-pull, PWM forward, PWM push-pull, and PWM bridge converters.

## TRANSISTORS[6]

The initial selection of a transistor for a switcher is basically a problem of finding the one with voltage and current capabilities that are compatible with the application. For the final choice, performance and cost tradeoffs among devices from the same or several manufacturers have to be weighed. But before these devices can be put in the circuit, both protective and drive circuits will have to be designed.

Table 3-3 shows the evolution of high-voltage transistors developed for power supply design usage and the dramatic improvement in switching frequency. Here

---

[6]Ibid. 1.

## Table 3-2    Guidelines for Determining the Rating

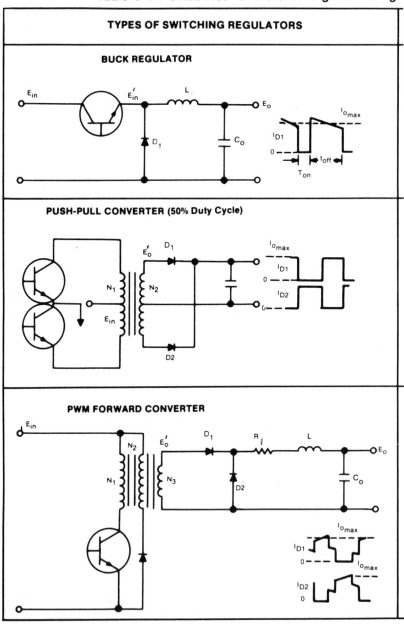

[5]Courtesy of Unitrode Corporation. Used by permission.

## of a Rectifier in a PWM Switched-Mode Converter[5]

| OUTPUT VOLTAGE | STEADY STATE — POWER DISSIPATION IN RECTIFIERS | MINIMUM DC BLOCKING VOLTAGE REQUIRED |
|---|---|---|
| $E_O = E'_{in} \times \frac{t_{on}}{\tau}$ <br><br> $E_O \approx E_{in} \times \frac{t_{on}}{\tau}$ | Power dissipation in Diode $D_1$ due to forward conduction: <br><br> $P_{D1_F} = I_{o_{max}} \times V_F \cdot \dfrac{E_{in_{max}} - E_O}{E_{in_{max}}}$ <br><br> Power dissipation due to leakage current, $I_R$: <br><br> $P_{D1_R} \leq I_R \times E_O.\quad I_R @ E_{in_{max}}$ | For Diode $D_1$: <br><br> $1.2 \times E_{in_{max}}$ |
| $E'_O = E_O + V_F$ <br><br> $E_O = E_{in} \times \frac{N_2}{N_1}$ | Power dissipation in Rectifier $D_1$ or $D_2$ due to forward conduction: <br><br> $P_{D1_F}$ or $P_{D2_F} = \dfrac{I_{o_{max}} \times V_F}{2}$ <br><br> Power dissipation due to leakage current, $I_R$: <br><br> $P_{D1_R}$ or $P_{D2_R} = 2.0 \times E_{in_{max}} \times \dfrac{N_2}{N_1} \times I_R$ | For $D_1$ or $D_2$: <br><br> $2.4\,(E_{in_{max}}) \times \dfrac{N_2}{N_1}$ |
| $E'_O = E_O + V_F + I_{o_{max}} \times R$ <br><br> $E_O = E_{in_{min}} \dfrac{N_3}{N_1 + N_2}$ <br><br> Where: <br> $E_O$ = dc Output Voltage <br> $E'_O$ = Output of Secondary Winding When $D_1$ is conducting | Power dissipation due to forward conduction in Rectifier $D_1$: <br><br> $P_{D1_F} = I_{o_{max}} \times V_F \dfrac{N_1}{N_1 + N_2}$ <br><br> Power dissipation in Rectifier $D_2$: <br><br> $P_{D2_F} = I_{o_{max}} \times V_F \left[ 1 - \dfrac{N_1}{N_1 + N_2} \cdot \dfrac{E_{in_{max}}}{E_{in_{min}}} \right]$ <br><br> Power dissipation due to reverse leakage current: <br><br> $P_{D1_R} = E_{in_{min}} \cdot I_R \cdot \dfrac{N_3}{N_1 + N_2}$ <br><br> $P_{D2_R} = I_R \times \dfrac{N_3}{N_1 + N_2} \times E_{in_{min}}$ | For $D_1$: <br><br> $1.2 \times E_{in_{max}} \times \dfrac{N_3}{N_2}$ <br><br> For $D_2$: <br><br> $1.2 \dfrac{E_{in_{max}} \times N_3}{N_1}$ |

**Table 3-3   Motorola High-Voltage Switching Transistor Technologies (Copyright of Motorola, Inc. Used by permission)**

| Family | Typical Device | Typical Fall Time | Approximate Switching Frequency |
|---|---|---|---|
| SWITCHMODE I | 2N6545 MJE13005 MJE12007 | 200–500 ns | 20 k |
| SWITCHMODE II | MJ13081 | 100 ns | 100 k |
| SWITCHMODE III | MJ16010 | 50 ns | 200 k |
| TMOS | MTP5N40 | 20 ns | 500 k |

Motorola's families of devices are used as illustrative examples. The original switchmode or SWITCHMODE I series of transistors were introduced in the early 1970s with data sheets that provided all the information that a designer would need including reverse bias safe operating area (RBSOA) and performance at elevated temperature (100°C). The SWITCHMODE II series is an advanced version of SWITCHMODE I that features faster switching. And yet another transistor family, the SWITCHMODE III, is a state-of-the-art bipolar transistor series with exceptional speed, RBSOA, and up to 1.5 kV blocking capacity. Here, device cost is somewhat higher, but system costs may be lowered because of reduced snubber requirements and higher operating frequencies. Similarly, Motorola TMOS Power FETs switch more efficiently at high frequencies (200 to 500 kHz) and are easier to drive, but at higher initial device cost.

Table 3-4 lists the transistor voltage requirements for the various off-line converter circuits. As illustrated, the most stringent requirement for single transistor circuits (flyback and forward) is the blocking or $V_{CEV}$ rating. Bridge circuits, on the other hand, turn-on and off from the DC bus and their most critical voltage is the turn-on or $V_{CEO(sus)}$ rating.

Most switching-power-supply transistor load lines are inductive during turn-on and turn-off. Turn-on is generally inductive because the short circuit created

**Table 3-4   Power Transistor Voltage Chart (Copyright of Motorola, Inc. Used by permission)**

| Line Voltage | Circuit | | | |
|---|---|---|---|---|
| | Flyback, Forward or Push-Pull | | Half or Full Bridge | |
| | $V_{CEV}$ | $V_{CEO(sus)}$ | $V_{CEO(sus)}$ | $V_{CEV}$ |
| 220 | 850–1 kV | 450 | 450 | 450 |
| 120 | 450 | 250 | 250 | 250 |

by output rectifier reverse recovery times is isolated by leakage inductance in the transformer. This inductance effectively snubs most turn-on load lines so that the rectifier recovery (or short circuit) current and the input voltage are not applied simultaneously to the transistor.

Sometimes primary interwinding capacitance presents a small current spike, but usually turn-on transients are not a problem. Turn-off transients due to this same leakage inductance, however, are almost always a problem. In bridge circuits, clamp diodes can be used to limit these voltage spikes. If the resulting inductive load line exceeds the transistor's reverse bias switching capability (RBSOA) then a different transistor should be selected or an RC network should be placed across the primary (see Figure 3-34) to absorb some of this transient energy. The time constant of this network should equal the anticipated switching time of the transistor (50 to 500 ns), and resistance values of 100 to 1000 ohms are generally appropriate. Trial and error will indicate how low the resistor has to be to provide the correct amount of snubbing. The RC snubber values may be calculated from the following:

$$C = \frac{It_f}{V} \tag{3-7}$$

where   $I$ = The peak switching current
$t_f$ = The transistor fall time
$V$ = The peak switching voltage (approximately twice the DC bus)

also $R$ = $t_{on}/C$ (it is not necessary to completely discharge this capacitor
in order to obtain the desired effects of this circuit)    (3-8)
where $t_{on}$ = The minimum on time or pulse width


and

$$P_R = \frac{CV^2F}{2} \tag{3-9}$$

where $P_R$ = The power rating of the resistor
and $f$ = The operating frequency

In many switching-power-supply designs, snubber elements are small or non-existent, and voltage spikes from energy left in the leakage inductance become a more critical problem depending on how good the coupling is between the primary and clamp windings and how fast the clamp diode turns on. MOSFETs often have to be slowed down to prevent self-destruction from this spike. (Bipolar and MOS power transistors are presented in detail in Chapter 5.)

## BASE DRIVE CIRCUITS

Figure 3-12 shows the relationship of the voltages and currents in the base and collector of a power transistor switching an inductive load. It is common practice to use a forced gain ratio ($\beta$) which is a function of the collector current and saturation characteristics of the switching transistor. The magnitude of the reverse drive, however, is determined by the turn-off characteristics and the energy capability of the device. Thus, as $I_{B2}$ is increased, storage and fall times are decreased, but as $V_{EB}$ is increased, $E_{SB}$ (secondary breakdown voltage) decreases also.

The base drive circuit should be designed to provide adequate $I_{B1}$ to minimize the saturation losses and sufficient $I_{B2}$ to minimize the storage time and switching losses, and, therefore, must exhibit low source impedance. In addition, the transition times between $I_{B1}$ and $I_{B2}$ should be fast. This section describes several base drive circuits which perform the above functions.

Typical methods for driving bipolar switching transistors are shown in Figure 3-13. Circuit (a) is the simplest: it utilizes no additional drive power or turn-off circuitry. In this circuit $Q_1$ is kept from deep saturation by $Q_2$'s saturation voltage. Circuit (b) utilizes a pair of diodes to reduce overdriving $Q_1$ when an

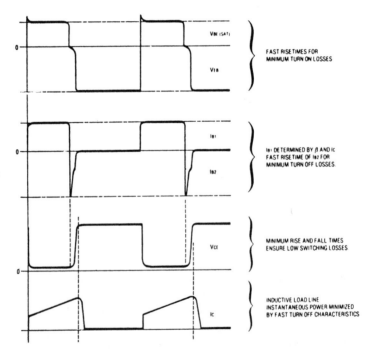

Figure 3-12   Inductive load switching waveforms.

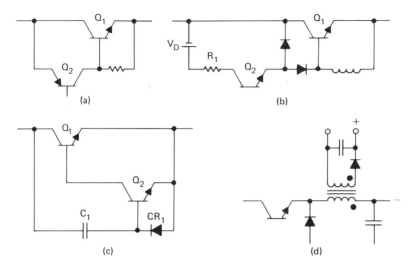

Figure 3-13   Methods of driving switching transistors.

additional drive voltage is used. Caution must be used in determining drive voltage $V_o$ and driver transistor $Q_2$ to prevent the possibility of $Q_2$ taking too great a portion of load current or even driving the base collector junction of $Q_1$ positive at light loads. Limiting resistor $R_1$ can do wonders here. Circuit (c) shows a dynamic means of turning off $Q_1$. This does not decrease storage time in the transistor but serves to assist other turn off methods as $Q_1$ comes out of saturation.

Circuit (d) shows a method of obtaining a semi-regulated DC voltage for use as a turn-on or turn-off bias. This is especially useful for post regulators with DC inputs.

The circuit shown in Figure 3-14 fulfills the requirements for low source impedance for $I_{B2}$ by means of transformer base drive. As shown by the waveforms, $I_{B1}$ has a magnitude of 1 A and $I_{B2}$, 1.5 A. $V_{BE(SAT)}$ is 0.8 V and $V_{EB}$ is $-3.5$ V.

Under these drive conditions, a typical high-speed planar switching transistor (2N6583) switches 300 V and 7 A peak with rise and fall times of 50 ns each. Storage time is approximately 1.5 $\mu$s under these conditions.

As shown by the curve of energy at turn-off measured at various values of $I_{B2}$ for a reverse drive of 1.5 A, the energy at turn-off is 120 $\mu$J. At 20 kHz, this yields a turn-off power loss of only 2.4 W per transistor.

Several other bipolar transistor base drive circuits are shown in Figure 3-15.

The Baker clamp (Figure 3-15d) is a circuit that keeps transistors from operating in hard saturation. Diodes $D_1$ and $D_2$ prevent $V_{CB}$ from going negative, avoiding hard saturation. $D_3$ provides negative off bias and diode $D_4$ compen-

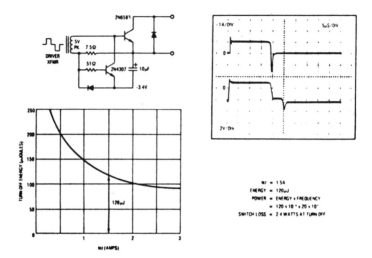

Figure 3-14   Base drive circuit.

sates for the negative potential at the actual collector-base junction during high current operation, preventing hard saturation.

By eliminating overdrive, the Baker clamp allows high-speed capabilities to be fully utilized. The Baker clamp is most effective where load variations are large, a simple drive is desired, and operating frequency is between 20 kHz and 100 kHz. At higher frequencies, the effectiveness of Baker clamps is limited by dynamic saturation—and in the audio range, by on-state losses. Just as was

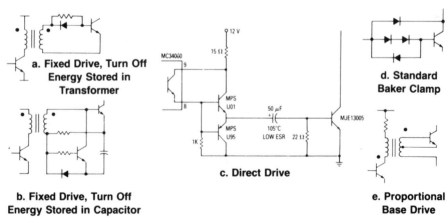

Figure 3-15   Additional bipolar transistor base-drive circuits. (Copyright of Motorola, Inc. Used by permission)

stated earlier in the section on rectifiers, the rectifier diode's characteristics are critically important to the operation of a Baker clamp. The following is a discussion of the Baker clamp requirements.

Beginning with the series base diodes ($D_2$, $D_3$, and $D_4$ in Figure 3-15d) the most important parameter is voltage. The breakdown voltage of these diodes does not have to be high—the lowest voltage parts available are normally more than sufficient. However, there can be a maximum voltage constraint. If the circuit topology requires the power transistor to turn on rapidly, a high-voltage series diode is a poor choice. Like power transistors, high-voltage diodes achieve a low forward drop by conductivity modulating a lightly doped voltage layer. Since the conductivity modulation process takes time, the high-voltage diode is notably slower than its low-voltage counterpart.

In series with the base of a power transistor, the high-voltage diode(s) can be slow enough to appreciably affect the transistor's rise time, and adversely affect turn-on SOA. Therefore, as a general rule, it is best to choose a low-voltage epitaxial type for the series base diodes. Fallout on "downgraded" devices from high-voltage series is to be avoided, since the breakdown voltages for such devices generally run considerably in excess of the specified minimum. For the flyback converter, the choice of the series diodes is generally not critical. Moreover high-voltage diodes can be used in flyback circuits, since the power transistor's rise time is not important.

Additional considerations for the series base diodes are the forward current rating and the reverse recovery time. Both are straightforward. The series diodes are required, in the worst case, to handle the full input base drive, and should be sized accordingly. As for reverse recovery time, it is not important. Bidirectional current flow to the base eliminates the need for an abrupt turn-off transition.

The collector feedback diode ($D_1$) has a different set of constraints. Its breakdown voltage should be compatible with either the collector-base or collector-emitter breakdown voltage rating of the transistor. Which one will depend upon whether or not voltage excursions beyond $V_{CEO}$(sus) are allowed by the circuit topology. Unlike the series diodes, turn-on time is not critical. Slow turn-on is actually somewhat desirable, since base current tends to peak at initial turn-on. Thus, one requirement is held in common. The collector feedback diode is also sized to handle the full input base drive.

The most important consideration is reverse recovery time. Diode $D_1$'s reverse recovery time can have a first-order influence on performance. Thus, it is recommended that ultrafast recovery diodes be used for the actual reverse-bias voltage seen by this diode in the Baker clamp application during the transistor's storage time.

Reference 2 discusses the operation and subtleties of the Baker clamp and the considerations involved in its use.

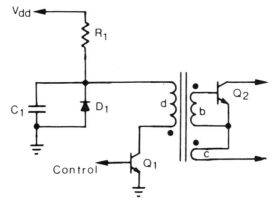

Figure 3-16   Proportional base-drive circuit. (Reprinted with permission from Unitrode Corporation)

Proportional base drive is a simple and effective method of achieving improved performance with high-voltage, bipolar-power-switching transistors in off-line applications. As shown in Figures 3-15 e and 3-16, a current transformer provides regenerative base-drive current whose amplitude is proportional to the collector current being switched. The drive current ratio is established by the turns ratio of the collector and base windings. The primary advantages of the proportional base-drive circuit include optimal performance under varying load current conditions and less drive power from the control circuit.

The proportional base-drive technique works in the following manner.[7] Referring to Figures 3-16 and 3-17, when drive transistor $Q_1$ is on, power switch $Q_2$ is off. Magnetizing current $I_{d1}$ in the control drive winding $N_d$ approaches a steady-state value equal to the drive circuit supply voltage $V_{dd}$ divided by $R_1$. Capacitor $C_1$ is discharged and there is zero voltage across all windings of $T_1$.

When the output of the control circuit turns on, driver $Q_1$ turns off and primary current $I_{d1}$ must cease. Energy stored in $T_1$ causes the voltage at the dotted ends of all windings to flyback in the positive direction. $I_{d1}$ multiplied by turns ratio $N_d/N_b$ becomes $I_{b1}$, the turn-on base drive current pluse to $Q_2$. Collector current $I_c$ starting to flow in winding $N_c$ causes a regenerative increase in base drive to $Q_2$ until it is switched fully on. The final value of $I_c$ induces a proportional base drive current, $I_b$, according to the turns ratio $N_b/N_c$.

During the time that $Q_2$ is on and $Q_1$ is off, capacitor $C_1$ charges through $R_1$ to supply voltage $V_{dd}$. At the end of this ''on'' period, driver transistor $Q_1$ is turned on again, applying the voltage on capacitor $C_1$ to the drive transformer primary. This drives the voltage on the base of $Q_2$ sharply negative. The turn-off base current pulse, $I_{b2}$, can be made larger than $Q_2$ collector current, resulting

---

[7]Courtesy of Unitrode Corporation.

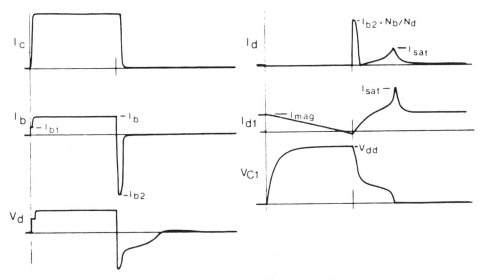

Figure 3-17 Waveforms of proportional base-drive circuit. (Reprinted with permission from Unitrode Corporation)

in very rapid turn-off of $Q_2$. After $Q_2$ is off and $I_{b2}$ ceases, any remaining voltage on $C_1$ across the drive transformer primary helps to rebuild the magnetizing current. Diode $D_1$ prevents the possibility of any underdamped ringing from driving the upper end of $N_d$ negative. At the end of the "off" period, magnetizing current $I_{d1}$ has been reestablished and the cycle repeats.

Using proportional base drive, high-voltage bipolar transistors can be operated at frequencies above 50 kHz with reasonable efficiency because of the large amplitude base drive pulses obtainable with this method. However, the circuit of Figure 3-16 is not capable of operation at frequencies above a few kilohertz. This is because capacitor $C_1$ must charge to $V_{dd}$ during the "on" period of $Q_2$, and the $R_1C_1$ charging time constant is far too long for this to be accomplished at 50 kHz.

This problem is solved by the addition of a rapid recharge circuit as shown in Figure 3-18. During the time that $Q_2$ is on and $Q_1$ is off, current through $R_1$ is multiplied by the current gain of $Q_3$, which significantly reduces the charging time of $C_1$. When $Q_1$ turns on, $C_1$ discharges through $D_2$. The base-emitter of $Q_3$ is reverse-biased, holding $Q_3$ off during the entire $Q_2$ "off" time.

In many cases, power MOSFETs are replacing bipolar transistors as the power switch in switchmode supplies. Compared to a circuit that uses MOSFET power switches (such as the one in Figure 3-19)[8], the drive circuit of a power supply

---

[8]Excerpted from Banfalvi, Stephen, "Seven Design Hints Help Improve Switcher Performance," *EDN*, May 30, 1985. © 1985 Cahners Publishing Company, a Division of Reed Publishing USA.

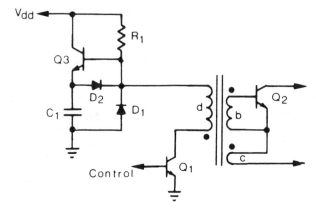

Figure 3-18   Improved proportional base-drive circuit. (Reprinted with permission from Unitrode Corporation)

that uses bipolar transistors typically has an extra stage that provides the current necessary to saturate the output transistors ($Q_1$ and $Q_2$ in the 200 W supply of Figure 3-19b). The resistors, 2N2222 transistors, and driver transformers of Figure 3-19b consume approximately 5 W. Because MOSFETs are voltage-controlled devices, their drive circuitry is simple. Note that the 500 W supply base drive circuit of Figure 3-19a does not require the large power resistors shown in Figure 3-19b. Furthermore, transformer $T_1$ is driven directly from the PWM control chip. Secondary resistors $R_1$ and $R_2$ prevent the ringing that the light load on the transformer would ordinarily cause. If the MOSFETs are switching too fast (causing oscillation), 20Ω resistors ($R_3$ and $R_4$ in Figure 3-19a) can be used to slow them down, thereby preventing oscillation.

The gate voltage of a power MOSFET should never exceed about 20 V. To ensure that the voltage remains below 20 V, the PWM IC should be powered with a regulated housekeeping supply. If the housekeeping supply varies, a 12 to 15 V zener diode can be used to clamp the supply voltage.

The MOSFETs in Figure 3-19a are switches that have very low on-resistance. For example, the data book for the IRF 250 MOSFET lists the on-resistance as 0.085Ω. However, the data book also shows that at 120°C the MOSFET's on-resistance rises to 0.2Ω. Thus, the MOSFETs in a 500 W supply with a 40 V input will generate 50 W that must be removed from the devices. To keep the MOSFETs in a safe temperature range that provides both low on-resistance and low junction temperature, a careful thermal design, including a thorough thermal analysis and proper heat-sink selection, is required.

The transformer-coupled MOSFET drive circuit of Figure 3-20 produces forward and reverse voltages applied to the MOSFET gate which vary with the duty cycle as shown. For this circuit, a $V_{GS}$ rating of 20 V would be adequate

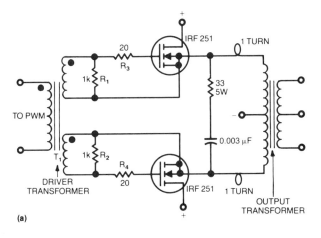

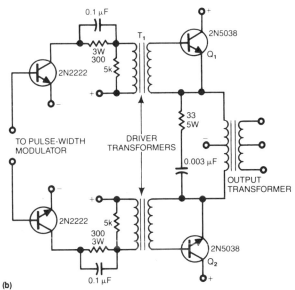

Figure 3-19  Compared to a MOSFET output section (a) of a power supply, a bipolar output section (b) contains many extra components. Excerpted from Banfalvi, Stephen, "Seven Design Hints Help Improve Switcher Performance," *EDN*, May 30, 1985. © 1985 Cahners Publishing Company, a division of Reed Publishing USA.

for the worst case condition of high logic supply (12 V) and minimum duty cycle. And yet, minimum gate drive levels of 10 V are still available with duty cycles up to 50%. If wide variations in duty cycle are anticipated, a semi-regulated logic supply should be considered.

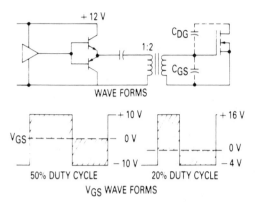

Figure 3-20   Typical transformer coupled MOSFET drive. (Copyright of Motorola, Inc. Used by permission)

One point that is not obvious when looking at Figure 3-20 is that MOSFETs can be directly coupled to many ICs with only 100 mA of sink and source capability, and still switch efficiently at 20 kHz. However, to achieve switching efficiency at higher frequencies, 1 to 2 amps of drive current may be required on a pulsed basis in order to quickly charge and discharge the gate capacitances. A simple example illustrates this point and also shows that the Miller effect, produced by $C_{DG}$, is the predominant speed-limiter when switching high voltages (see Figure 3-21). A MOSFET responds instantaneously to changes in gate voltage and will begin to conduct when the threshold is reached ($V_{GS}$ = 2 to 3 V) and be fully on with $V_{GS}$ = 7 to 8 V. The gate waveform shows a plateau at a point just above the threshold voltage which varies in duration depending on the amount of drive current available, thus determining both the rise and fall

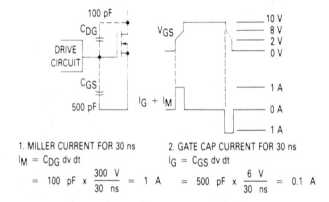

Figure 3-21   FET drive current requirements. (Copyright of Motorola, Inc. Used by permission)

times for the drain current. To estimate the drive current requirements, the following calculations are made:

$$1.\ I_M = C_{DG}\ dv/dt \text{ and} \tag{3-10}$$

$$2.\ I_G = C_{GS}\ dv/dt \tag{3-11}$$

$I_M$ is the current required by the Miller effect to charge the drain-to-gate capacitance at the rate desired so as to move the drain voltage (and current). And $I_G$ is usually the lesser amount of current required to charge the gate-to-source capacitance through the linear region (2 to 8 V). As an example, if 30 ns switching times are desired at 300 V where $C_{DG} = 100$ pF and $C_{GS} = 500$ pF, then

$$I_M = 100\ \text{pF} \times 300\ \text{V}/30\ \text{ns} = 1\ \text{A and}$$

$$I_G = 500\ \text{pF} \times 6\ \text{V}/30\ \text{ns} - 0.1\ \text{A}$$

This example shows the direct relationship between drive current capability and speed and also illustrates that for most devices, $C_{DG}$ will have the greatest effect on switching speed and that $C_{GS}$ is important only in estimating turn-on and turn-off delays.

Aside from its unique drive requirements, a MOSFET is very similar to a bipolar transistor. Currently available 400 V MOSFETs compete with bipolar transistors in many switching applications. They are faster and easier to drive, but do cost more and have higher saturation, or "on" voltages. The performance or efficiency tradeoffs are analyzed using Figure 3-22. Here, typical power losses for 5 A switching transistors versus frequency are shown. The losses were calculated at $T_j = 100°C$ rather than 25°C because on-resistance and switching

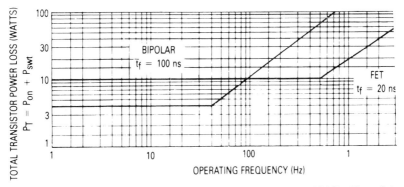

Figure 3-22 Typical switching losses at 300V and 5A $(T_j = 100°C)$. (Copyright of Motorola, Inc. Used by permission)

times are highest here and 100°C is typical of many applications. These curves, which are asymptotes of the actual device performance, are useful in establishing the "breakpoint" of various devices, which is the point where saturation and switching losses are equal. (The characteristics of MOS power transistors are presented in greater detail in Chapter 5.)

## CONTROL CIRCUITS

Over the past ten years, a variety of control ICs for switching power supply applications have been developed. Most control ICs have the following features:

- Programmable (to 500 kHz) fixed frequency oscillator
- Linear PWM section with duty cycle from 0 to 100%
- On-board error amplifiers
- On-board reference regulator
- Adjustable dead time
- Under voltage (low $V_{cc}$) inhibit
- Good output drive (100 to 200 mA)
- Option of single or dual channel output
- Uncommitted output collector and emitter or totem pole drive configuration
- Soft-Start capability
- Digital current limiting
- Oscillator sync capability

When it is necessary to drive two or more power transistors, drive transformers are a practical interface element and are driven by the conventional dual channel ICs. In the case of a single transistor converter, however, it is usually more cost effective to directly drive the transistor from the IC. In this situation, an optocoupler is commonly used to couple the feedback signal from the output back to this control IC.

### PWM Control Circuit

Several circuits that perform the pulse width modulation control timing portion of the switching regulator are shown in Figures 3-23 and 3-25.

A CMOS PWM circuit design that uses only two active components (the CD4047 CMOS multivibrator and $\mu A723$ IC voltage regulator) is shown in Figure 3-23. This circuit is capable of DC modulation and pulse width control from 10 to 100% of the duty cycle. The PWM segment of the circuit consumes only 30 mw but can be made to consume only 15 mW if the modulator stage is driven from a lower voltage source.

The $\mu A723$ serves as a reference, error amplifier and modulator. The CD4047

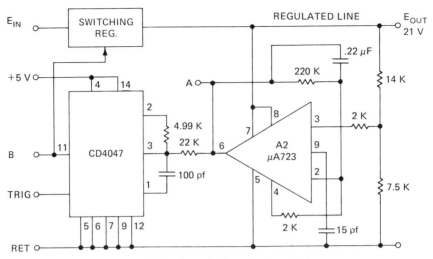

Figure 3-23   CMOS pulse-width modulator.

operates as a monostable multivibrator, which is locked to a certain repetition rate by accepting trigger signals from a given frequency source (50 kHz in this example). The pulse duration of the monostable multivibrator is determined by the amount of charge current supplied by the 22 K$\Omega$ resistor, $R_3$. The transfer function of pulse width versus modulating voltage $E_m$ at point A is shown in Figure 3-24. The $\mu A723$ is connected in such a manner that it is energized from the output of the switching regulator which is driven by the CD4047. It uses its own reference voltage and compares it to a voltage sample of the output. It then

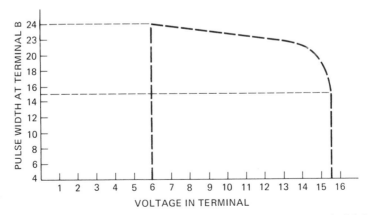

Figure 3-24   Empirical transfer function of pulse width vs voltage at terminal A for the PWM in Figure 3-23.

amplifies the error and provides to point A the required voltage to cause the pulse width of the CD4047 to be such that the output of the switching regulator remains fixed. For a $\pm 25\%$ line voltage variation, the above circuit regulates the output of the switching regulator to within $\pm 1\%$, and can be made to provide a tighter regulation.

Another PWM circuit with current foldback is shown in Figure 3-25.[9] This circuit uses the NE555 timer as the basic control element.

For astable operation, $R_1$, $R_2$, and $C_1$ are connected as shown. Capacitor $C_1$ charges to $2E_{in}/3$ through $R_1$ and $R_2$, and discharges to $E_{in}/3$ through $R_2$ when there is no external voltage at terminal 5. Thus the timer will retrigger itself, yielding a square wave output with:

$$t_{on} = 0.695 \ (R_1 + R_2)C_1 \tag{3-12}$$

$$t_{off} = 0.695 \ R_2 C_1 \tag{3-13}$$

$$\text{and:} \quad f = 1/(t_{on} + t_{off}) \tag{3-14}$$

This square wave is amplified by $R_3$, $Q_1$, $R_4$ and $R_5$, and fed to transistor $Q_2$. As long as the timer output is high, $Q_2$ will be on and driving current into $R_L$ and $C_2$ through inductor $L$. When $Q_2$ turns off, the energy stored in $L$ and $C_2$ is available to supply the load. Differing from the input level, the voltage thus generated is fed to a simple comparator formed by $Q_5$, $D_Z$, $R_{11}$, and $R_{12}$. $Q_5$ will not conduct unless the output voltage is less than the zener voltage. Therefore the voltage at the collector of $Q_2$ continuously changes depending on how it compares to the zener voltage. Since the collector voltage is fed to the modulating input (terminal 5), the pulse width of the generated square wave is modulated to provide the required output voltage. An approximate relation between $E_{in}$ and $E_{out}$ can be described as:

$$E_{out} = (t_{on} E_{in})/(t_{on} + t_{off}) \tag{3-15}$$

$R_7$ is the current sensing resistor. When the load current increases to a level such that the voltage drop across $R_7$ turns $Q_3$ on, $Q_4$ will be driven into saturation. The resultant low at pin 4 of the 555 resets the timer, bringing its output to zero. With the timer reset, no voltage develops across $R_8$ and $Q_4$ is turned off, enabling the timer, and bringing its output high. If an overload condition still exists, $Q_3$ and $Q_4$ will again be turned on and the timer reset. This closed-loop chain reaction continues as long as an overload condition exists. If the overload condition increases, the voltage and current will both decrease initiating the foldback action.

---

[9]"Put a 555 Timer in Your Next Switching Regulator Design," Chetty, P. R. K. Reprinted from *EDN*, January 5, 1976 by Cahners Publishing Company.

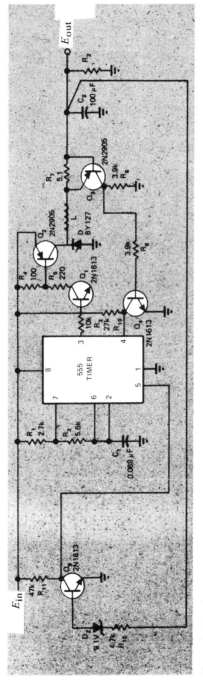

Figure 3-25  Complete with current foldback, this pulse-width modulated regulator takes advantage of the RESET and CONTROL VOLTAGE inputs of the 555 timer.

With a 15 V input, the circuit will deliver a 10 V, 100 mA output with line and load regulation figures of 0.5% and 1%, respectively. Foldback action will commence at a current value equal to $Q_3$'s $V_{BE(sat)}$ divided by $R_7$.

Chapter 6 presents and discusses the commercially available PWM and control ICs dedicated to power supply applications.

## SYMMETRY CORRECTION/PULSE SHAPING NETWORK

Failure of the output transistor when the power transformer goes into saturation is a common problem that can be cured by a pulse-shaping network (Figure 3-26). Power transistor failure results when the transistors of a push-pull driver have different conduction periods, and is especially likely to occur during load changes.

The resulting imbalance in drive current at the ends of a high-frequency ferrite transformer causes a DC component to flow through the transformer winding.

The cause of this imbalance is lack of symmetry—the pulse-width modulated signals feeding the transformer have slightly different on-times. The situation is corrected by a circuit that creates two out-of-phase signals which are duplicates of each other—in effect twins.

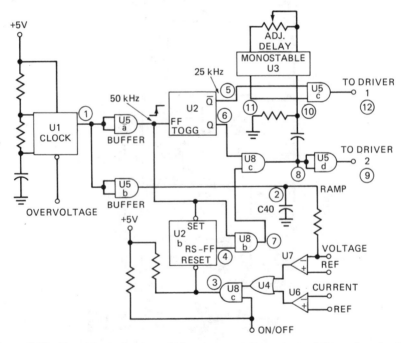

Figure 3-26   Two identical pulse widths appear at the output of this pulse shaping circuit.

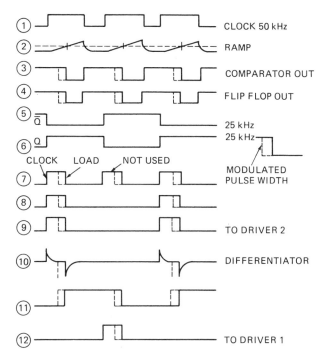

Figure 3-27    Waveforms 9 and 12 represent the identical signals required to operate a switching supply without saturation. Signals 5 and 6 ensure that the driver outputs are 180° out of phase.

Referring to Figure 3-27, which avoids transformer saturation, two pulses of identical width and 180° out of phase are generated by making a single clock source (the 50 kHz signal in Figures 3-26 and 3-27) responsible for both output pulses.

In the tuning circuit, $U2a$, a toggle flip-flop, produces gating signals (waveshapes 5 and 6) which make the two power output driver signals (waveshapes 9 and 12) 180° apart. The pulse-width modulated wave for driver 2 is produced through a long signal chain from the clock source and driver 1's pulse-width signal comes directly from a monostable multivibrator ($U3$) whose delay time is adjustable.

After the 50 kHz clock passes through buffer $U5b$, network $R8$ and $C40$ integrates it into a ramp which appears at the input of comparator $U7$. A reference voltage and the ramp are compared in $U7$ with the output fed to an OR gate. A similar comparison for current is performed in comparator $U6$. The net result is a pulsewidth-modulated output at gate $U4c$ (waveshape 3) which then passes through flip-flop $U2b$ to become the pulse signal for driver 2. Driver 1's waveshape, the so-called mirror signal, is generated from $U3$'s pulse and $\overline{Q}$

output of $U2a$. Because the monostable multivibrator has an adjustable delay, its pulse width can be made identical to that of driver 2.

Without such a pulse shaping scheme, transformer core saturation can induce high collector current spikes which eventually will cause transistor failure.

## SNUBBER CIRCUITS

In many applications, it is possible to add circuitry that alters the shape of the load line in a manner that reduces the peak power seen by the transistor and the total switching losses. In Chapter 2, Figures 2-33, 2-46, 2-69, 2-73, 2-74 and 2-82 through 2-88 contain such circuits which are often referred to by the following names: turn-off protection circuit, loading shaping network, energy snubber, slow rise circuit, and a snubber circuit or just plain snubber.

In a typical power supply, the ringing without a snubber can last for many cycles and generate substantial EMI (as shown in Figure 3-28a) and even damage the rectifiers. With properly designed snubbers, ringing can be damped out almost completely (Figure 3-28b) without interfering with circuit performance.

The following examples of load line shaping are presented for some of the more commonly used converters, but the general principles are widely applicable.

Snubbers for high current supplies are generally placed across each rectifier separately (as is shown in Figure 3-29a). For low current supplies, however, a single snubber across the transformer winding (as seen in Figure 3-29b) can suffice. Optimally designed, the snubber network will consume little power.

In the switching regulator of Figure 3-30 the load line is modified by inserting turn-on and turn-off snubber circuits. During turn-on, the 40 $\mu$H series inductor serves to isolate the transistor from the inductive load, thus permitting the transistor to turn on prior to conducting an appreciable amount of collector current. The rate of rise of collector current $dI_c/dt = E_{in}/L_{snubber}$. The diode resistor network around the snubber inductance dissipates the energy stored in the inductor during the transistor "off" period.

The "turn-off" snubber provides a path for filter choke current such that the transistor may cease conduction prior to full input voltage appearing across the collector-emitter. The 250 $\Omega$ resistor provides a charging path for the capacitor during the transistor on time. The diode serves to bypass the resistor for most efficient snubbing action.

In a flyback converter, a R-C-diode turn-off snubber can protect the switching power transistor from second breakdown by limiting $(dV_{CE}/dt)$ during turn-off as shown in Figure 3-31. It can also protect the transistor from collector-base avalanche breakdown by limiting the peak $V_{CE}$ which is imposed on the transistor after turn-off. In a transformer-coupled power converter, the snubber must cope

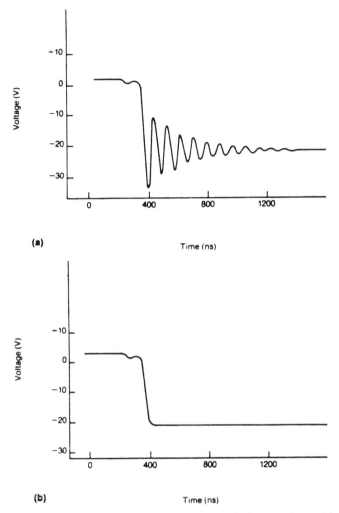

Figure 3-28    Voltage waveforms (a) without and (b) with the use of a snubber circuit.

with the transformer leakage inductance between the primary winding and the clamped secondary winding.

Referring to Figure 3-31, during the on time of $Q_1$ the $RC$ time constant is sufficiently small to allow $C$ to discharge, while $R$ limits the peak discharge current during the turn-on edge. When $Q_1$ turns off, the rate of rise of collector voltage is defined by the primary current in the transformer at that time and capacitor $C$; $C$ restricts the rate of rise of $V_{CE}$ during the fall time of the transistor

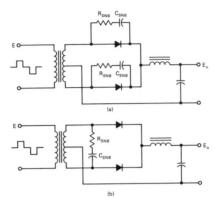

Figure 3-29   Typical snubber circuits for (a) high current power supplies and (b) low current power supplies.

collector current, $I_{CQ1}$. The larger the value of $C$, the slower $V_{CE}$ rises. A value of $C$ must therefore be chosen such that the $V_{CEO}$ rating of the transistor is not exceeded before the collector current has fallen to zero. It is necessary to almost fully discharge the capacitor when the power transistor turns on and to fully charge the capacitor to 2 $V_{CC}$ when the transistor turns off. As a result of this technique, efficiency will not be adversely impacted since a fair portion of the power dissipated in the snubber would have otherwise been dissipated in the transistor.

Additionally, this method will, in some cases, help to reduce EMI. References 3, 4, 5, and 6 provide detailed circuit analysis, circuit design methodology guidelines and design procedures for the turn-off snubber circuit of Figure 3-31.

In the switching regulator of Figure 3-32 the transistor is subjected to high currents during turn-on due to the reverse recovery of $CR_2$. This current can be reduced by adding an inductance between the transistor and diode. However,

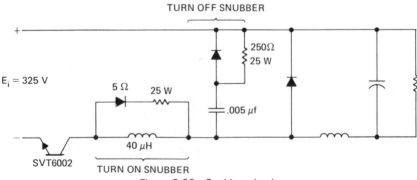

Figure 3-30   Snubber circuit.

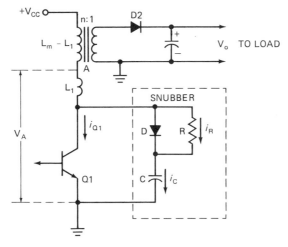

Figure 3-31  Snubber network in a flyback power converter.

during turn-off the energy due to the inductance must be prevented from being dissipated in the transistor.

A non-dissipative way of achieving this is to add the indicated secondary circuit and return this energy to the output filter capacitor.

Shown in Figure 3-33 is a buck-boost converter and associated snubber circuit consisting of $CR_1$, $CR_2$, $C_2$, and $L_1$. Without this network the transistor must dissipate the energy stored in the primary leakage inductance $L_1$ of $L_2$.

The circuit operates as follows: assume that $Q_1$ is on and that $C_2$ is charged to $-E_{in}$. When $Q_1$ is turned off suddenly at $t = 0$ the energy stored in leakage inductance $L$ causes a current equal to $I_{N1(pk)}$ to instantly begin flowing through a path formed by $C_2$, $CR_1$, and $N_1$. The voltage of $C_2$ is quickly reversed by this action. If the capacitor voltage rises to a value in excess of $E_{in} = E_{out(N_1/N_2)}$ a current path will be formed by $CR_2$, $L_1$, $C_2$, $N_1$, and $C_1$. $C_2$ will

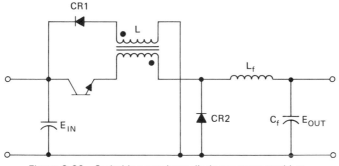

Figure 3-32  Switching regulator diode recovery snubber.

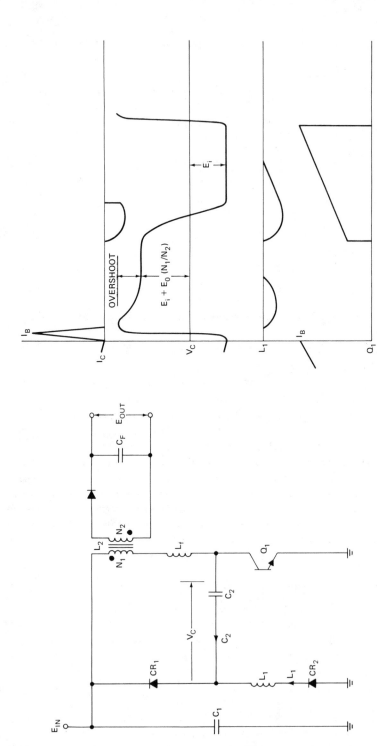

Figure 3-33   Non-dissipative snubber circuit.

then discharge to $E_{out(N_1/N_2)}$. If current flow in $N_2$ ceases before the next on interval of $Q_1$ the capacitor voltage will discharge to $E_{in}$ through the above-mentioned path.

During the next conduction cycle of $Q_1$, the voltage on $C_2$ begins to reverse due to the resonant path formed by $L_1$, $CR_2$, $C_2$, and $Q_1$. When the voltage across $C_2$ has reached $-E_{in}$ it is clamped to this level and the current in $L_1$ now flows through $CR_1$ and $CR_2$ back to input filter capacitor $C_1$, decaying at the rate of $E_{in}/L_1$, thus completing a full cycle.

The value of this circuit lies in the fact that it returns the leakage energy back to the input filter rather than dissipating it. Selection of component values begins by determining $L_1$. The capacitor voltage changes from $-E_{in}$ to $E_{in} = E_{out(N_1/N_2)}$ and by equating energies in $L_1$ and $C_2$ results in:

$$C_2 = \frac{L_1 I_p^2}{E_{in} E_{out(N_1/N_2)} + (E_{out(N_1/N_2)})^2} \tag{3-16}$$

$L_1$ is given by: $\pi \sqrt{L_1 C_2} = t_{on}$ minimum. $C_2$ should be a high quality AC capacitor.

In the push-pull regulated converter of Figure 3-34, the secondary filter inductance causes the device load line to be square in turn-off. This occurs because the voltage across the off going transistor must reach $E_1$ before current in $L_f$ can begin to free-wheel in $CR_3$ and $CR_4$. As a result the peak current and voltage occur simultaneously in the device. In practical circuits leakage inductance in the transformer will cause $V_{ce}$ to overshoot $E_1$. This undesirable situation is altered by adding the standard snubber circuit of $R_1$ and $C$. As can be seen, the peak and total energy absorbed by the transistor is substantially reduced.

Snubber circuit No. 2, as shown in Figure 3-34, is an alternate way of shaping the turn-off load line. In this circuit the load current is diverted through the diode to charge the snubber capacitor as voltage rises across the transistor during turn-off. The diode provides a low-impedance current path during turn-off and forces current through the current-limiting resistor during turn-on. This snubber results in a somewhat better load line than the first circuit and is the same as shown in Figure 3-31.

Both circuits require low-dissipation type capacitors and resistors with fairly large power ratings. The advantage of circuit No. 1 is the fewer parts it requires. Both circuits slow switch times somewhat. However, the power dissipated in the transistor will be reduced by as much as 60%, resulting in greater device reliability.

## TRANSIENT PROTECTION

The problem of high-voltage transients or spikes is a major cause of transistor failure in converters. When a transistor switches off an inductive load, a high-

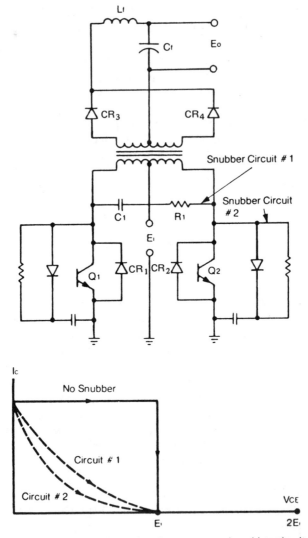

Figure 3-34 Push-pull regulated converter and snubber circuit.

voltage spike appears across its collector diode at the instant of switch-off, because of the collapse of the magnetic field. This effect is intensified by hole storage in the transistor. When an "on" transistor is abruptly turned off by making the base positive with respect to the emitter, the reverse resistance of the base-emitter diode does not change instantaneously to a high value. Hole storage keeps the reverse resistance low for a period of microseconds and a reverse base current pulse occurs. This induces a strong ringing and may break down the cutoff transistor. Spikes have two serious effects. First, even though of negligible energy, they may cause local breakdown at points in the junction, which leads to deterioration and ultimate failure of the transistor. Second, when the spikes are inadequately suppressed and very high, they may lead to catastrophic breakdown of the transistor because of overheating or breakdown of the junction.

Ringing-choke circuits are more prone to spikes than push-pull circuits. A push-pull square-wave converter is partially self-protecting (as compared with a single-ended circuit) because the voltage spike is reduced somewhat when first one end of the collector winding and then the other clamps itself to the negative supply by reverse conduction through the collector diode. Spikes may be minimized by using magnetic-amplifier-type transformer core materials with high remanence. With such a core, the field does not collapse so abruptly when the magnetizing force is removed.

Many DC supplies used for converters produce high-voltage transients with the switching on and off of loads. When transistor converters operate from such supplies, protective transient-suppressing networks are also necessary in the input. Even a simple clipping diode can prove effective.

Apart from careful transformer design, many circuit arrangements have been developed to reduce voltage transients to a safe level. A number of these are illustrated in Figures 3-35 and 3-36.

## CURRENT LIMITING

Many system requirements dictate the need that current-limiting circuitry be built into the power supply. In early switching power-supply designs, the current-limiting approach could shut down a cycle the instant that an overload was detected and would shift to reduced duty-cycle operation, thus allowing the circuit to recover on a low- or soft-start basis. This method, however, didn't accommodate high, transient peak power load demands. Reference 7 presents and discusses various current-limiting circuits for these early power supply designs.

Integrated circuit voltage regulators contain on-chip current-limiting and thermal shut-down circuitry, thereby providing flexibility to the designer. When MOSFETs are used in the power supply design, fast current-limiting is required, because if current-limiting is too slow, a short on the power supply's output can destroy the MOSFETS.[10] The slow rate of the current-limiting op-amp internal to some PWM control ICs is often too slow for MOSFET applications. To back up the on-board current-limiting function, one can use the circuit shown in Figure 3-37a. Current transformer $T_1$ has a 1-turn primary that is capable of

---

[10]Ibid. 8.

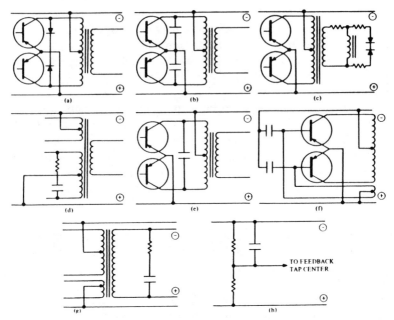

Figure 3-35 Transient spike suppression networks. (a) Clipping diodes from collectors to emitters. (b) Capacitors from collectors to emitters. (c) Diode clippers across inductive loads such as motor fields. (d) Capacitance-resistance series network across feedback windings. (e) Capacitor between collectors. (f) Capacitor from base to negative supply. (g) Capacitance resistance series network across transformer output. (h) Capacitor across part of base-bias network.

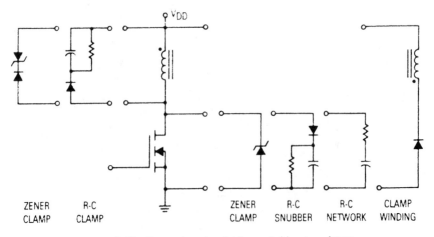

Figure 3-36 Protection circuits for switching transistors.

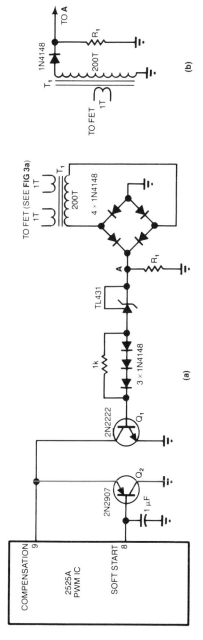

Figure 3-37  Current-limiting when using a PWM control IC for two-output transistor converters (a) and for single-output transistor converters (b). (Reprinted with permission from "Seven Design Hints Help Improve Switcher Performance," *EDN*, May 30, 1985. © 1985 Cahners Publishing Company, a Division of Reed Publishing USA)

carrying the current switched by the MOSFETs. The circuitry to the left of point $A$ is an extremely fast switch that turns $Q_1$ on when the voltage at point $A$ reaches approximately 2.5 V; $Q_1$ then shuts down the converter. Resistor $R_1$ is selected to make point $A$'s voltage equal 2.5 V at the designed current limit value. The circuit of Figure 3-37b is used for power supplies with a single power transistor.

In newer power supply designs, a sophisticated current-mode control system provides protection by sampling the converter output voltage and a current proportional to the load current. It then applies controls to:

(1) Sense an overload condition that would in time damage the switch or other components, shut down the switch for the remainder of its present switching cycle fast enough to prevent damage—without interrupting the switching operation, and then repeat the procedure on subsequent switching cycles;

(2) Maintain the above rapid switch turn-off mode long enough to "outwait" a minor transient overload of short duration, enabling the supply to deliver high-peak transient output power above the rated steady-state value;

(3) Shut down the switch completely for a prolonged period in the event of a severe and persistent overload lasting longer than several milliseconds that could damage both the supply and its load, but continue to recycle (sample) periodically until the overload ceases and normal supply operation resumes.

To current limit multiple output supplies, the switch current can be bounded to furnish no more than maximum rated total output power. The current limit is activated whenever the total supply output power exceeds this rated maximum value. This requires that each individual output be able to supply the total load power to overloads of short duration. Continuous overloads require additional protection. For example, a long-term overload imposed on any output might be sensed by a temperature rise in the output circuitry which, at a predetermined level, would shut the switch down.

Sensing individual output current levels in multiple-output supplies that contain no series-pass regulators is difficult. Generally, the transformer winding and rectifier of each output are made to withstand total load power. When some of the multiple outputs employ a series-pass regulator, protection becomes a simple matter of equipping each regulator with its own output current limiter. In this case, the main output—the one used to control overall supply regulation—must be able to handle the full output power and should be thermally protected.

## GENERATED EMI

All switching power supplies suffer from input ripple, which produces EMI, and from output ripple and spikes, which can affect load operation. The two major causes of generated EMI are the abruptly changing rectangular switching waveforms of both forward and flyback converters and imperfect diode turn-off characteristics. The Cuk converter, though, can be internally balanced to generate either no input ripple or no output ripple. However, cancelling ripple by balancing effectively cancels ripple only over a limited frequency range because of the harmonic-rich nature of the switching waveforms with their wide spectrum of component frequencies.

The flyback converter is often labeled as being noisy because the basic topology has the highest peak currents and the highest energy transfer at switch turn-off so that radiated interference is high. Also, the gapped transformers usually employed in flyback converters generate more magnetic interference than the ungapped transformers of forward converters. Thus, the hardware design should be checked carefully for radiated fields and then shielded as necessary.

All switching supplies require filtering of the fundamental and harmonics of the switching frequency, even the Cuk converter. While the forward converter has an inherent $LC$ filter in its output choke and capacitor, the flyback converter has only an output capacitor that must be augmented by an $LC$ filter. In general, practical considerations make the degree of filtering required by flyback, forward, and Cuk converters similar.

To control the generated EMI effects of diode turn-off characteristics, diodes are available that turn off fast or drift off in a slower and controlled manner. As discussed previously, the 50/60 Hz input rectifier diodes that are chosen to aid in generating the converter operating voltages usually have a soft turn-off characteristic. While the higher reverse leakage current of these soft turn-off devices reduces efficiency somewhat, they contribute little to interference problems. Conversely, fast turn-off is a desirable characteristic of output diodes in the secondary circuits of the forward and flyback converters. These fast turn-off diodes, which include conventional and Schottky types, induce oscillations at turn-off in the 5 and 12 MHz frequency bands. The noise and spikes that the action of these diodes generates can be minimized by use of lossy RC snubber networks, which, while trying to slow extremely high rates of change, also reduce switcher operating efficiency.

The flyback converter is not the worst "diode-induced" noise offender because the fast output diodes carry a linearly falling current that decays to zero before turn-off. The forward converter's high-speed output diodes, however, carry the full output current when they are abruptly switched off at the start and finish of the switch "on" cycle, thus generating significant diode noise.

All of these diode-switching, transient-induced-noisy emissions require $LC$ filtering in order to hold ripple and noise voltages to within about 1% of the nominal output.

## GROUNDING, BYPASSING, AND SHILEDING OF IC CONVERTERS

The quality of ground connections is key to the performance of DC/DC converters using high performance PWM ICs. Because the peak current in an inductor or switch (transistor) can reach several amperes, one must provide these points with very low impedance paths to the supply common. For example, in a flyback converter, the inductor current typically exceeds 1 A. For best results, separate paths to ground should be used for the high current paths so that they are separated from the chip's power and feedback connections. If this isn't possible, then a heavy single trace can be used to carry the high current back to the supply.

Loop instabilities, caused by interactive ground connections or stray capacitive pick up, can also severely limit the performance of an otherwise sound DC/DC converter design. Some of the symptoms of these problems are high ripple voltages at the output, efficiency that is lower than expected, and ''motorboating,'' or low-frequency oscillation. Motorboating occurs when the control loop of the DC/DC converter produces pulses in periodic clusters of 10 to 20 pulses rather than at more or less random intervals. In an IC PWM switching power supply, motorboating can be caused by one or more of the following phenomena: stray pickup at the feedback node, unwanted feedback to the reference, and feedback via the ground or power-input pin.

If the cause is stray pickup at the feedback node of the IC, a 100 to 1000 pF compensation capacitor should be added from the feedback terminal pin to the circuit output or the size of the connections at the feedback input should be reduced in order to reduce stray capacitance to ground. If unwanted feedback to the reference is the problem, the reference and power input pins should be bypassed to ground using a 0.1 to 1.0 $\mu$F capacitor. If the circuit is suffering from feedback via the ground or power-input pin, the power supply input should be bypassed with a 1.0 to 10.0 $\mu$F capacitor. High ground current connections should be separated from the reference, feedback, chip-ground, and chip-power connections.

## CONVERTER DESIGN PROBLEMS AND EXAMPLES

Converter design problems are best illustrated and easier to understand when explained in the form of practical examples. Several examples are presented in the following paragraphs which cover typical areas of oscillator design.

### A Simple Driven DC/DC Converter[11]

The switching-regulated DC-to-DC converter shown in Figure 3-38 combines switching regulation and voltage conversion by using only three signal and two power transistors and one voltage regulator IC. With an efficiency of about 83% and regulation to within $\pm 0.1\%$, power output is 5 W at +6.3 V and 4 W at +5 V over an input range of $-50$ to $-80$ V (nominally $-60$ V).

[11]Cowett, P. M., Switching Supply Converts $-60$ V to $+5$ V and $\pm 6.3$ V with 83% Efficiency. Reprinted with permission from *Electronic Design 2*, January 18, 1978, copyright Hayden Publishing Co., Inc. (1978).

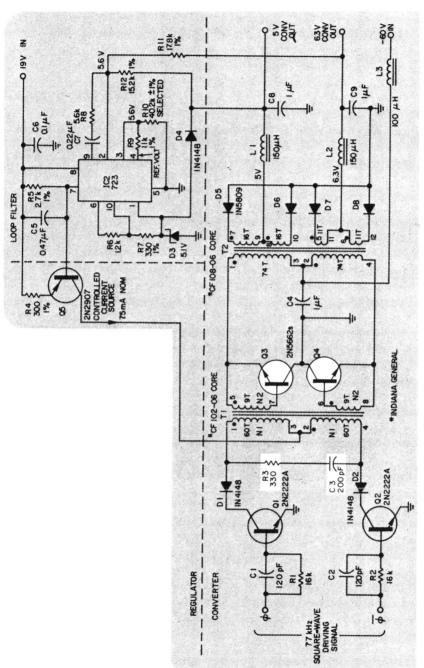

Figure 3-38 Driven converter schematic diagram. Reprinted with permission from *Electronic Design* 2, January 18, 1978, copyright Hayden Publishing Co. Inc. (1978).

The converter is a driven type, modified to be regulated by pulse-width modulation. Signal transistors $Q_1$ and $Q_2$ switch on and off alternately, driven by a 50% duty-cycle square wave. When $Q_1$ is on, diode $D_1$ is forward-biased and current flows through half the driver-transformer primary of $T_1$. The same thing happens for $Q_2$ and $D_2$. The voltage developed across half the primary equals approximately $(N_1/N_2)\ V_{BE}$, where $V_{BE}$ is the voltage across a converter power transistor, $Q_3$ or $Q_4$, and $N_1$ and $N_2$ are the number of turns in half the primary and secondary, respectively. Capacitor $C_3$ charges to twice the $(N_1/N_2)\ V_{BE}$ voltage.

Because the voltage across $C_3$ can't change instantaneously, the voltage across $T_1$ gradually shifts from the conducting power transistor to the other transistor. This action produces a "dead time" that not only prevents damaging cross-conduction of the power transistors, but also reduces RFI noise. Dead time, which occurs when both power transistors don't have enough base-emitter voltage for conduction, is proportional to $C_3$ and inversely proportional to the current from the controlled current source, $Q_5$.

Because inductor-input filters are used after the rectifier section, $D_5$ through $D_8$, the output voltages are proportional to converter "live" time. And since dead time is inversely proportional to the controlled source current, controlling source current controls the converter output voltage. A 723 regulator IC, loop filter and reference voltage before the current source complete the regulator loop.

The converter operates from $-60$ V, and the regulator operates from $+19$ V in the circuit shown. But, both can be made to operate from the same supply voltage. Resistor $R_3$ helps damp out ringing in $T_1$. A current-limiting feature in the 723 regulator limits the maximum current from current-source $Q_5$; resistor $R_7$ determines this current. Resistors $R_{11}$ and $R_{12}$ average the two final output voltages for comparison with a 5.6 V reference voltage, which is determined by a selected resistor, $R_{10}$. Diode $D_4$ helps minimize any overshoot occurring at turn on.

High efficiency is obtained with 1N5809 6A rated, fast-switching rectifiers, although the output current is only 700 to 800 mA. Their low forward drops produce low switching losses. Higher efficiency can be obtained with Schottky power rectifiers.

## DC/DC Converter Symmetry Correction Circuitry[12]

To ensure balanced operation of the power transformer in a push-pull converter supply, asymmetry can be corrected by using an MC3420 switch-mode-regulator control IC.

---

[12]Wurzburg, H. Regulator Performs Symmetry Correction in a Push-Pull Switching Power Supply. Reprinted with permission from *Electronic Design 19*, September 17, 1978, copyright Hayden Publishing Co. Inc. (1978).

Symmetry correction is always a problem in high-power push-pull switching supplies. Unlike a half-bridge or full-bridge circuit, the power transformer in a push-pull configuration can not be capacitively coupled to block net DC, which results from any asymmetrical operation in the primary.

In the correction circuit of Figure 3-39, the voltage impressed on the primary side $(P)$ of the transformer $T_1$ is sensed by the sensing secondary, $S_2$. The resulting secondary voltage, $V_{s2}$, is integrated by op amp $A_1$. As a result, the voltage on integrating capacitor $C$ represents the volt-second product impressed on $T_1$.

During transistor $Q_1$'s conduction period, the voltage on $C$ ramps-up from zero to some positive value—the output of $A_2$ is low this time. Drive for $Q_1$ is provided by the MC3420 (pin 11) whose duty cycle is controlled by output-voltage/current-error loops (not shown in figure).

Drive for transistor $Q_2$, from pin 13, is independently controlled by the symmetry-correction circuit.

When $Q_1$'s conduction period ends and $Q_2$ turns on, the voltage on $C$ ramps down until it reaches zero. Now the output of comparator $A_2$ goes high, which inhibits the drive signal from pin 13. But $Q_2$ receives a drive signal until a charge, equal and opposite to the charge built up on $C$, is removed from $C$— and asymmetrical operation results. Transistors $Q_3$ through $Q_5$ synchronize the conduction period of the drive signal to $Q_2$.

To use the circuit in a given application, one should apply the following design equations and operating restrictions:

1. Conduction time at pin 13 $(t_{ON2})$ can be greater than at pin 11 $(t_{ON1})$, but the difference is limited and given by

$$|(t_{ON2}) - (t_{ON1})| \text{ max} = \frac{1 - 2(DC)}{4f_o} - \frac{DT}{2} \qquad (3\text{-}17)$$

where $DC$ = duty cycle of waveform at pin 11
   $DT$ = dead time
   $f_o$ = output frequency.

2. The conduction period at pin 13 can be shorter than that at pin 11 by any amount.

3. Comparator $A_2$'s response time should be at least ten times better than the system switching frequency.

4. Soft-start circuitry should use pin 7, not pin 6 of the MC3420, for control. This minimizes asymmetry during start-up and after an inhibit.

5. Resistor $R$, at the input to integrator $A_1$, is calculated from

$$R \geq \frac{I_0}{|V_{s2}| \text{ max}}, \qquad (3\text{-}18)$$

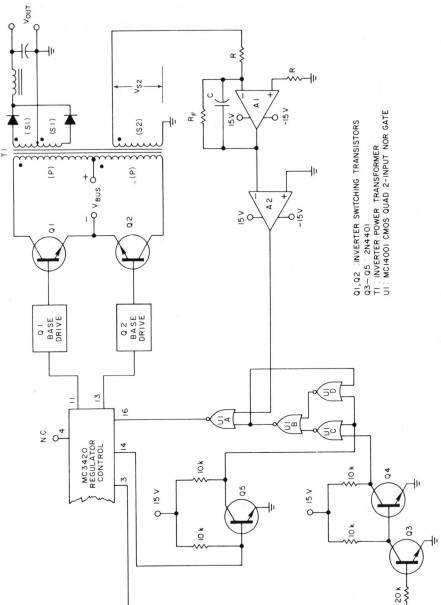

Figure 3-39 Symmetrical operation of a push-pull power supply comes from using an MC3420 regulator in the control circuit. Reprinted with permission from *Electronic Design 19*, September 17, 1978, copyright Hayden Publishing Co. Inc., (1978).

Q1,Q2 : INVERTER SWITCHING TRANSISTORS
Q3 – Q5 : 2N4401
T1 : INVERTER POWER TRANSFORMER
U1 : MC14001 CMOS QUAD 2-INPUT NOR GATE

where $I_o$ = maximum output current available from $A_1$.

Capacitor $C$'s approximate value is given by

$$C \simeq \frac{|V_{s2}| \; \max}{24Rf_o}. \qquad (3\text{-}19)$$

6. Op amp $A_1$ should have low input offset, low input bias current, and a slew rate given by,

$$SR \geq C\left(\frac{|V_{s2}| \; \max}{R}\right). \qquad (3\text{-}20)$$

7. Resistor $R_f$ resets $C$ during power-down and inhibits, and is found from

$$R_f \geq \frac{5}{\pi Cf_o}. \qquad (3\text{-}21)$$

### Five-Volt, Two-Watt, DC/DC Converter Oscillator Problems

Temperature tests on the original circuit shown in Figure 3-40 revealed two problem areas. First, the converter did not start satisfactorily at low temperature ($-30°$F) when fully loaded and with excitation voltage at 24 V DC.

Second, at high temperature ($+180°$F), input current increased and was not constant (the inverter seemed to have two modes of operation when fully loaded, depending upon excitation voltage). In addition, the base-emitter voltage waveform contained, negative pulses in excess of the reverse break-down voltage (Figure 3-41(a)). Also, the collector-emitter waveform, (Figure 3-42(a)) shows

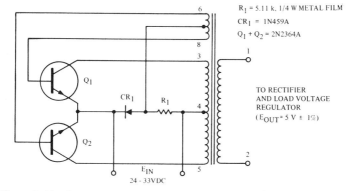

Figure 3-40   Original low-power; low-voltage DC/DC converter oscillator.

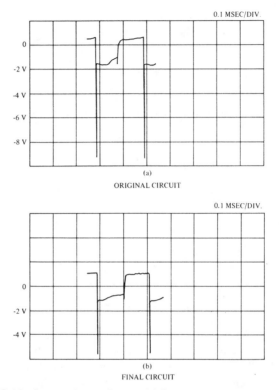

Figure 3-41  Base-emitter voltages. (a) Original circuit. (b) Final circuit.

a transient on the leading edge which reduced the margin of voltage rating on the transistor. Starting could be improved by reducing the value of $R_1$, but this would result in an increase in power dissipation. Reversing the polarity of $CR_1$ helped the starting characteristics by routing all starting current through the transistors. This also had some effect on the base-emitter waveform and allowed an increase in the value of $R_1$. The addition of $C_4$ had the greatest effect on switching and could result in almost perfect square waves without transients of any kind with a large enough capacitor value (0.1 $\mu$F). This improvement in switching, however, reduced starting and required a reduction of $R_1$. The resulting values ($R_1 = 5.1$ ohms, $C_4 = 0.01$ $\mu$F) represent a compromise between switching within the limits of the transistor's maximum ratings, satisfactory cold starting, and maximum efficiency. $R_6$ was added to limit the base-emitter drive in the forward direction. This is also solved the dual mode operation at high temperatures. The final circuit is shown in Figure 3-43. Figures 3-41(b) and 3-42(b) show the resulting waveforms of the final design.

The function of each of the components is as follows: Resistor $R_1$ supplies starting current to $Q_1$ and $Q_2$. Capacitor $C_4$ speeds up the transistor switching

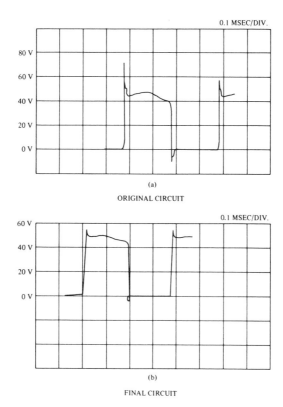

Figure 3-42   Collector-emitter voltages. (a) Original circuit. (b) Final circuit.

to increase efficiency by keeping the power dissipation in $Q_1$ and $Q_2$ to a minimum. Diode $CR_1$ also helps the efficiency by directing all of the starting current through $Q_1$ and $Q_2$. Resistor $R_6$ limits the drive current in the transistor base-emitter circuits, and $C_8$ prevents excessive voltage spikes from appearing on the leading edge of the square wave and exceeding the transistor breakdown voltage.

During low temperature ($-10°F$) thermal vacuum testing, the oscillator circuit in Figure 3-43 drew excessive input current; it was found that $Q_1$ was shorted collector-to-emitter. The failed transistor was removed from the circuit and replaced for troubleshooting. Waveforms of the collector-to-emitter voltage of $Q_1$ were displayed on an oscilloscope and are shown in Figure 3-44, depicting a ringing condition on the leading edge of the square wave. Expansion of the waveform indicated a tendency for the inverter to oscillate at a high frequency. This tendency increased with lower temperatures. Decreasing the ambient temperature to $-30°F$ increased the amplitude of the high-frequency oscillations. The existence of the oscillation on the leading edge of $Q_1$ indicates that the inverter tends to malfunction as $Q_2$ switches on.

By decreasing the number of feedback windings (to the bases of $Q_1$ and $Q_2$)

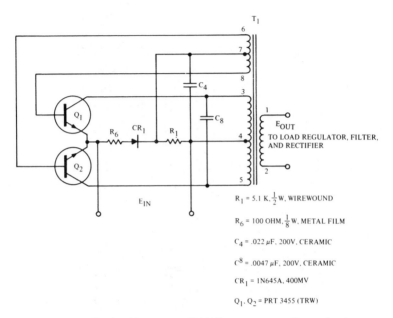

$R_1$ = 5.1 K, $\frac{1}{2}$ W, WIREWOUND

$R_6$ = 100 OHM, $\frac{1}{8}$ W, METAL FILM

$C_4$ = .022 $\mu$F, 200V, CERAMIC

$C_8$ = .0047 $\mu$F, 200V, CERAMIC

$CR_1$ = 1N645A, 400MV

$Q_1$, $Q_2$ = PRT 3455 (TRW)

Figure 3-43   Final low-power DC/DC converter oscillator circuit.

of the power transformer ($T_1$), the ringing was eliminated as was the possibility of stalling the oscillator in a DC mode and causing catastrophic transistor failure.

The trouble turned out to be that the transformer interwinding capacitance and leakage inductance formed a $LC$ tuned circuit which produced the ringing. The ringing probably confused transistor $Q_1$ as to whether it should come on or not, thus stalling the oscillator in a DC mode.

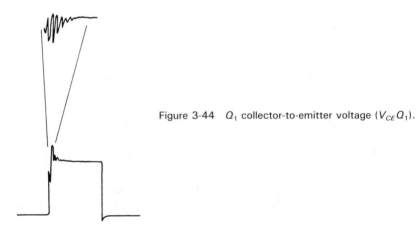

Figure 3-44   $Q_1$ collector-to-emitter voltage ($V_{CE}Q_1$).

## Transistor Matching

Problems of reduced power supply efficiency (and thus push-pull transistor thermal stress) are due to excessive power dissipation in the transistor's active region, slow transistor switching speed, and crossover distortion. The commutation or crossover effect has resulted in several power transistor catastrophic failures. The power transistors of push-pull oscillators (or power amplifiers) need to be matched for $h_{FE}$(beta), $I_{CBO}$, $BV_{CEO}$ $V_{CE}$(sat), $V_{BE}$(sat), rise and fall times. This ensures more balanced operation of the transistors; that is, the average power dissipated and junction temperatures are more evenly distributed for all of the power transistors and do not cause undue stresses or degradation.

In order to get current-driven converters to perform well, some degree of matching is required. Since $V_{BE}$ is a part of the base winding voltage which does the timing, it should be matched. Storage time, rise time, and fall time also influence the half-period length. The $h_{FE}$ influences how saturated the transistor will be and how long it will store charges. If all these parameters are matched, the transistor will become hard to obtain but the operation of the power supply will be enhanced.

## Peak Power Design Adequacy

The important switching transistor parameters mentioned before are switching speed, pulse safe operating area, high temperature collector leakage current, DC gain, $V_{CE}$(sat), and $V_{BE}$(sat). The predominant power dissipated in the transistor is that which occurs during switching time. This power loss is a function of frequency; therefore, the higher the frequency, the greater the switching interval portion of the total period becomes. The load line during the transition time determines the safe operating area required of the transistors. If the circuit elements force the transistor to exceed its safe operating area, then either a new transistor with the desired safe operation area must be chosen or circuit protection must be provided for realiable operation. The transistor can be protected from voltage spikes by connecting a capacitor or Zener diode between the collector and emitter junctions of the transistor as previously shown in Figure 3-35. If a Zener diode is used, it should have a breakdown voltage greater than twice the supply voltage plus a safety margin voltage of about 10 V for switching transients. If capacitors are used, the capacitance must not be too high as it may cause excessive charging currents and slow the transistor switching response.

The majority of power supply oscillator transistor failures have occurred as a result of high peak power. Experience shows that peak power failure occurs in power handling transistors under such conditions as turn-on, short circuit, or during switching under heavy-load conditions. In order to analyze the switching losses which result during high peak power conditions let us consider the simplified switching circuit in Figure 3-45.

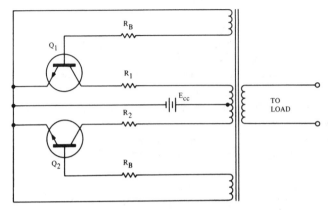

Figure 3-45   Simplified inverter switching circuit. (Figures 3-45 through 3-48 redrawn from James E. Comer, "Trends in Reliability of Space Power Conditioning Equipment," *Wescon Session 20*, 1965).

During normal operation, transistor $Q_1$ switches from a maximum current at saturation to an off condition with double the applied voltage. (The transformer action of $T_1$ doubles the applied voltage when $Q_1$ is off). During the time the transistor switches off it goes through a peak power condition. The switching losses can be estimated from switching time requirements but they must be checked in the unit by taking oscilloscope photographs that show the switching losses.

Photographs of switching losses may be taken by one of two methods, which are described as follows.

*Method 1* uses a small sampling resistor $R_1$ (approximately 0.10 ohm) in the collector circuit and displays the current and voltage on an oscilloscope. The $V - I$ oscilloscope photo will have large, bright areas for the on and off conditions; the waveform between conditions represents the important switching losses.

*Method 2* does not use a sampling resistor, but instead uses a magnetic pickup to display the current. The current and voltage are displayed on a dual-trace scope so that the time of each corresponds exactly. Both traces are displayed versus time. The current display will not reproduce the flat portion of the waveform due to the poor low-frequency response of the probe, but this portion does not represent switching losses. Typical waveforms are shown in Figure 3-46. The next step is to expand the rise and fall portions for both voltage and current and take suitable photos of these areas. Typical waveforms are shown in Figure 3-47.

A plot of these two curves at intervals of 0.04 $\mu$s is shown in Figure 3-48. This plot represents a typical load line for a power transistor switching a partially inductive load. This apparent disarray actually portrays the current and voltage

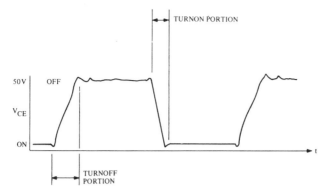

Figure 3-46 Typical $V_{CE}$ switching waveform for $Q_1$.

as seen by the transistor as it traverses from on to off conditions. If we integrate the area under this curve by sections, we can determine the amount of power loss during the switching time and average it over the total period to determine the effect on average power loss. Several interesting features should be noted. The shape of the curve seems to be more accurate than that produced by a sampling resistor. By eliminating the sampling resistor we eliminate the spikes and the pickup it causes. The low-frequency transistor curves have a very similar shape. The bright spots near the on time indicate the time required to remove the charge stored in the base-emitter region as the transistor starts to turn off. This charge must be reversed before the transistor switches off. The bright spots near the off period are the result of charging the collector base capacitance and the effect of frequency response. The switching loop can be separated into three distinct areas and analyzed individually, as follows:

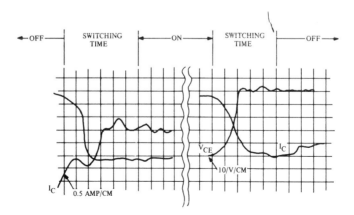

Figure 3-47 Voltage and current waveforms for $Q_1$ showing switching times.

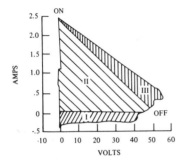

Figure 3-48   Power switching transistor operating area.

- *Area 1 (Figure 3-48).* A casual inspection of the load line shows that the current reverses when the power switches from off to on. This may appear erroneous but it is quite normal. The collector base circuit has become charged to twice the supply voltage.

  When $Q_1$ is on, $Q_2$ has double the supply voltage applied to the collector base. When $Q_2$ is first turned on, the 60 V charge of the collector base must discharge through the battery before $Q_2$ can conduct. Thus, the reverse current is present in Area 1. The power loss in Area I is equivalent to the charge of $C_{CB}$, where energy $J = 1/2 C_{CB} V_a^2$ with $C_{CB}$ = collector-base capacitance and $V_a$ applied voltage = 2 × battery voltage. The collector-base capacitance can be calculated from the formula

$$C_{CB} = \frac{KE_0 A}{dn + dp} \qquad (3\text{-}22)$$

where:     $A$ = area of collector,
           $K$ = relative dielectric constant,
           $E_0$ = dielectric constant,
  $dp + dn$ = depletion layer thickness of N and P region, respectively.

  For most transistors, this capacitance must be obtained from the manufacturer since the actual depletion layer is a function of the several variables in the N and P layer. For a 2N2658 the value has been estimated at 100 pF. The energy is $J$ is $J = 1/2 \times 100 \times 10^{-12} \times 60^2 = 0.18 \ \mu W$ s. This energy discharges in 0.16 $\mu$s and dissipates 0.18 $\mu$W $\mu$s/0.16 $\mu$s or 1.1 W.
- *Area II (Figure 3-48).* This area is equivalent to the amount of power lost in any resistive load line. Operating in Area II is considered Class A operation.
- *Area III (Figure 3-48).* When we switch off any magnetic element, the energy stored in that element must be absorbed by the transistor before the off state is finally reached. The flux in the magnetic element must reverse.

The energy stored in the inductance is $J = \int p\,dt = \int vi\,dt = L \int i\,di = 1/2\,Li^2$. For any inductor, $L$ can be calculated from the formula $L = N\phi/I$ since all the energy in the inductor will have to be dissipated in the transistor (assuming that the inductor is the only load). For a transformer that is loaded, this inductance is not calculated from the total flux. That flux, which links both the primary and secondary, is absorbed by the loaded secondary. The remaining flux, called the leakage flux $\phi_1$, is the major contribution to this inductance and is used to calculate energy.

A power circuit's peak power should be verified by some dynamic means similar to that described here. These peak power measurements should be performed under a worst-case condition. The measurements should give ample consideration to conditions such as input voltage, surge current at turn-on, and high-voltage transients.

## Peak Power Design Example

The previously described method for estimating switching losses (and determining if the power transistors were being operated within their safe operating area) was undertaken for an aerospace DC/DC power converter that had the following requirements.

| | |
|---|---|
| Input voltage: | 22 to 29.25 VDC |
| Outputs: | +28.3 VDC at 400 W |
| | −28.3 VDC at 20 W |
| Efficiency: | ≥ 65% |
| Output ripple: | 10 mV, peak to peak |
| Temperature range: | −30°F to +160°F |

Overload protection of outputs

The block diagram of this converter is shown in Figure 3-49 and the schematic diagram is shown in Figure 3-50. A brief description of the converter operation follows. The converter receives unregulated input voltage in the range of 22.0 to 29.25 V from a battery or a DC power supply. This voltage is converted to two regulated outputs, positive and negative 28.3 VDC $\pm$ 1%. These outputs are rated at 400 and 20 W respectively.

Referring to the block diagram in Figure 3-49, the converter consists of a multivibrator and boost rectifier, a simple rectifier, and two series regulators. The multivibrator is a basic circuit which converts an input DC to AC. It provides power for the boost rectifier and the simple rectifier. The series regulators provide low ripple and low dynamic impedance for good regulation under step load changes.

The multivibrator shown in Figure 3-50 operates in the following manner.

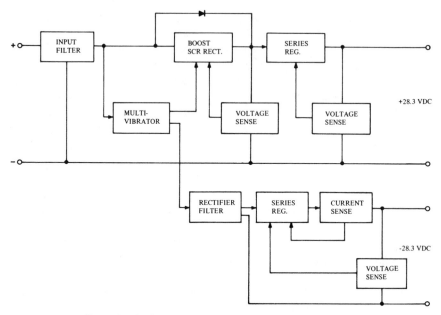

Figure 3-49   Block diagram of 400 W DC/DC converter.

Transistors $Q_1$ and $Q_2$ are in parallel driving one side of transformer $T_1$, and $Q_3$ and $Q_4$ are in parallel driving the other end of the $T_1$ primary. The circuit operates as a magnetically coupled multivibrator oscillator. The transistors switch in pairs alternatively from cutoff to saturation resulting in very low average power dissipation. Windings 10-11 and 11-12 provide regenerative drive to the bases of the transistor pairs. Winding 13-14 and SR-1 control the frequency of oscillation (2.5 kHz).

A forced-start circuit for low temperatures is included which uses SCR $Q_5$, $R_{10}$, and $R_{47}$ to ensure that the oscillator always starts when power is applied. The start circuit applies a turn-on base current to $Q_1$ and $Q_2$ as soon as the input voltage rises to a level sufficient to trigger $Q_5$.

The major portion of the converter power is supplied to the +28.3 VDC output which uses a boost principle to gain greater conversion efficiency (circuitry in upper portion of Figure 3-49). This is achieved by a direct connection from the input unregulated positive DC line (after the filter) to the midpoint of the full-wave rectifier. Thus, the converted DC output voltage fed to the positive regulator is the sum of the unregulated input and the rectified boost voltage. This connection reduces the power capacity required in the oscillator, transformer, and rectifiers. A magnetic amplifier preregulator is used to control the output of the full-wave rectifier. This permits the series regulator transistors to operate over a lower voltage range than would otherwise be required, resulting in a higher overall efficiency.

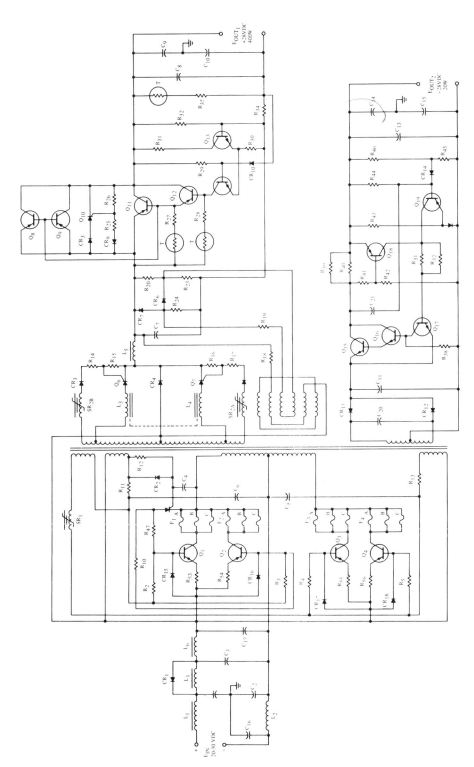

Figure 3-50   Schematic diagram of DC/DC power converter.

The input DC power is filtered and part of it is converted into a square wave AC by the multivibrator. The AC is then rectified by an SCR whose firing time is controlled by the average voltage appearing at the output of the rectifier. Note that the average voltage is the sum of the input voltage and the rectified boost voltage. If the input voltage decreases, the SCRs ($Q_6$ and $Q_7$) fire sooner and rectified boost voltage increases to maintain a constant average voltage. If the input voltage increases, the boost voltage decreases. The boost voltage decreases and reaches zero when the input voltage is greater than the average voltage by approximately one volt. At this time, the diode shunting the boost rectifier is forward-biased and acts as a voltage bypass.

The multivibrator and boost rectifier serve two purposes. They increase the input voltage and provide rough regulation. Since their response to step changes is dependent upon the firing time of the SCR rectifier, they are limited by the operating frequency of the controlling magnetic amplifier and the SCR recovery time. Closer regulation and quick response are provided by the series regulator. The voltage sense samples the regulated output and compares it to a reference voltage to develop an error signal. This signal is amplified and is used to control the conductivity of a transistor in series with the output line. Thus, if the output tends to rise, the error signal increases. This, in turn, decreases the conductivity and increases the voltage drop in series with the output to compensate for the rising tendency.

The negative output is supplied by a conventional rectifier and series regulator, with no preregulator or bypass circuit, but is current-limited.

## Switching Patterns

In order to optimize the efficiency of the power transistors in a switching mode, switching patterns may be produced on an oscilloscope and analyzed to determine regions of transistor operation with high dissipation. Circuit adjustments, filter and load changes may be made, and their effect on the transistor may be noted by means of switching patterns.

As was mentioned previously, in order to verify the peak power and the switching losses of this converter the following procedure was instituted: A 0.05 ohm resistor was inserted in the collector lead of transistor $Q_1$ to measure the collector current. A dual beam oscilloscope displayed the collector-to-emitter voltage versus time and collector current versus time waveforms under three load conditions: no load; half load, and full load. Also, the $I_C$ vs $V_{CE}$ characteristics of transistor $Q_1$ were displayed on the oscilloscope using a differential "plug-in" for each of the three loads.

The photographs that were taken of the collector-emitter voltage vs time and of collector current vs time were used to calculate the average power dissipated by oscillator transistor $Q_1$. Typical examples of these photographs are included as Figures 3-51 through 3-56. Table 3-5 shows that the calculated transistor

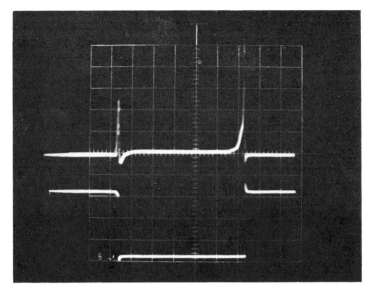

| $E_{in}$: | 24.0 V | Horizontal: 50 msec/cm |
|---|---|---|
| Upper: | $I_C$ at 0.4 A/cm | Frequency: 2174 Hz |
| Lower: | $V_{CE}$ at 20 V/cm | |

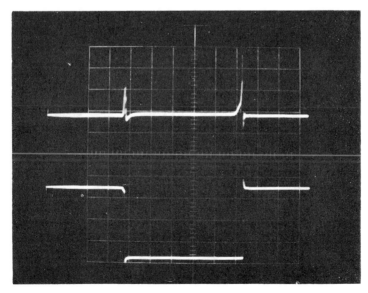

| $E_{in}$: | 26.0 V | Horizontal: 50 msec/cm |
|---|---|---|
| Upper: | $I_C$ at 1 A/cm | Frequency: 2340 Hz |
| Lower: | $V_{CE}$ at 20 V/cm | |

Figure 3-51   Photographs of $I_C$ and $V_{CE}$ for oscillator transistor $Q_1$ (no-load conditions).

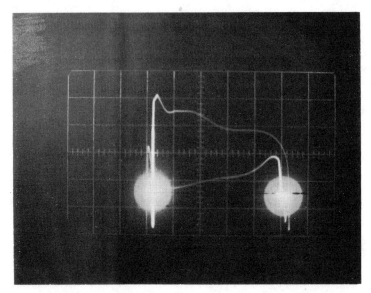

| $E_{in}$: | 24.0 V | Horizontal scale: | 10 V/cm |
| Vertical scale: | 0.2 A/cm | Frequency: | 2206.2 Hz |

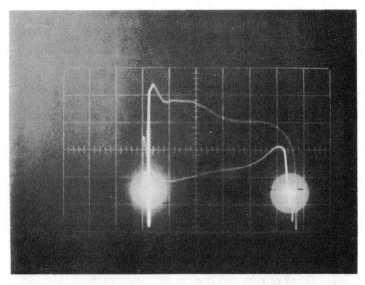

| $E_{in}$: | 26.0 V | Horizontal scale: | 10 V/cm |
| Vertical scale: | 0.2 A/cm | Frequency: | 2367.6 Hz |

Figure 3-52  Photographs of $I_C$ versus $V_{CE}$ for oscillator transistor $Q_1$ (no-load conditions).

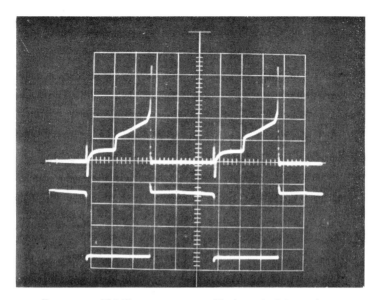

| $E_{in}$: | 25.0 V | Horizontal: 0.1 msec/cm |
|---|---|---|
| Upper: | $I_C$ at 1 A/cm | Frequency: 2178 Hz |
| Lower: | $V_{CE}$ at 20 V/cm | |

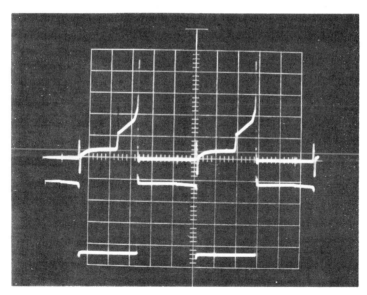

| $E_{in}$: | 27.0 V | Horizontal: 0.1 msec/cm |
|---|---|---|
| Upper: | $I_C$ at 1 A/cm | Frequency: 2351 Hz |
| Lower: | $V_{CE}$ at 20 V/cm | |

Figure 3-53   Photographs of $I_C$ and $V_{CE}$ for oscillator transistor $Q_1$ (one-half of full-load conditions).

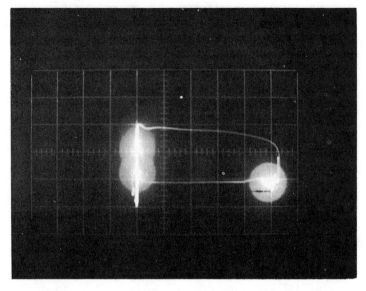

| $E_{in}$: | 25.0 V | Horizontal scale: | 10 V/cm |
|---|---|---|---|
| Vertical scale: | 1 A/cm | Frequency: | 2230.1 Hz |

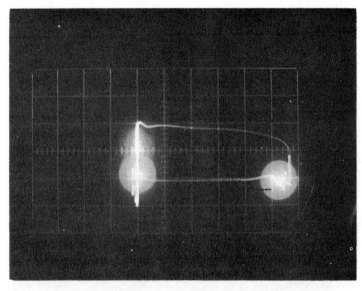

| $E_{in}$: | 27.0 V | Horizontal scale: | 10 V/cm |
|---|---|---|---|
| Vertical scale: | 1 A/cm | Frequency: | 2396.1 Hz |

Figure 3-54   Photographs of $I_C$ versus $V_{CE}$ for oscillator transistor $Q_1$ (one-half of full-load conditions).

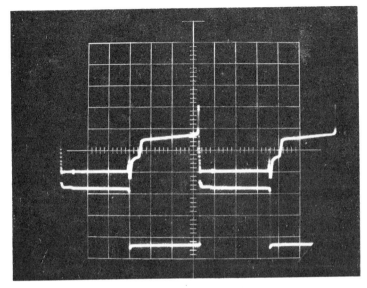

$E_{in}$:          24.0 V                  Horizontal: 0.1 msec/cm
Upper:        $I_C$ at 2 A/cm            Frequency: 1969 Hz
Lower:        $V_{CE}$ at 20 V/cm

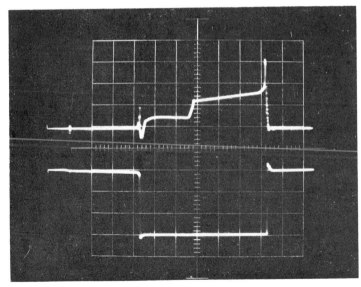

$E_{in}$:          26.0 V                  Horizontal: 50 msec/cm
Upper:        $I_C$ at 2 A/cm            Frequency: 2158 Hz
Lower:        $V_{CE}$ at 20 V/cm

Figure 3-55   Photographs of $I_C$ and $V_{CE}$ for oscillator transistor $Q_1$ (full-load conditions).

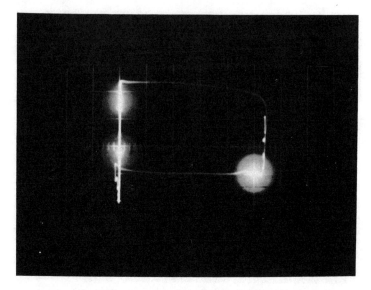

| $E_{in}$: | 24.0 V | Horizontal scale: 10 V/cm |
|---|---|---|
| Vertical scale: | 1 A/cm | Frequency: 2088.4 Hz |

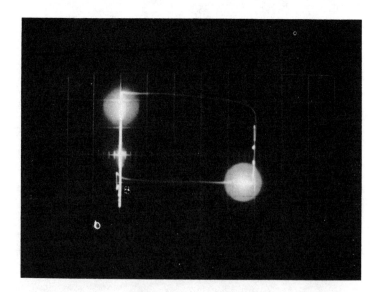

| $E_{in}$: | 26.0 V | Horizontal scale: 10 V/cm |
|---|---|---|
| Vertical scale: | 1 A/cm | Frequency: 2266.1 Hz |

Figure 3-56    Photographs of $I_C$ versus $V_{CE}$ for oscillator transistor $Q_1$ (full-load conditions).

### Table 3-5  Calculated Transistor and Oscillator Power Dissipation

| $E_{in}$ | $P_{diss} Q_1(W)$ | $P_{osc}(W)$ |
|---|---|---|
| *FOR NO LOAD* | | |
| 22.0 VDC | 0.0231 | 0.0462 |
| 24.0 VDC | 0.0199 | 0.0399 |
| 29.25 VDC | 0.0053 | 0.0106 |
| *FOR HALF-LOAD* | | |
| 22.0 VDC | 0.154 | 0.308 |
| 24.0 VDC | 0.119 | 0.276 |
| 29.25 VDC | 0.062 | 0.125 |
| *FOR FULL LOAD* | | |
| 22.0 VDC | 0.316 | 0.632 |
| 24.0 VDC | 0.278 | 0.556 |
| 29.25 VDC | 0.154 | 0.307 |

($Q_1$) power dissipation and total oscillator power dissipation are well within the power transistor's maximum rated value. Expanded scale photographs also depict $Q_2$'s crossover switching losses (Figures 3-57 and 3-58).

The average power dissipated by the transistor is equal to the frequency in hertz times the integral of the product of collector current and collector-emitter voltage, or the area under the curve in the current picture times the collector-emitter voltage corresponding in time to the nonzero portions of the current curves, times the time.

The two bright spots appearing in each photograph of $I_C$ vs $V_{CE}$ under no-load conditions (Figure 3-52) represent the two states of transistor $Q_1$: (a) when it is saturated to approximately 0.2 V and (b) when it is cut off and sees twice the supply (input) voltage. For one-half load and full load (Figures 3-55 and 3-57) there are three bright spots, the third being due to the duration that the magnetic amplifier is on. The $V_{CE}$ vs $I_C$ loop joining these points (bright spots) represents the transients when the transistor switches between saturation and cutoff and back into saturation.

The spikes of collector current which occur during turn-on and turnoff of an oscillator transistor as shown in the photographs can be easily explained. The storage time of a transistor, which is directly proportional to overdrive, causes increased losses in both the transistor and the transformer. Overdrive is required to keep $V_{CE}$(sat) losses low during on time, but an undesirable side effect is that the transistors want to stay on for a finite length of time after removal of drive.

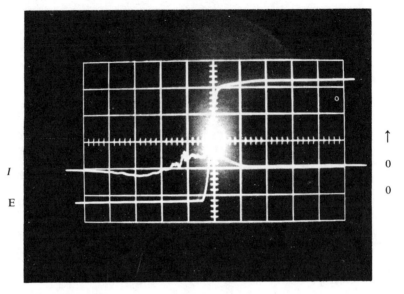

$I$

$E$

$Q_2$ Going Off        $0.5$ V/cm $= V_S$
$Q_2$ No Load          $10$V/cm $= V_{CE}$
.1 $\mu$sec/50 Expanded Scale

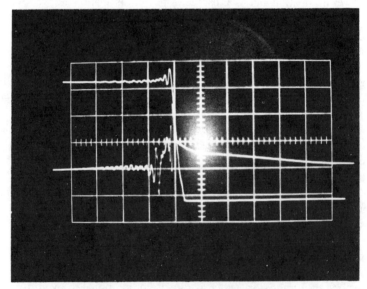

$Q_2$ Going On     .1$\mu$sec/50 Expanded Scale
$Q_2$ No Load

Figure 3-57     400 W DC/DC converter—expanded-scale photographs.

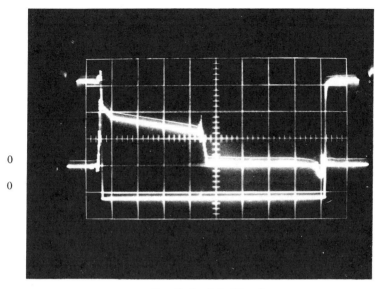

0

0

One Cycle of $Q_2$ Waveform
$Q_2$ ½ Load
$50\mu sec/2$ Expanded Scale

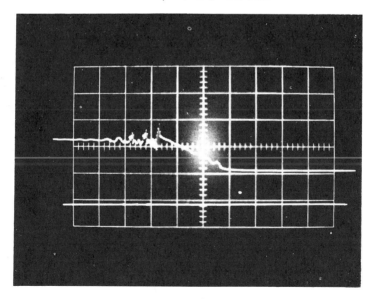

$Q_2$ On
$Q_2$ ½ Load

Figure 3-58 Transistor $Q_2$ switching waveforms.

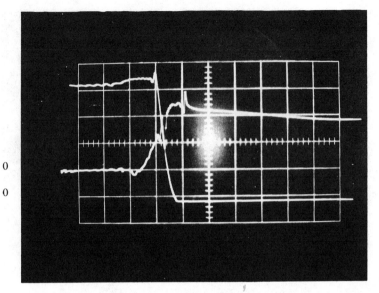

$Q_2$ Going On
$Q_2$ ½ Load

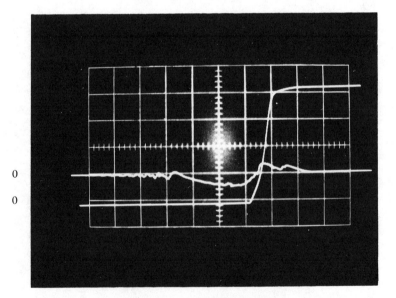

$Q_2$ Going Off
$Q_2$ ½ Load

Figure 3-58 (*Continued*)

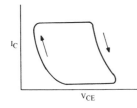

Figure 3-59   Typical safe operating area curve.

The oscilloscope photograph of $I_C$ vs $V_{CE}$ (Figures 3-52, 3-54, and 3-56) verify that the power transistors were not operating outside of their Safe Operating Areas (SOAR) as defined by the semiconductor manufacturer.

## Transistor Safe Operating Area

The Safe Operating Area (SOAR) curve defines the region that encloses all of the points representing simultaneous values of the collector current and the collector-to-emitter voltage which the transistor can safely handle during switching into any load under specified conditions for base current, output resistance of the driving circuit, switching time, junction temperature, and average power dissipation. SOAR is a complete graphical representation of the combined effect of the absolute maximum ratings that appear on device specification sheets. SOAR supplies guidelines for operating conditions that are beyond the scope of data sheet information. At any given time the transistor should be operated within its SOAR envelope.

When the transistor switches off an unclamped inductive load current, as in the DC/DC converter under discussion, the transistor must absorb the stored energy less the energy absorbed by the resistor in series with the inductance. The load line for this case is a function of the transistor voltage-current characteristic for the base condition when the load line is traversed. A typical curve of this is shown in Figure 3-59. Figures 3-52, 32-54, and 3-56 depict the actual operating curves. It was found that these operating curves are well within the SOAR curve for the transistors at hand under all line, load, and temperature conditions, thus verifying the design adequacy of this converter.

In addition to the waveforms presented previously to check the adequacy of the oscillator power transistor design, other converter waveforms must be checked to verify design accuracy and conformance to specification requirements. These waveforms are shown in Figures 3-60 through 3-62 for the DC/DC converter shown in Figure 3-50.

The design was improved in incorporation of a pulse-type oscillator start circuit which is shown in Figure 3-63 (the drive power is proportional to the load current). The start circuit was changed to a pulse-type circuit to prevent the oscillator from stalling in a DC mode and damaging the power transistors.

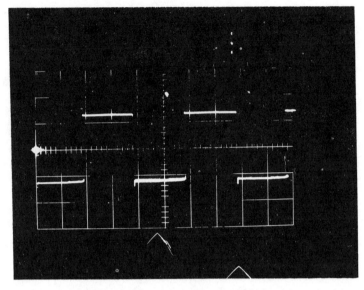

$T_1$: Waveform of Pin 2 to Pin 1 (Pin 2 is minus)
20 V/cm          ½ Load
100 $\mu$sec/cm      +28 VDC Input

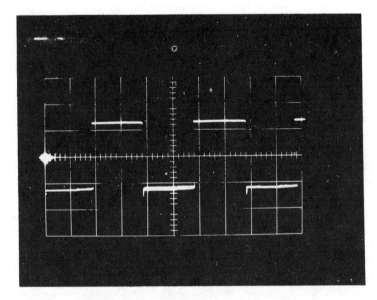

$T_1$: Waveform of Pin 2 to Pin 3 (Pin 2 is minus)
20 V/cm          ½ Load
100 $\mu$sec/cm      +28 VDC Input

Figure 3-60   Waveforms of transformer $T_1$ primary.

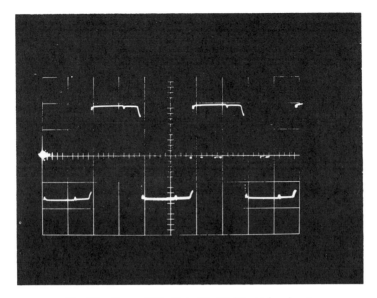

$T_1$: Waveform of Pin 11 to Pin 12 (Pin 11 is minus)
1 V/cm          ½ Load
100 $\mu$sec/cm     +28 VDC Input

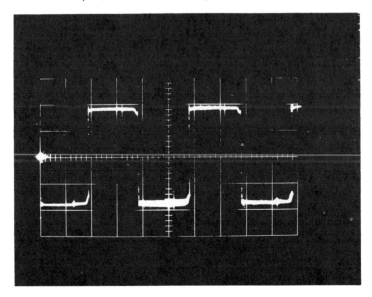

$T_1$: Waveform Pin 10 to Pin 11 (Pin 10 is minus)
1 V/cm          ½ Load
100 $\mu$sec/cm     +28 VDC Input

Figure 3-60   (*Continued*)

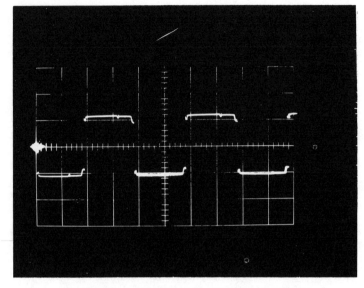

$T_1$: Waveform of Pin 14 to Pin 13 (Pin 14 is minus)
5 V/cm          ½ Load
100 $\mu$sec/cm     +28 VDC Input

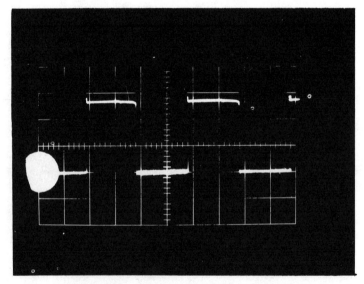

$T_1$: Waveform of Pin 1 to Pin 12 (Pin 1 is minus)
20 V/cm         ½ Load
100 $\mu$sec/cm     +28 VDC Input

Figure 3-61   Waveforms of transformer $T_1$ primary.

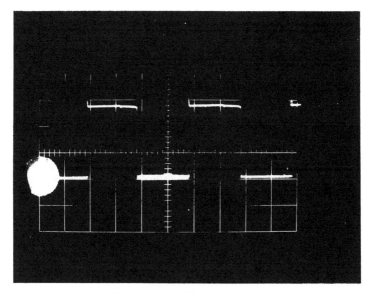

$T_1$: Waveform of Pin 3 to Pin 10
20 V/cm          $\frac{1}{2}$ Load
100 $\mu$sec/cm      $+28$ VDC Input

Figure 3-61   (Continued)

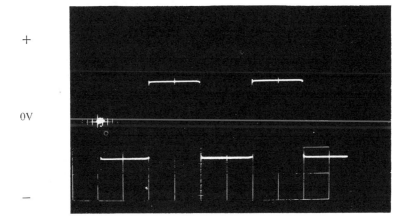

No Load
Pin #1 of $L_4$ and Pin #8 of $T_1$
Cal. 10 VDC/cm
0.1 msec/cm

Figure 3-62   Waveforms of transformer $T_1$ plus secondary (10 V/cm).

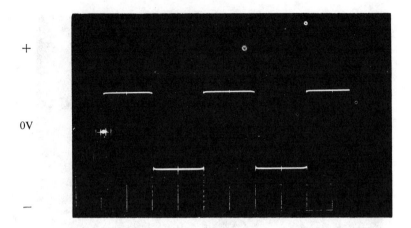

No Load
Pin #4 of $L_4$ and Pin #8 of $T_1$
Cal. 10 VDC/cm
0.1 msec/cm

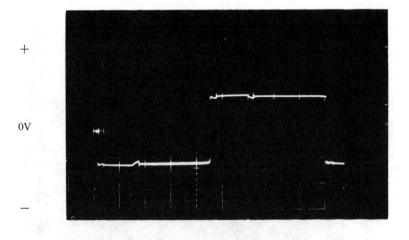

Load: 350 W
Pin #of $L_4$ and Pin #8 of $T_1$
Cal. 10 VDC/cm
50 $\mu$sec/cm

Figure 3-62 (*Continued*)

+

0V

−

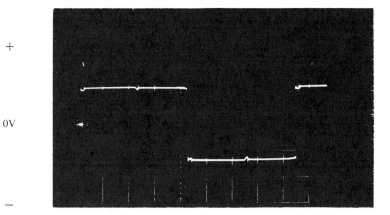

Load: 175 W
Pin #4 of $L_4$ and Pin #8 of $T_1$
Cal. 10 VDC/cm
    50 μsec/cm

Figure 3-62   (*Continued*)

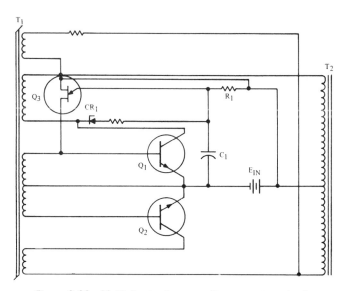

Figure 3-63   Multivibrator incorporating new start circuit.

The multivibrator start circuit consists of a double base diode $Q_3$, resistor $R_1$, and capacitor $C_1$: a low-frequency relaxation oscillator. The input voltage $E_{in}$ is applied directly across the series resistor-capacitor combination $R_1 C_1$ as well as across bases one and two of $Q_3$. When the input is applied, the voltage across $C_1$ increases exponentially until the peak point of $Q_2$ is reached and the emitter is forward-biased. Capacitor $C_1$ discharges through the emitter and the base-to-base resistance is lowered to drive a pulse of current into the base of $Q_1$. If the multivibrator does not start oscillating, the emitter voltage drops to its valley point to become back-biased and the charge and discharge of capacitor $C_1$ repeats. If the multivibrator begins oscillating, capacitor $C_1$ is shunted to limit its voltage buildup below the peak point of $Q_3$ by resistor $R_2$, diode $CR_1$, and the saturated transistor $Q_1$.

This type of start circuit offers maximum reliability through simplicity. The circuit provides a shunt pulse of base drive to start oscillation without affecting the normal off condition of transistors $Q_1$ and $Q_2$. This circuit permits the multivibrator to oscillate under all operating conditions, including external faults and when the start circuit is off.

## REFERENCES

1. Schaade, O. H., *Proc. IRE*, Volume 31, 1943.
2. Schultz, Warren, "Baker Clamps: Traditional Concepts Updated for Third Generation Power Transistors," *Powerconversion International*, July/August 1984.
3. McMurray, William, "Selection of Snubbers and Clamps to Optimize the Design of Transistor Switching Converters," IEEE Power Electronics Specialists Conference—1979, San Diego, California, June 18–22, 1979. *PESC'79 Record*, pp. 62–74, IEEE Publication 79CH1461-3 AES.
4. Domb, Moshe, "Optimum Design of Dissipative R-C-D Snubber Networks," *Design Automation*, Inc., 809 Massachusetts Avenue, Lexington, MA-02173, U.S.A.
5. Goncharoff, Nik, "Isolated Flyback Regulator Turn-off Current Snubber," *Powerconversion International*, May 1984.
6. Domb, Moshe and Sokal, Nathan O., "R-C-Diode Turn-off Snubber," *PCIM*, October 1985.
7. Hnatek, Eugene R., *Design of Solid State Power Supplies*, Van Nostrand Reinhold, 2nd Edition, 1981, Chapter 9.

# 4

## Transformer and Inductor Design

### INTRODUCTION

The heart of the DC converter is the power transformer. Since inductance does not readily lend itself to microminiaturization, transformers are, and are likely to remain, a fairly large block in power supply design.

The route to smaller power supplies uses higher operating frequencies and thus smaller magnetics. Tape-wound toroidal cores are very suitable for the frequencies at which power can be reasonably switched.

Two types of transformer are generally used in DC-to-DC converters. They are the saturable or square-loop core transformer, and the linear core transformer. Fabrication techniques for the coils and the selection of core material for these transformers are extremely important considerations.

Leakage inductances, self-capacitance of windings, and inter-winding capacitances become an ever increasing problem as the operating frequency is increased to the low-megahertz region. Whatever the core material, care should be taken to keep these effects to a minimum. Leakage inductance or stray capacitance, although in the microhenry and picofarad ranges, at these frequencies may prevent a circuit from operating properly due to various resonant effects.

Customarily, leakage inductance can be reduced by using bifilar windings to provide close coupling between coils and by avoiding an excessive number of turns; self-capacitance can be reduced by winding the coils in a progressive fashion so that the voltage between adjacent turns is held to a minimum; and inter-winding capacitance can be reduced by restricting the area of the coils to different segments or sections of the transformer core. The employment of bifilar sector windings using progressive-winding techniques in each core section is a good fabrication method for transformers to minimize all of the stray effects. For example, a toroidal magnetically coupled-multivibrator transformer for VHF operation can be constructed using bifilar collector windings on one sector with the output coil wound on top of the two collector windings, and bifilar base windings located on another sector of the core with progressive winding tech-

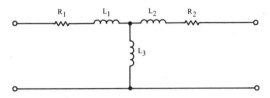

Figure 4-1    Low-frequency transformer model.

niques used for all windings in both sectors. For the progressive-winding approach to be practical, compromise often is necessary between the number of turns of each winding and the mean length of the magnetic path of the core. A model often used to represent a transformer is shown in Figure 4-1.

The winding resistances are $R_1$ and $R_2$; $L_1$ and $L_2$ are primary and secondary leakage inductances; and $L_3$ is the primary shunt inductance. $L_1$ and $L_2$ must be small and $L_3$ must be large if any power is to be transformed from primary to secondary. Leakage inductance is a function of both core geometry and winding configuration. Bifilar winding on a toroidal core is an excellent start for low leakage inductance; interleaving of windings and symmetrical voltage gradients reduce the leakage still further. The primary inductance is largely a function of the core material and the slope of its $B\text{-}H$ curve. A typical $B\text{-}H$ curve for a ferrous material appears as shown in Figure 4-2.

As more and more magnetizing force is applied, the flux density increases steadily at first and then begins to saturate. When magnetomotive force is reduced, the flux density follows a different path than it did when the magnetomotive force was increased. Square loop material has the steepest slope and thus the highest primary inductance. Tape-wound toroids have low eddy currents and thus lower core loss. The very square hysteresis loop of these square loop materials makes their saturation points dependable enough to use for timing.

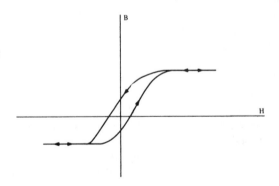

Figure 4-2    Typical $B\text{-}H$ curve for ferrous material.

This is a justification for the use of square loop tape-wound toroids as transformer cores.

For saturable transformers, square B-H loop materials are required which may be either of ferrite material or tape-wound cores. For both materials, the area of the dynamic B-H loop is large and the cyclic rate is very high, meaning high core loss. In the case of square-loop ferrites, the loss may be high enough to cause the core temperature to rise and approach the Curie temperature. Since the metallic square-loop materials have a considerably higher Curie temperature, one-eighth mil tape-wound cores such as Permalloy-80 bobbin cores appear preferable for most applications.

A variety of materials exhibit square loop properties, and they have a wide range of saturation flux, coercive drive levels, squarenesses, and temperature coefficients. For extreme environments and for high operating frequencies (up to 100 kHz), the nickel-iron alloys are most useful. (Above 50 kHz, ferrites, even with their poorer temperature coefficient, become more attractive.) Two types of cores have been widely used; 50% nickel-iron (orthogonal) and 80% nickel-iron (permalloy-80). The latter is now used almost exclusively because of its low coercive force, higher gain, and faster switching. Small size and high efficiency are two desirable parameters in a power converter. Small size is achieved by increased frequency which in turn increases core losses, and to avoid this, thin tape is used. One mil square permalloy is compatible with frequencies up to 20 kHz in transformers; in magnetic amplifiers, $\frac{1}{2}$ mil material should be used above 10 to 15 kHz. Ten kilohertz is a good operating frequency for most components and is used in most of the supplies to be discussed.

When a linear transformer is used for DC-to-DC energy conversion, the deleterious effects of winding capacitances usually are not as serious as those due to leakage inductances which are of the utmost importance. Excessive leakage inductance in the VHF region may easily introduce a high-voltage switching spike across a switching transistor and cause the transistor to break down.

The core with a linear B-H curve has a softer saturation and almost no hysteresis loop. Thus, the core has no memory, and B falls back to zero when H is removed. Molybdenum permalloy powdered cores have served very well as inductors, current transformers, and blocking oscillator transformers. They allow much higher drive levels than switching cores; however, their permeability is quite a bit lower. Lack of memory is the important parameter—even if saturated by a high mmf, powdered cores follow the original B-H curve without need of reset. Aside from the above mentioned effects related to the method of fabrication, the leakage inductance of a linear transformer can be decreased by selecting a high-permeability core material to increase the ratio of the self-inductance to the leakage inductance. High-permeability ferrite materials such as Ferroxcube 3D3 material or Magnetics, Inc. type A material are good choices for linear transformer cores due to their low core losses at high frequency and their reasonably high permeabilities. Ferrite cores, however, have the disadvantage of being relatively low-Curie-temperature materials and require that extra attention be given to this. As to the core shapes, pot cores as well as toroidal cores are desirable configurations to keep the electromagnetic radiation to a minimum.

The powdered cores are useful in applications which do not exceed about 1 MHz (at low inductance). Above this frequency, ferrites and carbon-alloy/iron cores are needed. The ferrites have an extremely wide range of frequency and permeabilities available. The carbon-alloy/iron cores, while considerably cheaper than moly-permalloy, have lower permeabilities.

## DESIGN PRINCIPLES FOR TRANSFORMERS AND INDUCTORS

### Core Materials

The flux density $B$ of a transformer is determined by the core material and operating frequency.

There are four types of materials used in cores that are manufactured for transformers and inductors. Each has characteristics peculiar to certain design requirements.

*Air Core*   Air cores are used if magnetic field distortions cannot be tolerated. The transformer or inductor windings are wound around a form, the sole purpose of which is to sustain the windings. Core forms may be purchased as preformed mountings or fabricated to meet specific requirements. The material used in the core forms (particularly at high frequencies) must have a low dielectric constant and loss which should not be greatly affected by variations of humidity and temperature (see Table 4-1).

*Laminated Iron*   This type of core is for inductors or transformers used in the low frequency ranges. It is fabricated from soft iron alloy stampings. The thick—ness and number of laminations used in a specific core are determined by the frequency and current requirements of the inductor. The laminations vary in thickness from 0.003 to 0.014 in. and are coated with an electrical insulation to restrict eddy currents through the core. The permeability and hysteresis loss of the core are controlled by air gap adjustment which varies the magnetic field within the core (see Table 4-1).

*Powdered Iron*   This type of core is used when high quality factor ($Q$) is required in the frequency range from 5 Hz to 500 kHz, depending on the core geometry. However, eddy current losses must be seriously considered when an inductor or transformer is to be used at frequencies above 30 kHz. The powdered iron core is composed of ferromagnetic particles coated with an insulating binder and compressed into a uniform mass. Permeability and core losses are controlled by the size and spacing of the magnetic particles, composition of the magnetic particles, and thickness of the insulating binder. Temperature and humidity also affect permeability and core losses (see Table 4-1).

*Ferrite*  This type of core possesses high magnetic permeability and high resistivity to eddy currents, permitting its use up to frequencies of 100 MHz. The saturation flux, however, is considerably lower than that of the laminated or powdered iron material, which somewhat restricts its use in high-current applications.

Ferrite cores can be used to replace laminated and powdered iron cores in many applications by use of controlled core geometry. This permits control of hysteresis losses and permits the use of this material in the low audio frequency range. Ferrite material is composed of nonconductive mixed crystals of manganese ferrite ($MnFe_2O_4$) and nickel ferrite ($NiFe_2O_4$). The binding process and heat treatment of these crystals provide a core which is ceramic in nature (see Table 4-1).

Table 4-2 summarizes the application considerations of ferrite material as a function of operating frequency.

Table 4-3 compares the properties of various soft magnetic materials.

## Core Types

Selection of the core shape and geometry is dictated by electrical and electromechanical requirements.

The equation for Area Product

$$A_p = W_A A_C = \frac{Pt}{KfBK_\mu J} \qquad (4\text{-}1)$$

Where  $A_p$ = area product ($cm^4$)
  $A_C$ = effective core area ($cm^2$)
  $W_A$ = window area ($cm^2$)
  $Pt$ = apparent power (watts, volt amp)
  $f$ = operating frequency (Hz)
  $B$ = flux density (tesla or webers/$m^2$)
  $K_\mu$ = window utilization factor
  $J$ = current density ($A/cm^2$)

permits the selection of a core on the basis of the area product being proportional to the power-handling capability of the transformer. In other words, the amount

**Table 4-1 Transformer and Inductor Design Parameters and Core Considerations**

| Type Core | Core or Material | Use | Frequency Range | Dielectric Constant | Dielectric Losses | Advantages or Disadvantages | Fabrication |
|---|---|---|---|---|---|---|---|
| Air core | Low dielectric and loss | Sustain winding | High | Low | Low | Inductance and frequency adjustable with adjustable tuning slug in core. | Purchased preformed or can be fabricated to requirements. |
| Laminated | Soft iron alloys | Low frequency ranges | Low | Adjustable | Adjustable | Permeability and hysteresis controlled by air gap adjustment which varies magnetic field. | Laminations stamped (0.003 to 0.014 in.) and coated with electrical insulation to restrict eddy currents. Thickness and number of laminations determined by frequency and current requirements. |
| Powdered iron | Ferromagnetic magnetic particles | Where high "$Q$" factor required | 5 Hz to 500 kHz (depending on core geometry) | Affected by humidity and temperature | Affected by humidity and temperature | Permeability and core losses controlled by size and composition of magnetic particles, and thickness of insulating binder spacing of the magnetic particles. Permeability and core losses also affected by temperature and humidity. | Ferromagnetic particles coated with an insulating binder and compressed into a uniform mass. |

| Ferrite | Nonconductive mixed crystals of manganese ferrite ($MnFe_2O_4$) and nickel ferrite ($NiFe_2O_4$) | Low audio frequency range to 100 MHz | Up to 100 MHz | Controllable by geometry at time of design | Controllable by geometry at time of design | High magnetic permeability and high resistivity to eddy currents. Saturation flux lower than that of laminated or powdered iron material which restricts its use in high-current applications. Can replace laminated and powdered iron core many times by controlling core geometry. | Binding process and heat treatment of crystals provide core that is ceramic in nature. |

## Table 4-2   Application Considerations: Ferrite Advantages and Disadvantages

| Application | Advantages | Disadvantages |
|---|---|---|
| Low frequency (<1 KHz)<br>High Flux Applications<br>  Generators<br>  Motors<br>  Power Transformers | • Ease of forming shapes allows possible use in inexpensive, high loss applications such as relays, small motors. | • Flux density low<br>• Relative cost high<br>• Limited size of parts |
| Medium Frequency (1-100 KHz)<br>Non-Linear High Flux Applications<br>  Flyback Transformers<br>  Deflection Yokes<br>  Inverters<br>  Wide Band Transformers<br>  Recording Heads<br>  Pulse Transformers<br>  Memory Cores | • Cost much lower than Nickel-Iron alloys, especially thin tapes<br>• Moderately high permeabilities available<br>• Low losses, especially in upper half of this range<br>• Inherent shielding in pot cores<br>• Good wear resistance<br>• Easily adapted to mass production | • Flux density lower than Nickel-Iron alloys<br>• Permeabilities lower than Nickel-Iron alloys<br>• Curie Temperature fairly low<br>• Good mating surface necessary for high inductance<br>• Smaller flux change than bobbin cores |
| Medium Frequency (1-100 KHz)<br>Low Flux, Linear Applications<br>  Loading Coils<br>  Filter Cores<br>  Tuned Inductors<br>  Wide Band Transformers<br>  Antenna Rods | • Permeabilities higher than powdered iron or Permalloy cores<br>• Gapped pot cores provide:<br>  1. Adjustability<br>  2. Stability · temperature, time, A. C. flux density, D. C. bias<br>  3. Self-shielding<br>• $\mu Q$ Products higher than other materials<br>• Wide choice of Inductance and Temperature Coefficient | • Low Curie point<br>• Need precision grinding of air gap<br>• Brittleness<br>• Mounting hardware needed |
| Higher frequencies (>200 KHz)<br>Low Flux, Linear Applications<br>  Filters<br>  Inductors<br>  Tuning Slugs | • Low losses (especially eddy current)<br>• Only Ferrites and powdered iron can operate at higher frequencies<br>• Medium frequency advantages apply | • Permeability decreases with frequency<br>• Medium frequency disadvantages apply<br>• Poor heat transfer |
| Microwave Frequencies (>500 MHz) | • Low dielectric losses<br>• Good gyromagnetic properties<br>• Only bulk material available | |

of copper (wire) and the amount of iron (ferrite or other appropriate core material) determines the total power capability of the transformer.

Figure 4-3 shows the cross sectional areas of several commonly used cores. The core types commonly used for transformer and inductor design are tube, rod, toroid, pot, and fabricated.

*Tube Core*   This type of core is ideally suited for air inductors where inductance values are low and the diameter, length, and spacing of turns are critical. The tube, made of nonmagnetic material, serves to maintain the coil shape and dimensions. Thread-like grooves in the tube maintain the desired turn spacing (see Figure 4-4). When space is critical, two or more windings may be wound on the same core, provided they are spaced so that a minimum of electrical or magnetic coupling is encountered. The tube core can also be adapted for use as

## Table 4-3   Properties of Soft Magnetic Materials

| Material | Initial Perm. $\mu_0$ | $B_{max}$ Kilogausses | Curie Temperature °C | Resistivity (ohm-cm) | $\mu_0 Q$ at 100 KHz | Operating Frequencies |
|---|---|---|---|---|---|---|
| Fe | 250 | 22 | 770 | $10 \times 10^{-6}$ | — | 60–1000 Hz |
| Si-Fe (unoriented) | 400 | 20 | 740 | $50 \times 10^{-6}$ | — | 60–1000 Hz |
| Si-Fe (oriented) | 1500 | 20 | 740 | $50 \times 10^{-6}$ | — | 60–1000 Hz |
| 50–50 Ni Fe(grain-oriented) | 2000 | 16 | 360 | $40 \times 10^{-6}$ | — | 60–1000 Hz |
| 79 Permalloy | 12,000 to 100,000 | 8 to 11 | 450 | $55 \times 10^{-6}$ | 8000 to 12,000 | 1 KHz–75 KHz |
| Permalloy powder | 14 to 550 | 3 | 450 | 1. | 10,000 | 10 KHz–200 KHz |
| Iron powder | 5 to 80 | 10 | 770 | $10^4$ | 2000 to 30,000 | 100 KHz–100 MHz |
| Ferrite-MnZn | 750 to 15,000 | 3 to 5 | 100 to 300 | 10 to 100 | 100,000 to 500,000 | 10 KHz–2MHz |
| Ferrite-NiZn | 10 to 1500 | 3 to 5 | 150 to 450 | $10^6$ | 30,000 | 200 KHz–100 MHz |
| Co-Fe 50% | 800 | 24 | 980 | $70 \times 10^{-6}$ | | |

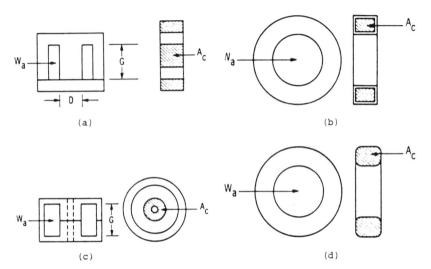

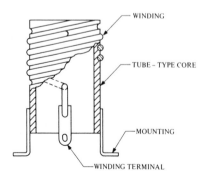

Figure 4-3   Core areas of various core types. (a) El lamination, (b) tape-wound toroidal core, (c) pot core, and (d) powder core.

an adjustable inductor by inserting and adjusting a magnetic material into the tube in the proximity of the winding (see Figure 4-5). All mountings and terminals used with this core must be composed of nonmagnetic materials.

*Rod Core*    This type of core, when used for air-type inductors, offers the same advantages as the tube type. However, the rod core has greater mechanical strength because of its rigidity and mounting capability (see Figures 4-6 and 4-7). Rod-type cores may be composed of either magnetic or nonmagnetic materials. Higher inductance values can be realized in smaller volumes if the rod is composed of magnetic material. This type of core lends itself more readily to machine winding.

Figure 4-4   Tube-type core.

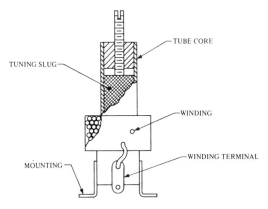

Figure 4-5   Adjustable inductor tube-type core.

*Toroid Core*   The toroid core (shown in Figure 4-8), when composed of powdered iron, is used in the frequency range from approximately 5 Hz to 50 kHz. The toroid core provides higher permeability and lower dielectric losses than rod-type cores of a similar iron composition because it contains magnetic flux in a continuous low-reluctance path. This type of core may be composed of magnetic or nonmagnetic material. However, the toroid type is ideally suited for powdered iron material since no physical air gap is required. An effective air gap, which is evenly distributed throughout the powdered iron material, results from the magnetic flux passing through the nonmagnetic insulating binder. Ferrite and laminated type cores are also available in this form for special applications. (See Table 4-4 for a comparison of the various core types.) Most converter and inverter designs use the toroid core. Generally, it can be said that toroid cores are used in high-power applications where EMI (electromagnetic interference) is not a prime consideration. Should it be desired to reduce EMI, a cup core (which is not discussed here) can be used in place of the toroid. However, cup cores cannot support high power loads.

*Pot Core*   This type of core is used to extend the frequency range of the powdered iron material from approximately 50k to 500 kHz. Ferrite cores in

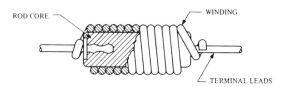

Figure 4-6   Rod-type core.

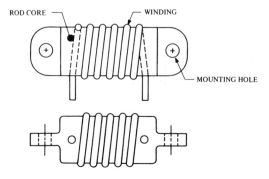

Figure 4-7   Rod-type core.

this form have a usable frequency range from 1 kHz to 20 MHz. The windings are prewound on a form or bobbin and assembled into the body of the core. This results in the magnetic flux being virtually contained within the body of the core, thus minimizing magnetic coupling to adjacent components. This geometry offers further advantages in that an air gap can be provided within the body of the core to reduce the hysteresis losses, reduce temperature coefficient, and control the permeability. This air gap lends itself to reducing the low-saturation problem of ferrite material. The pot core geometry is also more efficient at high frequencies which further favors the use of ferrite materials. Most ferrite pot cores have air gaps preset by grinding to a specific dimension. A tuning slug of magnetic material varied in the air gap of the core varies the permeability and consequently the inductance. This provides for adjustment of inductance up to ±10%. Seven sizes of ferrite pot cores conforming to international standards are made by all manufacturers. Special mountings are provided by the core manufacturers for printed circuitry and other electromechanical layouts. These should be considered during the initial design period if a vendor-purchased core is to be used.

*Fabricated Core*   Fabricated cores are used at very low frequencies when large inductance values and high current requirements are involved. They are either assembled laminations of soft iron alloy stampings or molded solid of powdered iron or ferrites. The laminations of the required thickness are either stacked to attain the desired core volume or the core is molded to size. The shapes of both the laminated and the solid-type cores resemble alphabetic forms such as E, I, U, F, and are designated as such. The "E-I" laminated and the "EE" molded cores are commonly used for inductor design. These forms allow various air gaps throughout the core to permit control of permeability and flux saturation. The air gap can be maintained by mechanically setting them in a form or by

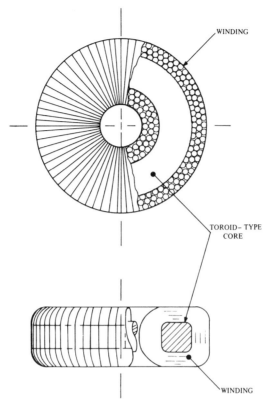

Figure 4-8   Toroid-type core.

adding high-reluctance shim stock, or both. Windings used in these cores are prewound to a specific turn requirement and mounted in the core.

Bobbin-wound transformers are used in power supply applications where voltage isolation is required between primary and secondary windings. Coils are wound on the bobbin and then assembled with the core to make the finished transformer. An example of this is the cross-section of an E-core, bobbin-wound transformer with a rectangular center post, shown in Figure 4-9. The empty spaces on each end of the bobbin are called end margins. Note that the insulation between windings (wrapper) extends out to the edge of the end margins. This technique is used to increase the creepage distance between adjacent windings so as to avoid high voltage breakdown due to surface contamination. In Figure 4-9, three windings are shown. The first and last are standard magnet wire windings, each having 30 turns. The middle one is a two turn foil winding. The coil has insulation between each successive layer, as well as wrappers, to isolate individual windings and to protect the last winding.

**Table 4-4  Comparison of Core Types**

| Type Core | Inductance Values | Materials | Adjustability | Special Mounting or Terminals Considerations | Advantages or Disadvantages | Fabrication |
|---|---|---|---|---|---|---|
| Tube | Low | Nonmagnetic | Adjusting magnetic material inserted in tube in proximity of winding | Nonmagnetic | (1) When space is critical two or more windings may be wound on same core.<br><br>(2) Two windings must be spaced so that a minimum of electrical or magnetic coupling is encountered. | (1) Core made of nonmagnetic material to sustain coil shape and dimensions.<br><br>(2) Thread-like grooves in the tubing to maintain the desired turn spacing.<br><br>(3) All mountings and terminals will consist of nonmagnetic material. |
| Rod | Higher inductance values in smaller volume | Magnetic or nonmagnetic | None | None | (1) Higher inductance values realized in smaller volumes (if rod is composed of magnetic material.)<br><br>(2) Better for machine winding.<br><br>(3) Greater mechanical strength due to rigidity and mounting capability. | (1) Rod-type cores composed of either magnetic or nonmagnetic material.<br><br>(2) Especially good when machine windings are used. |

| Toroid | High | (1) Powdered iron (2) Ferrite iron | Usually nonadjustable | None | (1) When powdered iron is used, frequency range is approximately 5 Hz through 50 kHz. (2) Core may be magnetic or nonmagnetic material. (3) No physical air gap is required. | (1) Core composed of either magnetic or nonmagnetic material. (2) No physical air gap required. (3) Nonmagnetic insulating binder. (4) Ferrite and laminated tape cores are available in this form. |
|---|---|---|---|---|---|---|
| Pot | Adjustable to approximately ±10% | Ferrite | Adjustable with tuning slug of magnetic material which varied in air gap of the core, varies the permeability and consequently the inductance up to ±10%. | Special mountings are provided by core manufacturers for printed circuitry and other electromechanical layouts. | (1) Used to extend frequency of the powdered iron material from approximately 50 K to 500 kHz. (2) Ferrite cores in this form have 1 k to 20 MHz useable frequency range. (3) Minimizes magnetic coupling to adjacent components. | (1) Windings previously wound on form or bobbin and assembled into the core. (2) Air gap can be provided. (3) Ferrite pot cores can have air gap preset by grinding to specific dimensions. (4) Tuning slug can be provided. |

**Table 4-4  (Continued)**

| Type Core | Inductance Values | Materials | Adjustability | Special Mounting or Terminals Considerations | Advantages or Disadvantages | Fabrication |
|---|---|---|---|---|---|---|
| Pot (Cont.) | | | | | (4) Air gap can be provided within the body of the core to: (a) reduce the hysteresis losses, (b) reduce temperature coefficient and control permeability, (c) reduce low-saturation problem of ferrite materials, and (d) be more efficient at high frequencies. Ferrite pot cores can have an air gap preset by grinding to a specific dimension.<br><br>(5) Seven sizes of ferrite pot cores conforming to international standards are provided by core manufacturers for printed circuitry and other electromechanical layouts. | (5) Seven sizes of ferrite pot cores conforming to international standards are provided by core manufacturers for printed circuitry and other electromechanical layouts. |

| Fabricated | Will fulfill large-inductance and high-current requirements | Soft iron stampings or molded solid of iron or ferrites | None, except by external circuitry or mechanically adding or removing shims during assembly | None | (1) Used at low frequencies when either large inductance values or high current requirements are involved.<br><br>(2) Shapes of both the solid and laminated-type resemble alphabetic forms such as E, I, U, F, and are designated as such.<br><br>(3) These forms allow various air gaps throughout the core, which permit control of permeability and flux saturation. | (1) Assembled laminations of soft iron alloy stampings, or<br><br>(2) Molded solid of powdered iron or ferrites.<br><br>  (a) Laminations of required thickness are stacked to attain the desired core volume, or (b) the core is molded to size.<br><br>(4) The E-I laminated and the EE molded are commonly used for inductor and transformer design.<br><br>(5) Air gap can be maintained by mechanically setting the core in a form, or by adding high-reluctance shim stock or both.<br><br>(6) Windings used in these cores are previously wound to a specific turn requirement and mounted in the core. |

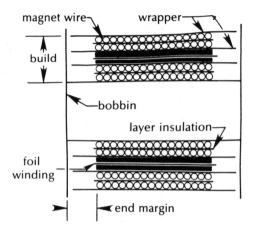

Figure 4-9 Typical cross section of a bobbin wound coil.

## Cores with Gaps

The energy storage capability of a core can be increased by "gapping" the core, whereby a significant portion of the total energy is stored in the air gap. The drawback, though, of a gapped core is that the effective permeability drops, requiring many more turns to achieve the required inductance. More turns require a larger winding window. The overall size of the inductor, however, can be considerably reduced with a properly gapped core, especially with a high-permeability core material. The formula for inductance with a gapped core is:

$$L = \frac{\mu \cdot A \cdot N^2 (0.4\pi \times 10^{-8})}{\ell \left(1 + \dfrac{\mu g}{\ell}\right)} \tag{4-2}$$

Inductance drops by the factor, $(1 + \mu g/\ell)$.

With a $\mu$ of 2000, $\ell = 2$ in., and $g = 0.02$ in., inductance will drop by 22:1, requiring that $N$ be increased by the square root of 22 to maintain constant inductance. Increase in energy storage is equal to the decrease in permeability.

$$\frac{E_{MAX}(\text{with gap})}{E_{MAX}(\text{no gap})} = 1 + \frac{\mu \cdot g}{\ell} \tag{4-3}$$

There are several practical limits on the amount by which the gap size may be increased. First, large gaps require many more turns to achieve the same inductance. This requires smaller diameter wire which increases copper losses from $I^2R$ heating. Secondly, with large gaps the *effective* gap size is considerably less than the actual gap because of fringing fields around the gap.

When using commercially available cores, data sheet information on $\ell$, $A$ and $\mu$ is usually given in *effective* values. The theoretical value of $\mu$, for instance, is the bulk value for the core material. The *effective* value for a single piece core may approach the bulk value, but with two-piece cores, the tiny air spaces left in the mating surfaces can reduce the *effective* permeability by as much as $2:1$. This may sound unreasonably pessimistic, but a core with bulk $\mu = 3000$, and $\ell = 1.5$ in., will use half its permeability for $g = 0.0005$ in. Data sheets for gapped cores list effective values of $\mu$ for each gap size to make calculations simple. They may also list a parameter, "inductance per (turn)$^2$" for each gap to further simplify inductance calculations.

There are two types of core material which are effectively self-gapped: (1) iron powder and (2) permalloy. These materials distribute the gap evenly throughout the core, allowing a gapless core to be constructed with much higher energy storage capability. The permeability of this material is much reduced, but if the winding window will accommodate the extra turns, the current-handling capability of the inductor will be much higher for the same inductance compared to a high-$\mu$ formulation.

Iron powder cores are cheaper than ferrite and can be custom tailored quickly, but high core-loss limits their application to low AC flux density applications such as inductors. A significant advantage of iron powder is that it saturates very "softly," preventing catastrophic total loss of inductance for large over-current conditions. Permalloy is more expensive, but has much lower core loss.

## Sealants, Encapsulants, and Containers

The purpose of sealants, encapsulants, and containers is to protect the transformer electrically and mechanically. In some cases, sealant alone may be sufficient to accomplish this. Design considerations, however, may require that all three be used.

*Sealants*  Sealants are applied directly to transformer windings to prevent moisture penetration. It is important (electrically) that the sealant maintain a low dielectric loss with changes of temperature and humidity environments. When the sealant is to be used to impregnate windings, the dielectric constant of the sealant must be considered to minimize the effects of distributed capacity. Polystyrene dope, silicone rubber coatings, and certain epoxy resins are excellent sealants. The characteristics of epoxy sealants are such that when used to coat frequency-determining elements (chokes), they change the designed frequency of operation and should thus be avoided. Usually soft rubber sealants are used first, then the entire assembly is encapsulated in epoxy. This prevents a frequency shift and yet provides protection from vibration and shock environments.

*Encapsulants* Encapsulating compounds must display the same electrical characteristics (dielectric losses) as sealants. Epoxy compounds that do not develop high exothermic temperatures or internal stresses during curing should be used to avoid transformer or inductor damage. In some instances, as mentioned above, soft rubber compounds are used to coat the components first. This protects them from both thermal and mechanical potting (encapsulating) stresses during the epoxy curing cycle.

*Containers* Plastic or metal containers are recommended when feasible, since they improve the mounting capability of the transformer or inductor and provide a container to hold the encapsulating material. Metal containers provide the added advantage of magnetic and electrical shielding. They also provide for the possibility of hermetic sealing.

## Integrated Magnetics

Integrated magnetics, in which the inductor (choke) and main transformer are physically combined (integrated), are used to reduce the size of the magnetic components, and thus power supply size. Integrated magnetics are used whenever the choke and transformer have the same waveforms, such as in the Cuk converter. However, they present some significant disadvantages: they contain high output ripple voltages due to the complex physical construction and gap "tweaking" required; they do not attenuate high-frequency harmonics; and they present a source of noise due to both capacitive and inductive coupling of source noise directly to the output. Thus, currently, the disadvantages for using integrated magnetics outweigh the advantages, thereby precluding their widespread use.

## High Operating Frequency Conversion Considerations

The trend toward higher operating (switching) frequencies in the 200 kHz to 1 MHz range presents some problems to power transformer designers that do not exist at lower frequencies: flux densities in magnetic materials become loss-limited instead of saturation-limited. Thus, more attention must be given to the core and core materials used. For example, many ferrite materials work well at 100 kHz, but not at 250 kHz, where the losses start to increase rapidly with frequency. As a result, new low-loss ferrites have been developed for the inductors and transformers used in megahertz frequency range power supply applications. Such high-frequency magnetics are substantially smaller than their low-frequency counterparts.

At high frequencies, the AC resistance of magnet wire, and thus the apparent resistance of a winding, increases dramatically, resulting in high losses. The high-frequency AC resistance of a winding is related to two phenomena: (1) the penetration depth of current in the wire, and (2) the proximity effect of current flowing in the surrounding wires. Several equations and a myriad of curves have been published to help designers predict this increase in resistance.

In general, the apparent resistance of a wire increases significantly as frequency increases, as wire diameter increases, or as the number of layers increases. Therefore, techniques which reduce the thickness of the wire or decrease the number of layers will reduce the ratio of AC to DC resistance.

One well-known method for reducing the apparent resistance is the use of Litz wire, consisting of a twisted bundle of very fine wires. However, this is not necessarily an ideal solution. The DC resistance of Litz wire for a given cross sectional area tends to be high, since so much of the cross section is used by the insulation of the strands. Also, due to the nature of the twisted bundle, the length of individual strands is somewhat longer than the bundle. Litz wire is more difficult to handle, that is, strip and terminate. Furthermore, for large sizes, the bundles are somewhat irregularly shaped and cannot be wound as tightly together as standard magnet wire.

Foil windings are often a good choice for high-current windings (see Figure 4-9). Foil is particularly effective in low-voltage, high-current applications where the number of turns is low. Foil is relatively thin for a given cross sectional area, compared to magnet wire. Therefore, foil windings have much lower AC resistance than large diameter magnet wire windings.

Interleaving the primary and the secondary windings significantly reduces the high-frequency resistance by effectively cutting the number of layers in half. Therefore, since resistance increases proportionally with the square of the number of layers, halving the number of layers can reduce the AC resistance by as much as 4 to 1.

Substituting two smaller wires for a large wire can often reduce the AC resistance of a winding. This is usually most effective when the substitution will not result in an increased number of layers. Of particular significance are applications where a single layer winding consumes less than 70% of one layer. In this case, two parallel wires having the same total cross-sectional area as the single large wire will fit onto a single layer. The total build will be reduced and the AC resistance will be improved, due to the smaller diameter of the wire.

Of equal importance is the problem of inductive losses caused by loose coupling between the primary and secondary windings. As the frequency rises, coupling becomes a critical factor. In DC/DC converters, close coupling can be achieved through the use of multifilar windings in which the primary and secondary windings are wound side by side.

High-frequency power chokes also require careful choice of core and attention to winding techniques to minimize stray fields. However, since energy is not being coupled to another circuit, the problems facing the manufacturer of high-frequency transformers are not much of a factor in designing inductors.

## TRANSFORMER REQUIREMENTS FOR CONVERTERS

In square-wave applications the core is rapidly switched from full magnetization in one direction to full magnetization in the opposite direction. Characteristic of such applications is the large class of converters in which the power transformer is not operated in a saturated mode but is driven nearly to saturation by a transistor switching circuit having an independent drive-signal source. The basic converter circuit indicated in Figure 4-10 is fairly well described in the available literature, but the detailed design and construction of the transformer are not as readily available.

Generally, the core material for the converter transformer is selected for maximum flux density and maximum squareness. Cores with higher flux capabilities enable smaller transformers to be built. Maximum squareness is by far the more important characteristic. This, of course, must be tempered by the operating frequency, allowable temperature rise, maximum power output, as well as cost considerations.

A lack of good *B-H* loop squareness results in spikes being generated on the output of the square wave. Theoretically, when the core is saturated, there is no flux change; however, in actual practice, due to imperfect saturation when the squareness ratio is less than 1.0, there is an unwanted flux change. The flux change, in addition to other possible causes, generates a spike which, even though of short duration, causes heating of the transistor junctions. This effect is minimized considerably with a square loop material. The ideal converter transformer should have a high squareness ratio $(B_r/B_m)$, low core loss, and

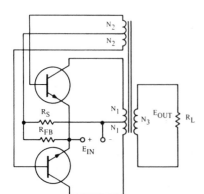

Figure 4-10 Conventional square-wave converter circuit.

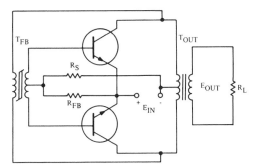

Figure 4-11   Two-transformer square-wave circuit.

high flux capabilities. The following nine steps are presented as a design summary for a one-transformer converter. Reference should be made to Figure 4-10 for this summary. These nine steps, however, also apply to the design of the main nonsaturating power transformer ($T_{out}$ of Figure 4-11) of a two-transformer inverter.

### Design Equations for a Single-Transformer Converter

Where power conversion is to be made, the load current dictates the size of the wire in the output windings $N_3$. The supply voltage is usually a fixed value so the turns ratio may be established between primary and secondary. Thus,

$$N_3/N_1 = E_{out}/E_{in} \qquad (4\text{-}4)$$

The size of the primary windings may then be found by determining the primary current. Thus,

$$I_{primary} = N_3/N_1 I_{load} \qquad (4\text{-}5)$$

If either $N_1$ or $N_3$ can be established, then the total cross section of all windings can be found, thus establishing the window area of the cores. The total flux capacity ($\phi_t$) of the core may be determined as follows: If one-tenth the frequency rating of the transistor is used as the operating frequency, and the supply voltage is known, the core flux $\phi_t$ may be determined from $E_{in} = 2\phi_t N_1 f \times 10^{-8}$. Thus,

$$\phi_t = \frac{E_{in}}{2N_1 \times f \times 10^{-8}}; \qquad \phi_t = 2B_m \times A_c \qquad (4\text{-}6)$$

*Note:* The value 2 is used here instead of 2.22 because average volts instead of rms volts are expressed.

For first approximation, use 10 turns for $N_1$. If, upon calculating $\phi_t$, a core is not available which has the required flux capacity, a new calculation should be made using a lower frequency or a smaller cross-section core. Notice that the core flux will vary inversely with frequency and turns.

Usually in power converter design, the frequency of the transistor dictates the maximum frequency of oscillation since the operating frequency of the converter should be one-tenth the frequency rating of the power transistor.

Steps used to select core size, turns, and wire size for converter design are:

(1) The input voltage, the output power and voltage must be known. The switching frequency or the output frequency of the converter must also be known or determined from the transistor ratings.
(2) Assume a minimum efficiency of 80% and calculate the input power and the input or primary current.
(3) Select the wire size for the secondary ($N_3$) using a 100% duty factor and the wire size for the primary ($N_1$) using a 50% duty factor.
(4) Determine the $W_a\phi_t$ product required from the formula:

$$W_a\phi_t = E_{in}A_w/2Kf \times 10^{-8} \qquad (4\text{-}7)$$

where: $W_a$ = case window area in circular mils,
$\quad\phi_t$ = total core flux in maxwells,
$\quad E_{in}$ = applied voltage in volts,
$\quad A_w$ = wire area in circular mils,
$\quad f$ = switching frequency,
$\quad K$ = winding factor (in power converters assume $\frac{1}{2}$ of the available winding area is used for both primary windings. Since the primary turns ($N_1$) are the turns on one side of a center-tapped winding, then $K$ equals $\frac{1}{4}$ of the nominal winding factor of 0.4. Therefore, $K = 0.1$).

(5) Select a core from the manufacturer's Core Data Section with a $W_a\phi_t$ product closest to the value calculated in step (4). The core $W_a\phi_t$ product can be determined from the formula:

$$W_a\phi_t = 2 \times W_a \times B_m \times A_c \qquad (4\text{-}8)$$

where: $B_m$ = flux density of core material,
$\quad A_c$ = effective core cross-sectional area.

In selecting the thickness of core material use 0.004-in. or 0.006-in. thickness for designs operating below 100 Hz. Use 0.002-in. thickness below

2000 Hz, 0.001-in. thickness below 4000 Hz. Ultra-thin materials such as 0.0005-in. thickness can be used up to 20,000 Hz while 0.00025-in. thickness can be used to 75,000 Hz. Above 75,000 Hz, 0.000125-in. thick material is suggested.

(6) Solve for $N_1$ using the formula:

$$N_1 = \frac{E}{2\phi_t f \times 10^{-8}}$$ (4-9)

(7) Determine the secondary turns:

$$N_3 = \frac{N_1 E_{out}}{E_{in}}$$ (4-10)

(8) Feedback turns $N_2$ should apply at least one volt to drive the bases of the power transistors.

$$N_2 = \frac{N_1 \times 1}{E_{in}}$$ (4-11)

If $N_2$ is less than five turns from the above formula, use five turns for $N_2$. Wire size for $N_2$ is determined from the transistor characteristics. Current flowing in feedback winding $I_F = I_C/\beta$. Refer to wire table for wire size.

(9) Check if core case will accommodate windings from formula.

$$K = \frac{(2N_1 A_{w1}) + (N_2 A_{w2}) + (N_3 A_{w3})}{W_a}$$ (4-12)

$K$ should be between 0.3 and 0.4.

### Design Equations for a Two-Transformer Converter

Referring to Figure 4-10, a high-power version of this circuit, operating at 60 Hz, would require a large and costly core. The two-transformer circuit in Figure 4-11 overcomes this by using a considerably smaller core of ideal characteristics (one whose size is determined by the power required by the base circuit of the transistors) for the saturating/feedback transformer, and a low-cost, that is, oriented silicon-iron alloy, nonsaturating power output transformer. This not only reduces the overall cost, but contributes to lower losses and temperature rise.

Of equal importance to core selection is winding technique. Ideally, the two halves of the center-tapped collector windings and the feedback winding should be wound simultaneously to maximize the coupling coefficients between windings. An asymmetrical winding technique, with resultant loose coupling, produces spikes on the output waveform. These spikes indicate improper switching which will cause additional heating of the transistor junction. A more practical method of winding would be to spiral one-half of the collector winding over the full periphery of the core and then place the other half of the collector winding in between the turns of this spiral. The feedback windings should be wound in a similar manner. If the center-tapped collector windings consist of many layers, then the simultaneous winding of the collector and feedback windings should be considered. As a minimum condition, the feedback winding must be directly next to the collector winding and should be wound in one 360° sweep.

Some special consideration must be given to the output winding. On the assumption that the DC supply is a low voltage, the turns on the primary and the feedback windings will be relatively few. However, output conditions may require many secondary turns. A large number of turns (the number depending upon the frequency of operation) can develop large inter-intrawinding capacitances, resulting in "ringing" or damped oscillations being superimposed upon the square-wave output.

It is recommended that the total number of turns be kept as small as practical by using a larger core area, but where a larger number of turns is unavoidable, they should be placed upon the core using a progressive or sector winding method.

### Design Notes

- The turns required for each half of the collector winding are calculated in a manner similar to that of an ordinary transformer. The only difference is the change of the factor 4.44 to 4.0, since it represents an average induced voltage rather than the rms value of a sine voltage which is considered:

$$N = \frac{E_{in}10^8}{4fA_cB_m} \tag{4-13}$$

- The turns required for the feedback winding are determined by the base-emitter voltage and the voltage required by the base resistor which limits base current.
- Wire sizes are determined in the conventional manner, but the collector circuit must supply the power to the base and feedback circuit in addition to supplying the load power.
- For high-power converters, consider the two-transformer type. For converters with highly reactive loads, which may upset the feedback/switching

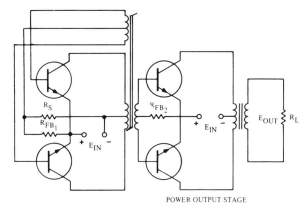

POWER OUTPUT STAGE

Figure 4-12  Square-wave oscillator using isolated square-wave oscillator and power output stage.

circuit, an isolated oscillator driving a power output stage, such as shown in Figure 4-12, can be employed.

■ For applications where supply voltages are higher than available transistor ratings, consider the circuit of Figure 4-13. The voltage impressed across each emitter-collector equals twice the source voltage divided by the number of pairs of transistors.

■ For higher frequencies, thinner tapes must be used. By referring to a core manufacturer's core-loss curves, an optimum tape material and thickness can be selected for every application.

In a nonsaturating push-pull DC/DC converter, large current spikes in the switching transistors can occur if the transformer core saturates when either of the transistors is conducting.

In an ideal converter, the volt-seconds applied to both the primary and the secondary windings average zero over one switching cycle. Also, the corresponding flux excursions of the transformer core will be symmetrical about the origin of its hysteresis loop. Thus, the core will not saturate as long as the maximum flux capability of the core is not exceeded. Practically, perfect volt-second balance is never achieved because the conduction interval of the transistor switches may not be exactly identical, and their conduction voltage drops may not be equal. Also, conduction voltage drops, as well as conduction-time intervals of rectifier diodes on the transformer secondary, may not be equal because of parasitic inductances.

There are several ways to prevent transformer core saturation:

1) Select a core material that has relatively low permeability and low residual flux density (a soft hysteresis curve similar to that of Figure 4-2) and which

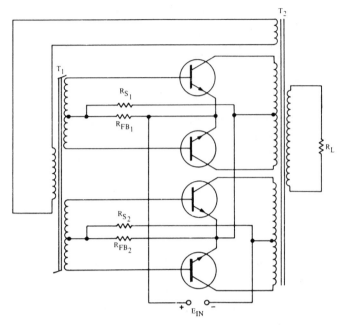

Figure 4-13 Square-wave converter for supply voltage ($E_{in}$) higher than transistor rating $V_{CE}$.

will thus tolerate some DC offset without saturation. But this results in high magnetizing current levels for small core area dimensions or low numbers of turns on windings. To reduce this high magnetizing current, the core area and winding turns must be increased, increasing the size and cost of the transformer.

2) Select a core that has a very high effective permeability in the normal operating region (about the origin of the B-H loop) and a lower (or sloped) permeability in the upper and lower regions, as shown in Figure 4-14. In this core, if some DC imbalance is present and the operating loop moves off to one side of the core characteristic, the core does not go into hard saturation, but moves into the upper region of the characteristic, where the permeability is lower. Thus, sufficient inductance exists to limit the peak switch currents to safe levels. From a practical viewpoint, there is no single core material that provides the shaped hysteresis loop of Figure 4-14. However, modified or composite core structures can be manufactured which closely approximate this curve, as shown in Figure 4-15, using toroidal cores. In Figure 4-15a, a partial air gap is placed on the outer part of the core so that nearly all of the core flux is contained in its inner ungapped section during normal operation. If the inner section should saturate, the core flux will move to the outer air-gapped portion, which has a lower but still significant permeability.

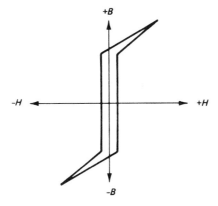

Figure 4-14   Shaped hysteresis loop.

In the stacked composite core shown in Figure 4-15b, two cores with the same inner and outer diameters are used but with different permeability materials.

3) Insert a DC blocking capacitor in series with the primary winding of full-bridge or half-bridge DC/DC converters. In the half-bridge DC/DC converter, the DC blocking capacitor function is provided by voltage division capacitors in parallel with the switching transistors.

4) Matching the switching transistors, both electrically and thermally.

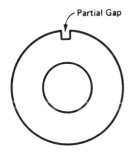

(a) Partial Air Gap (top view)

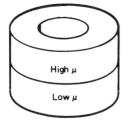

(b) Stacked Core          Figure 4-15   Composite core structures.

5) Use inductive primary current-limiting methods or active core state-flux sensing techniques to prevent turn-on core saturation caused by the remaining large residual flux density at power supply turn-off.

## CONVERTER POWER TRANSFORMER DESIGN APPLICATION

In order to select a core for a particular application, several relationships should be kept in mind. The first governs the length of time a core will support a voltage:

$$T = \frac{N_1 \phi_t}{E_{in}} 10^{-8} \text{ sec} = \text{turn-maxwells volt} \tag{4-14}$$

Maxwells can also be represented as Gauss-cm$^2$ ($\phi_t = A_c B_m$ = cross section × flux density). Both square waves and quasi-squares are covered by this relationship. The average value of any waveform over the period $T$ can also be used for $E$. A second relationship governs the amount of wire that will fit in a core: $\Sigma(N)(W) = (W_a)(K)$. ($W$) is the cross section of one wire in circular mils and $W_a$ is the window area of a core in circular mils. The winding factor $K$ is the ratio of total wire area to window area. A carefully wound core with wire very small compared to the window ($W_a \gg 500$ W) can be wound to $K$ approaching 0.7. This means that every wire has been placed exactly next to the previous one, is parallel to the axis of the toroid, and has been worked around the corners to snugly fit the core. Winding factors of 0.6 with just good winding are more practical and even leave a small window. A third useful relationship deals with the current required to magnetize a core $H_c = 0.4\pi NI/m\ell$: $I_{mag} = H_c m\ell/0.4\pi N$ where $H_c$ is the coercive force and $m\ell$ is the mean length of the core in centimeters. $H_c$ for permalloy-80 at 10 kHz is 0.14 Oe.

An example illustrates the use of these relationships. Consider that a transformer, which has a 20 V input (Figure 4-16), is switched not quite to saturation at 10 kHz in a push-pull manner, and has a single-output winding to drive a full-wave bridge with 100 V. The desired output power is 20 W. Since $T = (N\phi_t/E_{in}) \times 10^{-8}$ and $\phi_t = A_c B_m$ (this formula comes from $E = Nd\phi/dt$), $A_c$ (cross-sectional core area) = $(E_{in}T/N_1 B_m) \times 10^8 = [(20V)(50 \times 10^{-6} \text{ sec}) 10^8]/[(N)(14 \times 10^3 \text{ G})]$. The symbol $B_m$ is used because the core is not to saturate in 50 $\mu$sec. Fourteen kilogauss is the flux change necessary to go from residual flux in one direction to saturation in the other. If good efficiency can be assumed, then only 20 W are required. It is a reasonable first approximation to assume that the total cross section of wire in the primary is the same as in the secondary. Since each $N_1$ is half the primary, $W_a = 4N_1 W$. $W$ is the cross section of a primary wire which must handle 1 A on a 50% duty cycle. Again, 1000 circular mils per ampere is used to choose wire sizes. $N_1$ will require 500 circular mils

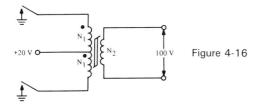

Figure 4-16

of insulation. The standard type of insulation used is a double-layer film of heavy formvar (105°C service). The double-layer films add about 1 to 2 mils diameter to the wire. Wire tables are published which take this insulation into account and give the current ratings of each wire size. A current of 500 mA at 1000 circular mils per ampere calls for AWG No. 23 wire which is 620 circular mils area. The required core window area is $W_a = (4N_1)(620)/0.6$. Two relationships both containing $N$ have now been formulated. The product of the cross-section area and the window area is a number which determines the core. Most core manufacturers make this product available along with the other data concerning each core.

$$A_c W_a = \frac{(20)(50)(10^{-1})}{(14)(N_1)} \times \frac{(4)(N_1)(620)}{0.6} = 0.0295 \times 10^6$$

Allowing the core to go to 90% of saturation is probably safe; $(A_c)(W_a) = 0.033$. From the Magnetics, Inc. catalog, there are several cores which would work. 52086-ID ($A_c = 0.038$), 52134-ID ($A_c = 0.076$), 52033-ID ($A_c = 0.151$). For the 52134-ID core the number of primary turns is $N_1 = 102$ turns, and the number of secondary turns is $N_2 = 510$ turns.

Had the output voltage been higher it would have been better to accept the core loss of a larger core in order to keep the turns down. The secondary current will be 200 mA with a 100% duty cycle. Number 27 wire is indicated here (see Tables 4-5 and 4-6). Since the copper loss is greater than the iron (core) loss, it would be better to use the next larger size core; that is, Magnetics, Inc., No. 52033. It has twice the cross-sectional area and therefore requires only half of the number of turns on the 52134 core.

In either case, (1) the loss could be reduced by putting the primary on first thus reducing its length on loss. Actually, the primary should go on first, for best coupling. The "neatest" winding results from putting the lightest wire on first. (2) Another way to reduce overall losses is to shuffle the wire sizes slightly: Increasing the primary to No. 22 (767 circular mils) gives a 23% increase in wire area; reducing the secondary to No. 28 is a decrease of 25%, so the two changes balance.

The secondary resistance is 65 mohms/ft, or 1.8 ohms. The power loss is

### Table 4-5   100 V, 20 W Transformer Core Design Data
### (Using a Magnetics, Inc. 52134-ID Core)[a]

| Primary Winding | Secondary Winding |
|---|---|
| $N_1 = 102$ turns | $N_2 = 510$ turns |
| Turn length $= 1.2$ in./$t = 123$ in. $= 10.2$ ft | Turn length $= 1$ in./$t = 510$ in. $= 42.5$ ft |
| Resistance $= 0.2$ ohm (where resistance of No. 23 wire is 20 mohm/ft) | Resistance $= 2.18$ ohm (where resistance of No. 27 wire is 51 mohm/ft) |
| Voltage drop in the wire $= 0.2$ V | Voltage drop in the wire $= 0.44$ V |
| Power loss $= 0.2$ W | Power loss $= 87$ mW |
| Copper loss $= 0.3$ W | |
| Core loss $= 0.1$ W | |
| Total power loss $= 0.379$ W | |

[a] Core weight $= 4.6$ g (not including the case).  The winding will occupy 0.8 to 1.4 in./turn  Core loss at 10 kHz (near saturation) for square permalloy $= 20$ mV/g.  Therefore the worst case loss is 92 mW.

now 72 mW. The primary resistance is now 16 mohms/ft or 0.11 ohm and the power loss is 110 W. Total loss is now 340 mW. While this is only a small change, the losses are better balanced.

Had the core loss been greater than the wire loss, it would have been worthwhile to reduce all wire sizes and try to go to a smaller core of the same cross section.

The transformer design for converting 20 to 100 V at 20 W has thus been optimized. The efficiency of the transformer was found to be 98%.

The packaging of a circuit is as important as the electrical design. Therefore,

### Table 4-6   100 V, 20 W Transformer Core Design Data
### (Using a Magnetics, Inc., 52033-ID Core)

| Primary Winding | Secondary Winding |
|---|---|
| $N_1 = 51$ turns | $N_2 = 255$ turns |
| Turn length $= 6.8$ ft | Turn length $= 27.6$ ft |
| Resistance $= 0.135$ ohm | Resistance $= 1.4$ ohms |
| Voltage drop in the wire $= 0.136$ V | Voltage drop in the wire $= 0.28$ V |
| Power loss $= 0.136$ W | Power loss $= 56$ mW |
| Core weight $= 7.9$ g | |
| Core loss $= 158$ mW | |
| The windings will occupy 1.1 to 1.8 in./turn | |
| Total power loss $= 352$ mW which is lower than that obtained using the 52134 core. | |

it is necessary to predict the finished size of a transformer. The two cores considered for the 20 W transformer would have the following finished sizes:

|   | 52134 | 52033 |
|---|-------|-------|
| *H* | 0.51 in. | 0.58 in. |
| *D* | 1.22 in. | 1.06 in. |

If the height is tolerable, the 52033 takes up much less room on a pc board.

## Winding Technique

As mentioned before, good winding can give winding factors approaching 0.7. Normally, a winding factor of 0.6 is used. Whether the winding is a single wire or bifilar the same rules hold. Consider a bifilar winding on a toroid. As the wire is placed around the core, it should be worked at the corners so that it lays flat along the surface it is crossing. Pulling on the wire will not get it to lay flat; it will only increase the danger of breaking the wire. Once the winding is completed, its finished end must be secured in some manner. A wrap of tape around the outside of the core will do this. But a better way is to use nylon string to tie the winding to the core. This also helps when starting a winding to keep the first turn from loosening.

When winding different-sized wires, the sequence of windings is at least a compromise. Starting with the lightest wire gives the neatest appearance. However, as the window gets smaller and the wire gets larger, positioning becomes difficult. Also, the length of the windings with the greatest currents increases and thus the total power loss increases. Placing light wires over heavy wires makes it difficult to avoid crossed turns as they fall into the cracks between windings. This must be done, however, if the core is to be filled completely and efficiently. Working light-to-heavy-to-light is probably best.

## Interwinding Capacitance

The transformer model shown earlier in Figure 4-1 included leakage inductance but not interwinding capacitance. The bifilar winding, which is very necessary to reduce leakage inductance and improve switching and symmetry, also adds distributed capacitance. The capacitance appears as a small generator between the windings which couples the transients that occur in switching a square wave. Two things can be done to reduce the distributed and interwinding capacitance. First, arrange the windings with symmetrical voltage gradients. This is difficult at best. The other way is to separate the windings by adding a toroidal layer of tape between primary and secondary windings. This acts as a weak faraday (electrostatic) shield. If a metal shield is placed between the windings, the capacitance is reduced still further. Using four-layer film insulation instead of two-layer film will also help separate the windings. Naturally, taping takes up window area which could otherwise be used by wire. Thus, the winding factor must be reduced accordingly. A factor of 0.4 to 0.45 seems to be realistic. Porous polyester webbing (5 mil reinforced) is excellent for taping when the transformer is to be encapsulated. Tables 4-7 and 4-8 are typical examples of core manufacturer's Core Selection Tables (these being from Magnetics, Inc.).

## POWER INDUCTOR DESIGN[1]

Not every inductor is appropriate for DC/DC conversion circuitry use. Even with the proper inductance value an "unknown" inductor may still saturate if it cannot handle the required current. The inductance of several types of coils, such as RF chokes, air core inductors, and noise filtering components frequently fall in the appropriate range for DC/DC converter applications (100 to 500 $\mu$H is common), but the coils saturate at only a few milliamps.

An inductor in saturation ceases to behave like an inductor, i.e., the mechanism which limits the rate of current rise breaks down. Energy is no longer being stored in the coil's magnetic field, so the mechanism that normally limits the inductor current no longer operates; all that limits the current is the series resistance. This resistance is quite low; consequently, the current can rise to an excessive, and possibly destructive, level and most of the input power is lost as heat.

An inductor doesn't saturate as long as its operating current is less than its rated maximum current. However, in most DC/DC converter designs the peak inductor current ($I_{pk}$) is significantly greater than the average output current, often by as much as 4 to 6 times. In the case of flyback converters, this peak current flows not just under peak load conditions, but each time the current switch turns on. $I_{pk}$ may be as high as several amperes in DC/DC converter circuits using external MOSFET switches and low inductance values ($<100$ $\mu$H). For this reason, one must give careful consideration to the current rating of the selected inductor.

To design an efficient filter power inductor in a short amount of time, several factors must be considered. One must first distinguish between the maximum incremental current, $I_m$, and the rated current at maximum temperature, $I_r$. The incremental current is the total instantaneous AC or pulse current ($I_{pk}$) and DC offset or bias ($I_{dc}$):

$$I_m = I_{pk} + I_{dc} \qquad (4-15)$$

[1]Manka, W. V., "Design Power Inductors Step-by-Step." Reprinted with permission from *Electronic Design*, **26**, December 20, 1977, copyright Hayden Publishing Co. Inc. (1977).

### Table 4-7   Selection Charts for Magnetics Incorporated Moly-Permalloy Cores

| $\mu_0 = B/H$ is assumed as follows: | | | | | |
|---|---|---|---|---|---|
| $\mu_0$ | 60 | 125 | 160 | 200 | 550 |
| $H_{sat}$ | 100 | 40 | 30 | 22 | 8 |
| for $H_{sat}$ at 90% $B/H$ | 54 | 112 | 144 | 180 | 495 |
| $H$ | 90 | 22 | 20 | 14 | 4 |

### Table 4-8   Magnetics Incorporated Core Selection Tables

| $Core$[a] | $\mu_o$ | $W_a$ | $\mathscr{L}_o$ | $A_c W_a \mu_o H$ |
|---|---|---|---|---|
| | | cir. mils | mH/1000 $turns$ | 90% $B/H$ |
| 55021 A2 | 60 | | 24 | 2.70m[b] |
| 55020 | 125 | 10K[b] | 50 | 2.25m |
| 55018 | 160 | | 64 | 2.16m |
| 55017 | 200 | | 80 | 1.98m |
| 55031 | 60 | | 25 | 7.20m |
| 55030 | 125 | 21K | 52 | 6.00m |
| 55028 | 160 | | 66 | 5.77m |
| 55027 | 200 | | 83 | 5.28m |
| 55041 | 60 | | 32 | 19.8m |
| 55040 | 125 | 36K | 66 | 16.5m |
| 55038 | 160 | | 84 | 15.9m |
| 55037 | 200 | | 105 | 14.6m |
| 55051 | 60 | | 27 | 54.0m |
| 55050 | 125 | | 56 | 45.0m |
| 55048 | 160 | 81K | 72 | 43.2m |
| 55047 | 200 | | 90 | 39.6m |
| 55046 | 550 | | 225 | 39.6m |
| 55121 | 60 | | 35 | 170m |
| 55120 | 125 | 150K | 72 | 142m |
| 55118 | 160 | | 92 | 136m |
| 55117 | 200 | | 115 | 125m |

[a] Magnetics, Inc. cores.
[b] $K = 10^3$; $m = 10^6$; $G = 10^9$.

### Table 4-8 (Continued)

| Core | $\mu_o$ | $W_a$ | $\mathscr{L}_o$ | $A_c W_a \mu_o H$ |
|------|---------|-------|-----------------|-------------------|
|      |         | cir. mils | mH/1000 $t$ | 90% $B/H$ |
| 55848 | 60  |       | 32  | 300m |
| 55206 | 125 |       | 68  | 250m |
| 55204 | 160 | 225K  | 87  | 240m |
| 55203 | 200 |       | 109 | 220m |
| 55202 | 550 |       | 320 | 220m |
| 55059 | 60  |       | 43  | 592m |
| 55310 | 125 | 302K  | 90  | 494m |
| 55308 | 160 |       | 115 | 474m |
| 55307 | 200 |       | 144 | 435m |
| 55894 | 60  |       | 75  | 1220m |
| 55930 | 125 |       | 157 | 1002m |
| 55928 | 160 | 320K  | 201 | 980m |
| 55927 | 200 |       | 257 | 895m |
| 55926 | 550 |       | 740 | 895m |
| 55071 | 60  |       | 61  | 2300m |
| 55548 | 125 | 590K  | 127 | 1920m |
| 55546 | 160 |       | 163 | 1840m |
| 55545 | 200 |       | 203 | 1690m |
| 55585 | 125 |       | 79  | 1820m |
| 55583 | 160 | 810K  | 101 | 1750m |
| 55582 | 200 |       | 126 | 1600m |
| 55076 | 60  |       | 56  | 2930m |
| 55324 | 125 | 730K  | 117 | 2440m |
| 55322 | 160 |       | 150 | 2340m |
| 55321 | 200 |       | 187 | 2150m |
| 55083 | 60  |       | 81  | 5360m |
| 55254 | 125 | 860K  | 168 | 4550m |
| 55252 | 160 |       | 215 | 4360m |
| 55251 | 200 |       | 269 | 4000m |
| 55716 | 60  |       | 73  | 11.3G[b] |
| 55715 | 125 | 1480K | 152 | 9.40G |
| 55713 | 160 |       | 195 | 9.00G |
| 55712 | 200 |       | 243 | 8.38G |
| 55110 | 60  |       | 75  | 16.8G |
| 55109 | 125 | 1870K | 156 | 14.0G |
| 55107 | 160 |       | 200 | 13.4G |
| 55106 | 200 |       | 250 | 12.3G |

## Table 4-8 (*Continued*)

| Core | $\mu_o$ | $W_a$ | $\mathscr{L}_o$ | $A_c W_a \mu_o H$ |
|------|---------|-------|-----------------|-------------------|
|      |         | cir. mils | mH/1000 $t$ | 90% $B/H$ |
| 55439 | 60  |       | 135 | 10.4G |
| 55438 | 125 |       | 281 | 8.68G |
| 55436 | 160 | 860K  | 360 | 8.32G |
| 55435 | 200 |       | 450 | 7.64G |
| 55090 | 60  |       | 86  | 985m |
| 55089 | 125 |       | 178 | 820m |
| 55087 | 160 | 1210K | 228 | 788m |
| 55086 | 200 |       | 285 | 720m |

Since these currents together change the choke's inductance, the minimum inductance is determined by $I_m$.

The rated current is the total effective AC, $I_{\text{eff}}$, and $I_{\text{dc}}$.

$$I_r = I_{\text{eff}} + I_{\text{dc}} \qquad (4\text{-}16)$$

These currents heat the inductor, so $I_r$ determines the minimum size of the magnet wire.

With a sine wave of $E_{\text{ac}}$, the maximum rms voltage across the choke, $I_{\text{eff}}$ depends on $L_m$, the incremental inductance at $I_m$, and the operating frequency, $f$:

$$I_{\text{eff}} = E_{\text{ac}}/2\pi f L_m \qquad (4\text{-}17)$$

also

$$I_{\text{pk}} = \sqrt{2}\, I_{\text{eff}} \qquad (4\text{-}18)$$

Be careful in assigning a value to $f$—it may not be the line frequency. For instance, with a 60 Hz three-phase full-wave bridge rectifier, $f$ is 360 Hz.

For pulse excitation of the choke, as in switching regulators, the voltage and current waveforms are shown in Figure 4-17. When a choke handles pulses rather than sine waves, $I_{\text{pk}}$ depends on the pulse excitation voltage ($E_{\text{pk}}$), the pulse duration ($D_p$), the $L_m$ and the core-loss current ($I_c$), which is due to hysteresis and eddies:

$$I_{\text{pk}} = (E_{\text{pk}} D_p/L_m) + I_c \qquad (4\text{-}19)$$

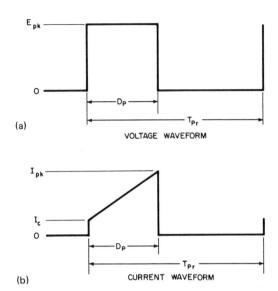

Figure 4-17   A trapezoid of current (a) results when a square wave of voltage (b) is impressed on an inductor. The initial jump of current, $I_c$, ramps up to the final value, $I_{pk}$, with a slope that depends on coil inductance and resistance (*Courtesy Electronic Design*).

Choose the proper core and $I_c$ will be less than 10% of $I_{pk}$:

$$I_c \leq 0.1\ I_{pk} \tag{4-20}$$

The pulse-repetition time, $t_{pr}$, comes into play for $I_{eff}$.

$$I_{eff} = \sqrt{D_p(I_p^2 + I_{pk}I_c + I_c^2)/3t_{pr}} \tag{4-21}$$

Once the values for the appropriate coil currents have been determined, a core material can be selected from the hundreds of magnetic materials available. Table 4-9 lists a representative cross-section of several popular core materials.

To pick the right core material, one needs to know the required $f$, maximum operating temperature, $T_m$, required core properties, cost, geometry and winding limitations, among other things. But whenever possible, and especially for switching regulators, gapped-ferrite cores should be used.

The core loss per unit volume, $C_v$, depends on the core-loss factor:

$$C_v = C_{vf}f \tag{4-22}$$

The crux of this entire procedure is that it determines optimum values for maximum operating flux density, $B$, and maximum operating magnetic-field strength, $H$, for the selected core material. Empirical analysis has shown that incremental effects begin approximately when the intrinsic magnetic-field strength, $H_i$, equals the saturation-flux density, $B_s$, minus the residual-flux density, $B_r$, all divided by twice the initial permeability, $U_i$, or

$$H_i = (B_s - B_r)/2U_i \qquad (4\text{-}23)$$

For inductors passing substantial DC, the maximum operating flux density must be lowered to the differential flux density, $B_s - B_r$, rather than just $B_s$:

$$B = B_s - B_r \qquad (4\text{-}24)$$

Maximum operating field strength depends not only on B, but also on incremental permeability, $U_m$:

$$U_m = U_i L_m/L_0 \qquad (4\text{-}25)$$

where $L_o$ is the choke's initial low-level AC value of inductance with no DC, and $L_m$ is the desired inductance value at $I_m$.

Now one can find H:

$$H = B/U_m \qquad (4\text{-}26)$$

Because of the large variations in the permeabilities of many magnetic materials, especially the electrical steels, specify $U_i$ at no more than 40 G. And expect variations in $U_i$, $B_s$ and $B_r$ at high temperatures.

Having found values for $B$ and $H$, the thinnest magnet wire that is required can be calculated.

The size of the magnet wire is limited by its current rating, $I_r$. However, the usual current rating for copper wire doesn't apply to inductors. Ratings based on a current density of 1000 circular mils/A—the loading that causes a 2% voltage drop per 100 ft. of standard house wiring—aren't realistic.

Instead current ratings for industrial and military inductors are based on a more practical consideration—maximum temperature rise. Since these ratings range typically from 5 to 20% of the wire's fusing current, rating the inductor wire at 10% of its fusing current is realistic, not to mention convenient for calculation. This rating applies at $T_m$.

At 20°C, the fusing current for bare-copper magnet wire of diameter $d_0$ is found by

$$I_f = 10{,}244 \, d_0^3 \qquad (4\text{-}27)$$

Table 4-9  Average Values of Common Core Materials (Courtesy Electronic Design)

| Material Designation | Freq. Range (Hz) | Temp. Range (°C) | $U_i$ (G/Oe) | $B_s$ (G) | $B_r$ (G) | $H_i$ (Oe) |
|---|---|---|---|---|---|---|
| **Electrical Steels**[4] | | | | | | |
| Silicon Iron | 20 to 1 M | −55 to +300 | 500 | 17.5 k | 12 k | 5.5 |
| Silectron | 20 to 10 k | −55 to +375 | 1.5 k | 16.5 k | 14 k | 0.83 |
| Alloy 48 | 20 to 8 k | −55 to +250 | 1 k | 12.5 k | 10 k | 1.25 |
| HY-MV 80 | 20 to 25 k | −55 to +230 | 10 k | 8 k | 4.4 k | 0.18 |
| Supermalloy | 20 to 25 k | −55 to +230 | 50 k | 7.3 k | 4 k | 0.033 |
| Supermendur | 20 to 2 k | −55 to +460 | 1.5 k | 21 k | 19 k | 0.67 |
| **Molypermalloy Powders** | | | | | | |
| MPP 14[2] | 400 k to 1 M | −55 to +250 | 14 | 6 k | 10 | 214 |
| MPP 26[2] | 400 to 650 k | −55 to +230 | 26 | 6 k | 10 | 115 |
| MPP 60 | 400 to 250 k | −55 to +200 | 60 | 6 k | 10 | 49.8 |
| MPP 125 | 40 to 100 k | −55 to +200 | 125 | 6 k | 200 | 23.2 |
| MPP 160 | 40 to 70 k | −55 to +200 | 160 | 6 k | 200 | 18.1 |
| MPP 200 | 40 to 30 k | −55 to +175 | 200 | 6 k | 200 | 14.5 |
| MPP 300 | 40 to 25 k | −55 to +150 | 300 | 6 k | 300 | 9.5 |
| MPP 550 | 40 to 20 k | −55 to +125 | 550 | 6 k | 650 | 4.86 |

| Powdered Irons | | | | | | |
|---|---|---|---|---|---|---|
| Carbonyl SF[2] | 2 to 50 M | −55 to +125 | 7.5 | 8 k | 10 | 533 |
| Carbonyl E[2] | 200 k to 10 M | −55 to +125 | 10 | 8 k | 10 | 400 |
| Carbonyl C[2] | 100 k to 2 M | −55 to +125 | 20 | 8 k | 10 | 200 |
| Carbonyl GQ4[2] | 50 k to 1 M | −55 to +125 | 35 | 8 k | 10 | 114 |
| Carbonyl HA | 1 to 100 k | −55 to +105 | 60 | 8 k | 2200 | 48.3 |
| 75 Powder | 400 to 50 k | −55 to +105 | 75 | 8 k | 2200 | 38.7 |
| 90 Powder | 400 to 10 k | −55 to +105 | 90 | 8 k | 2200 | 32.2 |
| **Ferrites[3]** | | | | | | |
| F40 | 10 to 80 M | −55 to +250 | 40 | 2400 | 750 | 20.6 |
| F125 | 200 k to 10 M | −55 to +250 | 125 | 2350 | 1200 | 4.6 |
| F175 | 100 k to 5 M | −55 to +230 | 175 | 2550 | 1400 | 3.28 |
| F250 | 50 k to 4 M | −55 to +200 | 250 | 2200 | 1100 | 2.2 |
| F400 | 10 k to 2.5 M | −55 to +175 | 400 | 2700 | 1100 | 2 |
| F750 | 1 k to 1.5 M | −55 to +125 | 750 | 4000 | 1800 | 1.47 |
| F1000 | 1 k to 1 M | −55 to +125 | 1000 | 4200 | 1700 | 1.25 |
| F1500 | 1 k to 650 k | −55 to +125 | 1500 | 4000 | 1100 | 0.97 |
| F2000 | 400 to 500 k | −55 to +125 | 2000 | 4400 | 1500 | 0.72 |
| F2300 | 400 to 300 k | −40 to +105 | 2300 | 4000 | 1200 | 0.61 |
| F2700 | 400 to 250 k | −30 to +105 | 2700 | 4700 | 2000 | 0.50 |
| F5000 | 400 to 100 k | −25 to +105 | 5000 | 4300 | 1200 | 0.31 |
| F10000 | 400 to 80 k | −25 to +90 | 10000 | 4300 | 1200 | 0.16 |

Notes:

(1) For specific data and tolerances, refer to individual suppliers. (2) Excessive heating occurs before incremental effects are observed. (3) There are no standard designations for equivalent materials between ferrite suppliers. Designation used is for convenience. (4) Frequency range of electrical steel depends on material thickness. The thinner the material, the higher the frequency range.

As a result, the current rating at 20°C is

$$I_{r0} = 0.1\, I_f \qquad (4\text{-}28)$$

Furthermore, for a required $I_r$, the magnet wire's minimum current rating at 20°C must be

$$I_{r0} = I_r\, [1 + 0.00393\, (T_m - 20)] \qquad (4\text{-}29)$$

Current ratings for copper magnet wire, based on Eq. 4-28, are tabulated in Table 4-10 along with other selected magnet-wire design data.

Using the values for $B$ and $H$, determine the minimum effective volume, $V_{em}$, that will sustain the inductor's operating conditions.

$$V_{em} = 0.4\pi \times 10^8\, L_m I_m^2\, (BH) \qquad (4\text{-}30)$$

For ungapped cores like toroids, the actual core volume must equal or exceed $V_{em}$. But meeting this requirement may make the core size excessive. Fortunately, however, an air gap in the magnetic circuit can reduce core size significantly. If the air-gap length, $\ell_g$, is small with respect to the effective magnetic-path length, $\ell_e$, the equivalent effective volume, $V_e$, is:

$$V_e = A_c \ell_e$$
$$= A_c(\ell_c + U_i \ell_g)$$
$$= V_c + (A_c U_i \ell_g), \qquad (4\text{-}31)$$

where $\ell_e$ is the magnetic path length of the core, $A_c$ is the core's cross-sectional area and $V_c$ is the actual volume.

One must ensure that

$$V_e \geq V_{em} \qquad (4\text{-}32)$$

to get the $L_m$ once the core gap has been selected.

Now with this minimum effective core volume, select the smallest core that can sustain the required inductance. With the help of core-size indexes pick a core from magnetic-core catalogs. They also provide data for the winding area, $W_a$, and for $A_c$ that you can use in

$$W_a A_c = 5.067 \times 10^8\, L_m I_m d_i^2/(kB) \qquad (4\text{-}33)$$

Here $d_i$ is the diameter of the insulated-wire (double-film insulation is recommended), and $k$ is the winding-utilization, or "fit" factor. For toroidal windings, $k$ is typically 0.4; for bobbin windings, 0.8.

The core must satisfy Eqs. 4-30 and 4-33. Often, the core's geometry is severely limited by mounting space and DC winding resistance, $R_{DC}$. (Other factors that affect core geometry are compared in Table 4-11.)

Even slug-type inductors can be designed with this step-by-step procedure, but one must empirically determine $\ell_g$ and $V_e$ for slugs.

Now that the material, size and geometry of the core is known, one can calculate the maximum value of the incremental inductance index, $A_{Lm}$, from

$$A_{Lm} = (BA_c)^2 \times 10^{-16}/(L_m I_m^2) \qquad (4\text{-}34)$$

Then with $A_{Lm}$, compute the number of turns, $N$, for the $L_m$ required:

$$N = \sqrt{L_m/A_{Lm}} \qquad (4\text{-}35)$$

Take $N$ and $H$, and determine $\ell_e$:

$$\ell_e = 0.4\pi N I_m/H \qquad (4\text{-}36)$$

For cores without air gaps, such as toroids, make sure that

$$\ell_c \geq \ell_e \qquad (4\text{-}37)$$

For cores with air gaps, compute the air-gap length from

$$\ell_g = (\ell_e - \ell_c)/U_i \qquad (4\text{-}38)$$

Now the effective permeability can be found:

$$U_e = U_i/[1 + (U_i \ell_g/\ell_c)] \qquad (4\text{-}39)$$

For core geometries in which an air gap interrupts the magnetic circuit twice, the inserted material thickness should be half that computed in Eq. 4-38.

The air gap affects the initial-inductance index, $A_{Lo}$, as follows:

$$A_{Lo} = 0.4\pi \times 10^{-8} A_c/[(\ell_c/U_i) + \ell_g] \qquad (4\text{-}40)$$

Calculate the initial inductance, $L_o$, from

$$L_o = A_{Lo}N^2 \qquad (4\text{-}41)$$

Table 4-10   Magnet Wire Design Data (Courtesy Electronic Design)

| AWG Size | Bare Wire | | | | Single Film | | | | | Double Film | | | | |
|---|---|---|---|---|---|---|---|---|---|---|---|---|---|---|
| | Max OD | Max I | Max Tensile Strength | Ω/ft | Max OD | ft/lb | Ω/lb | Tpi | Tpi² | Max OD | ft/lb | Ω/lb | Tpi | Tpi² |
| 10 | 0.1024 | 33.6 | 82.4 | 0.00100 | 0.1047 | 31.6 | 0.0316 | 9.551 | 91.22 | 0.1061 | 31.5 | 0.0315 | 9.425 | 88.83 |
| 11 | 0.0912 | 28.2 | 65.3 | 0.00126 | 0.0935 | 39.8 | 0.0501 | 10.70 | 114.4 | 0.0948 | 39.7 | 0.0500 | 10.55 | 111.3 |
| 12 | 0.0812 | 23.7 | 51.8 | 0.00159 | 0.0834 | 50.3 | 0.0800 | 11.99 | 143.8 | 0.0847 | 50.0 | 0.0795 | 11.81 | 139.4 |
| 13 | 0.0724 | 19.9 | 41.2 | 0.00200 | 0.0746 | 63.3 | 0.1266 | 13.40 | 179.7 | 0.0757 | 62.9 | 0.1258 | 13.21 | 174.5 |
| 14 | 0.0644 | 16.7 | 32.6 | 0.00252 | 0.0666 | 79.9 | 0.2013 | 15.02 | 225.5 | 0.0682 | 79.3 | 0.1998 | 14.66 | 215.0 |
| 15 | 0.0574 | 14.1 | 25.9 | 0.00318 | 0.0594 | 101 | 0.3212 | 16.84 | 283.4 | 0.0609 | 100 | 0.3180 | 16.42 | 269.6 |
| 16 | 0.0511 | 11.8 | 20.5 | 0.00402 | 0.0531 | 127 | 0.5105 | 18.83 | 354.7 | 0.0545 | 126 | 0.5065 | 18.35 | 336.7 |
| 17 | 0.0455 | 9.94 | 16.3 | 0.00505 | 0.0475 | 159 | 0.8029 | 21.05 | 443.2 | 0.0488 | 158 | 0.7979 | 20.49 | 419.9 |
| 18 | 0.0405 | 8.35 | 12.9 | 0.00639 | 0.0424 | 201 | 1.284 | 23.58 | 556.2 | 0.0437 | 199 | 1.272 | 22.88 | 523.6 |
| 19 | 0.0361 | 7.02 | 10.2 | 0.00805 | 0.0379 | 253 | 2.037 | 26.39 | 696.2 | 0.0391 | 251 | 2.020 | 25.58 | 654.1 |
| 20 | 0.0322 | 5.92 | 8.14 | 0.0101 | 0.0339 | 318 | 3.212 | 29.50 | 870.2 | 0.0351 | 315 | 3.182 | 28.49 | 811.7 |
| 21 | 0.0286 | 4.95 | 6.42 | 0.0128 | 0.0303 | 402 | 5.146 | 33.00 | 1089 | 0.0314 | 397 | 5.082 | 31.85 | 1014 |
| 22 | 0.0254 | 4.15 | 5.07 | 0.0162 | 0.0270 | 508 | 8.230 | 37.04 | 1372 | 0.0281 | 503 | 8.149 | 35.59 | 1266 |
| 23 | 0.0227 | 3.50 | 4.05 | 0.0203 | 0.0243 | 633 | 12.85 | 41.15 | 1694 | 0.0253 | 625 | 12.69 | 39.53 | 1562 |
| 24 | 0.0202 | 2.94 | 3.20 | 0.0257 | 0.0217 | 806 | 20.71 | 46.08 | 2124 | 0.0227 | 794 | 20.41 | 44.05 | 1941 |
| 25 | 0.0180 | 2.47 | 2.54 | 0.0324 | 0.0194 | 1013 | 32.82 | 51.55 | 2657 | 0.0203 | 990 | 32.08 | 49.26 | 2427 |
| 26 | 0.0160 | 2.07 | 2.01 | 0.0410 | 0.0173 | 1282 | 52.56 | 57.80 | 3341 | 0.0182 | 1260 | 51.66 | 54.95 | 3019 |
| 27 | 0.0143 | 1.75 | 1.61 | 0.0514 | 0.0156 | 1608 | 82.65 | 64.10 | 4109 | 0.0164 | 1580 | 81.21 | 60.98 | 3718 |
| 28 | 0.0127 | 1.47 | 1.27 | 0.0653 | 0.0140 | 2033 | 132.8 | 71.43 | 5102 | 0.0147 | 1990 | 129.9 | 68.03 | 4628 |
| 29 | 0.0114 | 1.25 | 1.02 | 0.0812 | 0.0126 | 2525 | 205.0 | 79.37 | 6299 | 0.0133 | 2470 | 200.6 | 75.19 | 5653 |
| 30 | 0.0101 | 1.04 | 0.801 | 0.104 | 0.0112 | 3215 | 334.4 | 89.29 | 7972 | 0.0119 | 3140 | 326.6 | 84.03 | 7062 |
| 31 | 0.0090 | 0.874 | 0.636 | 0.131 | 0.0100 | 4065 | 532.5 | 100.0 | 10000 | 0.0108 | 3950 | 517.4 | 92.59 | 8573 |
| 32 | 0.0081 | 0.747 | 0.515 | 0.162 | 0.0091 | 5000 | 810.0 | 109.9 | 12076 | 0.0098 | 4880 | 790.6 | 102.0 | 10412 |
| 33 | 0.0072 | 0.626 | 0.407 | 0.206 | 0.0081 | 6369 | 1312 | 123.5 | 15242 | 0.0088 | 6170 | 1271 | 113.6 | 12913 |
| 34 | 0.0064 | 0.524 | 0.322 | 0.261 | 0.0072 | 8064 | 2105 | 138.9 | 19290 | 0.0078 | 7870 | 2054 | 128.2 | 16437 |

| | Inches | Amperes | Pounds | Ω/ft | Inches | ft/lb | Ω/lb | $T_{pi}$ | $T_{pi}^2$ | Inches | ft/lb | Ω/lb | $T_{pi}$ | $T_{pi}^2$ |
|---|---|---|---|---|---|---|---|---|---|---|---|---|---|---|
| 35 | 0.0057 | 0.441 | 0.255 | 0.331 | 0.0064 | 10210 | 3380 | 156.2 | 24414 | 0.0070 | 9940 | 3290 | 142.9 | 20408 |
| 36 | 0.0051 | 0.373 | 0.204 | 0.415 | 0.0058 | 12760 | 5295 | 172.4 | 29727 | 0.0063 | 12440 | 5163 | 158.7 | 25195 |
| 37 | 0.0046 | 0.319 | 0.166 | 0.512 | 0.0052 | 15800 | 8090 | 192.3 | 36982 | 0.0057 | 15300 | 7834 | 175.4 | 30779 |
| 38 | 0.0041 | 0.269 | 0.132 | 0.648 | 0.0047 | 19920 | 12908 | 212.8 | 45269 | 0.0051 | 19300 | 12506 | 196.1 | 38847 |
| 39 | 0.0036 | 0.221 | 0.101 | 0.847 | 0.0041 | 26040 | 22056 | 243.9 | 59488 | 0.0045 | 25100 | 21260 | 222.2 | 49383 |
| 40 | 0.0032 | 0.185 | 0.0804 | 1.08 | 0.0037 | 33110 | 35759 | 270.3 | 73046 | 0.0040 | 32200 | 34776 | 250.0 | 62500 |
| 41 | 0.0029 | 0.160 | 0.0661 | 1.32 | 0.0033 | 40100 | 52932 | 303.0 | 91827 | 0.0036 | 39500 | 52140 | 277.8 | 77160 |
| 42 | 0.0026 | 0.136 | 0.0531 | 1.66 | 0.0030 | 51000 | 84660 | 333.3 | 111111 | 0.0032 | 49800 | 82668 | 312.5 | 97656 |
| 43 | 0.023 | 0.113 | 0.0475 | 2.14 | 0.0026 | 65800 | 140.8k | 384.6 | 147928 | 0.0029 | 63700 | 136.3k | 344.8 | 118906 |
| 44 | 0.0021 | 0.0985 | 0.0346 | 2.59 | 0.0024 | 79400 | 205.6k | 416.7 | 173611 | 0.0027 | 76300 | 197.6k | 370.4 | 137174 |
| 45 | 0.00176 | 0.0756 | 0.0243 | 3.62 | 0.00205 | 104k | 376.5k | 487.8 | 237954 | 0.00230 | 99600 | 360.6k | 434.8 | 189036 |
| 46 | 0.00157 | 0.0637 | 0.0194 | 4.54 | 0.00185 | 132k | 599.3k | 540.5 | 292184 | 0.00210 | 126k | 572.0k | 476.2 | 226757 |
| 47 | 0.00140 | 0.0536 | 0.0154 | 5.71 | 0.00170 | 162k | 925.0k | 588.2 | 346020 | 0.00190 | 153k | 873.6k | 526.3 | 277003 |
| 48 | 0.00124 | 0.0447 | 0.0121 | 7.29 | 0.00150 | 205k | 1.494M | 666.6 | 444444 | 0.00170 | 199k | 1.451M | 588.2 | 346020 |
| 49 | 0.00111 | 0.0379 | 0.0097 | 9.09 | 0.00130 | 258k | 2.345M | 769.2 | 591716 | 0.00150 | 252k | 2.291M | 666.6 | 444444 |
| 50 | 0.00099 | 0.0319 | 0.0077 | 11.4 | 0.00120 | 312k | 3.557M | 833.3 | 694444 | 0.00140 | 306k | 3.488M | 714.3 | 510204 |
| 51 | 0.00088 | 0.0267 | 0.0061 | 14.5 | 0.00110 | 416k | 6.032M | 909.1 | 826446 | | | | | |
| 52 | 0.00078 | 0.0223 | 0.0048 | 18.4 | 0.00100 | 555k | 10.21M | 1000 | 1.000M | | | | | |
| 53 | 0.00070 | 0.0190 | 0.0038 | 22.9 | 0.00085 | 667k | 15.27M | 1176 | 1.384M | | | | | |
| 54 | 0.00062 | 0.0158 | 0.0030 | 29.1 | 0.00075 | 859k | 25.00M | 1333 | 1.777M | | | | | |
| 55 | 0.00055 | 0.0132 | 0.0024 | 37.0 | 0.00070 | 1.090M | 40.33M | 1429 | 2.041M | | | | | |
| 56 | 0.00049 | 0.0111 | 0.0019 | 46.6 | 0.00065 | 1.380M | 64.31M | 1538 | 2.367M | | | | | |
| Units | Inches | Amperes | Pounds | Ω/ft | Inches | ft/lb | Ω/lb | $T_{pi}$ | $T_{pi}^2$ | Inches | ft/lb | Ω/lb | $T_{pi}$ | $T_{pi}^2$ |

Maximum ODs. Ω/ft and ft/lb are taken from Materials and Processes Handbook. Maximum rated current is 10% of the fusing current at 20°C. Maximum tension is based upon a tensile strength of 10,000 PSI. Ω/lb is derived from (Ω/ft) × (ft/lb). $T_{pi}$ = 1/Max OD. $T_{pi}^2$ = (1/Max OD)$^2$.

## Table 4-11   Core-Geometry Selection Criteria

| Selection Factor | E-U-I Cores | Pot Cores | Toroid Cores | Slug Cores | Other Shielded Cores |
|---|---|---|---|---|---|
| Core Cost | Low | High | Low | Low | High |
| Winding Cost | Low | Low | High | Low | Low |
| Winding Flexibility | Excellent | Fair | Good | Good | Fair |
| Mounting Flexibility | Good | Good | Fair | Fair | Good |
| Shielding | Fair | Excellent | Good | Poor | Good |

Next, compute the percent change in inductance $L$

$$\%L = [(L_m/L_o) - 1]\ 100 \tag{4-42}$$

Don't let the inductance change more than 25%. Next, compute $P_t$, the total power dissipated by the complete inductor—both the winding and the core. First determine the power dissipated in the windings, $P_w$.

The $d_i$ used for the magnet wire when finding the minimum-sized core is the minimum for the $I_r$. However, the $W_a$ of the actual core may be able to accept a wire with a large diameter and thereby lower the windings' power loss. The maximum diameter for the insulated wire is related to the winding density in turns per inch, $tpi^2$, by

$$d_i = \sqrt{1/tpi^2} \tag{4-43}$$

The winding density must conform to

$$tpi^2 = 6.452\ N/(kW_a) \tag{4-44}$$

Select a wire with the best size for lowest DC resistance per unit length, $p$. The DC winding resistance, for a particular mean length per turn, $\ell_{tm}$, is found with

$$R_{DC} = N\ell_{tm}p \tag{4-45}$$

Compute $P_w$, using $R_{DC}$, the reference temperature $T_o$, the wire's thermal coefficient of resistance ($r$), and the $T_m$ and $I_r$:

$$P_w = I_r^2 R_{DC} [1 + r(T_m - T_o)] \qquad (4\text{-}46)$$

If $T_o$ is 20°C, $r$ is 0.00393; if $T_o$ is 25°C, $r$ is 0.00385.

Now, determine the core loss, $P_c$, from the data for core loss per unit volume, $C_v$, which most core manufacturers either graph or tabulate. If necessary, one can extrapolate the value of $C_v$ for particular conditions in a given design:

$$P_c = C_v V_c \qquad (4\text{-}47)$$

Finally, the total power dissipated by the inductor is

$$P_t = P_w + P_c \qquad (4\text{-}48)$$

Since the inductor hasn't been fabricated yet one can only estimate its maximum operating temperature. Surface dissipation, $S_d$, is the wide-ranging variable that makes determining the $T_m$ fuzzy. The many material variations plus nonuniformity in construction and processing can significantly alter $S_d$. But, a value for $S_d$ that is close enough for a first-order estimate can at least be calculated.

Find $S_d$ at the ambient temperature, $T_a$, by

$$S_d = 0.0014 + 1.217 \times 10^{-6} T_a^{1.585} \qquad (4\text{-}49)$$

With the total surface area, $A_s$, and $S_d$, approximate thermal conductance:

$$G_t = A_s S_d \qquad (4\text{-}50)$$

Then the temperature rise:

$$T_r = P_t / G_t \qquad (4\text{-}51)$$

and finally the crucial variable, $T_m$:

$$T_m = T_a + T_r \qquad (4\text{-}52)$$

If the $T_m$ gets too high, reduce it by increasing $A_s$, $R_{DC}$, or both.

The design sequence just presented can be summarized by the 11 steps listed below.

1. Determine the total current.
2. Select the core material.
3. Determine the optimum flux density and magnetic-field strength.

4. Select the minimum-sized magnet wire.
5. Determine the minimum effective volume.
6. Select the minimum-sized core.
7. Determine the number of turns.
8. Determine the air gap.
9. Check the incremental effect.
10. Determine the total power loss.
11. Determine the maximum operating temperature the inductor must withstand.

A practical example is now presented integrating all of these steps. The objective is to design an inductor meeting the following requirements:

- $L_o$ = 5.6 mH at 0 A DC.
- $L_m$ = 4.7 mH minimum at 1.4 A pk and 20 kHz with a 34% duty cycle.
- $T_r$ = 60°C maximum at 20°C ambient, therefore $T_m$ = 80°C.
- $I_{DC}$ = 0.
- The inductor must be mountable on a PC board and shielded.
- The inductor must be small.

*Step 1:* Determine the total current. Because $I_{DC}$ is zero, from Eq. 4-15.

$$I_m = 1.4 + 0$$

$$= 1.4 \text{ A}$$

For a 34% duty cycle at 20 kHz,

$$t_{pr} = 1/20 \times 10^3$$

$$= 5 \times 10^{-5} \text{ s}$$

and

$$D_p = 0.34 \times 5 \times 10^{-5}$$

$$= 1.7 \times 10^{-5} \text{ s}$$

Assuming that $I_c$ is 10% of $I_m$,

$$I_c = 0.1 \times 1.4$$

$$= 0.14 \text{ A}$$

Although not specified in this design, $E_{pk}$ is always of interest, so from Eq. 4-19

$$E_{pk} = (1.4 - 0.14)4.7 \times 10^{-3}/(1.7 \times 10^{-5})$$

$$= 348 \text{ V pk}$$

From Eq. 4-21

$$I_{eff} = \sqrt{\frac{1.7 \times 10^{-5} [1.4^2 + (1.4 \times 0.14) + 0.14^2}{3 \times 5 \times 10^{-5}}}$$

$$= 0.497 \text{ A}$$

And from Eq. 4-16

$$I_r = 0.497 + 0$$

$$= 0.497 \text{ A}$$

*Step 2:* Select the core material. Choose F2000 ferrite for the core material because, as one can see from Table 4-9, it has the greatest operating flux density. From the table,

$$B_s = 4400 \text{ G}$$

$$B_r = 1500 \text{ G}$$

And from Table 4-10,

$$U_i = 2000 \text{ G}$$

$$T_m = +125°C$$

From the manufacturers' data, at 2900 G and 20 kHz.

$$C_{vf} = 12 \ \mu\text{W/cm}^3/\text{Hz}$$

Therefore, from Eq. 4-22

$$C_v = 12 \times 10^{-6} \times 20 \times 10^3$$

$$= 0.24 \text{ W/cm}^3$$

*Step 3:* Determine the optimum $B$ and $H$. From Eq. 4-22,

$$B = 4400 - 1500$$
$$= 2900 \text{ G}$$

From Eq. 4-25,

$$U_m = 2000 \times 4.7 \times 10^{-3}/5.6 \times 10^{-3}$$
$$= 1679 \text{ G/Oe}$$

From Eq. 4-26,

$$H = 2900/1679$$
$$= 1.73 \text{ Oe}$$

*Step 4:* Select the minimum-sized magnet wire. With an 80°C value for $T_m$ in Eq. 4-29

$$I_{r0} = 0.497[1 + 0.00393 (80 - 20)]$$
$$= 0.614 \text{ A}$$

From the data in Table 4-10, the thinnest wire that can accommodate $I_{r0}$ is AWG 33, whose

$$p = 0.206 \ \Omega/\text{ft}$$

and

$$d_i = 0.0088 \text{ in}$$

*Step 5:* Determine the minimum effective volume. From Eq. 4-22,

$$V_{em} = 0.4\pi \times 10^8 \times 4.7 \times 10^{-3} \times 1.4^2/(2900 \times 1.73)$$
$$= 231 \text{ cm}^3$$

To provide this effective volume, an ungapped core would have to fill a 6.13-cm cube—much too large for mounting on a PC board. Use a gapped core instead; and since shielding is required, make it a pot core.

*Step 6:* Select the minimum-sized core. From Eq. 4-33,

$$W_a A_c = \frac{5.067 \times 10^8 \times 4.7 \times 10^{-3} \times 1.4 \times 0.0088^2}{0.8 \times 2900}$$

$$= 0.1113 \text{ cm}^4$$

A search through core catalogs reveals the smallest standard pot core that one can use is $22 \times 13$ mm. For this core

$$A_c = 0.63 \text{ cm}^2$$

and

$$W_a = 0.292 \text{ cm}^2$$

Therefore,

$$W_a A_c = 0.292 \times 0.63$$

$$= 0.1840 \text{ cm}^4$$

which is large enough. From the catalogs, one obtains the following additional core data:

$$\ell_c = 3.15 \qquad \qquad W_o = 0.0453 \text{ in.}$$

$$V_c = 2 \text{ cm} \qquad \qquad \ell_{tm} = 0.145 \text{ ft}$$

$$A_s = 18.02 \text{cm}^2$$

*Step 7:* Determine the number of turns. From Eq. 4-34,

$$A_{Lm} = (2900 \times 0.63)^2 \times 10^{-16}/(4.7 \times 10^{-3} \times 1.4^2)$$

$$= 3.623 \times 10^{-8} \text{ H/turns}^2.$$

Then from Eq. 4-35

$$N = \sqrt{4.7 \times 10^{-3}/(3.623 \times 10^{-8})}$$

$$= 360 \text{ turns}$$

*Step 8:* Determine the air gap size. From Eq. 4-36,

$$\ell_e = 0.4\pi \times 360 \times 1.4/1.73$$

$$= 366 \text{ cm}$$

Then, from Eq. 4-38

$$\ell_g = (366 - 3.15)/2000$$

$$= 0.182 \text{ cm}$$

$$= 0.072 \text{ in}$$

The air gap interrupts the magnetic circuit twice, so the spacer should be 0.036 in. thick—half the computed value of $\ell_g$.

Check $V_e$ in Eq. 4-31:

$$V_e = 2 + (0.63 \times 2000 \times 0.182)$$

$$= 231 \text{ cm}^3$$

Therefore, $V_e$ complies with the $V_{em}$ required by Eq. 4-32.

*Step 9:* Determine the incremental effect. From Eq. 4-34,

$$A_{Lo} = 0.4\pi \times 10^{-8} \times 0.63/[(3.15/2000) + 0.182]$$

$$= 4.31 \times 10^{-8} \text{ H/turn}$$

Then, from Eq. 4-41,

$$L_o = 4.31 \times 10^{-8} \times 360^2$$

$$= 5.59 \text{ mH}$$

which is close enough to the required 5.6 mH. If $L_o$ isn't close enough, change the turns ratio appropriately—a simple process using Eq. 4-41.

*Step 10:* Determine the total power loss. From Eq. 4-44,

$$tpi^2 = 360/(0.8 \times 0.0453)$$

$$= 9934 \text{ turns/in}^2$$

From the wire data in Table 4-10, use AWG 32 heavy-film wire instead of AWG 33. Then,

$$p = 0.162 \ \Omega/\text{ft}$$

From Eq. 4-45

$$R_{DC} = 360 \times 0.145 \times 0.162$$

$$= 8.46$$

From Eq. 4-46,

$$P_w = 0.497^2 \times 8.46 \, [1 + 0.00393 \, (80 - 20)]$$

$$= 2.54 \text{ W}$$

From Eq. 4-47,

$$P_c = 0.24 \times 2$$

$$= 0.48 \text{ W}$$

Note that the core loss is much smaller than the winding loss. From Eq. 4-48.

$$P_t = 0.48 + 2.54$$

$$= 3.02 \text{ W}$$

*Step 11:* Determine the maximum temperature. From Eq. 4-49,

$$S_d = 0.0014 + 1.2717 \times 10^{-6} \times 25^{1.585},$$

$$= 0.0016 \text{ W/cm}^2/°C$$

From Eq. 4-50,

$$G_t = 0.0016 \times 18.02$$

$$= 0.02883 \text{ W/°C}$$

From Eq. 4-51,

$$T_r = 3.02/0.02883$$

$$= 105°C$$

From Eq. 4-52,

$$T_m = 20 + 105$$

$$= 125°C$$

This value of $T_m$ is too high. The open construction of the proposed inductor doesn't have enough surface to limit $T_r$ to 60°C max. From Eq. 4-51, the minimum thermal conductance required is

$$G_t = 3.02/60$$

$$= 0.0503 \ W/°C$$

Therefore, from Eq. 4-50, the minimum $A_s$ required is

$$A_s = 0.0503/0.0016$$

$$= 31.43 \ cm^2$$

The maximum diameter and height of the pot core are 0.866 and 0.536 inches, respectively. So one should be able to encapsulate it, with thermally conductive epoxy, in a round or rectangular plastic shell. The nearest suitably sized round shell has an outside diameter of 1.187 inches, a height of 0.75 inches, and a thickness of 0.03 inches. Therefore,

$$A_s = 5.01 \ in.^2$$

$$= 32.32 \ cm^2.$$

Combining Eqs. 4-50, 4-51 and 4-52,

$$T_m = 20 + 3.02/0.0016 \times 32.32$$

$$= 20 + 58$$

$$= 78°C$$

which meets the required design goal.

# 5

# Power Switch
# Considerations

## INTRODUCTION[1]

The rapid evolution occurring in power-switching device technology presents many challenges to, and opportunities for, the circuit or systems designer. The goal of a designer—to obtain the most satisfactory solution for the least total investment—is greatly complicated by these continual changes. Not only is the performance of bipolar and power MOS transistors improving, but new classes of devices such as MOS-gated four-layer devices and structures that combine power MOS and bipolar transistors are becoming available.

One basis for comparing power switching devices is shown in Figure 5-1, which compares devices as a function of power-handling capability and operating frequency. Figure 5-2 provides additional information, showing the current and voltage handling capability of these devices.

The large number of discrete power-switching devices indicates both the breadth and the rate of change in power electronics. The combination of majority and minority carrier devices promises to provide devices with a continuous spectrum of current, voltage, and switching characteristics. Table 5-1 compares the electrical parameters of a variety of power switching devices.

Power-switching device technology will continue to evolve as electrical and thermal limits of materials are better understood, and as a result, this evolution will provide designers with solutions to continually improve system performance.

## BIPOLAR POWER TRANSITORS

Bipolar power transistors needed for power supplies are special types and are critical components when considering the design of any power supply. Because

---

[1]Portions of this section used with permission from "Discrete Semiconductor Switches: Still Improving," by A. Cogan and R. A. Blanchard, *PCIM* January 1986.

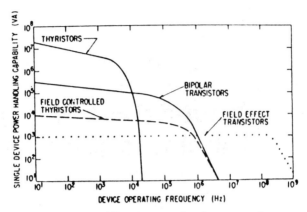

Figure 5-1   Power handling capability vs. operating frequency of power switching devices. (Courtesy Intertec Communications, Inc.)

the transistors produce square waves (push-pull power amplifiers), they must switch very quickly yet handle large amounts of power and have high collector current and breakdown voltage ratings.

Since efficient transistor operation is a requirement for power supply use, the saturation voltage must be low, but the collector-emitter breakdown voltage ($BV_{CEO}$) should be high for a push-pull square-wave circuit (approximately 2 × $E_{in}$ + 20 VDC).

$H_{FE}$(beta) is a parameter that varies with temperature, collector current, and from transistor to transistor due to manufacturing variations. Thus, the variation of beta must be compensated for by methods such as diode stabilization or use of resistors or thermistors.

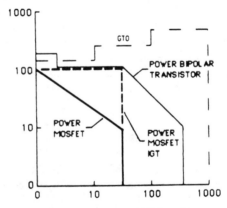

Figure 5-2   The current and voltage handling capabilities of commercially available power switching devices. (Courtesy Intertec Communications, Inc.)

## Table 5-1   A Comparison of Power Switching Device Characteristics (Courtesy Intertec Communications, Inc.)

| Device Characteristic | Power Bipolar Transistor | Gate Turn-Off Thyristor | Power MOSFET | Power JFET | Power MOS-IGT |
|---|---|---|---|---|---|
| Normally On/Off | Off | Off | Off | On | Off |
| Reverse Blocking Capability (volts) | < 50 V | 500–2500 V | 0 V | 0 V | 200–2500 V |
| Blocking Voltage Range (volts) | 50–500 V | 500–2500 V | 50–500 V | 50–500 V | 200–2500 V |
| Forward Conduction Current Density (A/cm$^2$) | 40 | 200 | 10 | 4 | 200 |
| Surge Current Handling Capability | 3x | 10x | 5x | 5x | 5x |
| Maximum Switching Speed | 50 kHz | 10 kHz | 20 MHz | 200 MHz | 50 kHz |
| Gate Drive Power | High | Medium | Low | Low | Very-Low |
| $dv/dt$ Capability | Medium | Low | High | High | High |
| $di/dt$ Capability | Medium | Low | High | High | High |
| Radiation Tolerance | Poor | Very Poor | Moderate | Good | Moderate |

Because a push-pull amplifier or oscillator is symmetrical, it stands to reason that optimum performance is obtained by using symmetrical or matched transistors. Most power transistors when used in push-pull aerospace power supplies are matched for $H_{FE}$ (beta), collector leakage current ($I_{CBO}$), collector-emitter breakdown voltage ($BV_{CEO}$), $V_{CE}$ (sat)-collector-emitter saturation voltage, rise times, and fall times. This matching ensures closer to balanced operation of the transistors, that is, the average power dissipation and junction temperatures are more evenly distributed.

The storage time of a power transistor is another important parameter. Transistors with small storage times are required to keep power dissipation low. In a push-pull inverter stage the on transistor stays on because of storage time while the off transistor is forward-biased and conducting collector current. This collector current is determined primarily by base current times beta. $V_{CE}$ is at its maximum at this time; therefore, transistor dissipation is extremely high until the stored charge is cleared in the opposite transistor. Since storage time is directly proportional to overdrive, these losses are even greater at lighter loads than at the rated load of the converter. These high collector currents also cause excessive voltage spikes at switchover which could destroy the transistors. For these reasons, it is desirable and necessary to reduce storage time and/or eliminate the losses due to storage time.

In power switching, the critical speed parameter is fall time ($t_f$). Although high turn-off drive current ($I_{b2}$) shortens $t_f$, excessive turn-on base drive ($I_{b1}$) lengthens it. Therefore, in switching applications, the base drive should be just enough to drive the transistor into saturation at a specified collector current.

A low $V_{CE(sat)}$ spec is desirable for efficient switching, but it must be specified under normal $I_{b1}$ drive conditions. A long $t_f$ is a major cause of power loss, which can easily cancel out any savings from an artificially lowered $V_{CE(sat)}$. Clearly, then, one should provide only just enough $I_{b1}$ to minimize saturation losses, but have high enough $I_{b2}$ to get a short $t_f$.

Another aspect of turn-off drive current that is often overlooked is its effect on the reverse-bias SOA (Safe Operating Area).

To get high $V_{CEO}$ in power transistors, current gain ($h_{fe}$) is usually sacrificed. To get both properties with high capabilities requires high-resistivity silicon material. But then current capability and power efficiency are drastically reduced.

Thus, trade-offs must be made based on the intended application and its attendant power semiconductor electrical parameter specification requirements.

Bipolar power transistors, though widely used as the power switch in DC/DC converters, suffer from two very important shortcomings. (1) complex drive requirements and (2) relatively long turn-off times. The problem in driving bipolar power transistors is that they are current amplifiers. This means that the driving circuit must supply power to the transistor, and as the transistor's output current increases, so too must the power supplied by the drive circuit. While designing drive circuits with the required power capacity is not a problem, it is an annoyance and adds to the cost of the converter.

Turn-on and turn-off times have not historically been a problem. But with high-frequency switching, they have become critical issues. In switching applications, the bipolar power transistor operates as a bistable device, being either on or off. When it is turned on, the transistor is fully saturated. Theoretically, this means that except for the small amount of power dissipated in the junction as a result of the forward junction voltage, no power is dissipated in the transistor. At the other extreme, the transistor is cut off, and since no current flows through it, there is no power loss. It is only during the period in which the bipolar transistor is switching between its two states that any appreciable power is dissipated. Thus, for given turn-on and turn-off times, also called "rise and fall times," there will be an instantaneous power loss.

At low frequencies, the turn-on/turn-off times represent only a small portion of the switching cycle time. As a result, the switching power losses represent only a small percentage of the power being handled. But as frequency rises, the turn-on and turn-off times, which are fixed by the transistor's geometry, become an increasingly larger percentage of the switching cycle time. Because of this, the power losses represent an increasingly larger percentage of the power being handled. Consequently, the overall efficiency of the converter is significantly reduced. This lower efficiency translates into a need for higher dissipation ratings for the transistors and increased thermal management problems.

Transition times not only affect efficiency, they also limit the maximum frequency at which a transistor can be used. The off-transition time of a bipolar transistor, moreover, includes not only the turn-off time, but also storage time— the time required for the carriers that facilitate conduction to migrate out of the base region. Storage time and turn-off time combine to determine how long it takes a bipolar transistor to complete its turn off. The transition times of today's bipolar power switching transistors are much shorter than those of previous generations of devices, but the fastest are still hard pressed to function in a power converter at frequencies above 100 kHz.

## Power Transistor Specifications[2]

For a user to completely specify a power transistor, several factors which must be considered are:

1. All maximum electrical and thermal stresses that can occur in the circuit.
2. All electrical parameters and conditions that affect the performance.
3. The capabilities of the transistor.
4. The interdependence of certain transistor parameters.

The performance that various circuits demand of a transistor can be evaluated with respect to voltage, current, switching time and power dissipation. Other, less obvious characteristics such as thermal runaway and peak turn-off-energy must also be considered.

### Sustaining Voltage

Inductive switching circuits (as shown in Figure 5-3) typically create a great variety of voltage stresses. For instance, in a switching power supply, the bias conditions on the base of the transistor are continually changing during the transition from on to off.

This is caused by removal of stored base-charge coupled with various well-known base drive techniques that reduce storage time, current fall-time and voltage rise-time.

---

[2]High Voltage Transistors for Switching Circuits by W. R. Peterson, *Proceedings of Power Con II*, October 1975.

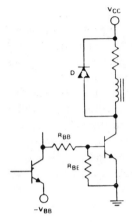

Figure 5-3   Inductive switching circuit.

Figure 5-4 shows inductive switching conditions during turn-off of the circuit shown in Figure 5-3. The greatest voltage that the transistor will see occurs just prior to the turn-on of the clamp-diode D at point A. However, the reverse base current $I_{B2}$ (of which the sustaining voltage is a function) is still near its peak value. Thus, the $V_{CEX}$(sus) rating should exceed the voltage at point A and should include $-V_{BB}$ and $R_{BB}$ when specified.

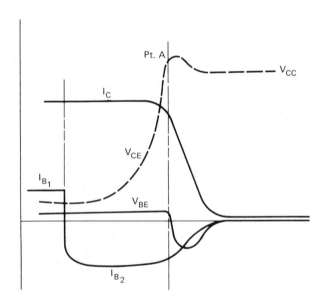

Figure 5-4   Inductive switching conditions during turn-off of the circuit shown in Figure 5-3.

If $-V_{BB}$ were to be applied to the base via $R_{BB}$ for the duration of the off cycle, no other sustaining voltage rating would be required. However, $-V_{BB}$ might be removed shortly after turn-off leaving $R_{BE}$ as the only path for $I_{B2}$. Here, the $V_{CER}(\text{sus})$ rating should exceed $V_{CC}$ and include $R_{BE}$ when specified.

Usually, the sustaining region is only encountered when an inductive load line sweeps to the sustaining value or a spike lifts the collector voltage above the blocking value.

## Leakage Current

Leakage current is normally specified at a voltage which exceeds the supply voltage plus foreseeable spike voltages. If $-V_{BB}$ is applied to the base during turn-off, $I_{CEX}$ should be specified along with $-V_{BB}$. Otherwise, $I_{CER}$ should be specified with $R_{BE}$.

Since $MI_{CBO}$ is dependent upon $V_{CC}$, the leakge current is a function of the base-bias conditions. $I_{CBO}$ is also temperature dependent, so that leakage ($I_{CEX}$, $I_{CER}$) should be specified at the maximum operating junction temperature.

Another implication of leakage current is the resultant power dissipation. For instance, with $V_{CC} = 400$ V and $I_{CEX} = 1.0$ mA, dissipation due to leakage at a 50% duty cycle is 200 mW (avg.). $I_{CEX}$ is usually small but can be significant if it is not controlled by the specification.

## Collector Current Fall Time

Traditionally, collector-current fall time $t_f$, is a major power transistor figure of merit for both designers and manufacturers. In switching applications, a major portion of a power transistor's losses occur during $t_f$. During $t_f$ the collector voltage is 90% of the supply (or clamp) voltage or higher, and the collector current falls from 90% of its maximum to 10%. Consequently, the power dissipated in the transistor is substantial (see Figure 5-5).

But often ignored is the "tail" that some transistors exhibit on the collector-current waveform, before the collector current falls to zero. The tail's duration, $t_t$, usually is measured between the 10% and 2% points. Energy loss during this interval can be significant, if the tail extends a long time, since full voltage appears between collector and emitter. Unfortunately, tail time is rarely mentioned on data sheets.

In addition, the emitter-collector voltage rise time, $t_r$, at turn-off also can contribute considerable loss, but this time usually is short. And since $t_r$ varies little among different power transistors, and depends heavily on the circuit it isn't usually cited as a figure of merit.

For maximum safety, therefore, include all the turn-off times—$t_f$, $t_t$ and $t_r$— in an over-all timing specification for your power transistors, from the 10% point of the collector-emitter voltage to the 2% point of the tail.

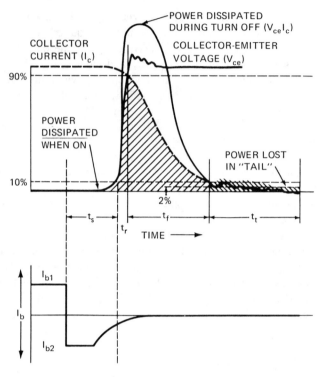

Figure 5-5   Power dissipated during transistor turn-off.

## Turn-Off Power Dissipation

There are two major sources of power dissipation in most switching circuits, saturation and turn-off losses.

Figure 5-6 shows typical turn-off waveforms with resistive and inductive loads. Since the rise-time of the voltage is equal to the fall-time of the current, the resistive turn-off energy dissipation is readily calculated from the equation:

$$E_R = \frac{I_c V_{CC} t_f}{6} \tag{5-1}$$

Total turn-off dissipation is then $P_{SW} = E_R \times f$. In contrast to resistive turn-off dissipation which is predictable from data sheet information, inductive turn-off dissipation is seldom symmetrical or predictable from present data sheet ratings. Since the power turn-off waveform is equal to $V_{CE} \times I_C$, its shape will necessarily follow the rising $V_{CE}$ ($I_C$ remains constant) and the falling $I_C$ ($V_{CE}$ remains constant) waveforms. It can be safely approximated by a triangular

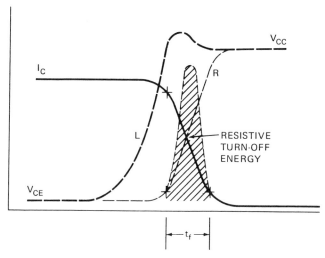

Figure 5-6  Typical turn-off waveforms with resistive and inductive loads.

waveshape. The peak value is approximately 80% of $V_{CE}I_C$ and the base of the triangle can be stated as the time from 10% of $V_{CC}$ to 10% $I_C$. This is shown in Figure 5-7 as $t_{VI}$. Thus:

$$E_L = 0.8 V_{CC} I_C \tfrac{1}{2} t_{VI} \quad \text{or} \quad E_L = 0.4 I_C V_{CC} t_{VI}. \tag{5-2}$$

Total turn-off dissipation is then

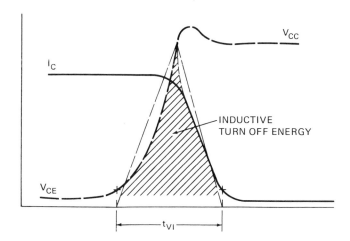

Figure 5-7  Waveforms showing inductive turn-off energy approximation.

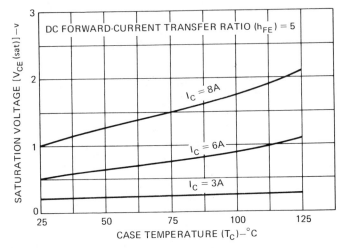

Figure 5-8   Case Temperature vs. $V_{CE(sat)}$.

$$P_{SW} = E_L \times f. \tag{5-3}$$

It is interesting to note that if $t_{VI} = 1.25\ t_f$, then $E_L = 3E_R$, indicating that inductive turn-off dissipation can be much larger than resistive turn-off dissipation. From Figure 5-8 it can be seen that $t_{VI}$ at 100°C is about twice the value at 25°C. This has been found to be true for many high voltage switching transistors from different technologies. Most of the increase with temperature is in the rise-time of $V_{CE}$.

There are several advantages that can be provided by $t_{VI}$ as a rated parameter. The first is that its measurement as a final factory test allows a limit to be established and elimination of the transistors with higher turn-off dissipation. The second is that the user can select transistors with a knowledge of their turn-off dissipation. A third is that a curve of $t_{VI}$ vs. case temperature, when combined on the data sheet with a curve of $V_{CE}$(sat) vs. case temperature, can be used to calculate the conditions that are necessary for thermal instability.

### Thermal Resistance vs Voltage

The voltage sensitivity of thermal resistance was known in early germanium days, but was not emphasized because of adequate safety margins. It was therefore not shown or implied on the data sheets. High voltage silicon power transistors, on the other hand, have significant thermal resistance change with voltage, and the change is evident in the DC margin of the safe operating graph. Figure 5-9 shows a safe operating graph of the DTS-424 transistor that is representative of high voltage silicon transistors. It can be seen that the DC margin

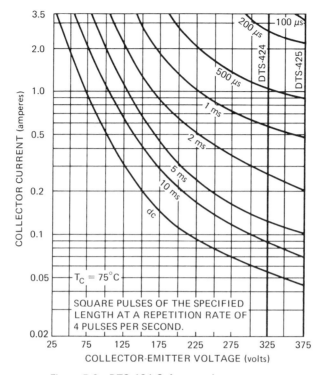

Figure 5-9   DTS-424 Safe operating curves.

ranges from 100 W at 40 V to 18 W at 300 V. This margin was established experimentally and was based upon the assumption that junction temperature is constant. Apparent thermal resistance therefore ranges from 0.75 °C/W to 4.2 °C/W, as shown in Figure 5-10. If a linear regulator for a variable power supply is required to operate with 200 V across the series pass element, the 200 V thermal resistance would be essential in calculating junction temperature. In this case, the 200 V thermal resistance would be 3.0°C/W.

## Thermal Stability

Curves of $t_{VI}$ and $V_{CE}$(sat) are shown in Figures 5-8 and 5-11, for a high voltage switching transistor at particular current and base drive conditions.

The classical equation for thermal stability is that:

$$\frac{\partial Pd}{\partial T} \leq \frac{1}{R\theta_{JA}}$$

(5-4)

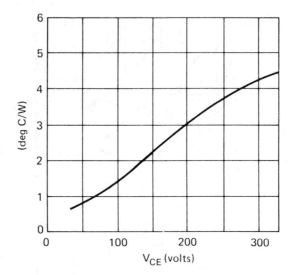

Figure 5-10   Thermal resistance variation

where $R\theta_{JA}$ is the thermal resistance from the transistor junction to the ambient atmosphere.

The contribution of $t_{VI}$ can be determined from:

$$\frac{\partial P_{SW}}{\partial T} = 0.4I_C \times V_{CC} \times f\frac{\partial t_{VI}}{\partial T} \qquad (5\text{-}5)$$

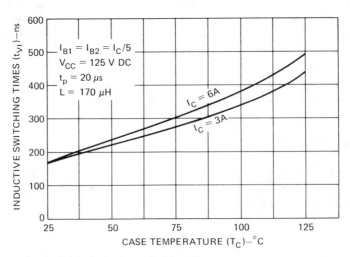

Figure 5-11   Inductive switching time vs. case temperature.

where $\partial t_{VI}/\partial T$ can be taken from Figure 5-8 at the design temperature. Likewise, the contribution of $V_{CE}(\text{sat})$ can be determined from:

$$\frac{\partial P_{on}}{\partial T} = 0.5 I_C \frac{\partial V_{CE}(\text{sat})}{\partial T} \tag{5-6}$$

and can be taken from Figure 5-11 at the design temperature. This assumes a maximum 50% duty cycle for half-bridge or push-pull converters, but in switching regulators and other inductive applications the duty cycle can be greater.

Thus, if the calculation shows that the dissipation increases faster with temperature than the heatsink can remove it; i.e.,

$$\frac{\partial Pd}{\partial T} \geq \frac{1}{R\theta_{JA}} \tag{5-7}$$

then some modifications must be made to reduce $R\theta_{JA}$ or reduce $\partial Pd/\partial T$.

### Understanding Device Power Ratings

Dissipation ratings are based upon thermal resistance and maximum allowable junction temperature. Derating of dissipation is based upon the approximate inverse of thermal resistance, with 100% derating applicable at the maximum junction temperature, as shown in Figure 5-12. Maximum junction temperature is usually determined experimentally in storage and reverse bias life tests with the criterion being leakage stability. Satisfactory dissipation levels in the customer's circuit depend upon heat sinking, ambient temperature, and fail rate goals. Life tests and field data above 25°C show that failure rates double for $T_J$ increments of approximately 20–25°C.

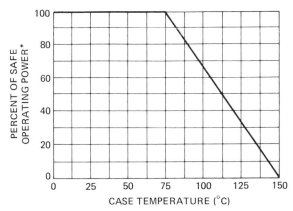

Figure 5-12   Safe operating power vs. case temperature.

For a complete understanding of power transistors, the designer must know how to interpret and use the device power ratings. These include maximum steady-state and peak power ratings, as well as the maximum allowable junction temperature.

The average or steady-state power dissipated in a transistor is:

$$P_D = I_B V_{BE} + I_C V_{CE} \tag{5-8}$$

which is the sum of the base and collector power. This expression can be better understood by writing it as the sum of the power dissipated in each junction:

$$P_D = V_{BE}(I_B + I_C) + V_{CB}(I_C) \tag{5-9}$$

If the maximum allowed junction temperature, $T_J$, is specified (100°C for germanium and 150°C for silicon), the maximum allowable DC power can be found from

$$P_{D(max)} = (T_{j(max)} - T_A)/\theta_{JA} \tag{5-10}$$

If the maximum temperature is exceeded, the transistor goes into secondary breakdown. Thus, both safe operating areas and allowable power limits must be observed when designing a power circuit.

The formulas are somewhat more complex for the transient case because the pulse power dissipation effects now depend on the thermal capacitance as well as thermal resistance.

## Secondary Breakdown

Transistors often fail during switching even though analytical calculations and predictions, such as power dissipation, based on parameters given by the manufacturer's data sheet should guarantee safe operation. This discrepancy arises because the majority of information contained in the data sheet is based on static measurements. Absolute ratings of current, voltage, power, and temperature may be specified, but the device may not be able to withstand these conditions if they occur simultaneously.

Transistor failures that occur for no apparent reason are a source of frustration for the design engineer. An analysis of the damaged transistor(s) seldom yields the exact failure mechanism, but most of these unexplained failures can be traced to second breakdown.

One of the most critical limitations of a transistor is the magnitude of voltage to which it can be subjected. The voltage at which a transistor breaks down is a function of both the individual transistor characteristics and its associated circuitry. None of the basic breakdown modes or any mechanism of voltage breakdown will, by itself, damage the transistor. It is the power dissipation and heat developed by the high current flow under breakdown conditions that result in permanent degradation of the transistor characteristics.

Voltage breakdown characteristics are most important in circuits which have inductive loads because the high peak powers that are generated influence both the transistor mode of operation and the system reliability.

If operation in the primary-voltage-breakdown mode is allowed to continue for too long, a thermal-runaway condition occurs resulting in the transistor entering second breakdown. Second breakdown is a major limiting factor in the power performance and power handling capability of a transistor.

The problem of second breakdown was first realized in 1958, even though it was existent since junction transistors were first made. Much progress has been made on a practical basis toward the solution of the problem. Second breakdown cannot be completely eliminated, but design steps must be taken to extend it beyond the normal operating range of the transistor.

### Characteristics and Causes of Secondary Breakdown

There exist three primary factors that cause collector current instability in transistors: They are temperature, voltage and time.

Secondary breakdown (SB) can occur in all transistors, and is known to be a function of the material from which the device is made. This is the true maximum rating of a power transistor. The symptoms of SB result in a derating of maximum allowable collector voltage as a function of collector current. When SB occurs it usually results in a catastrophic failure of the transistor.

Second breakdown is associated with the collector-to-base junction, and is controlled by the emitter-to-base junction. Second breakdown is triggered at low collector voltages as the base drive voltage is allowed to go from positive to negative in a pnp transistor, during which time the collector current increases.

Secondary breakdown is a condition that makes the transistor output impedance change instantaneously to a small positive value. This is shown in Figure 5-13 for various base-drive conditions.

The most obvious characteristic of secondary breakdown is an abrupt decrease in collector-to-emitter voltage and an increase in collector current. It is similar to avalanche breakdown* with two exceptions:

- The final limiting voltage is much less than the avalanche condition.
- Secondary breakdown is energy-dependent.

---

*Negative resistance region where large changes in current accompany small changes in applied voltage.

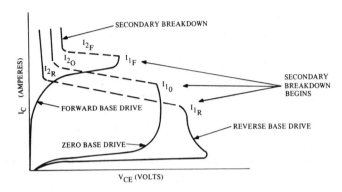

Figure 5-13  Secondary breakdown can occur at various base drives and is energy-dependent.

Figure 5-14 depicts the collector voltage and collector current of a transistor before, during, and after second breakdown occurs.

If conditions are right for SB, then $V_{CE}$ and $I_C$ remain constant during the time period prior to SB, called the delay time. Delay time is a function of the operating current, ambient temperature, and base drive. The delay time decreases as the operating current is increased. As the base drive changes continuously from reverse to zero and on to forward bias, the delay time increases. At the conclusion of this delay time, which may be many milliseconds in length, the voltage drops to a lower value and the transistor goes into SB. The current will have increased to a value determined by a current limiting resistor and the SB voltage-current characteristics of the transistor. There is always a delay time before the collector voltage decreases and the collector current increases.

The delay time that is observed before SB occurs has led to the idea that a certain critical energy, $E_m$, must be dissipated in the transistor before SB can be initiated. This energy exhibits the same kind of dependence on ambient temperature, base drive, and operating current as did the delay time. For the particular base termination used, the integration of the volt-ampere product with time at the boundary where SB occurs is the energy limit, $E_m$. This energy limit

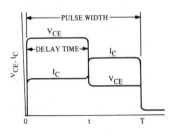

Figure 5-14  $V_{CE}$ and $I_C$ before, during, and after onset of second breakdown.

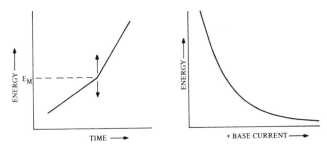

Figure 5-15   Second breakdown energy limit.

is inversely proportional to the reverse base current as shown in Figure 5-15. The rate of change of voltage with time has been observed to be a function of the thermal characteristics of the device, and the final limit of a failure is a function of the collector-to-base diode.

Figure 5-16 shows a sketch of the voltage-current characteristics of a transistor operating in the reverse breakdown mode. Note that at low collector currents, the voltage across the device exceeds the open base breakdown rating.

The peak of the curve and the negative resistance region is the first breakdown, or the normal breakdown, and is a result of avalanche action in the transistor. However, as current in the avalanche mode is increased to higher values, a critical current $(I_m)$ is reached at which the voltage across the device drops to a very low level. This behavior is aptly called Second Breakdown.

It was subsequently discovered that transistors need not be operating in ava-

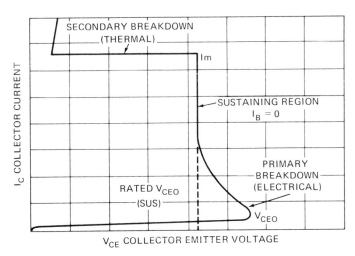

Figure 5-16   Manifestation of second breakdown in a transistor.

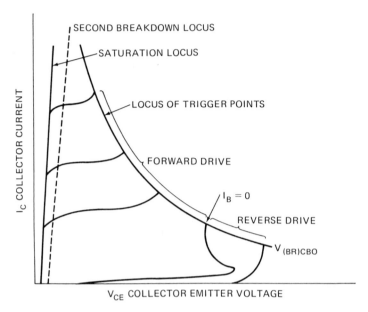

Figure 5-17   Locus of second breakdown trigger points.

lanche breakdown in order to encounter SB. Figure 5-17 shows a family of collector curves and the locus of critical or trigger currents at which the transistor enters SB. Note that as collector voltage is increased, $I_m$ occurs at lower currents and becomes extremely low as the emitter-base junction becomes reverse biased. It has also been found that the amount of time a power pulse is applied at a particular operating point also determines whether or not SB will occur. The observed behavior is a result of "hot spots" forming in the device as a result of nonuniform current density.

The effect of SB is a direct result of excessive current being concentrated in a relatively small area of the semiconductor chip. This effect causes the device to lose the ability to sustain the rated collector-to-emitter voltage with the base open. The voltage applied to the device then drops to a low value and will generally cause the external circuit to apply a very high current to the device; this causes a direct collector-to-emitter short and destroys the device.

The transistor can be protected from SB by restricting the time that the transistor spends in operating conditions where SB is possible. The energy dissipated, $E_m$, refers to a DC condition in which the voltage and current of the transistor are constant. The value of $E_m$ that can be tolerated is highly $V_{CE}$-dependent. As $V_{CE}$ increases, the critical energy ($E_m$) required to cause SB failure decreases. $E_m$ is reduced when the current distribution over the area of the collector junction is altered to produce a localized region of higher current density, and thus a localized region of higher temperature. $E_m$ is also reduced when the transistor case temperature is increased. Increasing the power dissipation for a given base drive decreases $E_m$.

Defects in a transistor can be responsible for the current variations, and these significantly increase the susceptibility of the device to SB by virtue of the reduced $E_m$ for a given operating condition.

Physically, then, secondary breakdown appears to be caused by a local thermal runaway induced by severe current concentrations. These current concentrations may be caused by:

- Defects in the transistor structure.
- Application of a reverse base drive.

That is, when the base current is flowing in the reverse direction, potential gradients may reverse-bias the periphery of the base but forward-bias its center. The injected current from the emitter flows in a small area toward the center of the base and causes a concentration of current forming a hot spot.

Temperatures of such a hot spot are not high enough to cause extensive melting and failure of the device, but they can cause high local currents and thermal generation of carriers. One can expect reduced beta under conditions of hot-spot formation where most of the base current flows into the hot spot. Here the device operates at currents well above those for which the structure is designed. At such high current densities the current gain is well below the peak value obtainable in the device, therefore reducing the overall current gain of the transistor.

The low sustaining voltage that occurs in secondary breakdown characterizes this mode. Operation is sustained by a mesoplasma or a small molten region in the device. Most of the voltage drop is the result of resistance within the microscopic molten region.

## Hot Spots Can be Avoided

Thermal instability, which leads to hot-spot formation, can be demonstrated best by considering the relatively large area of a power device as a number of small areas. For sake of experiment, these small areas can be separated into individual devices on a common heat sink connected in parallel. Due to nonuniformity in diffusions, in geometries, and in thermal structure, not all of the small devices draw identical currents in the parallel system. Those with higher collector currents must dissipate more heat. The higher temperature on these base-emitter diodes would cause them to operate 2 mV/°C lower base-emitter voltage for a given base current, or at a somewhat higher base current in the parallel array. This again means a higher collector current and higher temperatures.

If current to the array is not limited by an external circuit, the device goes into second breakdown or short-circuits itself. The designer, therefore, needs to provide negative feedback within the transistor that counteracts the positive feedback.

### Discrete Emitter Resistors

Individual resistors in series with each of the transistor emitters in a parallel array stabilize the system very effectively. The device that tends to draw the highest current also develops the largest voltage drop in the emitter resistor. This negative feedback voltage subtracts from the input voltage of the device, thereby limiting the amount of collector current that the device will conduct. Inclusion of 10 resistors in series with each of the emitters of an experimental 20-transistor array has resulted in complete stabilization at 40 W, compared to instability at 8 W without resistors. Built-in emitter resistors for the prevention of thermal instability are used in advanced silicon power devices.

### Avalanche Failures

A closer look at the inside of the device helps to explain reverse-biased avalanche failures. Under forward bias, base current within the device flows from the base contact to the region under the emitter. The voltage drop caused by the lateral base current in the sheet resistance of the base tends to "debias" the area under the center of the emitter. Maximum conduction in the device, therefore, occurs close to the emitter periphery. When the base is reverse-biased and the collector voltage is increased toward avalanche, base current caused by leakage and avalanche multiplication is first distributed uniformly over the collector base junction area. Current flowing from the center area of the emitter through the base sheet resistance to the base contact develops a voltage drop. The base voltage is more positive under the center of the emitter and, when large enough, causes the transistor to turn on at this point. The transistor, therefore, conducts at the geometric center of the emitter, which can be a point or at best a line. In reverse-biased avalanche conditions a hot spot forms, leading to failures exactly like those resulting from lateral thermal instability.

### "Pinch-In" Failure

"Pinch-in" (Figure 5-18) occurs when current crowds under the center of the reverse-biased emitter-base junction. In devices constructed with thin, high-resistivity collector regions—such as transistors with thin epitaxial films or triple diffused devices—"pinch-in" can lead to another failure.

With high base currents, such as those that form at the "pinched-in" con-

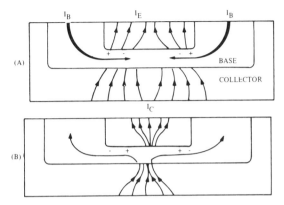

Figure 5-18 (A) Current distribution with forward bias. Here base current in the device flows from base contact to region under the emitter. (B) Pinched-in current distribution with reversed base current.

duction points, the transistor's base width tends to increase. Since the collector width is fixed by the thickness of the high-resistivity region, an increase in base-width decreases the available collector depletion width and therefore lowers the maximum sustaining voltage. As current builds up in the pinched-in conduction spot, the sustaining voltage decreases. If current to the transistor is limited in the external circuit, capacitance discharging into the pinched-in spot may cause a sudden oscillation. This is not the same problem as lateral thermal instability (overheating of the pinched-in conduction spot). Thermal mechanisms are characterized by time constants of the temperature build-up in the device material, and are typically in the microsecond or millisecond range. The time constant associated with base widening is much smaller, usually measured in nanoseconds.

## Preventing "Pinch-In"

The circuit designer can avoid pinch-in failures in two ways. Reverse-biasing a power transistor's base-emitter junction causes the depletion region to extend from the emitter into the base. At high collector voltages, a depletion region also extends from the collector-base junction into the base. The electrical base width is significantly reduced and the sheet resistance of the base is substantially increased over the normal value of several hundred or a few thousand ohms per square. The device thus goes into pinched-in operation at relatively small values of avalanche current. If one avoids reverse-biasing, the emitter-base junction pinch-in at low currents is prevented from occurring. Also, operation above $LV_{CEO}$ cannot occur and power dissipation takes place in the active region of the device.

Clamping the collector voltage to a value of less than $LV_{CEO}$ also prevents failures. If a transistor input is reversed-biased and collector voltage builds up due to discharge of an inductor in the collector circuit, the inductive current is bypassed into the clamp without turning the transistor on.

The designer can also do several things to make the device dissipate power under reverse-biased avalanche conditions. Reducing the sheet resistance of the base under the emitter helps to conduct more collector-base avalanche current before turn-on occurs. Diffusion structures that avoid thin base widths also substantially lessen the danger of pinch-in. Failures caused by the thin, high-resistivity collector regions can be avoided by using thick epitaxial structures, nonepitaxial devices, or triple diffused devices with thick regions of high-resistivity materials. Changes in the lateral geometry of the transistor can minimize the path length of the high sheet resistance base under the emitter, thereby also improving the amount of current that can be conducted before pinch-in occurs. Manufacturers usually supply safe-operating-area curves that indicate the energy ($I_C \times V_{CE} \times t$) limits below which the device will not go into secondary breakdown.

### Switching Inductive Loads

Since secondary breakdown occurs most often when switching inductive loads (such as is the case in most DC/DC converters), this area is examined in detail. Figure 5-19 shows the characteristic output curve of a transistor when the base is reverse-biased.

In region I, the transistor is cut off and has a high resistance. At voltage $V_a$, the transistor exhibits a negative resistance (avalanche breakdown condition). When the current reaches $I_1$, a sharp breakdown occurs, and the characteristics

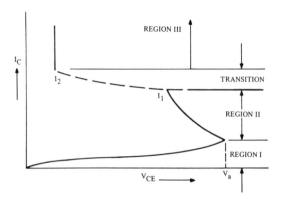

Figure 5-19 Characteristic output curve of power transistor with base reverse-biased pin-points safe operation areas.

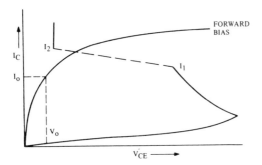

Figure 5-20   Characteristic output curve with base forward-biased.

jump into region III, the low-positive-resistance region. This is referred to as
the second breakdown. If reduced, the current jumps back into region II at the
point $I = I_2$.

On the same set of curves, the forward-bias output characteristic of the transistor can also be plotted (Figure 5-20). $I_o V_o$ is the quiescent operating point of
the device. Point $I_o$ is below breakover point $I_2$. If the transistor is suddenly
reverse-biased and switches off an inductive load, the turnoff path resembles
that in Figure 5-21.

The turnoff path depends on the $L/R$ time constant of the load and transistor,
and the amount of energy that was stored in the inductor. The initial jump in
voltage occurs because the inductor current cannot change instantaneously. For
the case shown in Figure 5-21, no transistor damage occurs if the turnoff path
is within the safe operating area. With a different turnoff path, as in Figure
5-22, the energy stored in the inductive load is so large that the initial jump in
voltage causes the turnoff path to hit the negative-resistance region. If the energy

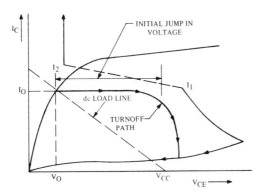

Figure 5-21   Safe turnoff path for inductive load must lie within the forward- and reversed-biased curve.

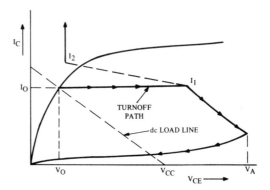

Figure 5-22 Maximum safe turn-off path follows reverse-bias curve. If energy delivered by inductive kick exceeds the transistor dissipation, the transistor will be destroyed.

delivered by the inductor is greater than the maximum energy that can be dissipated by the transistor, the device will be destroyed.

A second condition that can cause catastrophic failure appears in Figure 5-23. In this instance, point $I_o$ is above breakover point $I_2$. When the transistor is turned off, the load line hits the low positive-resistance region. The current then increases until it is limited by the DC resistance of the load. In this state, energy dissipation may again cause transistor destruction. The higher the limiting voltage in the negative-resistance region, the more rapid the transistor destruction. And in any case, destruction usually results in a collector-to-emitter short circuit.

Thus, when operating a power transistor into an inductive load, the inductive kick is often high enough to exceed the voltage rating of the transistor—a problem most troublesome with high-speed switching transistors. Wherever possible, a fast turn-on, free-wheeling diode should be placed across the inductor to absorb

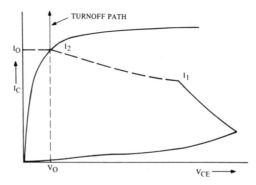

Figure 5-23 Secondary breakdown will occur during the turnoff when the quiescent current $I_o$ is near the positive-resistance region.

the inductive energy, in some cases, an RC network will suffice to keep the voltage spike within the transistor voltage rating. Such measures are not always acceptable, however, because they slow the decay of current through the coil, and in the case of a fast transistor driving a long wire line, the inductor is not accessible. In these cases, the transistor must absorb the inductive energy.

It is important that a transistor that is to be used in this mode of operation have data supplied which guarantees its energy-sustaining ability in the avalanche region. There is no relationship between active region (forward base current) and avalanche region (reverse base current) second breakdown capability; however, it is reported that in the avalanche mode considerably less power is required to destroy the device than in the normal operating mode.

The usual explanation is that under reverse base current, the IR drops in the base are such that severe current concentration occurs near the center of the emitter. The current is thus restricted to a much smaller area than would be the case when operating in the normal active region where current tends to concentrate near the emitter periphery.

It has also been observed that avalanche operation produces a wide variation in SB capability among devices of the same type. Sometimes avalanche data are available on a data sheet and is usually in the form of a plot of maximum allowable inductance as a function of collector current with the electrical conditions at the base as a parameter.

Both blocking voltage and sustaining voltage are important in switch mode applications. The basic push-pull converter requires high blocking capability since the transistor is subjected to a substantially higher voltage than $V_{CC}$ after turn-off.

For inductive loads, high voltage and current must be sustained simultaneously during turn off, in most cases, with the base-to-emitter junction reverse biased. Under these conditions, the collector voltage must be held to a safe level at or below a specific value of collector current. This can be accomplished by several means such as active clamping, RC snubbing, load line shaping, etc. The safe level for these devices is specified as $V_{CEX(sus)}$ at a given high collector current, and represents a voltage-current condition that can be sustained during reverse biased turn-off. This rating is verified under clamped conditions so that the device is never subjected to an avalanche mode.

As shown on the reverse bias SOA curve of Figure 5-24, two voltage levels are specified: One at the maximum continuous current level and one near the recommended operating level so that both normal and fault/transient conditions can be taken into consideration.

In most applications, a large percentage of total device power dissipation occurs during turn-off time and $t_f$ is normally used as a figure of merit. There are, however, two portions of the turn-off waveform that can add losses and in some cases can be significant. (See Figure 5-25). The interval $t_v$ is part of the

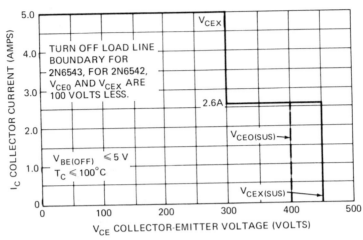

Figure 5-24   Reverse biased safe Operating Area. (Courtesy of Motorola Semiconductor)

total storage time $t_s$ and is defined as voltage switching time. During $t_v$, the $V_{CE}$ voltage changes from saturation to clamp voltage while collector current has only decreased by 10%. The time $t_t$ occurs after fall-time and appears as a "tail" on the collector current waveform. Significant dissipation occurs during the total period $t_v + t_f + t_j$.

Figure 5-26 is a typical example of the safe area information now seen on manufacturer's data sheets. The solid lines show second breakdown limitations while the dotted lines represent thermal limitations. For this transistor family,

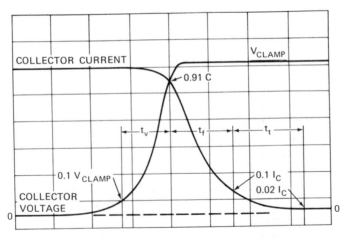

Figure 5-25   Inductive switching turn-off waveform.

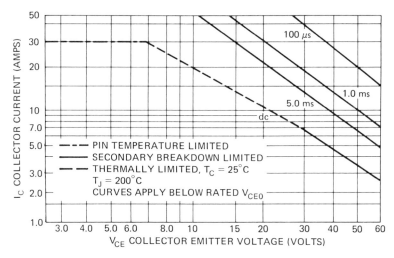

Figure 5-26  Example of an Active Region Safe Operating Area.

DC currents above 30 A cause excessive emitter pin temperatures*; therefore, operation above 30 A DC is not recommended. Above 6.5 V, the allowable DC current is junction temperature limited. If the case temperature ($T_C$) is 25°C, then power dissipation ($P_D$) must not exceed 200 watts. The DC curve also shows that the DC power level must be lowered from the 200 W level as voltage increases above 30 V if second breakdown is to be avoided. At case temperatures higher than 25°C, thermal resistance information must be taken into account to derate the power dissipation limit in order to keep the junction temperature ($T_J$) below its limits, $T_{J(MAX)}$. The second breakdown limitation curve is valid when $T_J \leq T_{J(MAX)}$; therefore, the curve is not derated with increases in case temperature. That is, the curve is dependent on junction temperature only.

The various pulse curves given show no thermal limitations for currents to 50 A; the transistor power is limited solely by second breakdown. (At higher case temperatures, the transient thermal resistance must be used to determine if operation is within $T_{J(MAX)}$.) The given pulse curves are used if the duty cycle is 10% or less. For higher duty cycles, the curves gradually degrade toward the DC curve.

It is common industry practice to rate power devices at a case temperature of 25°C, even though it is very unlikely that the case would ever be held at 25°C in a practical operating situation. Figure 5-27 is a safe operating curve for the devices of Figure 5-26 at 125°C.

The SB curves are unchanged; they are already based upon $T_J = T_{J(MAX)}$.

---

*On low level devices $I_C$ is limited by the bonding wire at low values of $V_{CE}$.

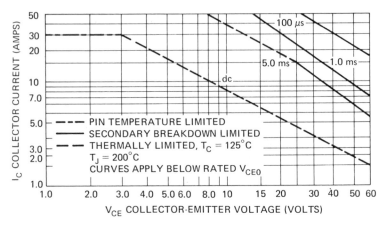

Figure 5-27   Constructed safe operating area data at $T_C = 125\,^\circ C$ for transistor of Figure 5-26.

Note that DC operation is entirely thermally limited and a thermal limitation appears on the lower voltage range of the 5 ms pulse. However, at 1 ms and 100 $\mu$s, no thermal limitations appear below 50 amperes. Such behavior is typical of most transistors, i.e., DC power becomes thermally limited at fairly low case temperatures while under short pulse operation, power dissipation is limited by second breakdown, even at fairly high case temperatures.

The curves of Figures 5-26 and 5-27 show the limits of power dissipation which may be sustained for a time interval as shown by the designated line, i.e, a 20 V, 30 A power pulse may be sustained for 1 ms without encountering Second Breakdown. In switching applications, the load line may traverse along one of the limit lines for the time indicated.

The specification of safe operating area curves on manufacturer's data sheets has provided the power supply designer with the necessary information to select and use power transistors in a reliable manner. The importance of these curves to effectuate a viable and trouble-free power supply design cannot be overstated.

### Preventing Secondary Breakdown

Based on the present knowledge of second breakdown the following guidelines are presented in applying power transistors to power supply circuits.

1. Design the power supply using the manufacturer's Safe Operating Area Curve.
2. In switching circuits, the driving circuit output resistance should be as small as possible.
3. The base current must be increased to reduce the total decay time energy;

but the base current must be reduced to increase the energy capability of the transistor. Thus, the base current must be of such value as to yield a "happy medium."

4. The reverse base current and voltage must exceed the collector junction leakage current and emitter floating potential, respectively, to refrain from operating in a negative beta region.
5. Keep the transistor power dissipation for a given base drive as small as possible. Increasing the power dissipation reduces $E_m$.
6. The transistor itself should have:

- Large ratios of emitter peripheral length to emitter area.
- Highest base conductivity in keeping with other device critical parameters.
- Minimum base width in keeping with other parameters; probably a device with split emitters is best.
- An $I_2$ breakover point higher than maximum collector current $I_0$ at the maximum temperature. The maximum temperature is specified because breakover point $I_2$ decreases with increasing temperature. At room temperature, $I_2$ may be above $I_0$, but at maximum junction temperature, $I_2$ will have fallen below $I_0$ and will cause the transistor to be destroyed when it is turned off.
- A large enough first breakover point $V_a$ for the inductive load line not to hit this part of the characteristic. This point is also temperature-dependent and moves to the left with increasing temperature.
- The transistor should be designed so that the "punch through" voltage is less then the collector base breakdown voltage. If punch through occurs prior to collector base breakdown, the effect of base bias on the device's SB characteristic is minimized, if not eliminated.
- Use appropriate active clamping, RC snubber and load line shaping circuits to maintain collector voltage at a safe value.

## Avoiding Transistor Problems

The problems encountered in using power transistors in converter applications are best illustrated by examples. Seven common pitfalls are described in the form of commonly accepted, but false, statements. The discussion that follows each statement explains why it is incorrect.

1. *A transistor can dissipate 50 W if it has a maximum power dissipation rating of 50 W.*

The above statement is false because the maximum power dissipation rating of a power transistor is based on its maximum permissible junction temperature and its thermal resistance.

## Example A

A manufacturer will say that a transistor has a maximum dissipation rating of 50 W at a case temperature of 25°C if it has a maximum junction temperature of 100°C and a maximum thermal gradient of 1.5°C/W. Fifty watts flowing through a thermal resistance of 1.5°C/W develops a thermal potential difference of 75°C between the junction and the case of the transistor. Subtracting this 75°C thermal potential from the 100°C maximum junction temperature results in a case temperature of 25°C.

An additional factor is introduced when the transistor is mounted on a heat sink, which may have a thermal resistance of 1 to 3°C/W. When added to the inherent thermal resistance of the transistor, the heat-sink resistance reduces the maximum permissible dissipation to a value considerably below the manufacturer's rated maximum power dissipation.

## Example B

If a transistor ($R_T$ = 1.5°C/W) is mounted on a heat sink having a thermal resistance of 2.5°C/W in an ambient temperature of 60°C, the original 50 W rating of the transistor is reduced to 10 W. When 10 W flow through a thermal resistance of 4°C/W, a thermal potential difference of 40°C is developed.

The maximum safe ambient temperature is determined by subtracting the thermal potential difference from the maximum safe junction temperature. Thus, 100° − 40° = 60°C, which is the maximum safe ambient temperature.

*2. A transistor can be cut off by a short circuit between its base and emitter.*

This discussion concerns the bias required to place the transistor in the cutoff or nonconducting condition. At normal room temperatures, a short circuit or small resistance connected from emitter to base is usually sufficient to establish cutoff. However, at higher temperatures the situation is quite different. Due to the shift in the input characteristic of the transistor, a short circuit between emitter and base will no longer assure a completely conducting condition. To compensate for the effect of the floating potential which develops at the emitter at elevated temperatures, sufficient reverse bias must be applied to overcome this potential at the temperature and voltage in question.

The emitter floating potential ($V_{EB}$) is the potential developed between the base and emitter leads of a transistor when voltage is applied between its base and collector. In Figure 5-28, a transistor is cross-sectioned to indicate how this floating potential is developed. The collector leakage current flowing in the base lead produces a transverse voltage drop in the base region. When a high-impedance meter is connected between the base and emitter leads, it reads the sum of the emitter-to-base potential barrier $V_D$ and the transverse voltage drop, $V_R$. This potential barrier voltage is about 0.1 V at room temperature and has a negative temperature coefficient of approximately 1.8 mV/°C. The voltage drop across

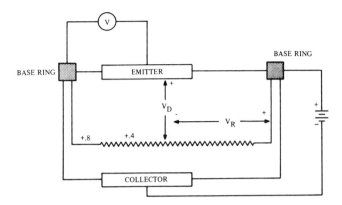

Figure 5-28    Transistor cross-section schematic diagram.

the base resistance, $V_R$, increases with increasing temperature because $I_{CO}$ increases.

3. *Collector diode reverse leakage current will double with each 13°C temperature rise.*

Collector diode reverse leakage current has two major components. One of these, identified as the bulk leakage current, doubles (with increasing temperature) with approximately each 13°C rise and is independent of collector voltage. The second component, identified with surface effects, varies markedly with collector voltage but in most transistors is essentially independent of temperature.

At room temperatures and at high voltages, surface effects predominate. Similarly, at high temperatures and low voltages, the component of the leakage current due to bulk effects is much larger than the current due to surface effects. For other values of voltage and temperature, the situation may not be quite as definite, and the leakage currents due to each effect may actually have similar values. For this reason, the collector leakage current doubles with each 13°C only for values of voltage and temperature for which the surface component can be neglected, specifically, at low to moderate voltages and temperatures in excess of approximately 55°C.

Figure 5-29 indicates the variation of collector leakage current with temperature and collector voltage for a typical germanium power transistor.

4. *The maximum circuit voltage is equal to twice the battery voltage.*

When a transformer is used in a balanced (push-pull) transistor circuit, the autotransformer effect of the primary causes a doubling of the collector-to-

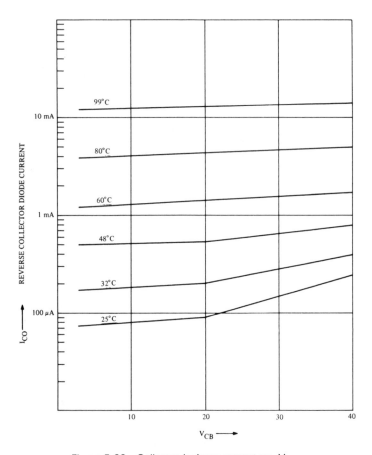

Figure 5-29   Collector leakage current vs. $V_{CB}$.

emitter voltage which is applied to each transistor. However, the collector-to-base voltage is the sum of the collector-to-emitter voltage and the emitter-to-base voltage. For example, if the maximum emitter-to-collector voltage is 28 V in a Class B amplifier and the transistor is cut off by a negative bias of 5 V, the total collector to base voltage is 33 V.

In push-pull switching (square-wave oscillator) circuits, the maximum collector-to-base voltage is (for the same reason as in balanced circuits) equal to twice the battery voltage plus the emitter-to-base cutoff voltage supplied by the base winding of the transformer. Switching transients and voltage drop across the base bias resistor contribute additional potentials which must be taken into account.

A similar problem is encountered in a Class A single-ended circuit which drives an inductive load. At full output, just below the level where clipping occurs, the maximum collector-to-emitter voltage is equal to twice the supply voltage. The collector-to-base voltage is a little less than, but about the same as, $V_{CE}$ because a Class A circuit is always forward-biased unless overdriven. When it is overdriven, the base voltage of a Class A amplifier goes into the cutoff region. This sharp cutoff of collector current (due to clipping) gives rise to voltage transients as high as 60 to 70 V, even though the supply voltage is low (approximately 25 V).

5. *Voltage transients of short duration are not harmful to transistors.*

The application of voltage transients (spikes) to the collector diode of a transistor must be prevented if such transients result in exceeding the rated maximum collector diode voltage rating, regardless of the duration of the transient. These transients may occur as the result of transformer leakage inductance which is due to poor oscillator transformer design, or they may be reflected to the collector circuit from the output circuit, as a result of an inductive load being coupled to the oscillator. Although the amount of energy in this spike may be small, experience indicates that any prolonged repetition of such spiking tends to cause an internal short circuit between the emitter and the collector. Voltage spikes can be eliminated or minimized by means of appropriate despiking circuitry and careful transformer design.

6. *A transistor will block current flow if all voltages are reversed.*

In a power transistor with a "ring" emitter configuration, the collector can act as an emitter and consequently a large current may flow if the polarity of the collector voltage is reversed. However, most transistors are designed to have high gain in only one direction. For example, at a collector current of 5 A, a power transistor which has a gain of 33 in the normal direction will have a reverse current gain of 3. At a collector current of 0.5 A, the same transistor will have a reverse current gain of 10, compared to a forward gain of 80.

On the other hand, this reverse conduction can be desirable under specific circuit conditions. For instance, a push-pull square-wave oscillator is partially self-protected from the effect of switching transients by reverse conduction. The amplitude of the transient is reduced, as first one end of the collector winding and then the other clamps itself to the power supply potential by reverse conduction through the respective collector diode. By transformer action, this also reduces the spike at the other collector.

Figure 5-30 shows how such transient voltages normally appear across the collector leads of an oscillator. Figure 5-31 shows how the insertion of blocking diodes in the collector leads increases the amplitude of the transients because the collector windings can no longer clamp to the supply voltage.

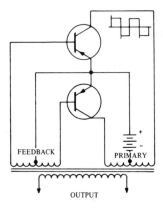

Figure 5-30   Push-pull converter.

7. *Rated $V_{CB}$ is the maximum voltage permissible in a voltage regulator circuit.*

At a voltage lower than the rated maximum collector diode voltage, which is commonly known as the "open-base" or "alpha-equals-unity" voltage, the sign of the current gain in the grounded emitter configuration reverses. Because voltage regulation depends on feedback, operation above this open-base voltage leads to difficulties due to the instability that occurs when the normally negative circuit gain becomes positive gain.

An additional and more serious problem also arises when reverse base current flows during operation in the negative resistance region. This problem is the increasing concentration of emitter current in the center of the junction area. Figure 5-32 illustrates the lateral voltage gradient which is developed in the base due to the reversed flow of base current. The reverse base current concentrates the flow of emitter current toward the geometric center of the emitter,

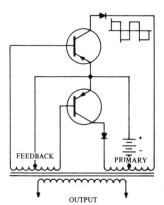

Figure 5-31   Push-pull converter with blocking diodes.

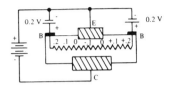

Figure 5-32  Internal transistor voltage gradients.

causing a very localized heating—a "hot spot." It is possible that only one-thousandth of the emitter area may be conducting under certain conditions of reversed bias.

Most of the problems discussed have been concerned with the effects of temperature and voltage. These are, indeed, the two areas where 90% of transistor pitfalls are encountered. Consequently, it is well worth the effort for anyone using transistors to completely familiarize himself with the effects of temperature and voltage on transistor operation.

## The Effect of Transient Voltages on Transistors

During transient conditions, the maximum junction temperature is dependent on the energy delivered to the transistor. For resistive and capacitive loads, transient energy is delivered during the turn-on period because the internal space-charge capacity induces a forward current in the base, which is amplified. For an inductive load, energy is delivered during the turnoff period because of the high induced voltage.

Good transistor circuit design requires an understanding of the voltage limitations of the transistor. Transistors have been rated by many systems with regard to maximum voltage, but the important dynamic considerations have often been overlooked. Transistors are quite susceptible to voltage failures during current periods of turning on or off. The criteria depend on the transistor characteristics and circuit conditions.

## Collector Characteristics

Figure 5-33 shows the typical collector characteristics of a pnp junction power transistor with the base current as a running parameter. It may be noted that independent of biasing, all characteristics are ultimately asymptotic to a certain limiting voltage $V_p$, the avalanche sustain voltage. This voltage is a function of the collector-base junction avalanche breakdown voltage $V_{bd}$.

$$V_p = \frac{V_{bd}}{(H_{FEO})^{1/n}} \qquad (5\text{-}11)$$

where $n$ is a number which depends on the resistivity and the type of material

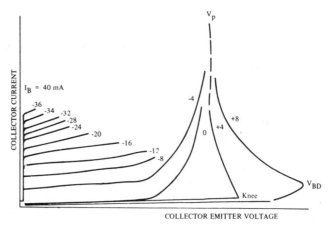

Figure 5-33   Power transistor collector characteristics (common emitter).

(germanium or silicon: 3 to 4 for pnp, 4 to 7 for npn), and $H_{FEO}$ is the common-emitter current amplification factor at low collector voltages.

Note that beyond $V_p$ there is a region that exhibits negative-resistance characteristics. The magnitude of negative resistance is dependent upon the base-emitter reverse bias: the greater the reverse bias, the greater the negative resistance. This negative resistance is also a function of the collector-base-leakage and saturation current. For transistors with a ''soft'' reverse collector-base characteristic, it takes a greater reverse bias to obtain negative resistance.

### Transient Energy and Junction Temperature

If the junction temperature of a transistor exceeds its maximum limit, the transistor can be destroyed. In steady-state operation, the power dissipation $P$ determines the junction temperature rise $\Delta T$. If $\theta$ is the thermal resistance in degrees centigrade per watt, then

$$\Delta T = P \cdot \theta \qquad (5\text{-}12)$$

During transient conditions, the time interval of power dissipation is usually less than the thermal time constant $\lambda$. Assuming a square-wave pulse, the instantaneous temperature rise can be expressed approximately as

$$\Delta T \simeq P \cdot \theta(1 - e^{-t/\lambda}) \qquad (5\text{-}13)$$

where $t$ is the time at any instant measured from the start of the square-wave pulse. Expanding $e^{-t/\lambda}$ and neglecting the insignificant higher-power terms

$$e^{-t/\lambda} \simeq 1 - \frac{t}{\lambda} \tag{5-14}$$

From equations (5-13) and (5-14)

$$\Delta T = \frac{(Pt)\,\theta}{\lambda} = \frac{J\theta}{\lambda} \tag{5-15}$$

Therefore, the temperature rise is directly proportional to the transient energy $J$ delivered to the transistor.

## Protection of Transistors

A transistor may be protected by limiting the voltage excursion $V_m$ or by avoiding the existence of a negative-resistance collector characteristic. $V_m$ may be reduced by using a transistor of high $C_{bc}$, or a circuit of low $L$ and $R_c$. $V_m$ may also be limited by using a nonlinear element (Zener diode) or any combination of these elements. When the base is reverse-biased, the negative-resistance characteristic usually exists unless the transistor has considerable leakage and saturation current. The base can be prevented from reverse biasing by the use of a diode. Chapter 3 discusses various transient spike suppression networks.

## Paralleling Power Transistors

In situations where greater current carrying capability of DC/DC converter power transistors is required than that available, the power transistors can be paralleled as was shown in Figures 2-32, 2-49, and 3-50. However, the implementation of this paralleling scheme involves a great deal of care.

The main consideration for operating switching transistors in parallel is equalization of collector current through the parallel paths during both the saturation and switching states.

Matched transistor parameters and circuit techniques may become purely academic if variation in the magnetic flux linked by each parallel path is overlooked. A symmetrical arrangement will minimize current imbalance resulting from rapidly changing magnetic fields. Mismatch of collector current due to mechanical arrangement can be the predominant factor. Collector circuit imbalance of as much as 10 to 1 can result from voltage induced in the base circuit. Good circuit layout techniques with attention to short lead lengths, symmetry, and use of twisted pair leads are essential.

Parallel devices being driven from the same impedance source can be closely matched for $V_{BE(sat)}$ so that base drive will be evenly distributed to each device. This, however, is not the most practical method because it requires different groupings of matched $V_{BE}$ devices. Differences in base-emitter voltages can be overcome more easily by the use of a series base resistor.

## MOS POWER TRANSISTORS

### Introduction

This section presents the characteristics of MOSFETs and compares them with bipolar power transistors for power supply applications.

Bipolar power transistors have long been considered as the primary power switch in switching power supplies. But with new developments in manufacturing technology, power MOSFETs have risen from obscurity in the 1970s to the point where they have now become a major force as a discrete power component. They offer unique characteristics and capabilities that are not available with bipolar power transistors. By taking advantage of these differences, overall systems cost savings can be achieved without sacrificing reliability.

The application of field-effect technology to power transistors was propelled by two of its operating characteristics: (1) a very high input impedance and (2) very fast switching times. Because the input impedance of a MOSFET is so high, it is for all practical purposes voltage driven. This means the driving circuit need not provide any power. As a result, the driving circuit can be made very simple. It should be noted, however, that a MOSFET's input acts as if it were a small capacitor, which the driving circuit must charge at turn on. Even so, except for high current devices, most power MOSFETs can be driven directly from an IC.

Typically, on-resistance values for power MOSFETs range from .005 to .1 ohm at current levels between 10 and 50 A. Although values for on-resistance of comparable bipolar power transistors are still lower than those of MOSFETs, the gap is closing rapidly.

Even though its high input impedance alone makes the MOSFET desirable, its fast switching speeds make it indispensable for high-frequency applications. Not only are switching power losses minimized, but the maximum usable switching frequency is considerably higher. The response time of a power MOSFET is 50 to 100 ns, compared with the relatively slower speeds of power bipolar transistors, generally specified in microseconds. This has resulted in considerably lower power losses and higher switching efficiencies in applications using MOSFETs.

Unlike small-signal types, bipolar power transistors are limited to typical values of about 100 kHz in normal operation, whereas power MOSFETs are capable of operating above 1 MHz (Figure 5-34). This is particularly useful in regulated switching power supplies in which higher frequency capability produces higher switching speeds, lower power losses, and higher power efficiency.

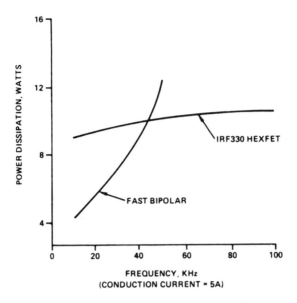

Figure 5-34  Switching Frequency versus Power Dissipation.

Also, at higher frequencies, the use of smaller and lighter components (transformer, filter choke, and filter capacitors) reduces overall component costs, while using less space for more efficient packaging at lower weight.

Because of its simple drive requirements and fast switching speed, the power MOSFET is widely used in low-voltage switchmode power converters. But the bipolar transistor still dominates at high voltages. The determining factor is the forward conduction loss. Unlike switching losses due to transition times, conduction losses occur during the period the transistor is saturated. The source of the loss, though, is different for the two transistor types. In a bipolar transistor, it is the product of the forward current multiplied by the junction voltage. This voltage is a function of the underlying physics upon which the transistor operates, and is typically 0.7 V for a silicon transistor.

The conduction loss in a power MOSFET is also the product of the forward current multiplied by the junction voltage. But whereas the junction voltage in a bipolar transistor is determined by the kind of semiconductor material it is made of, the voltage in a MOSFET is determined by the forward current and the MOSFET's forward on-resistance ($r_{DS}$ [on]). The value of this resistance is proportional to the MOSFET's blocking-voltage capability. Thus, as the voltage rating increases, so too does the on-resistance, and so too does the conduction loss. MOSFETs with voltage ratings of more than a few hundred volts can exhibit on-resistances of several ohms. If the drain current were 10 A and the on-resistance 2 $\Omega$, for example, the conduction loss would be 200 W. The same current flowing through a silicon bipolar transistor would result in a conduction loss of only 7 W.

Besides using simpler drive circuitry—resulting in easier design, fewer parts, less space, lower weight, and lower costs—power MOSFETs feature improved breakdown characteristics. Bipolar power transistors suffer from secondary voltage breakdown, reducing the breakdown voltage capability at certain current levels and resulting in the need for extra circuitry to ensure that the bipolar transistor operates only in its safe operating area (SOA).

Because MOSFETs act as resistors once they are turned on, their breakdown voltage is independent of operating current level, and they do not exhibit the secondary breakdown voltage effect. Although a power MOSFET's voltage rating provides a truer picture of its actual voltage capability, the device's power capability is still limited by its absolute maximum ratings and the need to remove heat to minimize an excessive increase of temperature.

A power MOSFET also offers an advantage over a power bipolar transistor by eliminating two related bipolar transistor problems: "thermal runaway" and "current hogging."

The output current of bipolar transistors increases for a given amount of input current as temperature increases. This causes a further increase in temperature, which, in turn, causes a further increase in output current, resulting in a runaway condition—called "thermal runaway"—and eventual device failure. To prevent this condition, additional components are required in the transistor circuit. Power MOSFETs, on the other hand, exhibit no thermal runaway condition, since their output currents decrease as temperature increases.

To increase current-handling capability in a power bipolar circuit, two or more bipolar power transistors are generally connected in parallel. If one of the parallel bipolar transistors is drawing more current than the other(s), the transistor operating at the higher current level will heat up and eventually hog all the current instead of sharing it proportionately. Components can be added to prevent current hogging, resulting in increased circuit complexity. Again, though, MOSFETs don't exhibit this problem and can easily be connected in parallel without additional components.

Another advantage is that power MOSFETs have more uniform operating characteristics over a broad temperature range. Transconductance—the equivalent of current gain in the bipolar transistor—varies less than $\pm20\%$ over the entire military temperature range ($-55°C$ to $+125°C$), while the current gain of a bipolar transistor will vary as much as 4 to 1 over the same temperature range.

The trend in power MOSFET technology toward higher frequencies, lower on-resistance and higher power capabilities is leading to the development of new, low-cost plastic and hermetic packages. These new MOSFET packages provide a more efficient approach to heat dissipation than used for bipolar power transistors, thereby giving a power handling and surge current capability that results in a more reliable component.

All these advantages, however, come at a price. Power MOS transistors are effectively equivalent to an array of several thousand very small MOS transistors wired in parallel. In order to manufacture these individual transistors efficiently, the manufacturing facility that produces power MOSFET devices must operate like a MOS IC manufacturing facility, rather than like one making bipolar power devices. In order to get good yields, the MOSFET fab requires a Class 100 or better environment with lithography registration of less than 10 microns. This compares to the Class 10,000 or 1,000 fab with lithography registration of 1 mil or greater used for the equivalent bipolar power transistor. These differences equate to a much more expensive silicon chip for power MOS than for power bipolar.

Table 5-2 compares the key features of bipolar and MOS power transistors. Figures 5-35 and 5-36 provide yet another comparison of these power switches when used in the same system, in this case a high-voltage flyback converter. For this application, a peak output voltage of about 700 V driving a 30 kΩ load $[P_{o(pk)} = 16\ W]$ was required. With the component values and timing shown, the inductor current required to generate this flyback voltage would have to ramp up to about 3.0 A.

Figure 5-35 shows the MOS transistor version. Because of its high input impedance, the FET, an MTM2N90, can be directly driven from the pulse-width modulator. However, the PWM output should be about 15 V in amplitude, and for relatively fast FET switching be capable of sourcing and sinking 100 mA. Thus, all that is required to drive the FET is a resistor or two. The peak

### Table 5-2   MOSFET versus Bipolar Power Transistor Comparison

| MOS | BIPOLAR |
|---|---|
| MAJORITY CARRIER DEVICE | MINORITY CARRIER DEVICE |
| No charge-storage effects | Charge stored in the base and collector |
| High switching speed; less temperature sensitive (10–20 ns switching time readily achieved) | Low switching speed; temperature sensitive<br>• Transition times of 50–200 ns with careful base drive design<br>• Storage time of 500–1500 ns (reducible to 150–300 ns with anti-saturation circuits |
| Drift current (fast process) | Diffusion current (slow process) |
| VOLTAGE DRIVEN | CURRENT DRIVEN |
| Purely capacitive input impedance; no DC current required | Low input impedance; DC current required |
| Simple drive circuitry | Complex drive circuitry (resulting in high base-current requirements) |

## Table 5-2  (Continued)

| MOS | BIPOLAR |
|---|---|
| PREDOMINANTLY NEGATIVE TEMPERATURE COEFFICIENT ON DRAIN CURRENT | POSITIVE TEMPERATURE COEFFICIENT ON COLLECTOR CURRENT |
| No thermal runway | Thermal runway |
| Devices can be paralleled with some precautions | Devices cannot be easily paralleled because of $V_{BE}$ matching problems, local current concentration, and current hogging |
| More uniform operating characteristics over a broad temperature range | |
| Less susceptible to secondary breakdown | Susceptible to secondary breakdown |
| SQUARE-LAW I-V characteristics at low current; linear I-V characteristics at high current | EXPONENTIAL I-V characteristics |
| Greater linear operation and fewer harmonics | More intermodulation and cross-modulation products |
| HIGH ON-RESISTANCE AND, THEREFORE, LARGER CONDUCTION LOSS | LOW ON-RESISTANCE (low saturation voltage) because of conductivity modulation of high resistivity drift region |
| Drain current proportional to channel width | Collector current approximately proportional to emitter stripe length and area |
| Low transconductance | High transconductance |
| High breakdown voltage as a result of a lightly doped region of a channel-drain blocking junction | High breakdown voltage as a result of a lightly doped region of a base-collector blocking junction |
| Contains inherent parasitic bipolar transistor which could destroy device—suffers from turn off $dv/dt$ | Does not suffer from this problem |
| Limited selection of power devices | Wide variety of speed, voltage and current ratings available |
| More expensive due to complex processing | Lower cost, simpler processing |

drain current of 3.2 A is within the MTM2N90 pulsed-current rating of 7.0 A (2.0 A continuous), and the turn-off load line of 3.2 A, 700 V is well within the switching SOA (7.0 A, 900 V) of the device.

Now let's compare this circuit with the bipolar version of Figure 5-36. To

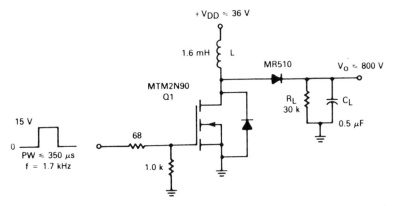

Figure 5-35 High-voltage flyback converter using a MOS output stage. (Copyright of Motorola, Inc. Used by Permission)

achieve the output voltage, using a high-voltage Switchmode MJ8505 power transistor, requires a rather complex drive circuit for generating the proper $IB_1$ and $IB_2$ currents. This circuit uses three additional transistors (two of which are power transistors), three Baker clamp diodes, eleven passive components and a negative power supply for generating an off-bias voltage. Also, the RBSOA capability of this device is only 3.0 A at 900 V and 4.7 A at 800 V, values below the 7.0 A, 900 V rating of the MOSFET. Thus, it is seen that since the drive requirements are not the same, it is not a question of simply replacing the

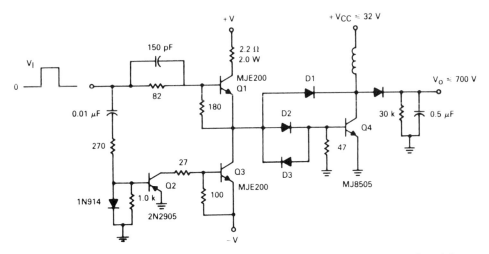

Figure 5-36 High-voltage converter using a bipolar driver and output stage. (Copyright of Motorola, Inc. Used by Permission)

bipolar transistor with the FET, but one of designing the respective drive circuits to produce an equivalent output, as depicted in Figures 5-35 and 5-36.

## Device Characteristics

Now lets look at some of these device characteristics, differences, and issues in greater detail.

### Speed

A conventional *npn* bipolar transistor is a current-driven device that contains an emitter-base *pn* junction as the source of current and a base-collector *pn* junction as a collector of current. Bipolar transistors are minority carrier devices: current flow is controlled by minority electrons across the emitter-base junction into the base region. Bipolar transistors have a fundamental drawback in switching speed because the speed is limited by the charge storage mechanism associated with minority carriers. Furthermore, because of its current-driven, base-emitter input, a bipolar transistor presents a low impedance load to its driving circuit. In most power circuits, this low impedance input requires somewhat complex circuitry to drive properly.

By contrast, a power MOSFET is a voltage-driven device whose gate terminal (see Figure 5-37a) is electrically isolated from the body (silicon) by a thin layer of silicon dioxide ($SiO_2$). As a majority carrier semiconductor, the MOSFET operates at a much higher speed than a bipolar transistor because there is no charge storage mechanism. A positive voltage applied to the gate—for an *n*-type MOSFET—repels positive charge from the silicon surface, thereby con-

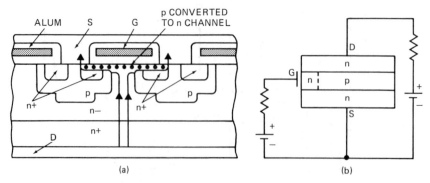

Figure 5-37   The MOSFET uses majority carriers to move current from source to drain (a). The key to MOSFET operation is the creation of the inversion channel beneath the gate when an electric charge is applied to the gate (b).

verting the channel region directly beneath the gate from $p$ type to $n$ type (see Figure 5-37b). This so-called "surface inversion phenomenon" allows current to flow between drain and source through an all $n$-type material. In effect, the MOSFET ceases to be an $npn$ device in this state, and the region between the drain and source can be represented as a resistor, although it does not behave in the linear manner of conventional resistors. Because of surface inversion, the $pn$ junctions that border the channel are effectively inoperable, and minority carrier injection does not occur. This is not the case in the bipolar transistor, which always retains its $npn$ character.

The high switching speeds allow efficient switching at higher frequencies, which in turn reduces the cost, size, and weight of reactive components. MOSFET switching speeds are primarily dependent on charging and discharging the device capacitances and are essentially independent of operating temperature.

Though higher speeds are an easily understood advantage of MOSFETs, their better efficiency needs some explanation. A MOSFET's usually higher on-state voltage drop might indicate greater overall power loss than that exhibited by comparable bipolar devices. However, that is generally not the case because the total dissipation, which includes both the loss in the on-resistance and the switching losses, must be considered.

Switching loss in a bipolar transistor is generally the major loss component— much higher than the switching loss in a MOSFET. Therefore, although the on-resistance loss may be high at DC, at high switching frequencies the total dissipation of the power MOSFET will at some frequency become less than that of a bipolar device.

The bipolar transistor also requires appreciable base drive, which means added dissipation in the external drive circuits. In addition, the bipolar transistor always needs a snubber circuit (collector-emitter diode) to shape the dynamic load line and keep it within the safe operating area when switching.

A comparison of the losses associated with the power MOSFET and an equivalent power bipolar device is a key factor in the choice of which to use and at what frequency. A comparison between the two types, both operating at 270 V and 2.5 A is shown in Figure 5-38. It takes into account the on-state saturation loss, the drive circuit power, and the switching losses of both devices and clearly indicates the superior efficiency of the MOSFET.

## Input Characteristics

To permit the flow of drain-to-source current in an $n$-type MOSFET, a positive voltage must be applied between the gate and source terminals. Because the gate is electrically isolated from the body of the device, theoretically, no current can flow from the driving source into the gate. In reality, however, a very small current does flow—in the range of tens of nA—and is identified on data sheets as a leakage current called $I_{GSS}$.

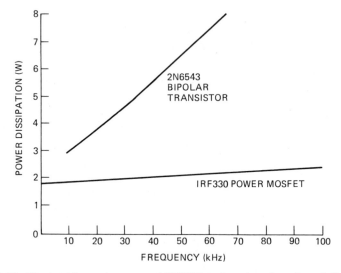

Figure 5-38   The total losses in a power MOSFET are less than those in a similar bipolar device, because switching losses are lower. These include both input drive requirements and saturation (on) losses.

Because the gate current is so small, the input impedance of a MOSFET is extremely high—typically greater than 40 MΩ—and is much more capacitive than resistive because of the gate terminal's isolation. Thus, a MOSFET's input circuit can be represented as being composed of equivalent resistance and capacitance elements (see Figure 5-39). The capacitance, called $C_{iss}$ on MOSFET data sheets, is a combination of the device's internal gate-to-source and gate-to-drain capacitance. Resistance $R$ represents the resistance of the material in the gate circuit. Together, the equivalent $R$ and $C$ of the input circuit pretty much determine the upper frequency limit of operation. The resistive portion

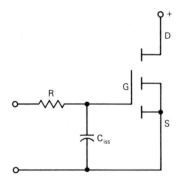

Figure 5-39   A MOSFET's switching speed is determined by its input resistance R and its input capacitance Ciss.

depends on the sheet resistance of the polysilicon gate overlay structure, and a value for $R$ is not found on data sheets.

On the other hand, $C_{iss}$ appears on virtually all data sheets. The value of $C_{iss}$ is closely related to chip size—the larger the chip, the higher the value. Since the driving circuit must charge and discharge the input $R/C$ combination and since the capacitance dominates, larger chips will have slower switching times than smaller chips—and are thus less suitable for higher frequency circuits. In general, the upper frequency limit of most power MOSFETs spans a fairly broad range, 1 to 10 MHz, depending upon chip size.

An important point comes up when considering the $V_{GS}$ level required to operate the MOSFET. In Figure 5-42, the device is not turned on—no drain current flows—unless $V_{GS}$ is greater than the threshold voltage level. This threshold voltage must be exceeded before an appreciable increase in drain current can occur. Generally, $V_{GS}$ for many types of DMOS devices is at least 2 V, an important consideration for devices or circuits that drive MOSFET gates. The gate drive circuit must provide at least the threshold voltage level, but preferably a much higher level. In fact, a MOSFET needs a fairly high voltage—on the order of 10 V or more—for maximum saturated drain current flow. However, ICs such as TTL types can not provide the necessary voltage levels unless they are modified with external pull-up resistors. Even with the pull-up to 5 V, a TTL driver can not fully saturate most MOSFETs. Thus, TTL drivers work best when the current to be switched is far less than the rated current of the MOSFET. CMOS ICs can run from supplies of 10 V, so these devices can drive a MOSFET into full saturation. On the other hand, a CMOS driver will not switch the MOSFET gate circuit as fast as a TTL driver. For best results, whether TTL or CMOS ICs provide the drive, the designer should insert special buffering chips between the IC output and gate input to match the needs of the MOSFET gate.

Since the gate is isolated from the source, the drive requirements are nearly independent of the load current. This reduces the complexity of the drive circuit and results in overall system cost reduction.

### Safe Operating Area

Power MOSFETs, unlike bipolar transistors, do not require derating of power handling capability as a function of applied voltage. The phenomena of second breakdown does not occur within the ratings of the device. Depending on the application, snubber circuits may be eliminated or a smaller capacitance value may be used in the snubber circuit. The safe operating boundaries are limited, though, by the peak current rating, breakdown voltages, and the power capabilities of the devices.

## On-Voltage

The minimum on-voltage of a power MOSFET is determined by the device on-resistance $r_{DS}$(on). For low voltage devices the value of $r_{DS}$(on) is extremely low, but with high voltage devices the value increases. $r_{DS}$(on) has a positive temperature coefficient which aids in paralleling devices.

## Output Characteristics

As shown in Figures 5-40 and 5-41, the output characteristics of the power MOSFET and the bipolar transistor can be divided similarly into two basic regions. The figures also show the numerous and often confusing terms assigned to those regions. To avoid possible confusion, this section will refer to the MOSFET regions as the "on" (or "ohmic") and "active" regions and bipolar regions as the "saturation" and "active" regions.

One of the three obvious differences between Figures 5-40 and 5-41 is the family of curves for the power MOSFET is generated by changes in gate voltage and not by base current variations. A second difference is the slope of the curve in the bipolar saturation region is steeper than the slope in the ohmic region of the power MOSFET, indicating that the on-resistance of the MOSFET is higher than the effective on-resistance of the bipolar transistor.

The third major difference between the output characteristics is that in the active regions the slope of the bipolar curve is steeper than the slope of the MOSFET curve, making the MOSFET a better constant current source. The limiting of $I_D$ is due to pinch-off occurring in the MOSFET channel.

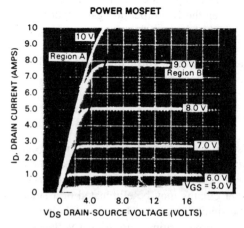

**POWER MOSFET**

Figure 5-40   $I_D$-$V_{DS}$ transfer characteristics of MTP8N15. Region *A* is called the ohmic, "on," constant resistance or linear region. Region *B* is called the "active," "constant current," or "saturation region." (Copyright of Motorola, Inc. Used by Permission)

**BIPOLAR POWER TRANSISTOR**

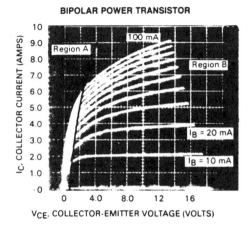

Figure 5-41  $I_C$-$V_{CE}$ transfer characteristics of MJE15030 (NPN, $I_C$ continuous = 8.0 A, $V_{CEO}$ = 150 V). Region $A$ is the saturation region. Region $B$ is the linear or active region. (Copyright of Motorola, Inc. Used by Permission)

## On-Resistance

The on-resistance, or $r_{DS}$(on), of a power MOSFET is an important figure of merit because it determines the amount of current the device can handle without excessive power dissipation. When switching the MOSFET from off to on, the drain-source resistance falls from a very high value to $r_{DS}$(on), which is a relatively low value. To minimize $r_{DS}$(on) the gate voltage should be large enough for a given drain current to maintain operation in the ohmic region. Data sheets usually include a graph, such as Figure 5-43, which relates this information.

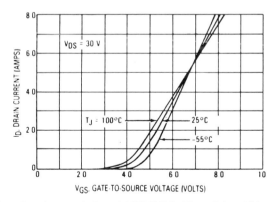

Figure 5-42  Transfer characteristics of MTP4N50. (Copyright of Motorola, Inc. Used by Permission)

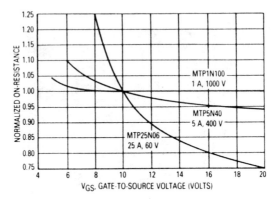

Figure 5-43   The effect of gate-to-source voltage on on-resistance varies with a device's voltage rating. (Copyright of Motorola, Inc. Used by Permission)

As Figure 5-43 indicates, increasing the gate voltage above 12 V has a diminishing effect on lowering on-resistance (especially in high voltage devices) and increases the possibility of spurious gate-source voltage spikes exceeding the maximum gate voltage rating of 20 V. Somewhat like driving a bipolar transistor deep into saturation, unnecessarily high gate voltages will increase turn-off time because of the excess charge stored in the input capacitance.

As the drain current rises, especially above the continuous rating, the on-resistance also increases. Another important relationship, which is addressed later with the other temperature dependent parameters, is the effect that temperature has on the on-resistance. Increasing $T_J$ and $I_D$ both effect an increase in $r_{SD}(\text{on})$ as shown in Figure 5-44.

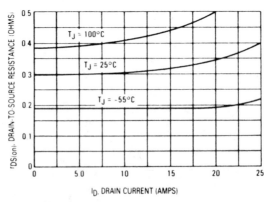

Figure 5-44   Variation of $r_{DS}(\text{on})$ with drain current and temperature for MTM15N45. (Copyright of Motorola, Inc. Used by Permission)

Most manufacturers of power MOSFETs use a vertical double-diffused process (DMOS). The DMOS MOSFET is a single silicon chip with many closely packed, hexagonal cells. The number of cells varies according to the dimensions of the chip. For example, a typical 120 $mil^2$ chip contains about 5000 cells, whereas a 240 $mil^2$ chip has more than 25,000 cells. One purpose of the multiple-cell construction is to minimize $r_{DS}$(on), thereby providing excellent power switching performance because the voltage drop from drain-to-source is also minimized for a given value of drain-to-source current.

To minimize $r_{DS}$(on), a large number of cells are manufactured in parallel on a chip. That is,

$$r_{DS}(on) = R_N/N$$

where $N$ is the number of cells. . .

Figure 5-45 shows the relationship between breakdown voltage, $r_{DS}$(on) and chip size. Using a larger chip results in a lower value for $r_{DS}$(on) because a larger chip has more cells. A larger chip also increases breakdown voltage capability. However, $r_{DS}$(on) increases with increasing breakdown voltage capability and low $r_{DS}$(on) must be specified if the MOSFET is to withstand higher breakdown voltages.

The penalty for using a larger chip, however, is an increase in cost, since chip size is a major cost factor. And because chip area increases exponentially, not linearly, with voltage, the additional cost can be substantial. For example, to obtain a given $r_{DS}$(on) at a breakdown voltage twice as great as the original,

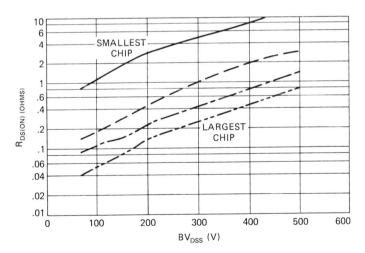

Figure 5-45  As chip size increases, $r_{DS}$(on) decreases, and voltage handling capability increases.

the new chip requires an area four or five times larger than the original. Although the cost does not rise exponentially, it is substantially more than the original cost.

### Transconductance

Since the transconductance, or $g_{FS}$, denotes the gain of the MOSFET, much like beta represents the gain of the bipolar transistor, it is an important parameter when the device is operated in the active, or constant current, region. Defined as the ratio of the change in drain current corresponding to a change in gate voltage ($g_{FS} = dI_D/dV_{GS}$), the transconductance varies with operating conditions as shown in Figure 5-46. The value of $g_{FS}$ is determined from the active portion of the $V_{DS}$-$I_D$ transfer characteristics where a change in $V_{DS}$ no longer significantly influences $g_{FS}$. Typically the transconductance rating is specified at half the rated continuous drain current and at a $V_{DS}$ of 15 V.

For power supply designs that switch the power MOSFET between the on and off states, the transconductance is often an unused parameter. Obviously when the device is switched fully on, the transistor will be operating in its ohmic region where the gate voltage will be high. In that region, a change in an already high gate voltage will do little to increase the drain current; therefore, $g_{FS}$ is almost zero.

### Threshold Voltage

Threshold Voltage, $V_{GS(th)}$, is the lowest gate voltage at which a specified small amount of drain current begins to flow. For example, Motorola specifies $V_{GS(th)}$

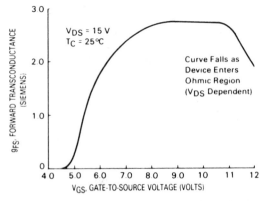

Figure 5-46   Small-Signal Transconductance versus $V_{GS}$ of MTP8N10. (Copyright of Motorola, Inc. Used by Permission)

at an $I_D$ of 1 mA. Device designers can control the value of the threshold voltage and target $V_{GS(th)}$ to optimize device performance and manufacturability. A low threshold voltage is desired so that the MOSFET can be controlled by low-voltage circuitry, such as CMOS and TTL. A low value also speeds switching because less current needs to be transferred to charge the parasitic input capacitances.

A low threshold voltage is undesirable for high power MOS devices for a number of reasons. High power MOS transistors generally operate at higher chip temperatures for optimum efficiency. Since the threshold voltage is temperature-dependent (a coefficient of approximately $-5$ mV/$°$C) a high threshold is mandated to assure operation in the enhancement region. But threshold voltage can be too low if noise can trigger the device. Also, a positive-going voltage transient on the drain can be coupled to the gate by the gate-to-drain parasitic capacitance and can cause spurious turn-on of a device with a low $V_{GS(th)}$. Furthermore, high-power devices have large input capacitance which necessitates a substantial drive. The wisdom of a high threshold precludes the possibility of driver noise causing false triggering of the MOS transistor. This noise immunity is especially important when working in switching power supplies.

## Temperature Dependent Characteristics

$r_{DS}(on)$.  High operating temperatures are a frequent cause of bipolar transistor failure because current tends to concentrate in areas around the emitter, thus creating hot spots. The result is thermal runaway and eventual destruction of the device.

MOSFETs, however, since current flow is in the form of majority carriers, operate in a completely different manner. The mobility of majority carriers in silicon depends on temperature. As temperature increases, mobility decreases, and as the chip gets hotter, the carriers slow down. In effect, this slowdown increases the resistance of the silicon path and prevents the sort of concentration of current that leads to hot spots. In fact, if hot spots start to form, the local resistance increases—rerouting the currents to cooler portions of the chip. Thus, the MOSFET has a positive temperature coefficient of resistance as shown in Figure 5-47. This curve, which is shown on most data sheets, shows that the majority carriers tend to slow down as the temperature increases, increasing $r_{DS}(on)$. Thus, a MOSFET is inherently stable with temperature and provides its own protection against thermal runaway and secondary breakdown.

Figure 5-43 shows that the temperature coefficient of $r_{DS}(on)$ is greater for high voltage devices than for low voltage MOSFETs.

Another benefit, not possible with bipolar transistors, is that MOSFETs can be operated in parallel without current hogging. If any device begins to overheat, its resistance increases, and the current is rerouted to the cooler chips.

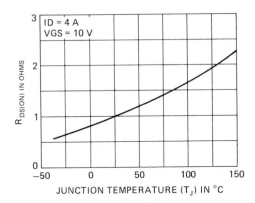

Figure 5-47   MOSFETs have a positive temperature coefficient of resistance, which greatly reduces the possibility of thermal runaway as temperature increases.

*Switching Speed.*   High junction temperatures emphasize one of the most desirable characteristics of the MOSFET, that of low dynamic or switching losses. In the bipolar transistor, temperature increases will increase switching times, causing greater dynamic losses. On the other hand, thermal variations have little effect on the switching speeds of the power MOSFET. These speeds depend on how rapidly the parasitic input capacitances can be charged and discharged. Since the magnitudes of these capacitances are essentially temperature invariant, so too are the switching speeds. Therefore, as temperature increases, the dynamic losses in a MOSFET are low and remain constant. Thus, speed remains constant, while in the bipolar transistors the switching losses are higher and increase with junction temperature.

*Drain-To-Source Breakdown Voltage.*   The drain-to-source breakdown voltage is a function of the thickness and resistivity of a device's *n*-epitaxial region. Since that resistivity and $V_{(BR)DSS}$ vary with temperature, as Figure 5-48 indicates, a 100°C rise in junction temperature causes $V_{(BR)DSS}$ to increase by about 10%. However, it should also be remembered that the actual $V_{(BR)DSS}$ falls at the same rate as $T_J$ decreases.

*Threshold Voltage.*   The gate voltage at which the MOSFET begins to conduct—the gate-threshold voltage—is temperature-dependent. The variation with $T_J$ is linear as shown in Figure 5-49. Having a negative temperature coefficient, the threshold voltage falls about 10% for each 45°C rise in the junction temperature.

## Importance of $T_{J(max)}$ and Heat Sinking

Without heat-sinking, the power ratings of packages used to encase power transistors (both bipolar and MOS) are meaningless. Because long-term reliability

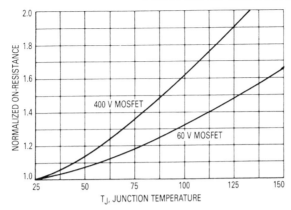

Figure 5-48   The influence of junction temperature on on-resistance varies with break-down voltage. (Copyright of Motorola, Inc. Used by Permission)

decreases with increasing junction temperature, $T_J$ should not exceed the maximum rating of 150°C. Steady-state operation above 150°C also invites abrupt and catastrophic failure if the transistor experiences additional transient thermal stresses. However, excluding the possibility of thermal transients, operating below the rated junction temperature can enhance reliability. A $T_J$(max) of 150°C is normally chosen as a safe compromise between long-term reliability and maximum power dissipation.

In addition to increasing the reliability, proper heat-sinking can reduce static losses in the power MOSFET by decreasing the on-resistance, $r_{DS}$(on). With $r_{DS}$(on), its positive temperature coefficient, can vary significantly with the quality of the heat sink. Good heat sinking will decrease the junction temperature, which further decreases $r_{DS}$(on) and the static losses.

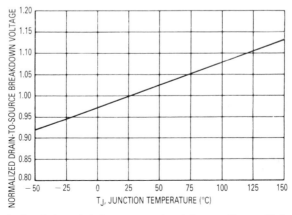

Figure 5-49   Typical variation of drain-to-source breakdown voltage with junction temperature. (Copyright of Motorola, Inc. Used by Permission)

## Drain-Source Diode

The ideal majority carrier MOSFET has no second breakdown failure mechanism, but an *npn* parasitic bipolar transistor is formed in the fabrication of an actual device as shown in Figures 5-50 and 5-51. Unfortunately, this bipolar device does break down, and catastrophic failures of MOSFETs can result from turn-on of this parasitic transistor.

As mentioned before, a MOSFET is a chip containing a large number of closely packed cells. Over 25,000 such cells can be found in a 240 × 240 mil device, each containing its own parasitic bipolar transistor. If just a single parasitic device turns off, its cell begins to conduct or "hog" current flowing through the device. Unable to handle the excess current, the cell heats and melts, destroying the chip.

In the vertical DMOS FET of Figure 5-51, the parasitic bipolar device has an $n+$ emitter, $p+$ base, and $n$ collector. Its base-emitter junction is shorted by the source metallization, but the effectiveness of the short in keeping the transistor out of conduction depends on design and process control. If carriers generated by high electric fields, as created by high drain-source voltages in the drain (collector) region, are allowed to cross the channel (base) region into the source (emitter) area, bipolar transistor action occurs. The breakdown mechanism is $V_{CER}$, as shown in Figure 5-52.

$V_{CER}$ breakdown occurs at a much lower voltage than $V_{DSS}$ breakdown. The latter would occur if the parasitic transistor's breakdown could be completely suppressed, providing a greater margin of safety. To make matters worse, the locus of the $V_{CER}$ curve exhibits a negative resistance characteristic as the gain of the bipolar transistor increases with increasing current. If the parasitic transistor is allowed to become active, the classic second breakdown failure mechanism can occur. This induces current hogging on both a macroscopic (among cells) and microscopic (within cells) level. Local heating results, which increases the parasitic transistor's gain, further restricts current, and eventually leads to MOSFET failure.

The measure of a power MOSFET's ruggedness is its ability to survive circuit

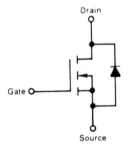

Figure 5-50   *N*-channel power MOSFET including parasitic bipolar drain-source diode. (Copyright of Motorola, Inc. Used by Permission)

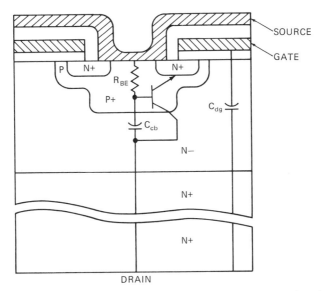

Figure 5-51 Cross-section of vertical DMOS power MOSFET construction showing the integral parasitic bipolar transistor.

conditions that tend to turn on its parasitic transistor. While it is difficult to entirely suppress parasitic transistor conduction, effective suppression by design and strict process control is vital to fabricating rugged devices virtually immune to second breakdown.

Since the current-hogging phenomenon is related to the bipolar transistor's gain, one approach to curing the second breakdown problem is to reduce gain.

In many applications, the drain-source diode is never forward-biased and does not influence circuit operation. Although the parasitic diode is fast, it is not as fast as the MOS transistor. Often used like a free-wheeling diode for transient suppression or energy recovery, the parasitic device may be too slow to protect the transistor. In addition, to drive their outputs toward either the supply voltage or ground, many circuits connect two transistors in series across the supply, as in totem-pole, push-pull, complementary pair, and half-bridge outputs. However, when both transistors turn on at the same time, they create a short circuit, destroying the devices. The speed with which MOS devices switch on poses the danger that the control circuit will not be fast enough to prevent both transistors from turning on simultaneously. Also, MOS speed requires a control circuit that is fast enough to protect itself, such as by means of a very fast feedback loop.

A unique role of the parasitic diode is shown in the totem-pole circuit of Figure 5-53. In this circuit, each transistor is protected from excessive flyback voltages—not by its own drain-source diode—but by the diode of the opposite

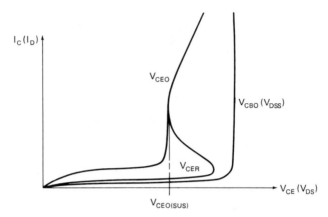

Figure 5-52   The parasitic transistor shown in Figure 5-51 deteriorates the second break-down effect in a power MOSFET, from $V_{DSS}$ to $V_{CER}$, a much lower voltage.

transistor. As an illustration, assume that $Q2$ of Figure 5-53 is turned on. $Q1$ is off and current is flowing up from ground, through the load and into $Q2$. When $Q2$ turns off, current is diverted into the drain-source diode of $Q1$ which clamps the load's inductive kick to V+. By similar reasoning, one can see that $D2$ protects $Q1$ during its turn-off.

## *dv/dt Limitations in Power MOSFETs*[3]

Under actual operating conditions, a MOSFET may be subjected to transients—either externally from the power bus supplying the circuit or from the circuit itself due, for example, to inductive kicks going beyond the absolute maximum ratings.

Power MOSFET performance is eventually limited by extremely rapid rates of change in drain-source voltage. These very high $dv/dt$'s can disturb proper circuit performance and even cause device failure in certain situations. High $dv/dt$'s occur during three conditions, each having its own $dv/dt$ threshold before problems arise. The first is what is termed "static $dv/dt$" and occurs when the device is off. For example, a voltage transient across the drain and source can be coupled to the gate via the gate-to-drain parasitic capacitance, $C_{rss}$. Depending on the magnitude of the gate-to-source impedance and the displacement current flowing into the gate node ($i = C \, dv/dt$), the gate-to-source voltage may rise above $V_{GS(th)}$, causing spurious turn-on. For this case, the $dv/dt$ immunity of the device depends to a large extent on the gate-to-source impedance. This underscores the importance of proper gate termination to promote good noise

---

[3]Courtesy of Motorola Semiconductor Products Inc.

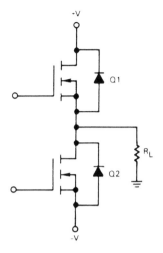

Figure 5-53 MOS totem-pole network with integral drain-source diodes. (Copyright of Motorola, Inc. Used by Permission)

immunity and is one reason why operation of power MOSFETs with the gate open-circuited is not recommended.

If the gate-to-source impedance is high and a voltage transient occurs between drain and source, spurious turn-on is much more likely than device failure. Typically, the transient will be coupled to the gate and cause the MOSFET to begin its turn-on. But as $V_{GS}$ rises, $V_{DS}$ falls, and the $dv/dt$ is reduced. Thus, the phenomena is self-extinguishing and generally is not destructive to any circuit elements.

The second mode in which $dv/dt$ may be of concern occurs when the MOSFET is turned off and an extremely rapidly-rising flyback voltage is generated. Since all loads appear inductive at high switching speeds, the device experiences simultaneous stresses imposed by a high drain current, a high $V_{DS}$ and large displacement currents in the parasitic capacitances. Problems associated with this "dynamic $dv/dt$" (so named because the device is being switched off and is generated its own $dv/dt$) are evidenced by device failure.

Unless extraordinary circuit layout techniques are used (for example, hybridized circuits that eliminate package and lead impedance) maximum attainable $dv/dt$'s in the dynamic mode range from 10 to 45 V/ns, depending on the $V_{DSS}$ rating of the device. Among the various available device types, maximum turn-off speeds do not differ widely, and the attainable $dv/dt$ is mainly determined by the magnitude of the voltage that the drain can be switched through. Consequently, a 1000 V MOSFET can generate a greater dynamic $dv/dt$ than a 60 V device, regardless of die size.

The third condition in which rapidly rising drain-to-source voltage may cause problems is the most stressful. It occurs in bridge configurations where the drain-source diode is allowed to conduct current. Failures are usually catastrophic and

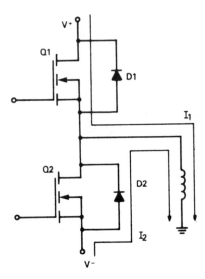

Figure 5-54 Totem-pole circuit in which *dv/dt* may be a concern. (Copyright of Motorola, Inc. Used by Permission)

are limited to a specific set of conditions. The circuit in Figure 5-54 serves as an illustration.

Assume the inductive load is being pulse-width modulated by $Q1$ and $Q2$ and that the intrinsic drain-source diodes $D1$ and $D2$ provide the current conduction path after the MOSFETs are turned off. When $Q1$ is turned on, it establishes current $I_1$. The load current ($I_2$) is commutated to $D2$ when $Q1$ is turned off. If $Q1$ is rapidly turned on again while $D2$ is still conducting, *dv/dt* considerations are in order. $Q2$ may then suffer damage because its diode is conducting while it experiences a rapid rise in $V_{DS}$.

One method of circumventing this problem is through employing the topology shown in Figure 5-55. The intent of this circuit is to eliminate the problem by not allowing the intrinsic diode to conduct. However, the greater number of parts, the additional cost, and the voltage drop due to the diode in series with the MOSFET are all undesirable. Thus, another solution is to limit the *dv/dt* by using snubbers or by slowing the MOSFET turn-on.

The optimum solution, however, for the device manufacturer is to make the devices more rugged, that is by making a MOSFET more forgiving to over-voltage transients. The difference between a ruggedized MOSFET and its conventional counterpart is that the ruggedized version is rated to withstand a specific amount of unclamped avalanche energy when operated at voltages above its maximum drain-to-source breakdown voltage ($BV_{DSS}$). In effect, the manufacturer guarantees that the transistor will not fail catastrophically up to a specified amount of avalanche energy. Thus, ruggedizing can enhance the reliability of power MOSFETs.

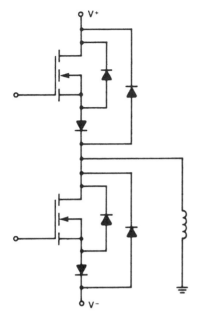

Figure 5-55 Circuit to eliminate current conduction in the MOSFET drain-source diode. (Copyright of Motorola, Inc. Used by Permission)

## Load Considerations

Freedom from second-breakdown limitations makes driving highly inductive or capacitive loads a natural application for MOSFETs.

In common with bipolar transistors, MOSFETs can be damaged if their voltage ratings are exceeded. Although their avalanche energy capability is much better than that of bipolar transistors, it is not good design practice to have the MOSFET absorb inductive energy unless the part is rated for this type of service. The spikes generated from inductive loads may have tremendous energy content, so usually some means of limiting their amplitude must be provided.

In addition, the transient power generated during the turn-on and turn-off intervals must be determined in order to check for excessive channel temperatures. Highly inductive loads may generate significant power at turn-off, whereas capacitive-type loads cause power surges at turn-on.

Usually with inductive loads, the peak voltage spike is limited to a value below the breakdown rating of the transistor by free-wheeling diodes, peak clipping, and snubbing, as was discussed in Chapter 3. No auxiliary circuitry is usually required with capacitive loads. Although the MOSFET is not subject to secondary breakdown, it is necessary to observe the safe-area curves of the MOS transistor in order to avoid the excessive temperature excursion during current inrush. When inrush power is excessive, increasing the gate drive will usually reduce it, and may hold it within bounds.

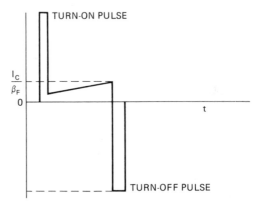

Figure 5-56 The ideal bipolar base-drive signal turns the device on and off with a fast rise-time current pulse.

## Providing Drive Current to Power Transistors[4]

Fast switching time in bipolar transistors is achieved by selecting fast devices— not by driving them properly. Often, manufacturers' switching-time test circuits do not drive the transistor on and off hard enough. Therefore, switching times published in data sheets are not always indicative of the true capabilities of a specific device. A gain-bandwidth ($f_t$) figure on the other hand is a more reliable indication of high-frequency usefulness. Generally, $f_t$ must be 10 MHz or greater, and the devices should be tested in the actual circuit.

An idealized base-current drive waveform is shown in Figure 5-56. At turn-on, the base must see a large, fast rising current pulse. With the transistor fully on, the turn-on pulse is removed, and sufficient base drive—in proportion to the collector current, if possible—should be provided to keep the device in or near saturation. At turn-off, a large, fast-rising negative pulse is applied to minimize both the storage and turn-off transition times. During the off-time, the transistor's base should be clamped to the emitter through a low impedance. This minimizes false turn-on resulting from noise and capacitively-coupled currents induced by high $dv/dt$'s in other parts of the circuit.

Several methods for properly driving bipolar transistors (using a DS0026 clock driver) are illustrated in Figure 5-57. The DS0026 sources and sinks 1 A in under 20 ns, besides providing an excellent interface between the logic and switch driver. (A more detailed discussion of bipolar transistor base drive circuits is presented in Chapter 3.)

To reduce bipolar storage time, an antisaturation circuit is required to prevent the collector voltage from falling below the base voltage. One simple way to

---

[4]Portions of this section used with permission of Siliconix.

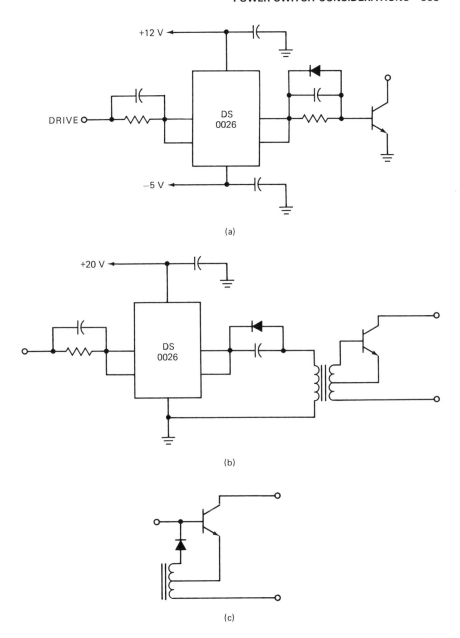

Figure 5-57 Drive circuits for bipolar transistors take many different forms depending on the application. A DS0026 clock driver makes a good interface element because it handles high current (1 A) and switches quickly (< 20 ns).

do this is by overdriving with a Baker clamp (see Chapter 3 for discussion). While this circuit reduces storage time by four or five times, rise and fall transition times are virtually unaffected. However, there is a limit beyond which overdrive offers no improvement. While the limit varies from one device to another, the forced-beta need not be less than 2 or 3. In fact, excessive overdrive leads to current crowding, which reduces the transistor's second breakdown energy capability.

The gate-current-drive waveform for a MOSFET is very similar to that of a bipolar transistor except that during the on time the current is essentially zero.

A MOSFET gate presents a capacitive load to the driver, and turn-on speed is limited primarily by the driver's resistance and the circuit's parasitic inductance. However, because a MOSFET has an *npn* transistor inherently built into its structure (Figure 5-58), triggering this transistor imposes another special limitation: turn-off *dv/dt*.

When a MOSFET's drain is rapidly pulled positive, current is injected into the base of the *npn* through capacitor *C*. If too much current is injected, the *npn* turns on, possibly destroying the device. The *dv/dt* limitations of a particular MOSFET depend greatly on geometry, which varies considerably between devices. And since data sheets do not contain information on *dv/dt*, designers must contact the manufacturer to learn of the limitations.

Several other characteristics must be kept in mind when designing with MOSFETs. Since MOSFETs are high-frequency devices, the MOSFET circuit can oscillate with improper circuit design. This oscillation can be eliminated by minimizing lead and PC board trace lengths, especially leads associated with the gate of the MOSFET, or by placing a ferrite bead on the gate lead or a small resistor in series with the gate.

Because of the extremely high input impedance of a MOSFET (in excess of

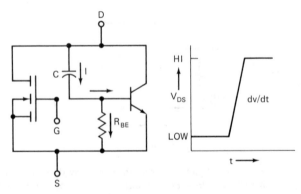

Figure 5-58 Parasitic *npn* switchback is inherent in the structure of MOSFETs. If too much current flows through capacitor *C*, the *npn* turns off, destroying the device.

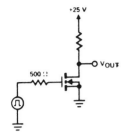

Figure 5-59    A typical MOSFET switching circuit. (Illustration courtesy of Siliconix Incorporated, Santa Clara, California)

$10^{12}$ Ω) drive circuits may be designed which are very high impedance. Under these conditions, it is possible for the gate node to get enough positive feedback from the gate-to-drain capacitance or just from stray fields in the circuit to cause oscillation.

When driving a MOSFET, it must be kept in mind that the dynamic input impedance is very different from the static input impedance. Since the input of a MOSFET is capacitive, the DC input impedance is very high, but the AC input impedance varies with frequency. Because of this effect, the rise and fall times of a MOSFET are dependent on the output impedance of the circuit driving it. The first approximation of the rise or fall time is simply:

$$t_r \text{ or } t_f = 2.2 \times R_{out} \times C_{iss} \qquad (5\text{-}17)$$

where $R_{out}$ is the output impedance of the drive circuit. This equation is valid only if the drain load resistance is much larger than $R_{out}$. Because of this, and because there is no storage or delay time with a MOSFET, it is very easy to calculate the rise and fall times and set them to any desired value. For example, to calculate the 10 to 90% rise or fall time for the circuit shown in Figure 5-59, using Equation 5-17, the rise time is equal to:

$$t_r = (2.2)(500)(50 \times 10^{-12}) = 55 \text{ nsec}$$

Lastly, when putting a positive voltage on the gate of a MOSFET, with respect to the source, the maximum voltage rating of the input protection Zener diode should not be exceeded. More importantly, the Zener diode must not be forward-biased by putting a negative voltage on the gate while the device is operating in a circuit. The reason for this is most easily explained by referring to Figure 5-60. As can be seen in the figure, the Zener diode is actually the base-emitter junction of a bipolar transistor. If a negative voltage greater than 0.6 V is placed on the gate, the base-emitter junction of the bipolar transistor will be forward-biased, which will then turn on the bipolar transistor. Consequently, when the bipolar transistor is turned on, current will flow from the drain through the

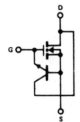

Figure 5-60 A parasitic *npn* transistor in Zener-protected MOSFETs. (Illustration courtesy of Siliconix Incorporated, Santa Clara, California)

bipolar transistor and out the gate, and this operating condition is very likely to be destructive.

Of all operating modes, the common-source configuration is the simplest to drive. Because of the high input impedance of a MOSFET, it can be driven directly from many logic families. When driving from a CMOS gate as shown in Figure 5-61, rise and fall times of about 60 nsec can be expected due to the limited source and sink currents available from the CMOS gate. If faster rise and fall times are required, there are several ways to obtain them. For example, if there are extra gates in the package that is driving the MOSFET, the designer can simply parallel the extra gates with the gate already being used. The additional current available will reduce the rise and fall times. If no extra gates are available, an emitter-follower buffer can be used, as shown in Figure 5-62. With this circuit, the current available to the MOSFET will be the output current of the CMOS IC multiplied by the beta of the bipolar transistors. Because the bipolar transistors are operating as emitter-followers, there will still be no storage time concerns, and the frequency limit will be determined by either the CMOS gate or the $f_t$ of the bipolar transistors, whichever comes first.

The MOSFET can also be driven directly from TTL gates. Because the output voltage of TTL is limited, the output current of the MOSFET will be limited to some value less than its maximum rated current. The expected output current can be determined from the transfer characteristic of the device being used. For example, if a TTL gate is driving a MOSFET such as the Siliconix VN46AF, the minimum output current of the MOSFET will be approximately 250 mA. This value was obtained by using the minimum output voltage of the TTL gate (3.2 V) for a high-level output and referring to the transfer characteristic for the

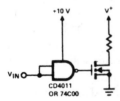

Figure 5-61 Driving a MOSFET with a CMOS gate. (Illustration courtesy of Siliconix Incorporated, Santa Clara, California)

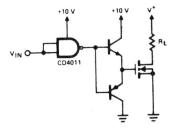

Figure 5-62   An emitter-follower circuit decrease MOS rise and fall times. (Illustration courtesy of Siliconix Incorporated, Santa Clara, California)

VNAZ which is the MOSFET geometry used in the VN46AF. If more than 250 mA is required, the output of a standard VMOS gate can be pulled up to the 5 V rail, as shown in Figure 5-63. With a full 5 V on the gate, the VN46AF will typically sink 600 mA.

For very high speeds, a capacitive drive such as the DS0026 clock driver can be used, as shown in Figure 5-64. With this drive configuration, typical rise and fall times are less than 10 nsec.

When operated in the common-drain mode, the MOSFET is somewhat more difficult to drive than when in the common-source mode. Again, because of the MOSFET's high input impedance, it is considerably easier to drive common-drain than a bipolar transistor would be when operated common collector. Moreover, common-drain circuits can be used when the load needs to be connected to ground, when an active pull-up and pull-down is required (totem-pole circuit), or in bridge-type circuits. However, for the purpose of this discussion, all examples are shown with totem-pole circuits.

The difficulty with common-drain circuits occurs because as the voltage across the load increases the enhancement voltage of the common-drain device decreases. Referring to Figure 5-65, as the voltage across $R_L$ approaches $V_2$ the enhancement voltage for the upper VN66AF decreases. If $V_1$ is not greater than $V_2$, then the voltage across $R_L$ can never reach $V_2$. For this reason, whenever a common-drain circuit is used, it is always necessary to have or to generate a voltage that is greater than the voltage which is desired to be impressed across the load. The amount the voltage has to be above the desired drain voltage is dependent upon the current the MOSFET must source and can be determined

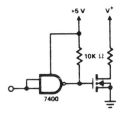

Figure 5-63   Pulling up a TTL output increases the sink current of the MOSFET. (Illustration courtesy of Siliconix Incorporated, Santa Clara, California)

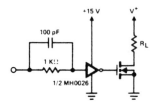

Figure 5-64   Using a MOS clock driver to drive a MOS-FET. (Illustration courtesy of Siliconix Incorporated, Santa Clara, California)

from the transfer characteristic of the MOSFET being used. If no supply voltage is available other than the one the load is to be pulled up to, one can be generated relatively easily because of the very low drive current requirements of the MOSFET.

One way of generating the required gate voltage is by using the bootstrap circuit shown in Figure 5-66. In the circuit, when $Q_1$ and $Q_3$ are on, $C_1$ is charged to the supply rail through $D_1$. When $Q_1$ and $Q_3$ are turned off, the gate voltage on $Q_2$ goes to the supply rail. As the source of $Q_2$ begins to pull $R_L$ up, the voltage across $C_1$ will be maintained; therefore, the gate-to-source voltage of $Q_2$ will be maintained. The size of $C_1$ should be large enough so that when it charges the gate capacitance of $Q_2$ a minimum voltage equal to the required enhancement voltage of $Q_2$ will be maintained across it. A good rule of thumb is to make $C_1$ equal to ten times the $C_{iss}$ of the FET. Figure 5-67 shows the same bootstrap circuit with some added components to improve the rise and fall times. In this circuit $Q_2$ acts as an emitter-follower to increase the peak gate current to $Q_3$. $D_2$ will be forward-biased when $Q_1$ turns on and serves as a low impedance path to discharge the gate of $Q_3$.

Another method to drive a common-drain MOSFET is shown in Figure 5-68. Rather than charging a capacitor and then feeding a signal back from the output as was done in the bootstrap circuit, this circuit stores the required charge in an inductor. When $Q_1$ is turned off, a flyback voltage is generated across the inductor. This voltage is used to maintain an enhancement voltage equal to the voltage of Zener diode $D_2$ across the MOSFET. Once $Q_2$ has been fully turned on and the voltage on $R_L$ is at the rail, a negligible amount of energy is required

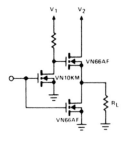

Figure 5-65   MOSFET totem-pole configuration. (Illustration courtesy of Siliconix Incorporated, Santa Clara, California)

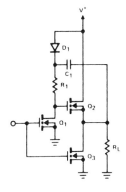

Figure 5-66 MOSFET bootstrap circuit. (Illustration courtesy of Siliconix Incorporated, Santa Clara, California)

to keep $Q_2$ on. $Q_2$ will remain on until $Q_1$ is turned on, or until the leakage currents of $Q_1$ and $D_2$ discharge the gate capacitance of $Q_2$.

Another method that can be used to drive a common-drain MOSFET is by using transformer drive, as shown in Figure 5-69. In this circuit, the transformer is used in the flyback mode when turning on the upper FET. $R_1$ and $R_3$ are used to suppress ringing, and $R_2$ and $R_4$ are used to assist with turn-off of the FETs. When driving with a transformer, care must be taken to design the transformer so that the secondary inductance in conjunction with the input capacitance of the FET does not create ringing or oscillation problems.

A MOSFET can be driven directly from an IC switching mode controller regulator chip without intervening power amplification. Moreover, the MOSFET lends itself to the trouble-free design of a family of supplies by requiring only minor adjustments in drive circuitry when changing from one output voltage and current rating to another. A bipolar power device, on the other hand, may require not only greatly different component values but also a totally different drive circuit for different variations of the same power supply family.

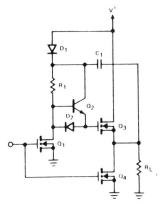

Figure 5-67 Bootstrap circuit with emitter-follower for improved rise times. (Illustration courtesy of Siliconix Incorporated, Santa Clara, California)

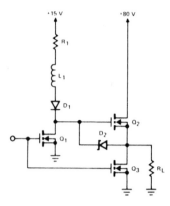

Figure 5-68 Inductive kickback drive circuit. (Illustration courtesy of Siliconix Incorporated, Santa Clara, California)

Although the power MOSFET's gate consumes virtually no current under steady-state conditions, it does require power while switching on and off, because the gate-to-drain and gate-to-source self-capacitance must be charged and discharged fast enough to obtain the desired switching speed. The drive circuit must therefore have a sufficiently low output impedance to supply the required charging and discharging current. But even under those conditions, a MOSFET's drive circuit needs negligible average current when compared with an equivalent bipolar transistor.

### Protecting the Gate

The gate of the MOSFET, which is electrically isolated from the rest of the die by a very thin layer of $SiO_2$, may be damaged if the power MOSFET is handled or installed improperly. Exceeding the 20 V maximum gate-to-source voltage rating, $V_{GS(max)}$, can rupture the gate insulation and destroy the FET. MOSFETs are not nearly as susceptible as CMOS devices to damage due to electrostatic discharge (ESD) because the input capacitances of power MOSFETs are much larger and absorb more energy before being charged to the gate breakdown

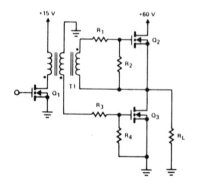

Figure 5-69 MOSFET transformer drive circuit. (Illustration courtesy of Siliconix Incorporated, Santa Clara, California)

voltage. However, once breakdown begins, there is enough energy stored in the gate-source capacitance to ensure the complete perforation of the gate oxide. To avoid the possibility of device failure caused by ESD, precautions similar to those taken with small signal MOSFET and CMOS devices apply to power MOSFETs.

The devices should be shipped only in antistatic bags or conductive foam, and when removing them from the packaging, careful handling procedures should be adhered to. Those handling the devices should wear grounding straps, while devices not in the antistatic packaging should be kept in metal tote bins. MOSFETs should be handled by the case and not by the leads, and when testing them, all leads should make good electrical contact before voltage is applied. As a final note, when placing a MOSFET into the system it is designed for, soldering should be done with a grounded soldering iron.

The gate of the power MOSFET could still be in danger after the device is placed in the intended circuit. If the gate experiences voltage transients which exceed $V_{GS(max)}$, a 20 V Zener should be placed across the gate and source terminals to clamp any potentially destructive spikes. Using a resistor to keep the gate-to-source impedance low also helps damp transients. It also serves another important function: voltage transients on the drain can be coupled to the gate through the parasitic gate-drain capacitance. If the gate-to-source impedance and the rate of voltage change on the drain are both high, then the signal coupled to the gate may be large enough to exceed the gate-threshold voltage and turn the device on.

## Available MOSFETs

Although both $n$- and $p$-channel power MOSFETs are available, $n$ channel dominates in most applications. $P$ channel MOSFETs cost more to manufacture than $n$ channel devices because the resistivity of $p$ type silicon is almost twice that of $n$ type. Thus, a $p$ channel MOSFET with the same on-resistance as an $n$ channel device would be about twice as large, driving up the cost of $p$ channel devices considerably. Because of these size and cost disadvantages, $p$ channel device technology has not experienced the rapid advances occurring in $n$ channel devices.

Steady advances in $n$ channel performance have been made in three major parameters: (1) on-resistance and current capability, (2) power handling capability of the package, and (3) cost. The most dramatic improvement in power MOSFET performance has occurred in the low voltage ($\leq 100$ V) high current area, which has been achieved by reducing the individual size of the paralleled cells (transistors) that make up a power MOSFET. The reduction in cell dimensions increases low voltage performance per unit area, resulting in lower on-resistance and higher current.

In addition to providing a wide product offering of power bipolar transistors, Motorola Semiconductor Products Division, GE/RCA Solid State Division, Solitron Devices Inc., Solid State Devices, and SGS-Thomson Microelectronics also provide an extensive offering of MOS power transistors. A brief sampling of the wide diversity of available device offerings is now presented.

GE/RCA Solid State Division offers virtually hundreds of power MOSFET $BV_{DSS}$, $I_D$, $r_{DS(on)}$ and package-style options as follows: $BV_{DSS}$ ratings range from 30 to 500 V, $I_D$ ratings from 13 A to 40 A; $r_{DS(on)}$ values from 0.055Ω to 5.6Ω in TO-204 (75–150 W), TO-220 (20–125 W), TO-247 (150 W), TO-39 (6.25–25 W), TO-237 (2 W) and 4-pin DIP (1 W) package options.

Motorola Semiconductor Products Division also has a broad product offering. For example, Motorola's plastic encapsulated TMOS power MOSFETs provide drain-source breakdown voltages of 50, 60, 80, and 100 V with the 60 and 80 V versions having on-resistances of 0.009 and 0.012Ω, respectively. The package's low inductance construction allows operation up to 500 kHz.

In a hermetic TO-3 package, Siliconix VN series devices combine the lowest on-resistance and highest power dissipation in a sealed package. VN devices are capable of 250 W dissipation, with an on-resistance of only 0.035Ω at 100 V and 0.300Ω at 500 V.

Advanced Power Technology has introduced a family of medium- to high-voltage power MOSFETs that exhibit on-resistance as low as 0.20Ω and switching at up to 1 MHz with minimal loss and encased in TO-3 packages. These devices are rated at up to 600 V drain-to-source voltage ($V_{DS}$) and handle up to 22 A in drain current ($I_D$). Previously introduced parts are rated at 450 and 500 V, 16 and 18 A, and 0.25Ω. Because of a unique cell design, these devices have an input capacitance that is less than one half that of conventional MOS-FETs, thus doubling frequency and reducing the drive current needed for a specific output current. A large (102,400 mil$^2$) die is used to achieve low on-resistance, handle higher currents, and improve the power MOSFET's ability to withstand transient overloads.

A growing trend in the packaging of power MOSFETs is the inclusion of both $n$ and $p$ channel types in the same package. The VC series from Supertex, Inc. provide two $n$-type and two $p$-type chips in a 14-lead ceramic DIP. Each of these low voltage devices has all three leads bonded out, allowing the devices to be used in a variety of ways.

Siliconix also has a family (the VQ family) of low voltage quad devices that feature a 2Ω on-resistance for the $p$ channel device and 1Ω for the $n$ channel device.

In the high power area, International Rectifier encases six power MOSFETs in a single package.

Advances in power MOSFET technology have yielded a family of bilateral MOSFET switches from Siliconix.[5] In terms of construction, the switches can

---

[5]Used with permission of Siliconix Inc.

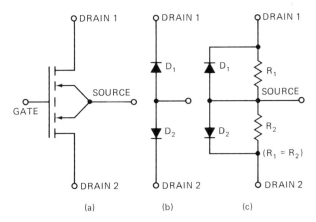

Figure 5-70  The schematic for the BLS 100 bilateral MOSFET (a) shows that it is a pair of $n$-channel MOSFETs with their sources and gates tied together. When off, it looks like a pair of back-to-back diodes (b), the switch resembles a voltage divider with clamp diodes across each resistor (c). (Illustration courtesy of Siliconix Incorporated, Santa Clara, California)

be thought of as two $n$ channel double-diffused MOS (DMOS) power FETs with common source and gate leads (Figure 5-70). As $n$ channel enhancement mode devices, they have no intrinsic offset voltage and can transmit AC or DC signals with equal facility. Furthermore, their impedance when turned on is resistive, which produces very low harmonic distortion.

In operation, when a positive voltage greater than the threshold voltage $V_{GS(th)}$, is applied between the gate and the source, both FETs turn on and conduct in series through their channels. When the gate signal is removed, both channels turn off and stop conducting. The source-to-drain parasitic diodes of the individual FETs, $D_1$ and $D_2$ (Figure 5-70b), do not conduct in reverse as they do in a unidirectional MOSFET, since either $D_1$ blocks $D_2$ or $D_2$ blocks $D_1$. Either of the diodes also may conduct (depending on the drain-to-drain voltage polarity), but only when the voltage drop across half the channel resistance exceeds the forward voltage of the diode (Figure 5-70c). When one of the diodes turns on, it clamps the voltage across its associated half of the channel, reducing power loss (lowers on-resistance) at high power levels and ensures surge protection in signal applications. The diodes, though, will conduct from the source to their respective drains if the source is biased above the drain potential.

Parasitic diodes make the device nonlinear whenever the drain-to-drain voltage exceeds about 1.4 V (peak) in either direction (approximately 2 V rms). Distortion is limited, though, since only one diode turns on at a time, and some resistance remains in the rest of the circuit. In AC signal switching, the distortion may be detectable during the switching interval.

With the proper gate-drive circuit, the switching intervals can be kept so short that the distortion is virtually undetectable. Moreover, in view of the low on-resistance of the switches, the 1.4 V drop required to initiate such distortion represents a signal that is more than 90 dB above the off leakage level. That figure is for a small device suitable for audio switching. In larger devices or for devices selected for low leakage, the dynamic range before distortion occurs is even wider.

Building gate-drive circuits for the bilateral MOSFETs involves some new techniques. The gate-to-source terminal pair may be viewed as a voltage-sensing capacitive load, just as in a unidirectional MOSFET. Like its unidirectional counterpart, the input capacitance $C_{iss}$ of the switch will vary with chip size and, hence, with voltage and current ratings. The bilateral devices have approximately double the input capacitance of equally rated unidirectional FETs. Moreover, the available range of threshold voltages is the same as it is for unidirectional devices; however, it must be well controlled on the bidirectional MOSFET to ensure symmetry.

Despite such similarities, there is a striking difference between the bilateral switch and any other device. When it is on, the source terminal (gate return) is the electrical center of a resistive divider formed by the two gate channels. When the device is off, the source is one diode voltage drop above the more negative drain terminal. Thus, in many cases, it is necessary to isolate the input signal from the MOSFET's gate in order to avoid common mode effects.

The isolation of gate drives presents few problems, partly because of the proliferation of devices meant to accomplish that function and partly because the bilateral FETs demand so little static drive power. A designer has the choice of transformer, optical, or capacitive isolation, as well as combinations of the three. Each method has its own distinct attributes, and the selection will depend on the application requirements.

## MOSFET Power Supplies[6]

Several power supplies using a MOSFET as the power switch are now presented.

A schematic diagram of a 5 V 10 A output offline flyback regulator is shown in Figure 5-71. The primary goals in this design were simplicity and low cost. Some tradeoffs had to be accepted to meet these goals, but overall performance of the supply is still excellent.

In operation, the supply was designed to maintain a constant on-time for the DMOSFET. The operating frequency of the supply is varied to change the duty cycle. On-time for the MOSFET is approximately 7 microseconds and the operating frequency varies from 5 to 100 kHz.

---

[6]Ibid. 5.

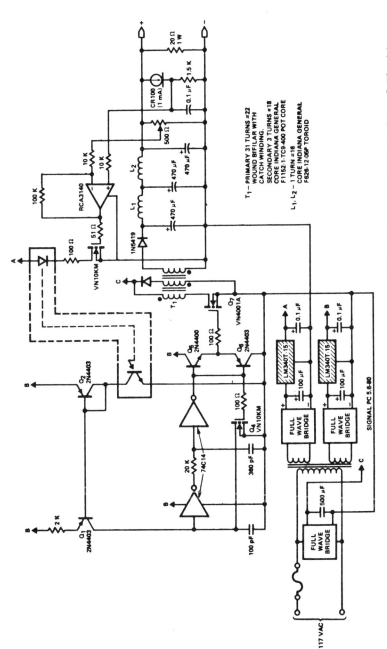

Figure 5-71  A 5 V 10 A Flyback Regulator. (Illustration courtesy of Siliconix Incorporated, Santa Clara, California)

This circuit configuration was chosen because of its ease of implementation and low parts count. However, there are two disadvantages associated with this circuit. First, the filter requirements are much more stringent because of the wide range of operating frequencies. Second, as with all flyback regulators, a larger transformer core than usual must be used because of increased energy storage requirements. These disadvantages are far outweighed, though, by the simplicity offered by this configuration for a low-power, low-cost switching supply.

In the circuit, two inverters from a hex Schmitt inverter package are used to form a constant on-time variable-frequency oscillator. The frequency of the oscillator is controlled by the amount of current flowing from the current mirror consisting of $Q_1$ and $Q_2$. The output of the CMOS oscillator is buffered by a pair of emitter-followers to drive the DMOS power transistor. The emitter-followers create a low-output impedance drive to charge and discharge the DMOS gate capacitance rapidly. The output transformer, $T_2$ is a pot core with three windings. Because of the high frequency operation of this regulator, care must be taken in the construction of the transformer to avoid excessive leakage inductance.

The transformer is wound using fifteen turns of #22 bifilar wire followed by the 3-turn secondary and then sixteen more turns of the bifilar wire. The two wires of the bifilar pair form the primary and the catch windings. By interleaving the secondary in the primary, losses are reduced to a minimum and excellent performance is obtained.

Because the operating frequency of this power supply varies with load, the output filter is more complex than is commonly used. However, the frequency response obtained from the supply is more than adequate. Figure 5-72 is a plot of the closed loop gain of the supply versus frequency. The drain voltage and current waveforms for the output transistor are shown in Figure 5-73. Switching times are very good, and waveforms are close to ideal.

The use of the bilateral switch in a Jensen saturating converter is shown in

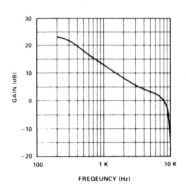

Figure 5-72 Closed-loop frequency response of the 50 Watt flyback regulator. (Illustration courtesy of Siliconix Incorporated, Santa Clara, California)

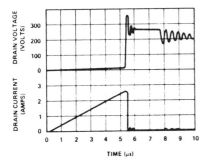

Figure 5-73 Operating waveforms of the 50 Watt flyback regulator. (Illustration courtesy of Siliconix Incorporated, Santa Clara, California)

Figure 5-74. MOSFET $Q_1$ is connected across the control winding of the saturating drive transformer core ($T_2$), regulating the converter's frequency. The floating drive, high gain, and true bidirectionality of the device mean that few parts are required to build a simple frequency stable converter.

The Jensen converter oscillates because the positive feedback from the power transformer ($T_1$) repeatedly saturates the core of the drive transformer in opposite directions. The frequency is determined by the flux capacity of the drive transformer's core and the voltage applied to its primary. In this case, the MOSFET functions as a linear shunt regulator, controlling the AC voltage present on the primary of the drive transformer.

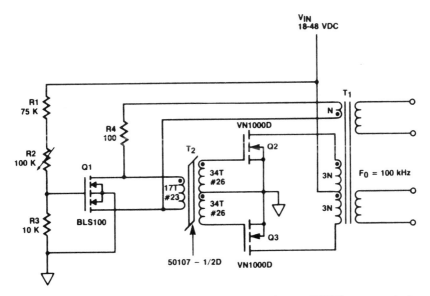

Figure 5-74 Using a bilateral MOSFET to control a Jensen DC/AC converter's frequency. Varying $R_2$ changes the voltage on $T_2$'s primary, changing $f_{out}$. (Illustration courtesy of Siliconix Incorporated, Santa Clara, California)

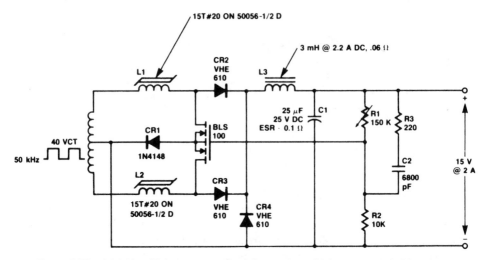

Figure 5-75   A highly efficient postregulator for use in multiple-output switching power supplies uses two 50 kHz magnetic amplifiers ($L_1$ and $L_2$) and employs a bilateral switch in a feedback loop to keep the output regulated to within $\pm 0.5\%$. $R_1$ adjusts the output voltage. (Illustration courtesy of Siliconix Incorporated, Santa Clara, California)

The regulator shown in Figure 5-75 delivers the reset current to a pair of magnetic amplifiers used for postregulating the auxiliary output of a switching mode power supply.

In a multiple-output switching converter, usually only the main (highest power) output is directly regulated by a pulse-width modulator IC. The smaller auxiliary outputs are regulated by transformer coupling to the main output. When the accuracy achieved with this approach ($\pm 5\%$ at best) is insufficient, post-regulation with magnetic amplifiers is a low-cost, simple, and reliable way to obtain finer regulation. What's more, it does so without degrading system efficiency as would a three terminal linear regulator.

In operation, the two saturating core inductors, magnetic amplifiers $L_1$ and $L_2$, function as switches (they are unsaturated at a high impedance, saturated at a low impedance), controlling the flow of power from the secondary of the switching power supply's transformer to the output. On any given half-cycle of the input, one magnetic amplifier is timing out and eventually saturates, conducting power to the output. Simultaneously, the other magnetic amplifier is being reset a controlled amount by running current backwards through it. (The current is generally taken from the output.) The amount the reverse-biased magnetic amplifier is reset determines the length of time it blocks before it conducts on the following half-cycle.

The reset current in the circuit is controlled by the bilateral MOSFET, which replaces an op amp, a bipolar transistor current source, and two steering diodes.

Since the MOSFET is an AC device, it can be connected ahead of the rectifiers ($CR_1$, and $CR_2$) rather than after them, as is normally the case, thus using the leakage current through one magnetic amplifier that is timing out to reset the other.

For that reason, neither the reset-current-steering diodes nor the reset current supply are necessary. The high gain and wide gain bandwidth product of the bilateral MOSFET permits a fast, high-gain regulator loop to be built without additional amplification. Also, because the set current in one magnetic amplifier is equal to the reset current in the opposing core, such a postregulator operates down to zero output current without requiring a dummy load.

## INSULATED GATE BIPOLAR TRANSISTORS

There are primarily two reasons for using power MOSFETs: (1) their fast switching speeds and (2) their high impedance inputs. When switching frequency is a critical factor, there is no substitute for a power MOSFET. But if the MOSFET is desirable primarily because of its high impedance input, there is an emerging technology that may be better: the insulated-gate bipolar transistor (IGBT).

Although IGBTs function as transistors, they are more closely related to MOS input SCRs. Like an SCR, they can be thought of as a complementary pair of bipolar transistors cross-connected, collector-to-base. And like an SCR, the emitter of the *pnp* transistor serves as the IGBT anode while the emitter of the *npn* serves as the cathode. The major difference between an IGBT and an SCR is that the IGBT is fabricated so that a shunting resistance appears between the base and emitter of the *npn* transistor. This shunting resistance enables the IGBT to function as a transistor by preventing the SCR-like structure from latching on.

What makes the IGBT desirable are its low forward power loss, high-impedance input, and relatively small die size for a given power level. Because IGBT conduction is through bipolar structures, on-resistance is not directly tied to blocking voltage. As a result, forward power losses are significantly less at high voltages than they are in a conventional power MOSFET. Although primarily a bipolar device, it is fabricated with a FET connected between the input terminal and the base of the IGBT's input *pnp* transistor. As a result, the IGBT exhibits the same high input impedance as does a conventional power MOSFET, which means that it can be driven by the same simple drive circuit. And since the IGBT die is considerably smaller than that of an equivalent power MOSFET, it is less expensive.

A side-effect of the IGBT's SCR heritage is that it does not contain the parasitic diode that is inherent in a power MOSFET. Whether this is an advantage or not, though, depends on the application. For example, its absence means that an additional component must be used in applications requiring a free-wheeling diode. On the other hand, reverse avalanche is not a problem.

The IGBT, however, is not without its own shortcomings, the most significant of which is its limited speed. Because its conduction relies on minority carriers, the IGBT suffers from the same storage time phenomena that plagues conventional bipolar transistors. The result is that IGBTs are relatively slow devices. In practical switching circuits, the fastest IGBTs can barely make 50 kHz with most topping out at 20 kHz.

## PACKAGING

The power transistors typically used in power converters are rated at under 2000 W and encased in metal TO-3 packages. But TO-3 packages have two significant disadvantages: (1) the baseplate serves as the collector terminal and (2) installation is labor-intensive. With the outer surface of the case serving as a terminal, a TO-3 transistor must be electrically isolated from its heat sink. The traditional solution has been the insertion of a mica washer between the TO-3 and the heat sink and an application of thermal grease to the metal surfaces. Although special electrically-isolated thermal pads are now available that replace the washer and thermal grease, mounting a TO-3 is still time consuming and not well suited to automatic assembly. Moreover, because of its metal construction, TO-3 cases are relatively expensive.

Over the past several years, there has been a move towards replacing the TO-3 with less expensive packages that are easier to mount. The most popular of these have been the TO-220 and the slightly larger TO-218. But neither of these is able to handle the power dissipation of the TO-3. To meet that need a plastic version called the TO-3P was developed. All three are rectangular molded plastic packages in which the three transistor terminals appear on pins. In many applications, these plastic packages are used on PC boards drilled so as to directly accept their leads. Often a heat sink is also bonded to the PC board. In these situations, the heat sink can itself be electrically isolated from the circuits and chassis, eliminating the need to isolate the transistor from it. This approach has been used effectively in power sources built around single-power transistors, such as low-power flyback and feedforward converters.

As power levels rise, topology shifts from single transistor converters to multiple-transistor, half- and full-bridge converters. As transistor count rises, the use of individual PC mounting heat sinks becomes increasingly costly. As power levels rise, increased power dissipation also complicates thermal management. The solution usually rests on the use of larger, more efficient heat sinks physically separated from the PC board. But when all of the transistors in a bridge converter are mounted on a common heat sink, each must be electrically isolated from the others, using mica washers of special isolating thermal pads.

Recently, a growing number of completely isolated plastic packages conforming to the TO-220, TO-218, and TO-3P footprints have become available. There have also been a number of hermetically-sealed packages conforming to the footprint of these plastic packages.

There is also a growing interest in surface mounting and an increasing number of lower power transistors are becoming available in surface mountable packages. In most cases, lead-formed TO-220s and the newer DPAK are being used. However, some innovative proprietary packages are also beginning to appear. There are limitations, however, to the amount of power that can be dissipated in a surface-mounted package. As a result, most commercial surface mounting applications will involve relatively low power.

# 6

# IC Voltage Regulators and Power Supply ICs

## HISTORICAL PERSPECTIVE

The first widely accepted IC regulators were (1) the μA723 precision voltage regulator introduced in 1968,—intended for use with positive or negative supplies—and providing an adjustable output voltage from 2 to 37 V, with up to 150 mA of load current; (2) the LM105, positive voltage regulator, introduced in 1969, that provided an adjustable output voltage from 4.5 to 45 V and between 12 and 45 mA of load current; and (3) the LM104 negative voltage regulator that supplied an output voltage from 0 to −40 V at load currents up to 12 mA operating from a single regulated supply.

These industry standards were then followed by a group of fixed-output, high-current voltage regulators, beginning with the innovative LM109. The introduction of the LM109 fixed 5 V regulator in 1971 (and subsequent fixed-output voltage regulators) provided a new set of easy to use design blocks for the power supply designer. The LM109 performs a complete regulation function with a minimum of external components, as shown in Figure 6-1. It is not necessary to bypass the output (because of internal frequency compensation), although this does improve transient response somewhat and reduces high-frequency output impedance. Input bypassing is needed, using a 0.22 μF capacitor on the output of the DC supply to prevent oscillation, under all conditions if the regulator is located an appreciable distance from the filter capacitors. Stability is also achieved by methods that provide very good rejection of load or line transients, as are usually seen with TTL logic.

The LM109 was designed to be essentially blowout proof, using new and unique circuit innovations that formed the basis for subsequent generations of IC regulator designs. Current-limiting is included to limit the peak output current to a safe value. In addition, thermal shutdown is provided to keep the IC from overheating. If internal dissipation becomes too great, the regulator will shut down to prevent excessive heating. The output also contains a crowbar clamp

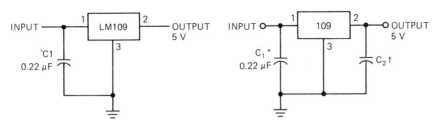

NOTES:
*REQUIRED IF REGULATOR IS LOCATED
AN APPRECIABLE DISTANCE FROM
POWER SUPPLY FILTER.
†ALTHOUGH NO OUTPUT CAPACITOR IS
NEEDED FOR STABILITY, IT DOES
IMPROVE TRANSIENT RESPONSE.

(a) INPUT BYPASSED         (b) BOTH INPUT AND OUTPUT BYPASSED

Figure 6-1   Fixed 5 V regulator connections.

to protect the load from damage. The regulator is designed so that it is not damaged in the event the unregulated input is shorted to ground when there is a large capacitor on the output.

At the time of its introduction, the internal reference that was developed for the LM109 advanced the state of the art for regulators. Not only did it provide a low voltage, temperature-compensated reference for the first time, but it was also expected to have better long-term stability than conventional Zeners. Also, noise is inherently much lower, and it can be manufactured to tighter tolerances.

In the T0-5 header, the LM109 develops 200 mA of output current, while 1 A is available in the T0-3 power package.

Next came several families of 1 A fixed-output voltage regulators: the $\mu$A7800 series (or LM140 series) of three terminal positive-voltage regulators providing thermal overload protection and internal short circuit current-limiting with fixed-output voltages of 5, 6, 8, 8.5, 12, 15, 18 and 24 V supplied by input voltages of from 5 to 24 V; the $\mu$A7900 series (or LM120 series) negative-voltage regulators with the same protection features as the $\mu$A7800 series but operating from an input of 5 to 18 V and supplying fixed negative-output voltages of 5, 8, 12 and 15 V with an output current of 1 A; and a host of other fixed 3 and 4 terminal voltage regulators (such as the $\mu$A78MG and $\mu$A79MG families).

These device families were followed by Silicon General's revolutionary series (and generation) of regulating pulse-width modulators. The series consists of the SG1524/3524; the SG1526/3526 and the SG1525A/3525A/SG1527A/3527A devices. These are briefly discussed below, beginning with the legendary and pioneering SG1524/3524 fixed-frequency, pulse-width-modulator voltage regulator control circuit, which greatly simplified power supply circuit design.

The first IC controller, the SG1524/3524, was introduced in 1976 and rapidly became an industry standard. It was primarily as a result of the ease of use of the SG1524/3524 that switching power supplies became economically feasible.

From the block diagram shown in Figure 6-2, it can be seen that the SG1524 contains the elements necessary to implement either single-ended switching regulators or DC/DC converters of several different configurations. The device includes a voltage reference, error amplifier, constant-frequency oscillator, pulse-width modulator, pulse-steering logic, dual alternating-output switches, and current-limiting and shutdown circuitry.

The regulator operates at a frequency that is programmed by one timing resistor ($R_T$) and one timing capacitor ($C_T$). $R_T$ establishes a constant charging current for $C_T$ which is fed to the comparator, providing linear control of the output pulse width by the error amplifier. The SG1524 contains an on-board 5

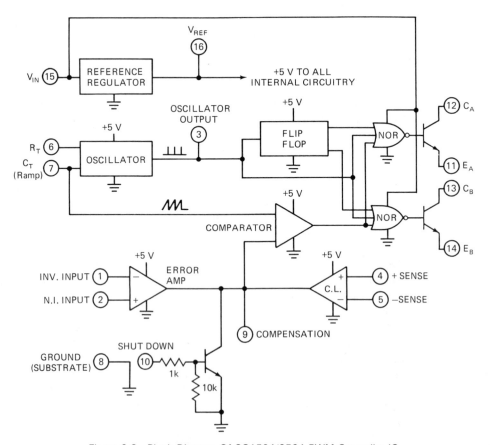

Figure 6-2  Block Diagram Of SG1524/3524 PWM Controller IC.

V regulator that serves both as a reference and for powering internal control circuitry. The regulator is also useful in supplying external support functions.

The reference voltage is lowered externally by a resistor divider to provide a reference within the common mode range of the error amplifier. An external reference may also be used. The power supply output is sensed by a second resistor divider network to generate a feedback signal to the error amplifier. The amplifier output voltage is then compared to the linear voltage ramp at $C_T$. The resulting modulated pulse out of the high-gain comparator is then steered to the appropriate output pass transistor ($Q_1$ or $Q_2$) by the pulse-steering flip-flop, which is synchronously toggled by the oscillator output. The oscillator output pulse also serves as a blanking pulse to assure that both outputs are never on simultaneously during the transition times. The width of the blanking pulse is controlled by the value of $C_T$. The outputs may be applied in a push-pull configuration in which their frequency is half that of the base oscillator, or paralleled for single-ended applications in which the frequency is equal to that of the oscillator. The output of the error amplifier shares a common input to the comparator with the current-limiting and shutdown circuitry and can be overridden by signals from either of these inputs. This common point is also available externally and may be employed to control the gain of, or to compensate for, the error amplifier; it can also provide additional control to the regulator.

Although complete controllers, such as Motorola's MC3420 and Texas Instrument's TL494, were subsequently introduced, the SG1524/3524 was still favored due to its versatility and generalized architecture. While the SG1524/3524 provides complete operational control, practical supplies often require other additional circuitry to interface with other elements in a system, to protect against fault conditions, to adjust for inaccuracies, or to improve control during power sequencing. The introduction of the "A" and "B" versions of the SG1524 and the Unitrode UC1843 addressed this problem by incorporating these functions "on-chip" and thereby provided improved performance with the same pin configuration as the SG1524.

Silicon General's SG1524A was introduced in 1982. Its improvements included a precision bandgap reference, $+5.00$ V trimmed to $\pm 1\%$ initial accuracy, which not only provides a standard for the regulation loop, but also powers most of the internal control circuitry. This eliminates adverse effects caused by fluctuating supply voltage. Under-voltage lockout circuitry further prevents spurious turn-on commands to external power transistors due to low supply voltages, and holds output devices off when the reference voltage drops below $+4.5$ V. When the reference rises to $+7$ V, the output drivers are enabled. During power up, this allows the controller's internal circuitry to be stabilized in the appropriate state before the turn-on voltage is reached, thus preventing start-up glitches. Other improvements included extending the common-mode range of the error amplifier from $+2.5$ V to $+V_{ref}$, improving the input voltage range and response time of the current limit amplifier, enhancing oscillator design to allow up to 400 kHz operation, adding double pulse suppression logic with last pulse memory, and modifying the output transistors to operate at continuous collector currents of 100 mA.

The SG1525A/1527A and SG1526 represent significant advancements to the original SG1524 PWM controller IC. The SG1525A/1527A series of pulse-width modulator integrated circuits are designed to offer improved performance and lowered external parts count when used to implement all types of switching power supplies. Key device features include the following:

- 8 to 35 V operation
- 5.1 V reference trimmed to $\pm 1\%$
- 100 Hz to 500 kHz oscillator range
- Separate oscillator sync terminal
- Adjustable deadtime control
- Pulse-by-pulse shutdown
- Internal soft-start
- Input undervoltage lockout
- Latching PWM to prevent multiple pulses
- Dual source/sink output drivers
- 16-pin DIP

The output stages are totem-pole designs capable of sourcing or sinking in excess of 200 mA. The SG1525A output stage features NOR logic, giving a LOW output for an off state. The SG1527A utilizes OR logic which results in a HIGH output level when off.

Completing the family is the SG1526 series of devices. The 1526 is a high-performance monolithic pulse-width modulator circuit designed for fixed-frequency switching regulators and other power control applications, and featuring the following:

- 8 to 35 V operation
- 5 V reference trimmed to $\pm 1\%$
- 1 Hz to 350 kHz oscillator range
- Dual 100 mA source/sink outputs

Included in an 18-pin dual-in-line package are a temperature-compensated voltage reference, sawtooth oscillator, error amplifier, pulse-width modulator, pulse-metering and steering logic, and two low-impedance power drivers. Also included are protective features such as soft-start and undervoltage lockout, digital current limiting, double pulse inhibit, a data latch for single pulse metering, adjustable deadtime, and provision for symmetry correction inputs. For ease of interface, all digital control ports are TTL and 4000B-series CMOS compatible. Active LOW logic design allows wired-OR connections for maximum flexibility. This versatile device can be used to implement single-ended or push-pull switching power supplies of either polarity, both transformerless and transformer coupled. The difference between the various device part numbers are as follows:

- the SG1526 operates from −55°C to +150°C
- the SG2526 from −25°C to +150°C
- the SG3526 from 0°C to +125°C.

This initial series of PWM Regulators was followed by a higher speed second generation series of PWM controller ICs (the Unitrode UC1840, UC1842, and UC1846/1847). The UC1840 was optimized for high-efficiency boot-strapped primary side operation in forward or flyback power converters, thus reducing system costs. The following key features are included:

- All control, driving, monitoring, and protection functions included
- Low-current, off-line start circuit
- Feed-forward line regulation over 4 to 1 input range
- PWM latch for single pulse per period
- Pulse-by-pulse current-limiting, plus shutdown for overcurrent fault
- No startup or shutdown transients
- Slow turn-on and maximum duty-cycle clamp
- Shutdown upon over or undervoltage sensing
- Latch off or continuous retry after fault
- Remote, pulse-commandable start/stop
- PWM output switch usable to 1A peak current
- 1% reference accuracy
- 500 kHz operation
- 18-pin DIP

In addition to startup and normal regulating PWM functions, these devices offer built-in protection from overvoltage, undervoltage, and overcurrent fault conditions. This monitoring circuitry contains the added features that any fault will initiate a complete shutdown with provisions for either latch off or automatic restart. In the latch off mode, the controller may be started and stopped with external pulsed or steady state commands.

The UC1840's PWM output stage includes a latch to insure only a single pulse per period and is designed to optimize the turn off of an external switching device by conducting during the off-time with a capability for both high peak current and low saturation voltage.

Other switching regulator ICs appearing concurrently with the UC1840 included Fairchild's μA78S40 and Texas Instrument's TL494, 496, and 497 circuits. The μA78S40 contains a temperature-compensated reference, a duty-cycle controllable oscillator with active current-limit circuit, an error amplifier, a high-current high-voltage output switch, a power diode, and an uncommitted operational amplifier. The output is adjustable from 1.3 to 40 V with output current to 1.5 A without external transistors and an overall efficiency of 75%.

The TL494 pulse-width modulator control circuit eliminates the many IC and discrete components used to regulate switching power supplies, and it contains an on-chip 5 V regulator, two error amplifiers, adjustable oscillator, dead-time control comparator, pulse-steering flip-flop, and output control circuitry. The uncommitted output transistor provides either common-emitter or emitter-follower output capability. Push-pull or single-ended output operation may be selected through the output control function.

The TL497A incorporates on a single monolithic chip all the active functions required in the construction of a switching voltage regulator: a precision 1.22 V reference, a pulse generator, a high-gain comparator, current-limit sense and shut-down circuitry, a catch diode, and a series pass transistor—as shown in Figure 6-3. The TL497A was designed to offer versatility and to optimize the ease of its use in the various step-up, step-down, and voltage inversion applications requiring high efficiency.

Introduced in 1983, the UC1842 family of high-speed (500 kHz operation) control ICs provides in an 8-pin DIP the necessary features to implement off-line, fixed-frequency current-mode control schemes with a minimum of external parts. The superior performance of this technique can be measured in improved line regulation, enhanced load response characteristics, a simpler and easier to design control loop, and inherent pulse-by-pulse current limiting.

Protection circuitry includes built-in undervoltage lockout and current-limit-

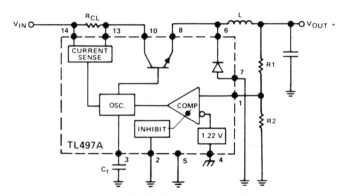

Figure 6-3 TL497A block diagram.

ing. Other features include fully latched operation, a 1% trimmed bandgap reference, and start-up current of less than 1 mA.

These devices feature a totem-pole output designed to source and sink high peak current from a capacitive load, such as the gate of a power MOSFET. Consistent with $n$-channel power devices, the output is LOW in the off state.

The UC1846/1847 family of control ICs provide the same basic performance features as the UC1842 family. The 1846/1847 is encased in a 16-pin DIP, and contains a shutdown function, which can initiate either a complete shutdown with automatic restart or latch the supply off; it also provides 200 $\mu$A totem pole outputs. The UC1846 features LOW outputs in the off state, while the UC1847 features HIGH outputs in the off state.

Currently produced IC switching regulators use pulse-width modulation to control switching; nearly every supplier complements its alternate source parts with proprietary regulator designs, with most betting that MOSFETs will be the dominant power switch within a few years; and since supplies handle extremes in current and voltage, protection schemes are found on-board nearly every IC.

In addition to pulse-width modulators and controllers, a variety of various supervisory sense and driver circuits was also developed to aid the power supply designer. Several of these are now presented.

The SG1543/3543 "Power-Supply Output Supervisory Circuit" contains, in a 16-pin DIP, all the functions necessary to monitor and control the output of a sophisticated power supply system. Overvoltage sensing with provision to trigger an external SCR "crowbar" shutdown; an undervoltage circuit which can be used to monitor either the output or sample the input line voltage; and a third op amp/comparator usable for current sensing are all included in this IC, together with an independent, accurate reference generator.

Both over- and undervoltage sensing circuits can be externally programmed for minimum time duration of fault before triggering. All functions contain open collector outputs which can be used independently or wire-ORed together, and although the SCR trigger is directly connected only to the overvoltage sensing circuit, it may be optionally activated by any of the other outputs, or from an external signal. The overvoltage circuit also includes an optional latch and external reset capability.

The current sense circuit may be used with external compensation as a linear amplifier or as a high-gain comparator. Although nominally set for zero input offset, a fixed threshold may be added with an external resistor. Instead of current-limiting, this circuit may also be used as an additional voltage monitor.

The reference generator circuit is internally trimmed to eliminate the need for external potentiometers, and the entire circuit may be powered directly from either the output being monitored or from a separate bias voltage.

The 18-pin DIP SG1544/3544 was designed to provide all of the operational features of the SG1543/3543 but with the added advantage of uncommitted inputs to the voltage sensing comparators. This allows monitoring of voltage levels less than 2.5 V by dividing down the internal reference supply.

The SG1542/3542 and 3523/3423 voltage sensing circuits provide the control functions necessary to protect sensitive electronic circuitry from overvoltage transients or the effects of voltage regulator failure. They are designed for use with an external SCR "crowbar" for immediate shutdown of the power supply but additionally provide logic level outputs for regulator turn-off and/or operator or system out-of-tolerance indication.

The SG1542/3542 and SG3523/3423 contain an accurate and stable 2.6 V reference which allows the sensing threshold to be set predictably without the need for potentiometers. Only on the SG1542/3542 does the uncommitted availability of both polarity inputs to the sensing comparator allow a wide flexibility of use, including the ability to sense voltages less than the reference voltage. The SG1542/3542 comes packaged in a 14-pin DIP; whereas the SG3523/3423 is encased in an 8-pin DIP.

The SG1549/3549 Current Sense Latch was designed to provide pulse-by-pulse current-limiting for switch-mode power supply systems, but many other applications are also feasible. Its function is to provide a latching switch action upon sensing an input threshold voltage, with reset accomplished by an external clock signal. This device can be interfaced directly with many pulse-width modulating control ICs, including the SG1524, SG1525A, and SG1527A. With delays in the range of 200 ns, this latch circuit is ideal for fast reaction sensing to provide overall current limiting, short circuit protection, or transformer saturation control.

Several high-speed driver circuits are available to provide the current required to drive high-speed power switching transistors and to interface digital control logic with high current loads—such as the SG1629/3629 and SG1627/3627, respectively. The SG1629/3629 is intended to interface between the secondary of a drive transformer and the base of an *npn* switching device. Positive drive current can be made constant with an external programming resistor, or can be clamped with a diode to keep the switching device out of saturation. Negative turn-off current is derived from a negative voltage generated in an external capacitor. All operating power is supplied by the transformer secondary, and these devices can be floated at high levels with respect to ground for off-line bridge converters.

The SG1627/SG3627 contains two independent drivers, which will either source or sink up to 500 mA of current. The sink transistor is designed as a saturating switch while the source transistor can be used either as a switch or as a constant current generator with external resistor programming. Each half of this device contains both inverting and noninverting inputs, which have two-volt thresholds for high noise immunity. Either input can be used alone to switch the output, or one input can be strobed with the other. Both device families have been designed to directly interface with the SG1524 PWM regulator control IC.

## ADVANCED IC REGULATOR CIRCUITS

The bywords in integrated circuit design and development are denser chips, higher speed operation, greater on-chip functionality (integration of system functions), and higher reliability. Power supply ICs (regulators, control circuits and single chip DC/DC converters) are no exception. Since the second edition of *Design of Solid State Power Supplies* was published in 1981, several generations of power supply ICs have appeared:

1. Pulse-width modulators and controllers, supervisory, and sense circuits; examples of which were just presented, and
2. High-speed subsystem ICs, which will be discussed next.

The driving trends affecting the design of switching power control ICs are as follows:

1. Expansion of the use of current-mode techniques in new designs with operation at frequencies up to one megahertz.
2. The continued trend toward higher frequencies to achieve higher power densities and higher efficiencies.
3. The developing use of variable frequency to achieve increased power supply stability in hysteretic control, as well as achieving higher power density in resonant-mode.
4. The increasing use of custom circuits to gain added specialized performance in specific applications. Essentially, the designer pays for only what he needs to use.
5. Integrated on-chip power and control circuitry to reduce board space and design complexity. This is currently being done only with bipolar, while mixed processes such as CMOS-bipolar (BICMOS) will be utilized as it becomes economically justified. The use of multichips—a control chip plus a separate power switch—for single-ended high-current, high-voltage applications, each optimized in performance and yield, within a single IC package, can provide a more cost-effective approach in some cases than the more glamorous, but much more complex, use of mixed processes requiring large silicon areas.

PWM current-mode techniques simplify the power supply designer's design task while improving supply performance. In current-mode, the power switch duty-cycle is directly controlled by inductor current instead of output voltage. As a result, current-mode design eliminates the inductance effect of the output LC filter from the voltage regulating loop. In doing so, the current-mode loop establishes a one-pole output filter whose phase shift cannot exceed 90°. In voltage-mode, the output filters exhibit two poles with a phase shift of up to 180°, requiring careful phase compensation circuit design to avoid possible long-term instability.

Further relevant characteristics gained from current-mode control include inherent current-limiting, automatic feed-forward, automatic symmetry correction, improved load response over voltage-mode PWM, and improved paralleling capability.

Since the initial introduction of current-mode ICs in 1983, acceptance of current-mode techniques in new designs has been growing. The continuing trend toward current-mode is not simply due to more IC manufacturers bringing out revisions of the industry accepted circuits but to the effort of IC designers to transform more of the power supply designers' concerns into low-cost silicon solutions.

There are three current-mode control techniques: (1) constant frequency with turn-on at clock time, (2) hysteretic, and (3) constant off-time. All current-mode control involves sensing, directly or indirectly, inductor current. Hysteretic control senses both peak and valley currents, while the remaining two control techniques sense only the peak inductor current. By providing feed-forward at both the input and output of the power supply, hysteretic control can provide the designer, for the first time, with truly unconditional stability.

The highest speed IC regulators, such as Siliconix's Si9100 (see Figure 6-4), use current-mode control almost exclusively, and they generally handle higher voltages than their voltage-mode counter parts. The 1 MHz Si9100 accepts either a low voltage DC supply or a 10 to 70 V unregulated DC supply that it then steps down to ±5 V. Notably, it drives its integrated power DMOS FET at 2 A—one of the highest current ratings available.

The Unitrode UC 3825 was one of the first current-mode control PWM ICs to operate in the megahertz range. It has a typical propagation delay of 50 ns and accepts a maximum input voltage of 30 V, which it steps down to 13.5 V. The UC3825 combines both current- and voltage-mode techniques, handling the former's two feedback loops and the latter's single. Through its dual totem-pole output, it delivers 1.5 A, enough to drive the input capacitance of larger power MOSFETs. The UC3825 operates at a constant frequency with turn-on at clock time.

The International Rectifier IR2100 power converter IC accepts 100 to 450 V of unregulated DC and generates 15 V DC (±1 V) with 500 mA of drive. Separate sense and power feedback pins enable engineers to raise the output voltage to, say, 24 V, while feedback and feed-forward loops tighten the IR2100's control. The feedback loop, as is customary, revolves around an error

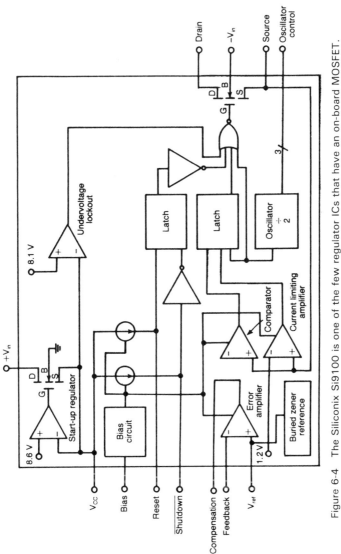

Figure 6-4   The Siliconix Si9100 is one of the few regulator ICs that have an on-board MOSFET.

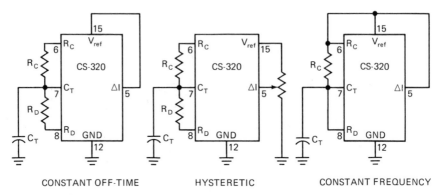

CONSTANT OFF-TIME          HYSTERETIC          CONSTANT FREQUENCY

Figure 6-5   The three current-mode control operating modes of the CS320.

amplifier whose output sets the switching point of the PWM comparator. The feed-forward loop produces a duty-cycle proportional to the ratio of output to input voltage, providing increased efficiency.

Cherry Semiconductor's 1 MHz CS320 uses all three current-mode control methods, as shown in Figure 6-5: hysteretic, constant off-time, and constant frequency—selectable by the power supply designer (usually current-mode control is characterized by constant frequency control). Because hysteretic control (implemented here for the first time on an IC) senses both the peak and valley inductor currents, it affords the tightest regulation and renders the chip unconditionally stable.

The CS320 also provides programmable undervoltage lockout (eliminating the requirement to stock different part numbers for different UVLO requirements) and built-in synchronization, which allows paralleling of power supplies without assigning a master/slave status. The CS320, like the UC3825, is designed to drive power MOSFETs directly.

Motorola's 300 kHz PWM current-mode MC34129 IC regulator is designed to interface with current sensing power MOSFETs. On the input side, the MC34129 pulls DC power from the telecommunication line; on the output side, it delivers about 8 to 9 V at 1 A to the external power device. The lower leg of the totem-pole output stage contains an SCR instead of a transistor—lowering supply current, conserving die area, and raising efficiency.

Whether an IC regulator uses current- or voltage-mode control, pulse-width modulation prevails, especially for high speed, heavy load applications. But in systems handling light loads, PWM faces difficulties, as the PWM duty-cycle always must meet the requirement for extremely narrow pulses.

Frequency-modulation, the common alternative, responds more closely to changes in the load. It reduces the duty-cycle by manipulating the interval between pulses, not by manipulating the width of the pulses. Thus, efficiency rises, although the lower operating frequency—a function of longer intervals—sometimes causes filtering problems.

GE Solid State's CA1523 was the first IC to combine pulse-width modulation (PWM) with pulse-interval modulation (PIM). Along with incorporating over-current sensing, soft-start, and the ability to operate up to 200 kHz, the CA1523 uses PIM to offset the low load problems inherent in PWM, where minimum pulse-width constraints can create design problems, reducing overall design efficiency. By using several external resistors and capacitors, the design engineer can tune the maximum width, rise and fall times, and frequency of the output pulses.

Achieving a significant increase in power density requires going well beyond the 100 kHz level. As power supply operating frequencies are driven higher, external components increasingly become the limiting factor in supply operation, with commutation losses in high-speed rectifiers contributing to a major loss of efficiency, and with loss of efficiency, heat sinking becomes a concern, offsetting to a varying degree potential size/weight reductions. Obviously, nothing is achieved if enhanced IC performance cannot be directly reflected in power supply performance.

If high power density is a key design objective, then zero-current sine-wave switching using resonant-mode techniques is a means of meeting it. Resonant-mode can deliver up to three to five times the power density of squarewave switchers. Linear Technology's (Canada, now called Gennum) 16-pin DIP LD405 (Figure 6-6) combines frequency modulation with resonant-mode switching to gain speed. Resonant-mode switching, which is based on half sine-waves, turns transistors on and off only when circuit current is zero, in other words, only at the zero crossings. This technique, dramatically reduces switching power losses, permitting the chip to operate at switching rates of 300 kHz to 1 MHz. Advantages include less stress on the power switch, less conducted and radiated EMI, and significantly less power dissipation.

All shutdown modes, overvoltage, and overload, are synchronous (the last output pulse is completed), defaulting to soft-start once the fault condition is removed. This important feature ensures that the shutdown condition always occurs at zero current, rather than at an unknown combination of voltage and current, which may risk power supply reliability.

Utilizing the same control philosophy as the LD 405/CS3805, Unitrode developed the UC1860 resonant controller in a 24-pin DIP, with frequency variation capability from 1 kHz to 3 MHz (see Figure 6-7). The UC1860 uses a flexible-fault-disposition capability, where, under a shutdown condition, the restart delay pin on the chip can be programmed to shut the supply down permanently after a fault, restart after a delay, or restart immediately once the fault has been removed.

Regulator ICs are clearly handling the high voltages that off-line power sup-

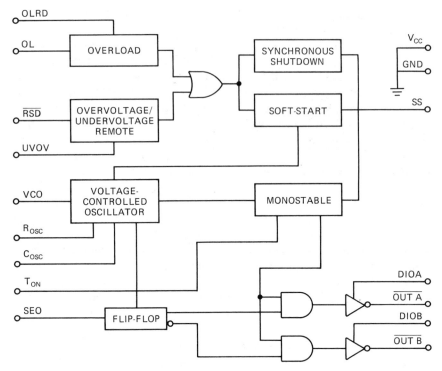

Figure 6-6  LD 405 functional block diagram.

plies demand, but isolating these systems from the AC line voltage strongly challenges designers. Safety is paramount; yet system tolerances, regulation, and transient response must not be compromised either. Integrated Power Semiconductor has developed an isolation scheme that splits regulation between two chips. The first chip, the largely digital IP1P00, controls the primary winding, and the second, the mostly analog IP1P01, serves the secondary winding. Between the two devices lies the isolation interface—a simple coupling transformer that remains virtually transparent to system designers (see Figure 6-8). This configuration is unique in that it contributes no phase shift to the regulating loops, giving designers more useable phase margin.

The Maxim Integrated Products MAX641 10 W step-up switching regulator is a revolutionary single-chip DC/DC converter that takes input voltages from 2 to 16 V and delivers a fixed output of +5, +12, or +15 V. Since the chip contains an $n$-channel power MOSFET (325 mA peak) and a catch diode at its output, it can serve either as the switching element in a low-power system (less than 250 mW) or simply drive an external high power MOSFET or bipolar transistor, as shown in Figure 6-9.

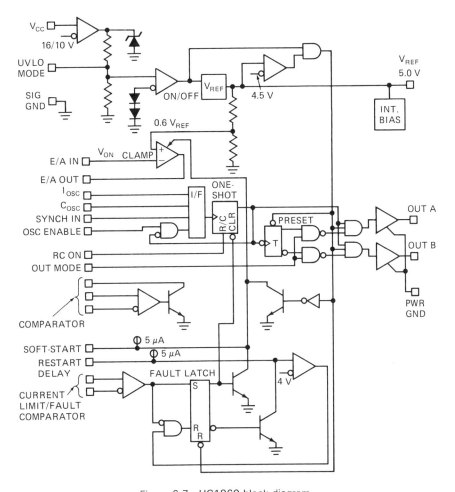

Figure 6-7   UC1860 block diagram.

The MAX 641's intended operation is as follows. When the output voltage drops below the preset (or externally set) value, the error comparator switches HIGH and connects the internal oscillator to the $L_x$ and EXT outputs. EXT is typically connected to the gate of an external $n$-channel power MOSFET (although the external MOSFET isn't necessary for many low power applications). When EXT is activated, the MOSFET turns on and off at the oscillator frequency. When EXT is HIGH, the MOSFET switches on, and the inductor current increases linearly, storing energy in the coil. When EXT switches the MOSFET off, the coil's magnetic field collapses, and the voltage across the inductor changes polarity. The voltage at the catch diode's anode then rises until

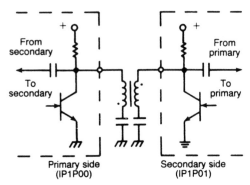

Figure 6-8 To alleviate concerns about safety and performance in off-line switching supplies, Integrated Power Semiconductors uses a coupled transformer to isolate its unique primary-winding and secondary-winding control ICs.

the diode is forward-biased, delivering power to the output. As the output voltage reaches the desired level, the error comparator inhibits EXT until the load discharges the output capacitor to a point at which the error comparator connects the oscillator to $L_x$ and EXT generates an output once again.

The MAX641 doesn't have a $V_{in}$ pin. Input power to start the DC/DC converter is supplied via the external inductor (and external diode, if used), to the $V_{out}$ pin. If an external catch diode is used, the cathode should be connected to $V_{out}$. Once the converter is started, it is powered from its own output voltage.

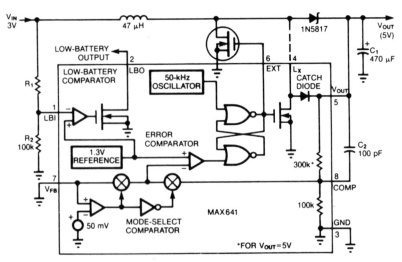

Figure 6-9 Equipped with a low-power MOSFET and a catch diode, the MAX641 can stand alone in DC/DC converters delivering 5 mW to 10 W.

This boot-strap design ensures that the external MOSFET has the maximum gate drive and, consequently, the minimum $R_{on}$. One external component that must be selected is the inductor.

Siltronic's S424 PWM regulator is designed to operate from very low alkaline or nickel-cadmium input voltages, below 1 V. The S424 works from a 0.9 to 1.7 V DC source and boosts it to either 3 V at 15 mA or 5.5 V at 10 mA. The device is encased in a surface-mounted plastic package.

Rather than being designed and optimized for a specific application or power supply topology, Linear Technology Corp's (California) LT1070 current-mode switching regulator integrates power and control circuitry on one chip and can be used in many different types of systems. Oriented toward simplifying switching power supply design, the LT1070 operates at up to 45 kHz with input voltages of 3 to 60 V and provides an output voltage of up to 75 V while supplying from 5 A to 15 A of load current.

Referring to the block diagram, of Figure 6-10, the switch is turned on at the start of each oscillator cycle, and it is turned off when switch current reaches a predetermined level. Control of the output voltage is obtained by using the output of a voltage sensing error amplifier to set the current trip level. This technique has several advantages. First, it provides immediate response to input

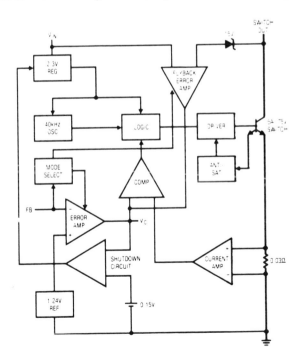

Figure 6-10   LT1070 block diagram.

voltage variations, unlike ordinary switchers, which have notoriously poor line transient response. Second, it reduces the 90° phase shift at midfrequencies in the energy storage inductor, which greatly simplifies closed-loop frequency compensation under widely varying input voltage or output load conditions. Finally, it allows simple pulse-by-pulse current-limiting to provide maximum switch protection under output overload or short circuit conditions.

A low dropout internal regulator provides a 2.3 V supply for all internal circuitry on the LT1070. This low dropout design allows input voltage to vary from 3 to 60 V, with virtually no change in device performance. A 40 kHz oscillator, which is the basic clock for all internal timing, turns on the output switch via the logic and driver circuitry. Special adaptive antisaturation circuitry detects the onset of saturation in the power switch and adjusts driver current instantaneously to limit switch saturation, minimizing driver dissipation and providing rapid turn-off of the switch.

A 1.2 V bandgap reference biases the positive input of the error amplifier. The negative input is available for output voltage sensing. This feedback pin has a second function: when pulled LOW with an external resistor, it programs the LT1070 to disconnect the main error amplifier output and connects the output of the flyback amplifier to the comparator input. The LT1070 will then regulate the value of the flyback pulse with respect to the supply voltage. This flyback pulse is directly proportional to the output voltage in the traditional transformer-coupled flyback regulator. By regulating the amplitude of the flyback pulse, the output voltage can be regulated with no direct connection between input and output. The output is fully floating up to the breakdown voltage of the transformer windings, and multiple floating outputs are easily obtained with additional windings. A special delay network inside the LT1070 ignores the leakage inductance spike at the leading edge of the flyback pulse to improve output regulation.

The error signal developed at the comparator input is brought out externally. This pin ($V_C$) has four different functions. It is used for frequency compensation, current-limit adjustment, soft-starting, and total regulator shutdown. During normal regulator operation, this pin sits at a voltage between 0.9 V (LOW output current) and 2.0 V (HIGH output current). The error amplifiers are current-output ($g_m$) types, so this voltage can be externally clamped for adjusting current limit. Likewise, a capacitor-coupled external clamp will provide soft-start. Switch duty cycle goes to zero if the $V_C$ pin is pulled to ground through a diode, placing the LT1070 in an idle mode. Pulling the $V_C$ pin below 0.15 V causes total regulator shutdown, with only 50 $\mu$A supply current for shutdown circuitry biasing.

The device can be effectively utilized in most power supply topologies, such as buck, boost, buck-boost, Cuk, flyback, forward, and isolated output. The LT1070 comes packaged in a standard 5-pin TO-220. The LT-1071 and LT-1972 are 2.5 A and 1.25 A versions, respectively.

Thus, one can see that there have been significant advances from the relatively simple multi-use variable-output voltage regulator and fixed-output voltage regulator to today's power supply systems on a single chip. Since the introduction of the PWM IC controller chip, a virtual plethora of power supply ICs has become available, and one thing is certain more advanced ICs will be forthcoming in the future.

## INTEGRATED CIRCUIT REGULATOR APPLICATION EXAMPLES

In this section, many different examples of using IC regulators are presented, beginning with the $\mu$A723 and ending with the advanced ICs. This discussion is not meant to be exhaustive in nature, but rather illustrative of the many possibilities available using ICs in power supply design.

### $\mu$A723 Application Examples

Figures 6-11 and 6-12 show the basic low and high voltage regulator configurations respectively. In Figure 6-11 $V_{out} = R_1/(R_1 + R_2)$ 7 V where $R_1 + R_2$ > 1.5 K. In Figure 6-12, a resistor attenuator network is required to regulate any voltage above the internal reference voltage. The frequency compensation uses feedback from the $R_1$, $R_2$ resistor divider output. In this circuit $V_{out} = (R_1 + R_2)/R_2$ 7 V.

Figure 6-13 shows how the current capability of the basic regulator configuration is increased by use of an external pass transistor. In this circuit $V_{out} = (R_1 + R_2)/R_2$ 7 V.

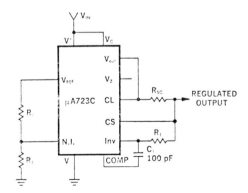

TYPICAL PERFORMANCE

| | |
|---|---|
| Regulated Output Voltage | 5 V |
| Line Regulation ($\Delta V_{IN}$ = 3 V) | 0.5 mV |
| Load Regulation ($\Delta I_L$ = 50 mA) | 1.5 mV |

Figure 6-11  Basic low voltage regulator ($V_{out}$ = 2 to 7 V).

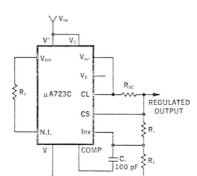

| | |
|---|---|
| Regulated Output Voltage | 15 V |
| Line Regulation ($\Delta V_{IN}$ = 3 V) | 1.5 mV |
| Load Regulation ($\Delta I_L$=50 mA) | 4.5 mV |

Note: $R_3 = \dfrac{R_1 R_2}{R_1 + R_2}$ for minimum temperature drift.

$R_3$ may be eliminated for minimum component count.

Figure 6-12  Basic high voltage regulator ($V_{out}$ = 7 to 37 V).

Negative output voltages may be regulated using the circuit of Figure 6-14. Since the regulator in this configuration sees only the output voltage applied across the device, the input voltage is limited by the breakdown voltage of the *pnp* series pass elements. In this circuit $V_{out} = 3.5 \, (R_2 + R_1)/R_1$.

In many cases, the dissipation which occurs under current-limited short circuit conditions is excessive and, if accounted for, results in excessive heat sink area. The use of positive feedback as shown in Figure 6-15 in the current limit

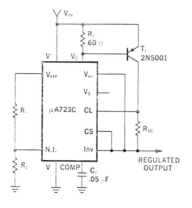

Figure 6-13  Positive voltage regulator with external *pnp* pass transistor.

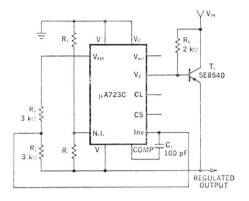

TYPICAL PERFORMANCE

Regulated Output Voltage        −15 V
Line Regulation ($\Delta V_{IN}$ = 3 V)   1 mV
$I_L$ = 100 mA)   2 mV

Figure 6-14   Negative voltage regulator.

circuitry produces short circuit currents which approach zero. The short-circuit-current knee can be calculated from the following formula.

$$I_{lim} = (\text{Knee}) = \frac{1.0}{R_{SC}} \qquad (6\text{-}1)$$

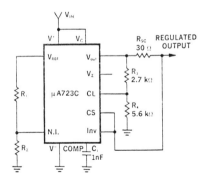

TYPICAL PERFORMANCE

Regulated Output Voltage        $+$5 V
Line Regulation ($\Delta V_{IN}$ = 3 V) 0.5 mV
Load Regulation ($\Delta I_L$ = 10 mA)   1 mV
Current Limit Knee                20 mA

Figure 6-15   Foldback current limiting.

### LM105 Switching Regulator

An example of IC switching regulator design is shown in Figure 6-16. This regulator was designed as the input series switching regulator of an on-board spacecraft computer. The electrical design parameters were:

| | |
|---|---|
| Input voltage: | 22 to 29.25 VDC |
| Output voltage: | +5 VDC ± 1% |
| Output ripple: | 50 mV p-p |
| Output power: | 2.5 W |
| Efficiency: | 75% |

Once the basic circuit was designed, according to the procedures outlined earlier, certain modifications were necessary. These are listed below:

- $C_{49}$ and $C_{50}$ minimize transient spikes.
- $C_{51}$ minimizes the impedance that the regulator sees.
- $C_{47}$ and $C_{48}$ slow down the switching speed in order to reduce the transient spikes being generated by the regulator switching action. This had a detrimental effect on the efficiency: it decreased from 75 to 60%. It was decided to accept this efficiency as a tradeoff for reduced EMI spikes.

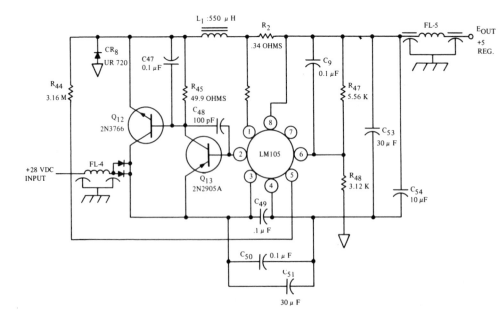

Figure 6-16   Practical +5 VDC voltage regulator using the LM105.

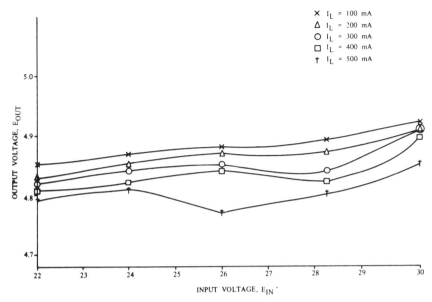

Figure 6-17   Output voltage versus input voltage.

- $L_1$ presented a problem: It was not possible to get the desired inductive resistance compatible with small inductor size. Thus $R_2$ was added.
- $C_{54}$ was added on the output for more EMI filtering, as was *FL-5*.
- *FL*-4 was added to the input to provide input protection to the regulator from EMI.

Figures 6-17 and 6-18 depict the results of electrical performance tests of this regulator. Figure 6-17 shows the output voltage variation as a function of the input voltage for given loads. The variation of output ripple voltage as a function of load current for a given input voltage is shown in Figure 6-18. In addition to these figures, the regulator of Figure 6-16 was found to have an efficiency of 67% over the input voltage range of 22 VDC to 31 VDC.

## Remote Shutdown

Electronic shutdown is used in some applications where, under certain conditions, the removal of power from the load is desired. This function can easily be achieved with multiterminal regulators, such as the $\mu$A723, LM105, and LM104, since these regulators are either equipped with shutdown capability or the non-inverting input of the error amplifier is accessible.

The $\mu$A723 regulator may be turned off by pulling down the compensation terminal, thereby shunting the drive current for the output stage to ground. The

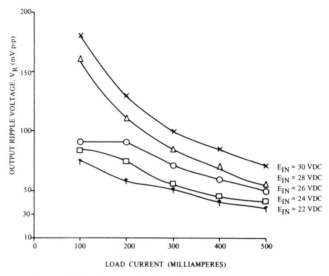

Figure 6-18   Output ripple voltage versus load current.

simplest method of achieving this in a positive regulator is shown in Figure 6-19a. When the current limiting function is required, an external transistor may be substituted ($Q1$ in Figure 6-19b). The logic input may be from any positive voltage source, for example, TTL or CMOS, capable of driving greater than 100 $\mu$A into the $CL$ terminal or $Q1$ base. Typically, $R3$ may be 3 k$\Omega$ from a 5 V TTL system, or 10 k$\Omega$ from a 10 V CMOS system. To protect the output stage from excessive reverse base-emitter voltage transients during the shutdown, $D1$ should be included when the output voltage $V_o$ is greater than 10 V. $R4$ reduces the peak current that flows when $Q1$ saturates.

Remote shutdown, when applied to a negative regulator, requires the additional circuitry to the right of the dashed line in Figure 6-19c. In operation, a logic LOW input, $V_{IL}$(max), holds $Q3$ off, disabling the shutdown circuit. A logic HIGH input, $V_{IH}$(min), from a TTL or CMOS gate turns $Q3$ on with the base drive limited by $R8$. Resistor $R5$ is calculated in the normal worst case manner for the series pass devices selected.

$$R_S = \frac{(V_{IN(min)} - V_o - 2V_{BE}{}^*)}{Q2I_{B(max)}} \tag{6-2}$$

When $Q3$ is turned on, $D1$ is forward-biased at a current limited by $R7$. The ratio of $R5$ and $R7$ is calculated such that the output of the supply is always at ground when the logic input is HIGH.

---

*This term becomes 3 $V_{BE}$ if $D2$ is included.

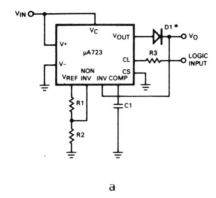

a

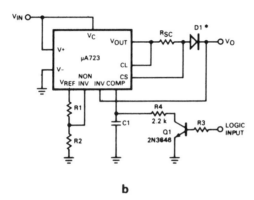

b

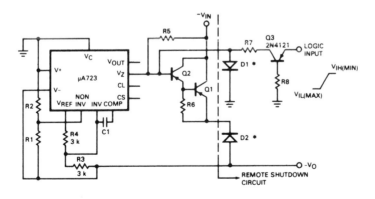

c

Figure 6-19   Remote shutdown.

$$\frac{R_7}{R_8} = \frac{V_{IH(min)}}{V_{IN(max)}} \tag{6-3}$$

This formula guarantees that the junction of $R5$ and $R7$ is always positive during shutdown ($R8 = 10\ R7$) to give a forced beta of 10 for $Q3$, fully saturating that device. Diode $D2$ protects the output devices from reverse base-emitter transient voltages, and should be included when the output voltage is greater than the combined base-emitter breakdown voltages of the series pass devices.

With the 4-terminal regulators, the control terminal is the inverting input, and therefore some external parts are necessary to turn off these devices. The same applies for the 3-terminal devices. The 3-terminal regulator circuit of Figure 6-20 has a remote shutdown feature. Under normal conditions, $Q2$ is on and provides the base current of $Q1$.

Figure 6-21 shows the SG1549 Current Sense Latch being used for current sensing in the input of a simple buck converter using the SG1524 PWM control IC. The value for $R_{SC}$ is determined by dividing the 100 mV input threshold by the peak current desired. High-frequency noise, or switching transients, can usually be eliminated by a small capacitor between pins 3 and 4.

The current-shutdown command can be coupled into the SG1524 by either connecting HI-OUT (pin 6) to the SG1524's shutdown pin or, as shown in Figure 6-21, by using the LO-OUT pin to pull the compensation terminal low. In either case, activation of the current sense latch will tend to discharge the compensation capacitor, $C_C$ which may cause slow recovery from pulse limiting. Keeping the value of $C_C$ as small as possible within the requirements of voltage loop stability will minimize this effect; however, slow turn-on from current limit is often desirable and can be optimized by using the LO-OUT signal to discharge a soft-start network, instead of coupling directly into the SG1524.

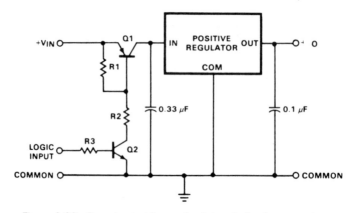

Figure 6-20   Remote shutdown of a 3-terminal voltage regulator.

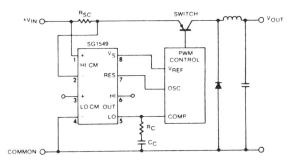

Figure 6-21   Using the SG1549 to sense input current to a simple buck converter and interface directly to the SG1524. (Reprinted with permission from Silicon General *1986 Product Catalog,* Application Note SG1549)

## Low Line Sensing[1]

In many types of feed-forward or push-pull converters, current protection may be provided by sensing through an emitter-resistor referenced to ground on the primary side of an output transformer. The fast reacting SG1549 can easily sense secondary overload as reflected back to the primary and, additionally, provide protection from unbalanced transformer saturation. When using the LO CM inputs as shown in Figure 6-22, the HI CM inputs should be shorted together.

While the LO CM inputs may be connected directly across a sense resistor, $R_{SC}$, a small low-pass filter, $R1$-$C1$, is often helpful in removing high-frequency transients. It must be remembered that the 500 ohm input impedance to the LO CM terminals will cause the use of $R1$ to increase the effective threshold; however, this also offers the possibility of an easily adjustable threshold by incorporating a potentiometer at the input to the SG1549.

Coupling the current-shutdown command back to the control circuit may be done in several ways as described above: but, again, the fastest approach is to go directly to the output switches. Figure 6-23 shows such an approach by adding two external shutdown transistors, $Q1$ and $Q4$. In this circuit, these transistors perform double-duty by the use of $C2$, $R3$ and $R4$ to generate a positive pulse when the main power switches, $Q2$ and $Q3$, are commanded off by the SG1524. Turn-off signals from either the PWM or the SG1549 are summed together through diodes $D2$ and $D3$ to $Q1$ and $Q4$.

One problem often experienced with using pulse-by-pulse current-limiting with a push-pull converter is half-cycling caused by limiting on one period without full recovery in time for the next. A maximum duty-cycle clamp, formed by $R1$, $R2$ and $D1$ in Figure 6-23, minimizes this effect by holding the error amplifier out of saturation when the feedback voltage begins to fall.

---

[1]Courtesy of Silicon General Corporation.

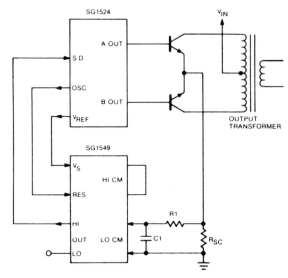

Figure 6-22 When sensing emitter current, a small input filter is often useful in eliminating transients to the SG1549. (Reprinted with permission from Silicon General *1986 Product Catalog,* Application Note SG1549)

Another convenient way to tie the output of the SG1549 into the PWM control in higher power applications is by using the SG1627 Dual Interface Driver and connecting the LO OUT terminal directly to the two non-inverting inputs of the

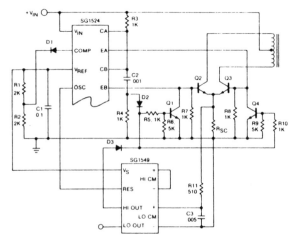

Figure 6-23 Fastest turn-off response is achieved by taking the output of the SG1549 direct to the power switches through turn-off transistors, Q1 and Q4. (Reprinted with permission from Silicon General *1986 Product Catalog,* Application Note SG1549)

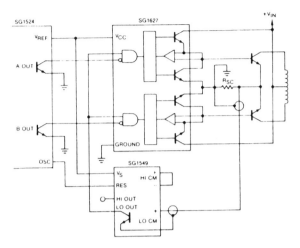

Figure 6-24   An SG1627 provides both power gain for the voltage control signal as well as a high-speed interface for the SG1549's current protection. (Reprinted with permission from Silicon General *1986 Product Catalog,* Application Note SG1549)

SG1627, as shown in Figure 6-24. The non-inverting inputs of the SG1627 will force the outputs off, regardless of the commands on the inverting inputs, and do so within 100 nanoseconds.

### Constant Current Regulator

Any 3-terminal regulator can be used as a constant current regulator as shown in Figure 6-25. The current $I_{out}$ which dictates the regulator type to be used is determined by the following equation.

$$I_{out} = \frac{V_o}{R1} + I_Q \qquad (6-4)$$

where $V_o$ is the regulator output voltage and $I_Q$ is the quiescent current.

The input voltage $V_{in}$ must be high enough to accommodate the dropout voltage at the low end, but must not exceed the maximum input voltage rating at the high end.

### Adjustable Dual Voltage Regulator

A comparator and a few external parts can transform two IC regulators into a dual-voltage regulator that tracks, has safe-area limiting, and provides protection

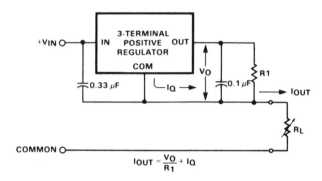

Figure 6-25 Positive output constant current regulator.

against short-circuits and thermal overload, as shown in Figure 6-26. The combination can deliver currents to 500 mA, and the positive and negative outputs of the two integrated regulators can be adjusted independently with variable resistors, $R_1$ and $R_2$.

The component values in the figure provide output voltages from the positive regulator—the $\mu$A78MG—over a range of 5 to 30 V and from the negative— the $\mu$A79MG—from $-2.2$ to $-27.2$ V. But care must be taken not to exceed the maximum voltage ratings of the $\mu$A741.

To achieve tracking, the common terminals of the two regulators are both tied to the output of the $\mu$A741. Thus a decrease in, say, the positive output because of temperature or a line or load variation would cause a reduction in the $\mu$A741 inverting input voltage. This, in turn, would raise the potential of the common terminals of the two IC regulators to reduce the output of the negative regulator.

In this way changes in one of the outputs tend to produce a corresponding change in the common regulator terminals, to force the other output to track. Since each regulator has its own reference, there is no slaving. The outputs and the degree of tracking are proportional to the ratio $(R_1 + R_3)/(R_2 + R_4)$.

## Multipurpose Regulator Circuit[2]

Figure 6-27 shows the schematic diagram of a multipurpose voltage regulator circuit that features both fixed and adjustable outputs. The basic component of this circuit is a three-terminal 5 V IC regulator (the $\mu$A7805).

[2]"Handy Supply Provides Fixed and Variable Outputs," by John Predescu. Reprinted from _Electronics_, June 27, 1974; Copyright © McGraw-Hill, Inc., 1974.

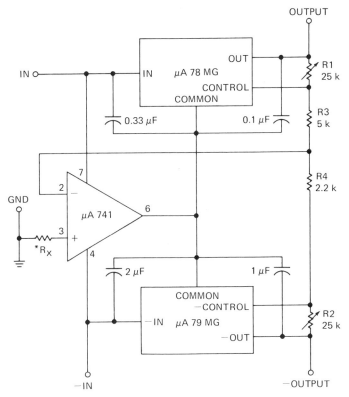

OUTPUT

IN o

GND

*R_X

−IN

−OUTPUT

*R_X = PARALLEL COMBINATION
OF (R1 + R3) AND (R2 + R4)

Figure 6-26  A tracking regulator for positive and negative voltages supplies independently adjustable voltages and load currents to 500 mA.

The variable (8- to 17 V) output is obtained by employing a conventional op amp in the regulator's output loop. For the fixed (5 V) output, only half the transformer secondary voltage is taken so that the regulator's power rating is not exceeded.

Output current is 200 mA. The double-pole double-throw switch permits the same IC regulator and diode bridge to be used for both the fixed and variable outputs. The switch bypasses the operational amplifier network and taps the transformer's secondary voltage at the halfway point. A 5 V output can then be obtained from the IC regulator without exceeding this device's power dissipation rating.

If a second supply circuit is built and the positive side of its output grounded, a complete ±15 V op-amp supply can be made. Regulation is better than 1%.

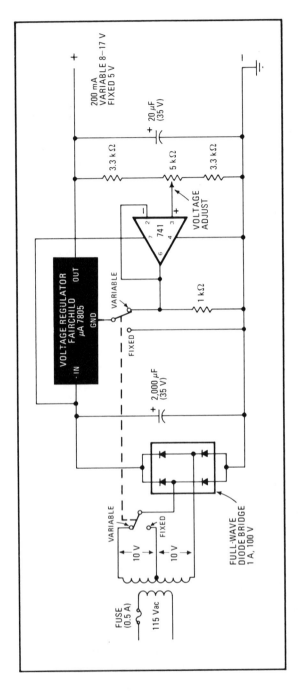

Figure 6-27  Multipurpose Voltage Regulator.

## Power Supplies That Regulate to Zero Volts[3,4]

Several examples are presented of power supplies that regulate to zero volts. In the first circuit shown in Figure 6-28 the LM723 voltage regulator, which provides 12 volts at 1 ampere, must be biased with a negative supply voltage at its $-E_{in}$ port (pin 5) for proper operation. This voltage is provided by the switching inverter shown within the dotted lines.

The LM111 voltage comparator is configured as an astable multivibrator that oscillates at a frequency of about 10 kHz. With the aid of the 1 mH inductor, which generates the counterelectromotive force required to produce a negative potential from a switched-source voltage, the inverter delivers a well-regulated $-7.5$ V to the $-E_{in}$ port of the 723.

The magnitude of this voltage is essentially equal to that of the regulator's internal reference voltage, $V_{ref}$, appearing at pin 4, and properly biases its voltage-reference amplifier. This condition in turn precipitates a condition in the amplifier whereby $V_{ref}$ clamps to ground potential. Thus the output voltage may be adjusted throughout its maximum possible range by potentiometers $R_1$ and $R_2$. Although the potential of $V_{ref}$ as measured with respect to ground has been changed, the circuit will retain the regulating properties of the 723. Both the line and the load regulation of the supply are 0.4%.

The second circuit using two 723 IC regulators, shown in Figure 6-29 allows the reference voltage to be adjusted all the way down to the offset voltage of the regulator's internal op amp. REGULATOR₁ and its associated circuitry form a bias supply that provides a voltage of about $-7$ V for the V-terminal of the main regulator (REGULATOR₂). Since the non-inverting input of this regulator is connected to the common ground of the circuit, its reference voltage appears to be $+7$ V with respect to this V-terminal.

There will be a 7 V drop across resistors $R_2$ and $R_3$. When $R_1$ is set to its minimum value, the circuit's output voltage will be equal to the reference voltage. If the output is measured with respect to the V-terminal of REGULATOR₂, it will be 7 V. But if it is measured with respect to the common ground, it will be zero.

The maximum voltage available at the output is determined by the value of resistor $R_2$. For the component values shown there, the maximum voltage may be set anywhere from 16 to 39 V. But voltages above 30 V will not be regulated very well because the supply is using a 24 V transformer ($T_2$).

The equation for the output voltage is:

---

[3]"Dc-dc Power Supply Regulates Down to 0 Volt," by P. R. K. Chatty and A. Barnabee. Reprinted from *Electronics*, January 19, 1978; Copyright © McGraw-Hill, Inc., 1978.

[4]"Regulating Supply Voltage all the way Down to Zero," by Brother Thomas McGahee. Reprinted from *Electronics*, June 27, 1974; Copyright © McGraw-Hill, Inc., 1974.

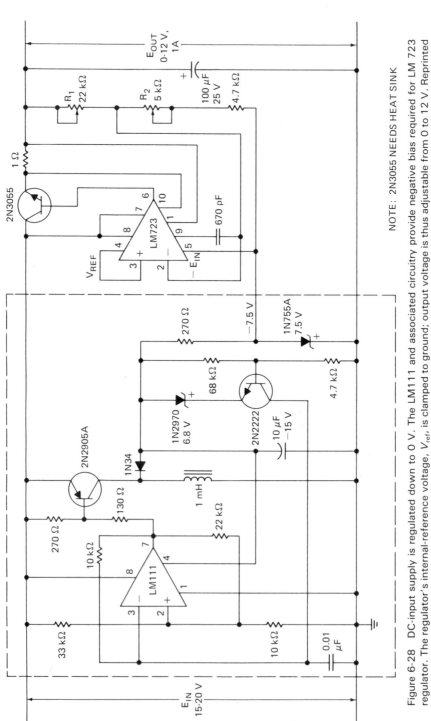

NOTE: 2N3055 NEEDS HEAT SINK

Figure 6-28 DC-input supply is regulated down to 0 V. The LM111 and associated circuitry provide negative bias required for LM 723 regulator. The regulator's internal-reference voltage, $V_{ref}$, is clamped to ground; output voltage is thus adjustable from 0 to 12 V. Reprinted from *Electronics*, January 19, 1978; Copyright © McGraw-Hill, Inc. 1978.

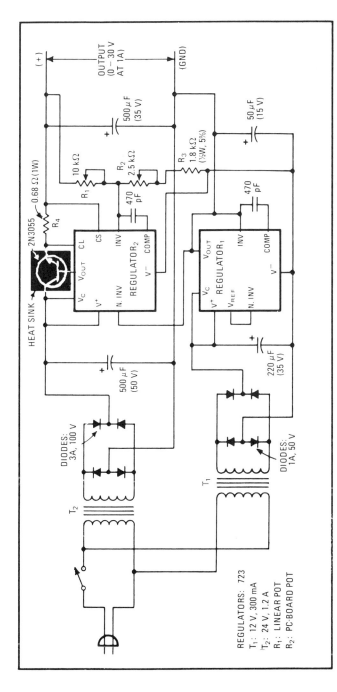

Figure 6-29 This power supply, which employs two IC voltage regulators, produces a regulated output voltage of between 0 and 30 V. REGULATOR₁ provides the bias voltage for REGULATOR₂ so that the latter device can operate with respect to a common ground. The lowest regulated output voltage, then, is approximately zero, rather than the reference voltage of REGULATOR₂. Reprinted from *Electronics*, June 17, 1974; Copyright © McGraw-Hill, Inc. 1974.

$$E_{\text{out}} = R_1 V_{\text{B}}/(R_2 + R_3) \tag{6-5}$$

where $V_{\text{B}}$ is the absolute value of the bias voltage (7 V in this case). The bias supply normally will be producing about 12 milliamperes of current. Under worst-case conditions, however, it may be required to provide a maximum of 40 mA. Transformer $T_1$, therefore, should be a 12 V unit capable of supplying at least 50 mA (since REGULATOR$_1$ will require some current itself).

The transistor at the output of REGULATOR$_2$ boosts the circuit's output current. Resistor $R_4$ acts as the current-limiting resistor.

## AC-to-DC Converter

Figure 6-30 shows an approach to AC/DC converter design that uses integrated circuits as the semiconductor components. Discrete transistors and diodes are not used because of their inherent lack of fail-safe features as well as due to the present availability of power regulator ICs at reasonable costs.

The converter is completely foolproof. Both the LM309 and LM320 voltage regulators have internal current limiting to limit peak output currents to a safe

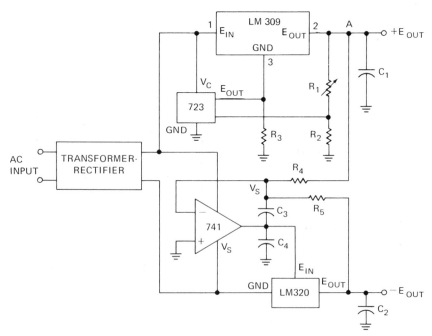

Figure 6-30   ICs replace the conventional discrete transistors and diodes in this AC-to-DC converter.

value. Also, thermal shutdown is provided to keep each IC safe from over-heating. Each regulator is thus "blow-out" proof.

The 741 op amp is also a foolproof device. It has internal overload protection on the inputs and output; it doesn't latch up when the common-mode range is exceeded; and it is also free from oscillation.

The circuit operates in the classical manner: an AC voltage is applied to a transformer and rectifier to obtain raw, but useable, positive and negative DC voltages. These voltages are fed, respectively, to the LM309 and LM320 series regulators to obtain the desired output voltages. In this circuit, both the LM309 and LM320 are used as positive and negative regulator pass transistors due to their high current capabilities and foolproof protection features. The 723 regulator is used only as a precision voltage reference which regulates the ground terminal of the LM309.

Resistors $R_1$ and $R_2$ are used to set up the positive output voltage and as a supply to the 723 reference. The 741 is used to invert the positive voltage to a negative voltage for the LM320. The connection point A on the positive output is fed back through $R_4$ to the inverting input of the 741 to enable the negative output to track the positive output. Capacitors $C_1$ and $C_2$ are used to smooth the output voltage.

With the following component values the circuit provides a +15 V, 1 A output and a −15 V, 1 A output with a total circuit efficiency of 80%.

$$R_1 = 5.6k\Omega, R_2 = 5.6k\Omega, R_3 = 300\ \Omega, R_4 = 20k\Omega, R_5 = 20k\Omega$$

$$C_2 = C_2 = 47\ \mu F, C_3 = 0.05\ \mu F, C_4 = 0.03\ \mu F$$

The positive output can be adjusted to any value (within the LM309's operating specifications) by means of $R_1$ and $R_2$ or by replacing the LM309 with an LM340 ($\mu$A7800) fixed voltage regulator of the desired output voltage rating. The negative output can also be changed by selection of a different output voltage rating LM320 regulator.

## Basic PWM IC Applications[5]

The circuit shown in Figure 6-31 is a low-current flyback converter that provides ±15 V at 20 $\mu$A from a +5 V regulated line.

In this application, the two output stages are connected in parallel and used as emitter-followers to drive a single external transistor. Since the currents in the secondary of a flyback transformer are out of phase with the primary current, current limiting is very difficult to achieve. In this circuit, protection is provided through the use of a soft-start circuit. If either output is shorted, the transformer

---

[5]Ibid. 1.

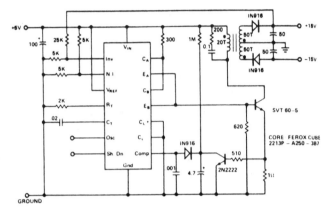

Figure 6-31   SG1524 used in a flyback converter application. (Reprinted with permission from Silicon General *1986 Product Catalog,* Application Note SG1524)

saturates, providing more current through the drive transistor. This current is then sensed and used to turn on the 2N2222 which resets the soft-start circuit and turns off the drive signal. If the short remains, the regulator will repetitively try to start up and reset with a time constant set by the soft-start circuit. Removing the short allows the regulating loop to re-establish control.

For higher current applications, the single-ended conventional switching regulator can be used, as shown in Figure 6-32. In this case, an external *pnp* darlington transistor is used to provide a 1 A current switch. The SG1524 has the two outputs in parallel, connected as a grounded emitter-amplifier for effective 0–90% duty cycle modulation. The current sense resistor is inserted in the

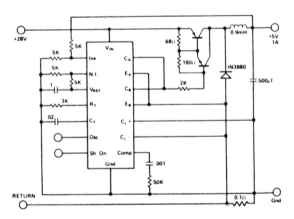

Figure 6-32   Single-ended IC switching regulator circuit. (Reprinted with permission from Silicon General *1986 Product Catalog,* Application Note SG1524)

ground line and the voltage across it used for constant current limiting. Note that in addition to the divider resistors and frequency setting $R_T C_T$, a phase-compensation resistor and capacitor are used to stabilize the loop now that an inductor has been added.

Push-pull outputs are used in the transformer-coupled DC/DC regulating converter of Figure 6-33. Here, the outputs of the SG1524 are connected as separate emitter-followers driving external transistors.

Current-limiting occurs in the primary circuit in this application for several reasons. First, it's easier to live within the ±0.3 V common-mode limits of the current-limit amplifier. Second, since this is a step-down application, the current—and, therefore, the power in the sense resistor—is lower. Third, and finally, if the output drive were to become nonsymmetrical causing the transformer to approach saturation, the resultant current spikes will shorten the pulse width on a pulse-by-pulse basis, providing a first order correction. The oscillator is set to run at 40 kHz to obtain a 20 kHz signal at the transformer; that is, the oscillator must be set at twice the desired output frequency since the SG1524's internal flip-flop divides the frequency by 2 as it switches the PWM signal from one output to the other.

The circuit illustrated in Figure 6-34 is a push-pull, +28 V to +5 V converter operating at 50 kHz using the SG1524B regulating pulse-width modulator. Capacitor C3 acts as a high-frequency bypass for the IC supply line; C6 is the high current reservoir for the power stage, and C4 and R5 set the oscillator frequency at 100 kHz. When divided by two by the action of the internal toggle flip-flop, this becomes 50 kHz at the power transformer. The +5 V reference is filtered against high-frequency noise pick-up by R2 and C1, and applied to the non-inverting input of the error amplifier. The inverting input is connected to the power supply output terminals to form the negative feedback loop required

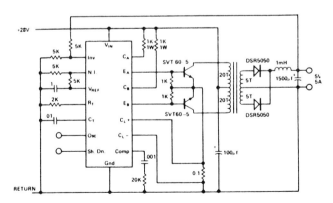

Figure 6-33   Push-Pull transformer-coupled circuit. (Reprinted with permission from Silicon General *1986 Product Catalog,* Application Note SG1524)

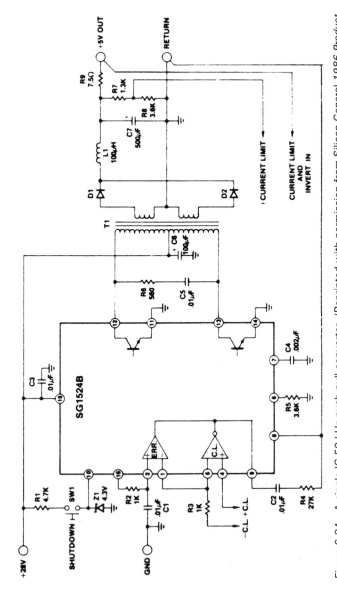

Figure 6-34   A single IC 50 kHz push-pull converter. (Reprinted with permission from Silicon General 1986 *Product Catalog*, Application Note SG1524)

for regulation. $R3$ minimizes the effects of input-offset-bias current by equalizing the source impedance seen by each error amplifier input terminal. Closed-loop stability is provided by frequency compensation components $R4$ and $C2$, using the common technique of cancelling one of the two poles of the $LC$ output filter with an open loop zero in the error amplifier.

In the power section of the supply, the two output transistors are used to directly drive a center-tapped transformer. A snubber network consisting of $C5$ and $R6$ modifies the inductive load line seen by each transistor, while the transformer itself is wound on a small ferrite core; the turns ratio is $3:1$. Rectifier diodes $D1$ and $D2$ are Schottky junction devices to maximize efficiency at $+5$ V output. Filtering is provided by $L1$, wound on a permalloy powder toroid, and $C7$.

Current sensing is done directly in the output line. A foldback ratio of 7.5 to 1 is obtained with the given values of $R7$, $R8$, and $R9$. The divider formed by $R7$ and $R8$ applies a back-bias of 1.3 V, or 6.5 times the current limit threshold, when the supply output is at $+5.0$ V. Peak output current before the onset of current limiting is 200 mA, while short circuit current is only 25 mA. Rapid turn-off of the control circuit is accomplished by closing $SW1$. $R1$ and $Z1$ limit the maximum voltage applied to the SHUTDOWN terminal to less than $+5$ V.

This design illustrates the controller's ability to perform all the major control functions required during start-up, normal regulation, and overload conditions.

Figure 6-35 shows the UC1840 used in an AC/DC converter application. In

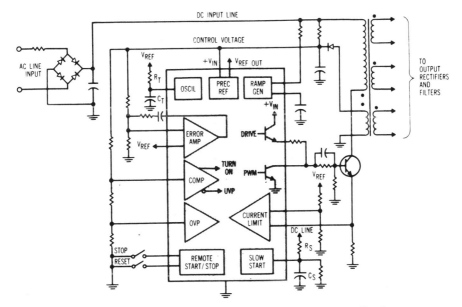

Figure 6-35   UC1840 PWM used in an AC/DC converter application.

this circuit, low-current start-up is accomplished through a sensing circuit which holds off all drive current to the power switch until the input voltage has risen to the point where there is adequate energy to accomplish a predefined start sequence. However, with the control on the primary side of the transformer, adequate regulation through feedback from the output is often a problem. This is circumvented, though, by feed-forward line sensing in the UC1840. This portion of the device accomplishes an open-loop line regulation by automatically and instantaneously varying the drive pulse width in a manner that is inversely proportional to the input voltage such that a constant volt-second product is always delivered to the output. The large dynamic range of this circuit allows the implementation of power supplies which operate equally well from either 110 or 220 V power sources.

The power supply of Figure 6-35 requires only the high-voltage power switch, the transformer, and a small number of passive components in addition to the IC. This particular configuration uses the high-voltage input line for starting, but receives more efficient low-voltage operating power from an auxiliary winding, $N_2$, on the transformer.

### Multiple Output Forward Converter[6]

The schematic diagram of a dual-output, high-efficiency forward converter using an SG1525A PWM IC is shown in Figure 6-36. The SG1525A's internal oscillator provides output pulses at A and B, alternately turning on power switches $Q1$ and $Q2$. The same pulses trigger flip-flop $IC_2$, which turns on $S_2$ (for A HIGH) and $S_1$ (for B HIGH). This action allows differential amplifier $IC_3$, for example, to provide a feedback output to the SG1525A IC when $Q_1$ is on.

The SG1525A compares differential amplifier $IC_3$'s (or $IC_4$'s) output with its internal reference; if the output exceeds the reference, the PWM terminates the pulse for the appropriate output transistor. When $IC_1$ generates the next power pulse (turning on the formerly off transistor), flip-flop $IC_2$ configures the feedback network so that the correct output controls the power switch's pulse width.

Current-limiting occurs pulse-by-pulse using the SG1525A's shutdown terminal. Here, the current-sense network develops appropriate signals to terminate a power pulse at a preset peak-current point. The next power pulse for the alternate power switch occurs at a later time (as set by $IC_1$'s internal oscillator), and therefore doesn't interfere with the other pulse's current setting.

The circuit of Figure 6-36 provides 75 to 85% efficiency over a 23 to 36 V input range (in a 5 V/10 V output configuration).

---

[6]Henderson, Ross, "PWM IC Controls Two Supply Outputs," *EDN*, June 23, 1983. Used with permission.

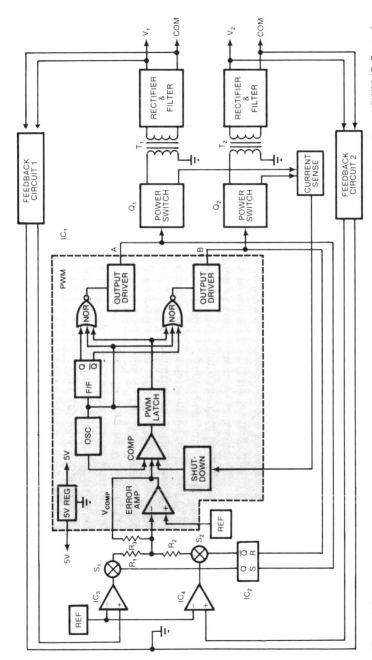

Figure 6-36 A dual-output switching supply using the SG1525A PWM IC. (Reprinted with permission from "PWM IC Controls Two Supply Outputs," *EDN*, June 23, 1983. © 1988 Cahners Publishing Company, a Division of Reed Publishing USA)

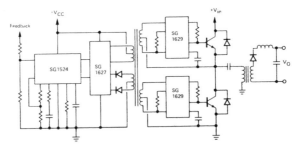

Figure 6-37   Block Diagram of highly integrated switchmode half-bridge converter. (Reprinted with permission from Silicon General *1986 Product Catalog,* Application Note SG1627)

## Half-Bridge Converter Design Considerations[7]

The use of power-supply support-circuit building blocks, such as the SG1627 and SG1629 High Current Power and Switch Drivers respectively, along with PWM IC controllers greatly simplifies the power supply design process, as shown by the block diagram of Figure 6-37. In this half-bridge converter circuit, the SG1524 drives the SG1627 directly which, in turn, provides the signal conditioning to develop the drive and reset commands to an interstage drive transformer. The secondary windings of that drive transformer are directly coupled to a pair of SG1629 floating drivers, which are then used to command the external 5 A switching transistors that form the high-power output stage of the power supply.

This converter utilizes ICs that have been designed to offer a maximum degree of flexibility, while incorporating what would otherwise be a substantial amount of discrete circuitry—and at the same time providing greater reliability.

Now, let's look at some of the details required to effect a viable design. In Figure 6-38, two SG1629's are used to provide the drive signals for the power transistors in a 5 A half-bridge switching supply. The drive transformer is shown with 10 V drive signals on the primary winding which, with a 2 : 1 transformer turns ratio, provides a 5 V peak signal on each half of the secondary. When the drive command is present on one secondary, it is translated into a constant current through the source transistor by the use of the sense resistor, $R_{CS}$, which, in this case, provides a constant 700 mA into the base of the external *npn* transistor. At the same time, the 20 microfarad external capacitor is being charged with a current through the rectifier in the SG1629 and the lower half of the secondary winding. While this is occurring, the opposite phase signal is being applied to the lower SG1629 circuit, which serves to further enhance the charge on its external capacitor while maintaining the power switch in the off state.

---

[7]Ibid. 1.

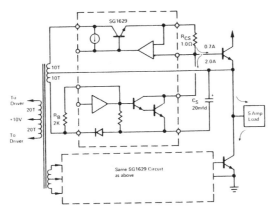

Figure 6-38 Use of the SG1629 in a 5 A, half-bridge converter. (Reprinted with permission from Silicon General *1986 Product Catalog,* Application Note SG1629)

When the drive command terminates, the voltage at both ends of the secondary winding goes to zero. Since there is approximately $-4$ V at the emitter of the sink transistor while its base is being driven through the external drive resistor, $R_B$, to zero volts, the sink transistor then immediately turns on and pulls a high current $Ib_2$ pulse out of the external transistor and through the capacitor. This current only flows as long as it is available from the stored charge within the base of the external transistor, since the source has been gated off. After that charge is depleted, the sink transistor remains on, thereby insuring a negative reverse voltage at the base of the switch transistor.

The performance of the SG1629 can be illustrated by the waveforms shown in Figures 6-39 through 6-41. In Figure 6-39, the command signal from one channel of the control circuitry and the waveform of the drive transformer pri-

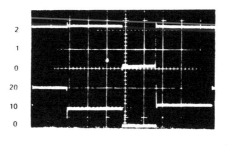

Time Base:     5 µSec/Division
Upper Trace:   P.W.M. Control Signal
Lower Trace:   Drive Transformer Primary

Figure 6-39 Input control to the drive transformer. (Reprinted with permission from Silicon General *1986 Product Catalog,* Application Note SG1629)

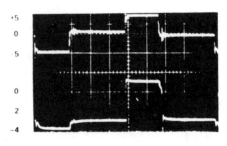

Time Base:      5 μSec/Division
Upper Trace:    Drive Transformer Secondary
Lower Trace:    Power Transistor Base Voltage

Figure 6-40   The base voltage delivered to the external power-switching transistor by the SG1629. (Reprinted with permission from Silicon General *1986 Product Catalog,* Application Note SG1629)

mary voltage are shown. The voltage on the secondary winding, referenced to the centertap and the power transistor emitter, is pictured in Figure 6-40. Also shown in this photograph is the input voltage at the base of the external *npn* transistor.

Note that at the very first portion of this waveform, when the opposite side is on, there is an additional negative charge supplied to the capacitor so that the maximum reverse base-to-emitter voltage is −4 V. During the off-time, the action of the sink transistor maintains a negative voltage bias of approximately −3 V on the base of the power transistor. When the drive command is given to turn on, the base voltage goes positive to the 0.7 or so volts necessary to turn it on and at turn-off, goes negative again. The important action is shown

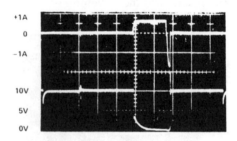

Time Base:      5 μSec/Division
Upper Trace:    Transistor Base Current
Lower Trace:    Collector Voltage with $R_L = 2\Omega$

Figure 6-41   Base-current and collector-voltage waveforms of the external switching transistor. (Reprinted with permission from Silicon General *1986 Product Catalog,* Application Note SG1629)

in Figure 6-41 which shows the actual base current of the power transistor with a scale of one amp per division. Both the constant turn-on $Ib_1$ of about 0.75 A and the peak $Ib_2$ of close to $-2$ A can be seen along with the collector voltage waveform with a 5 A resistive load. Remembering that the time base of all these waveforms is 5 $\mu$s per division, one can see approximately one $\mu$s delay between the turn-off signal at the base and the actual turn-off of the collector of the output transistor. Although this turn-off response is primarily a function of the transistor design, it is safe to say that any power switching transistor should perform faster with this form of base drive.

One can also see a soft knee in the collector waveform at turn-on where the power transistor is not saturated instantaneously. This is partially a result of the turn-on characteristics of the transistor and partially a result of the finite rise-time of the base current through the drive circuitry. This rise-time is primarily a function of the leakage inductance of the drive transformer which opposes a sudden change in current from zero to maximum value. Thus, the transformer should be designed to minimize, to the greatest extent possible, the leakage inductance, using a minimum number of turns and a maximum coupling between turns. In this example, a ferrite pot core with a diameter of approximately ¾ in. was used to configure the drive transformer.

One area of concern in the turn-off circuitry is circuit operation with a very narrow pulse command. Since the charge on the external capacitor is developed during the turn-on command, narrow pulse widths accomplish the transfer of a minimum amount of energy. As the drive-command pulse widths get narrower, there is a point where the voltage on the external capacitor begins to fall off. With the circuit components shown earlier, this loss of $Ib_2$ occurs at approximately 2 $\mu$s pulse widths. Figure 6-42 shows the base current and collector voltage waveforms at narrow pulse widths where $Ib_2$ has diminished from 2 A

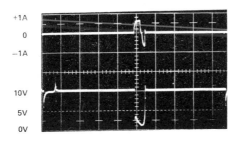

Time Base:      5 μSec/Division
Upper Trace:    Transistor Base Current
Lower Trace:    Collector Voltage with $R_L$ = 2 Ω

Figure 6-42  Operation with narrow pulse widths. (Reprinted with permission from Silicon General *1986 Product Catalog,* Application Note SG1629)

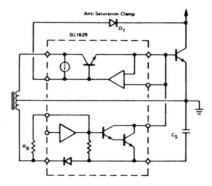

Figure 6-43   Use of the SG1629 in a load-dependent drive configuration. (Reprinted with permission from Silicon General *1986 Product Catalog,* Application Note SG1629)

down to approximately 0.75 A. Further reductions in pulse width reduce $Ib_2$ current ultimately to zero. This characteristic is, of course, a function of the time constants in the total circuit, and some compromise or optimization can be achieved by appropriate selection of capacitor values and secondary drive voltages.

The circuit of Figure 6-38 uses constant-current drive which is an advantage if the load current happens to be relatively constant. But in many applications this is not the case.

The SG1629 may also be used to provide a base drive proportional to load demand by adding an anti-saturation clamp diode, as shown in Figure 6-43. With the current sense terminals shorted, there are two $V_{BE}$ voltage drops between the clamp and source-emitter terminals. Therefore, clamp diode $D1$ will hold the collector on-voltage to approximately one diode drop above the base, keeping the switching transistor at the threshold of saturation, regardless of load current variations.

In applications where maximizing base-current rise-time is important—and secondary transformer inductance is a significant consideration—the use of the gating functions in the SG1629 can provide significant benefits. Figure 6-44 shows the addition of an external transistor, $Q1$, to drive the sink transistor's gate circuit. To explain the operation of this circuit, note that the sink transistor's base drive is now being generated with $R_B$ connected to the common or center tap of the drive transformer secondary, instead of the negative input terminal. This is essentially zero volts, and since the emitter of the sink transistor is attached to the capacitor which has a negative voltage on it, the sink transistor would normally be on continuously. Transistor $Q1$ is selected as a relatively slow non-gold-doped transistor which has a finite storage time. Its action is to turn the sink transistor off during the drive command signals, but to delay that off signal for some increment in time. Before the commencement of a drive

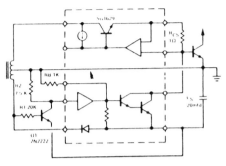

Figure 6-44   Improving base-current turn-on rise time. (Reprinted with permission from Silicon General *1986 Product Catalog,* Application Note SG1629)

command, the sink transistor is on by the action of $R_B$. Transistor $Q1$ is also saturated by its base resistor being connected to the end of the transformer which is also at zero volts prior to the command signal.

When the drive command is initiated, current begins to build up in the secondary circuit, and the first flow of current is through the source transistor, then through the current-limiting resistor, down through the sink transistor which is still conducting, and finally back to the negative terminal on the transformer secondary. The switching transistor is still back-biased while this occurs. Because the input to transistor $Q1$ is now at a negative voltage, it turns off; but since it has a finite storage time, that time is used to delay the rise of the input to the sink gate. Additional delay can be added with a small capacitor at the sink-gate input terminal. When the input to the sink gate goes high, its output goes low, forcing the sink transistor to turn off. Since the source current is already flowing, turn-off of the sink transistor diverts that current to the base of the output transistor, producing an $Ib_1$ rise time of less than 100 ns.

When the action is reversed at turn-off, there is negligible increase in delay between the turn-off signal and the actual turn-on of the sink transistor. This happens because $Q1$ is the only device with a long storage time, and it is turning on, so storage is not a factor.

Another method of speeding up the rise time of current into the base is the use of some reactive components to differentiate the drive-current signal. A simple approach is a capacitor bypass around the current sense resistor, $R_{CS}$, providing an initial boost in turn-on current.

## Gate Array Power Supply

The advent of gate arrays brings about the need of a good regulated low-voltage power supply. These arrays have from 800 inverters (or gates) to more than 50,000 gates per array. Typical power requirements are 20 V at about 200 mA

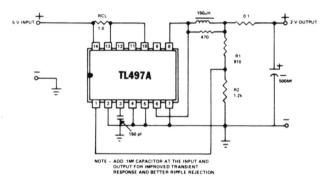

NOTE – ADD 1MF CAPACITOR AT THE INPUT AND
OUTPUT FOR IMPROVED TRANSIENT
RESPONSE AND BETTER RIPPLE REJECTION

Figure 6-45 TL497A Logic array power supply, step-down circuit.

per array for the 800-gate array. The input requirement is usually 5.0 V. The circuit of Figure 6-45 meets the above requirements at an overall efficiency of 72%.

Figure 6-46 shows another type of step-down regulator. With an input of from 7 to 12 V, it has an output of 5 V at 2.0 A. The TIP34 is a plastic TO-220 *pnp* transistor rated at 10 A. The IN5187D is a 3.0 A fast recovery diode.

## Telecommunication Regulated Power Supply[8]

Figure 6-47 shows a schematic diagram of a flyback converter configuration for a telecommunications application, using an LT1070 switching regulator IC. Telecommunications operate from an unregulated supply voltage of from $-40$ to $-60$ V. This high voltage requires protection of the $V_{in}$ pin of the LT1070 by means of $Q1$ and the 1N5936 30 V Zener diode, which drop the input voltage appearing at $V_{in}$ to about $-17$ V under all line conditions.

In this circuit, the "top" of the inductor is at ground potential, and the ground pin is at $-48$ V. The feedback pin senses with respect to the ground pin, so the circuit needs a level shift from the 5 V output. $Q_2$ accomplishes this function, and introduces only $-2$ mV/°C drift. This drift is normally not objectionable in a logic power supply, but if it is of concern, one can compensate with an appropriately scaled diode-resistor combination across the 1.2 k$\Omega$ resistor.

Frequency compensation is provided by the 2 k$\Omega$/0.22 $\mu$F combination at the $V_c$ compensation pin. This capacitor should have a low ESR resulting in less phase shift, and providing faster loop response because of reduced compensation time constant. The 68 V Zener diode clamps and absorbs excessive line transients that might otherwise damage the LT1070 ($V_{sw}$ is 75 V max).

---

[8]From "Regulator IC Speeds Design of Switching Power Supplies," Jim Williams, *EDN,* November 12, 1987. Used with permission.

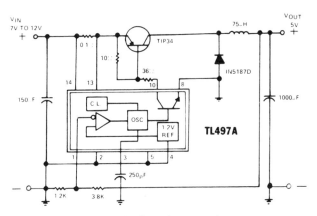

Figure 6-46   Step-down regulator.

## Low-Noise Converter[9]

Sensitive analog circuitry frequently requires a low-noise source of power. This can be obtained by controlling the power-supply output transistor's turn-on and turn-off (slowing the base transition), thus minimizing normal spikes due to a fast-switching oscillator. Figure 6-48 shows such a low-noise power supply that operates from a +5 V source and provides +15 V regulated outputs.

In this figure, the LM311 multivibrator clocks the SG3524 whose internal oscillator is disabled by grounding the timing-capacitor pin. While the LM311 output is HIGH, the SG3524 cuts the drive to $Q_1$ and $Q_2$, helping to minimize switching noise.

The main contributor to low-noise performance is the base-drive slowdown network used with $Q_1$ and $Q_2$. The 390 $\Omega$/0.1 $\mu$F time constant slows turn-on, and the diode forces base-emitter charge trapping to delay turn-off. The LM311's long on-time permits no current to flow in $Q_2$ until well after $Q_1$ has turned off. Furthermore, the current's rise and fall times are smoothly controlled and long, unlike those of the more common fast switching converters. Therefore, very little harmonic content appears in the transformer drive, so converter output noise is exceptionally low. In addition, the disturbance to the 5 V input is small.

This circuit's low noise comes at the expense of efficiency and available output power. During the slow base transitions, $Q_1$ and $Q_2$ dissipate power, reducing efficiency to about 50% and available output to approximately 50 mA. Heat-sinking $Q_1$ and $Q_2$ doesn't help because it involves the risk of secondary breakdown. The circuit is, however, short-circuit protected by the 0.1 $\Omega$ emitter-resistor and the SG3524's current-limiting circuitry.

---

[9]From "Conversion Techniques Adapt Voltages to Your Needs," Jim Williams, *EDN*, November 10, 1982. Used with permission.

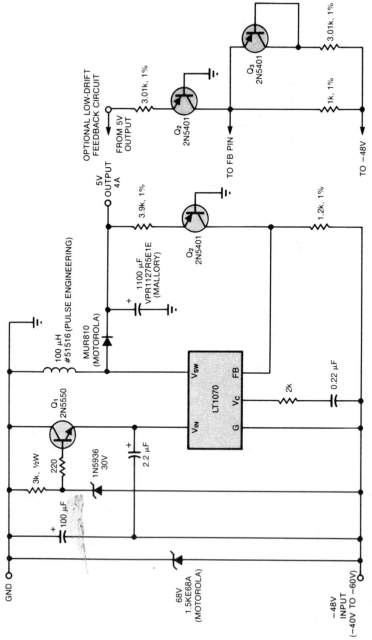

Figure 6-47 Using the LT1070 in a telecommunications application. (Reprinted with permission from "Regulator IC Speeds Design of Switching Power Supplies," *EDN*, November 12, 1987.)

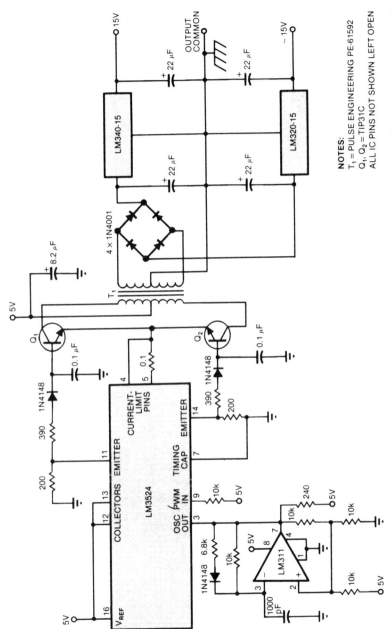

Figure 6-48  Low noise converter. (Reprinted with permission from "Conversion Techniques Adapt Voltages to Your Needs," *EDN*, November 10, 1982.)

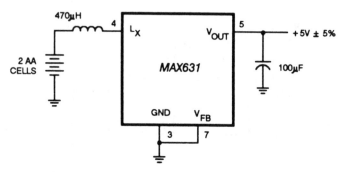

Figure 6-49   MAX631 3 to 5 V step-up converter.

## IC DC/DC Converter Usage Conditions[10]

The basic circuit diagram of the MAX631 DC/DC converter for use as a low voltage step-up converter is shown in Figure 6-49.

By connecting a second small inductor to the $L_X$ output of a MAX641 step-up DC/DC converter (as shown in Figure 6-50), the efficiency and power handling ability of many low-voltage converter designs can be dramatically improved. This circuit supplies 5 V at 40 mA with only a 1.5 V input.

The second coil (470 $\mu$H) works with $L_X$ to form a second step-up converter whose only function is to supply power to the MAX641 chip. An internal diode steers the coil's discharge current to a filter capacitor and Zener clamp at $V_{out}$. In this way, the MAX641 operates from 12 V and provides a larger gate drive signal to the MOSFET, which, in turn, has lower on-resistance than if 5 V were connected to $V_{out}$.

It should be noted that $V_{out}$ is actually the MAX641's voltage input, and not an output per se. The pin is labeled this way because it usually connects to the circuit output to provide power and the feedback signal back to the chip when the MAX641 is used in its standard configuration. When a MAX64X or MAX63X series device is used in other than the basic configurations, such as here, the $V_{out}$ pin is frequently NOT the output of the DC/DC converter.

Often, a digital system powered by a 5 V supply includes a few analog functions that require $\pm 12$ V. The circuit shown in Figure 6-51 uses two dedicated 8-pin converters—the MAX632 and MAX636—to derive 25 mA at 12 V and 15 mA at $-12$ V from a 5 V logic supply. One can configure the circuit for independently regulated outputs (Figure 6-51a) or for tracking regulation (Figure 6-51b).

The positive converter's efficiency is 85%; the inverter's is 75%. These efficiency figures can be slightly improved by using Schottky diodes rather than

---

[10]Courtesy of Maxim Integrated Products.

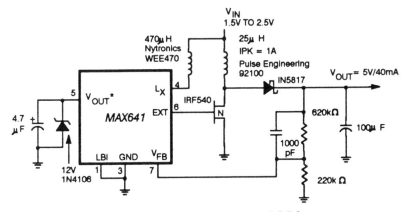

Figure 6-50  Low battery voltage to +5 V DC/DC converter.

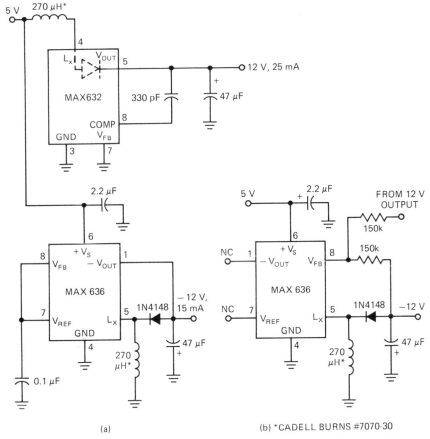

(a)

(b) *CADELL BURNS #7070-30

Figure 6-51  Boost converter provides either independently regulated outputs (a); or a tracking negative output (b).

the MAX632's internal diode and the 1N4148 signal diode connected to pin 5 of the MAX636. If a Schottky diode is used with the MAX632, it should be connected in parallel with the chip's internal diode (that is, between pins 4 and 5).

With several popular types of high-current rectifier diodes, such as the 1N4000 Series, efficiency and overall performance are poor for conversion rates greater than 10 kHz. Many of these diodes were designed to pass high current only at 120 Hz; therefore, they waste energy at 50 kHz and higher operating frequencies. In addition, these slow rectifiers might also allow the inductor's discharge voltage to reach excessive levels before the rectifier turns on and directs current to the load.

Small signal diodes, such as the 1N4148, are fast enough and work well in applications that require less than 50 mA. High-speed rectifiers, such as the 1N4935, are suitable in applications that require as much as 1 A. Schottky diodes, though, provide the best performance with respect to speed and forward voltage drop, and they can significantly improve efficiency in low-voltage, high-current applications.

For higher output current requirements, an external power MOSFET can be added as shown in Figure 6-52 to obtain 100 mA at 12 V and 60 mA at $-12$

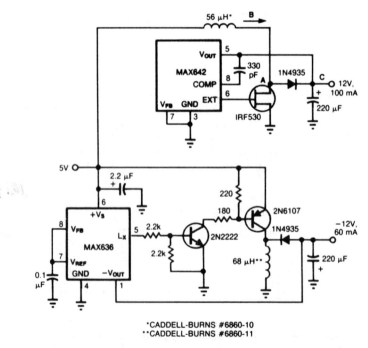

*CADDELL-BURNS #6860-10
**CADDELL-BURNS #6860-11

Figure 6-52  Boost converter with increased output current capability.

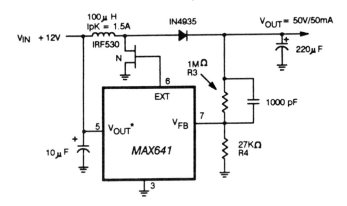

Figure 6-53   High voltage step-up converter.

V. The power MOSFET drops the 12 V converter's efficiency to 80%, but driving the power MOSFET doesn't require any additional parts.

The output voltage limits of the MAX6XX series DC/DC converters can be exceeded if an external FET or transistor with an adequate voltage rating is used as the switch. In Figure 6-53, a +12 V input is converted to +50 V at 50 mA by adding an IRF530 $n$-channel FET, which has a voltage rating of 100 V. Here, the circuit differs from the basic MAX641 circuit in that an external resistor-divider must provide the feedback signal to the $V_{FB}$ input and chip power comes from the +12 V input via the $V_{out}$ pin.

Boost converters are inadequate for some battery-powered applications. For example, when converting a 12 V sealed lead acid battery to a regulated +12 V output, the battery voltage may vary from a high of 15 V down to 10 V. Both step-up boost and step-down buck DC/DC converters have a common limitation in that neither can handle input voltages which may be both greater than or less than the output. Here, a buck-boost converter is best suited to handle the wide input voltage swing associated with the sealed lead-acid battery.

The circuit shown in Figure 6-54 is a buck-boost converter that delivers 100 mA at +12 V and accepts 8 to 16 V inputs. By using a MAX641 to drive separate $p$- and $n$-channel MOSFETs, both ends of the inductor are switched to allow noninverting buck-boost operation. A second advantage of this circuit over most boost-only designs is that the output goes to zero volts when SHUT-DOWN is activated. However, a drawback is that efficiency is not optimum because 2 MOSFETs and 2 diodes increase the losses in the charge and discharge path of the inductor. The circuit of Figure 6-54 has an efficiency of 70%.

The quality of ground connections is critical to the performance of most DC/DC converters. Since the peak current in an inductor or switch ground can reach 1 A, these points should have very low impedance paths to supply common. In the best case, separate grounds are used for high-current paths rather than for chip power or feedback connections.

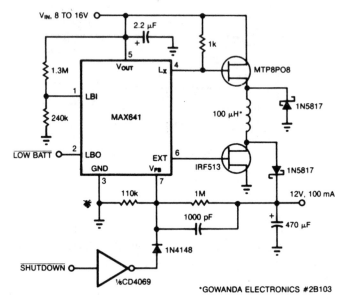

Figure 6-54   Buck-boost converter.

## Triple Output Battery Sourced DC/DC Converter[11]

The schematic diagram of Figure 6-55 depicts a DC/DC converter that uses a single integrated circuit (the MAX641) to provide three regulated output voltages (isolated $\pm 15$ V and non-isolated 5 V) from a lead-acid battery power source ($V_{in} = 8\text{–}16\ V_{DC}$) in a buck-boost topology. The MAX641 generates a 45 kHz signal that drives the gate of MOSFET $Q_1$.

As shown in Figure 6-55, $Q_1$ turns on when the gate voltage is high, causing a linear increase in $T_1$'s primary current, which stores energy in a magnetic field. The field then starts to collapse as $Q_1$ turns off, reversing the voltage polarity on all windings and causing the voltage on each secondary winding to increase. These secondary voltages then deliver energy to the outputs by forward-biasing Schottky diodes $D_1$, $D_2$ and $D_3$. When the 5 V output rises above a desired level, feedback to the chip causes an internal error comparator to turn off the gate signal to $Q_1$.

The secondary winding ratios set the output voltage levels, and close coupling between the trifilar windings assures good load regulation for the $\pm 15$ V supplies. For better regulation, the output voltage can be set higher and linear regulators can be added at the outputs. Inductors $L_1$ and $L_2$ block high-frequency ringing from the transformer that would otherwise boost the $\pm 15$ V outputs

[11]From "Derive $\pm 15$ V and 5 V from a 12 V Battery," by Andy Jenkins, *EDN*, February 18, 1988. Used with permission.

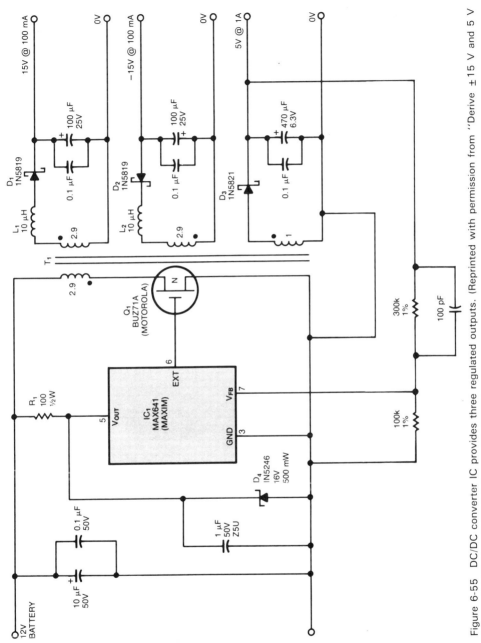

Figure 6-55  DC/DC converter IC provides three regulated outputs. (Reprinted with permission from "Derive ±15 V and 5 V from a 12 V Battery," *EDN*, February 18, 1988.)

above acceptable limits when lightly loaded. For best regulation, one should provide minimum loads of 10% for the 15 V supplies and 20% for the 5 V supply.

The protection network made up of resistor $R_1$ and Zener diode $D_4$ allows the circuit to withstand the classic 50 V, 1 msec overvoltage test that simulates the load dump of an automobile's alternator when the ignition is turned off. For an input change of 8 to 16 V, the 5 V output's line regulation is typically 0.2%.

Battery current is about 600 mA for nominal operation, but current peaks in the primary winding can be 4 A or more. Therefore, the circuit implementation should have good ground connections and short, low-impedance connections to the transformer and the MOSFET. Close decoupling using ceramic and electrolytic capacitors also reduces output noise. With proper circuit layout, the output noise is about 50 mV at the 5 V output and 30 mV at the $\pm 15$ V outputs.

The transformer, constructed with a ferrite pot core that offers low loss and minimal magnetic leakage, has a primary inductance of about 21 $\mu$H for the power levels shown. A core size and material must be chosen that will handle the 4 A peak currents without saturation. The 15 V secondaries have 2.9:1 turns ratios, which provide the desired 3:1 voltage ratio after covering the rectifier losses. Actual turns are as follows: the primary, $11\frac{1}{2}$ turns; the 15 V secondaries, $11\frac{1}{2}$ turns each; the 5 V secondary, four turns. High circuit efficiency (about 75% at full load with a 12 V input) eliminates any need for a heat sink on the MOSFET.

## DESIGN EXAMPLES

The following design examples are presented to illustrate the methodology of designing with IC switching regulators.

### Step-Down (Buck) Switching Regulator Design[12]

A schematic of a step-down regulator using the TL497A switching regulator is shown in Figure 6-56:

Conditions:

$$V_{in} = 15 \text{ V} \qquad I_{out} = 200 \text{ mA}$$

$$V_{out} = 5 \text{ V} \qquad V_{ripple} < 1\%$$

Calculations:

$$I_{pk} \geq 2I_{load} = 400 \text{ mA} \qquad (6\text{-}6)$$

---

[12]Courtesy of Texas Instruments, Inc.

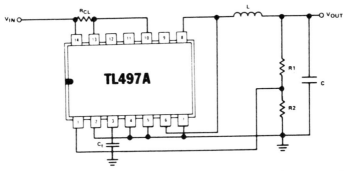

Figure 6-56   Basic step-down regulator. (Courtesy of Texas Instruments Inc.)

This is the limit condition for discontinuous operation. For design margin, $I_{pk}$ will be designed for 500 mA which is also the limit of the internal pass transistor and catch diode.

$$\therefore I_{pk} \rightarrow 500 \text{ mA}$$

$$L = \frac{V_{in} - V_{out}}{I_{pk}} t_{on}$$

$$L = \frac{10 \text{ V}}{500 \times 10^{-3}} t_{on} \qquad (6\text{-}7)$$

Recommended on time is: $19 \ \mu s < t_{on} < 150 \ \mu s$, thus the range of acceptable inductance is, 380 $\mu$H to 3 mH.

choosing $L = 390 \ \mu$H

$$t_{on} = \frac{390 \times 10^{-6} \times 500 \times 10^{-3}}{10} = 19.5 \times 10^{-6} \text{ sec.}$$

To program the TL497A for 5 $V_{out}$:

$$R_2 = 1.2 \text{ k}\Omega \text{ (fixed)}$$

$$R_1 = (5 - 1.2) \text{ k}\Omega = 3.8 \text{ k}\Omega$$

To set current limiting:

$$R_{CL} = 0.5/I_{limit}$$

$$R_{CL} = \frac{0.5}{500 \times 10^{-3}} = 1 \ \Omega \qquad (6\text{-}8)$$

For the on-time chosen above, $C_t$ can be approximated;

$$C_t(\text{pf}) \simeq 12t_{\text{on}}(\mu s)$$

$$C_t \simeq 240 \text{ pf} \tag{6-9}$$

To determine $C_{\text{filter}}$ for desired ripple voltage:

$$C = \frac{(I_{\text{pk}} - I_{\text{load}})^2}{V_{\text{ripple}} \, 2I_{\text{pk}}} \cdot \frac{t_{\text{on}} V_{\text{in}}}{V_{\text{out}}} \tag{6-10}$$

For constant $C$, $V_{\text{ripple}}$ increases as $I_{\text{load}}$ decreases.

$$C = 45 \ \mu F \ (\text{for 200 mA/1\% ripple})$$

The maximum operating frequency is encountered under maximum load conditions.

$$f_{\text{max}} = \frac{2I_{\text{load}}(\text{max})}{I_{\text{pk}}} \cdot \frac{V_{\text{out}}}{t_{\text{on}} V_{\text{in}}} \tag{6-11}$$

The minimum operating frequency occurs under minimum load conditions.

$$f_{\text{min}} = f_{\text{max}} \frac{I_{\text{load}}(\text{min})}{I_{\text{load}}(\text{max})} \tag{6-12}$$

Figure 6-57 illustrates the regulator with the above values applied to it. Waveforms at $C_t$ for indication of proper circuit performance are shown in Figure 6-58. For peak currents greater than 500 mA, it is necessary to use an external transistor and diode.

## Step-Up (Boost) Switching Regulator Design[13]

Figure 6-59 is the schematic diagram of a step-up regulator using the TL497A.

Conditions:

$$V_{\text{in}} = 5 \text{ V} \qquad I_{\text{out}} = 75 \text{ mA}$$

$$V_{\text{out}} = 15 \text{ V} \qquad V_{\text{ripple}} < 1\%$$

---

[13]Ibid. 12.

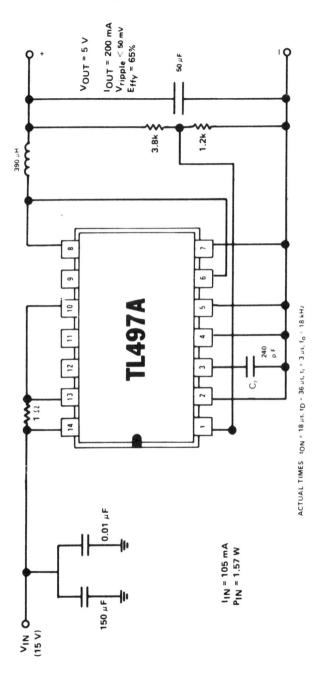

ACTUAL TIMES $t_{ON}$ = 18 $\mu$s, $t_D$ = 36 $\mu$s, $t_1$ = 3 $\mu$s, $f_0$ = 18 kHz

NOTE: 13 $\mu$s ON-TIME RESULTS IN $I_{pk}$ = 433 mA. RECALCULATING $t_D$, $t$, AND $f_0$ WILL CONCUR WITH ACTUAL TIMES OBSERVED.

Figure 6-57   15 V to 5 V switching regulator for output currents to 200 mA. (Courtesy of Texas Instruments Inc.).

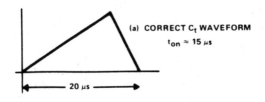

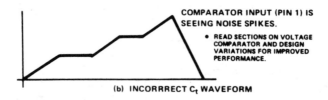

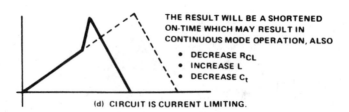

Figure 6-58   Circuit performance waveforms. (Courtesy of Texas Instruments Inc.).

Calculations:

$$I_{pk} \geq 2I_{load}\left[\frac{V_{out}}{V_{in}}\right]$$

$$I_{pk} \geq 450 \text{ mA} \tag{6-13}$$

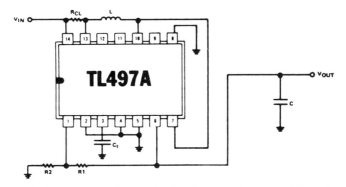

Figure 6-59    Basic step-up regulator using the TL497A. (Courtesy of Texas Instruments Inc.).

For design margin $I_{pk} \rightarrow 500$ mA

$$L = \frac{V_{in}}{I_{pk}} t_{on}$$

$$L = \frac{5}{500 \times 10^{-3}} t_{on} \qquad (6\text{-}14)$$

Recommended on-time is 19 $\mu$s $< t_{on} < 150$ $\mu$s, thus the range of acceptable inductance is; 190 $\mu$H to 1.5 mH

$$\text{choosing } L = 200 \ \mu\text{H}$$

$$t_{on} = 20 \ \mu\text{s}$$

To program the TL497A

$$R2 = 1.2 \text{ k}\Omega$$

$$R1 = (15 - 1.2) \text{ k}\Omega = 13.8 \text{ k}\Omega$$

To set the current limiting:

$$R_{CL} = 0.5/I_{limit}$$

$$R_{CL} = \frac{0.5}{500 \times 10^{-3}} = 1 \ \Omega \qquad (6\text{-}15)$$

For the on-time chosen above (20 $\mu$s) $C_t$ can be estimated:

$$C_t \text{ (pf)} \simeq 12t_{on} \text{ (}\mu s\text{)}$$

$$C_t \simeq 240 \text{ pf}$$

To determine $C_{filter}$ for desired ripple voltage

$$C = \frac{(I_{pk} - I_{load})^2 t_D}{V_{ripple} \, 2I_{pk}} \tag{6-16}$$

$$t_D = t_{on}\left(\frac{V_{in}}{V_{out} - V_{in}}\right) = 10 \ \mu s \tag{6-17}$$

$$C = 12.0 \ \mu F$$

The nominal operating frequency $f_o$ is:

$$f_o = \frac{1}{T} = \frac{2I_{load}}{I_{pk} t_D}$$

$$f_o = 30 \text{ kHz} \tag{6-18}$$

Applying these values to the TL497A results in the schematic diagram as shown in Figure 6-60.

### Isolated Flyback Converter Design[14]

A design example using the LT1070 Switching Regulator as a fully isolated flyback converter is now presented. The LT1070 has available an operating mode called "isolated flyback," as shown in Figure 6-61, which does not use the feedback pin to sense output voltage; instead, it senses and regulates the transformer primary voltage during switch off time ($t_{off}$). This voltage is related to $V_{out}$ by:

$$V_{out} = (N)(V_{pri}) - V_f \quad \text{(during } t_{off}) \tag{6-19}$$

$N$ = turns ratio of transformer
$V_f$ = forward voltage of output diode
$V_{pri}$ = primary voltage during switch "off" time

The secondary output voltage will be regulated if $V_{pri}$ is regulated. The LT1070 switches from normal mode to regulated primary mode when the current out of

---

[14]Courtesy of Linear Technology Corporation.

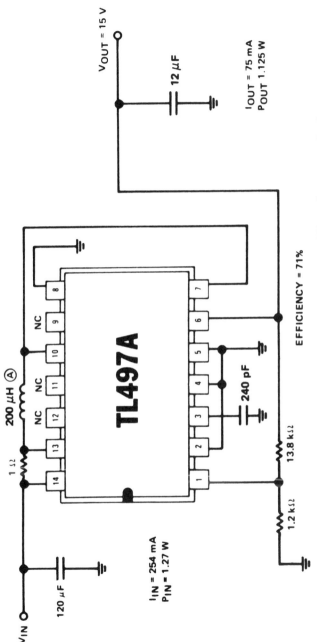

Figure 6-60  5 V to 15 V switching regulator. (Courtesy of Texas Instruments Inc.).

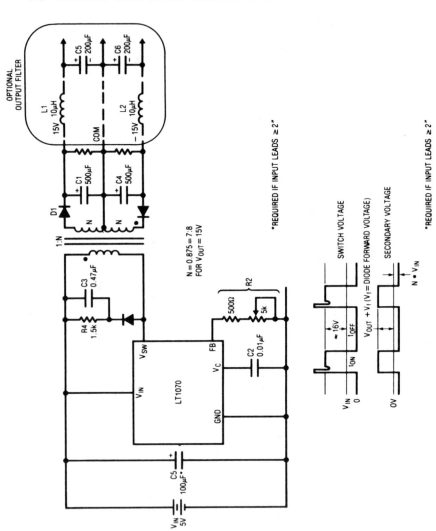

Figure 6-61 Totally isolated converter. (Reprinted with permission from "LT1070 Design Manual," Application Note 19. Courtesy of Linear Technology Corporation)

the feedback pin exceeds 10 $\mu$A. An internal clamp holds the voltage ($V_{FB}$) on this pin at approximately 400 mV. $R2$ is used to put the LT1070 in isolated flyback mode. It also doubles as an adjustment in the regulated output. $V_{pri}$ is regulated to 16 V + 7K ($V_{FB}/R2$), where $V_{FB}/R2$ is equal to the current through $R2$, and the 7K is an internal resistor. $V_{out}$ is therefore equal to:

$$V_{out} = N\left[16 + 7K\left(\frac{V_{FB}}{R_2}\right)\right] - V_f \qquad (6\text{-}20)$$

and the required transformer turns ratio is:

$$N = \frac{V_{out} + V_f}{16 + 7K\left(\dfrac{V_{FB}}{R_2}\right)} \qquad (6\text{-}21)$$

The term, 7K ($V_{FB}/R2$) is normally set to approximately 2 V to allow some adjustment range in $V_{out}$. Solving for $N$ in Figure 6-61 with $V_{out}$ = 15 V:

$$N = \frac{15 + 0.7}{16 + 2} = 0.872$$

The smallest integer ratio with $N$ close to 0.872 is 8:7 = 0.875. $T1$ is to be wound with this turns ratio for each output. The total number of turns is determined by the required primary inductance, ($L_{pri}$). This inductance has no optimum value; it is a tradeoff between core size, regulation requirements, and leakage inductance effects. A reasonable starting value is found by assigning a maximum magnetizing current $\Delta I$ of 10% of the peak switch current of the LT1070. Magnetizing current is the difference between the primary current at the start of switch on-time and the current at the end of switch on-time. This gives a value for $L_{PRI}$ of:

$$L_{pri} = \frac{V_{in}}{(\Delta I)(f)\left(1 + \dfrac{V_{in}}{V_{pri}}\right)} \qquad (6\text{-}22)$$

$\Delta I$ = primary magnetizing current
$V_{pri}$ = regulated primary flyback voltage

For $V_{in}$ = 5 V, $\Delta I$ = 0.5 A, $V_{pri}$ = 18 V:

$$L_{pri} = \frac{5}{(0.5)(40 \times 10^3)\left(1 + \dfrac{5}{18}\right)} = 196 \ \mu H$$

Again, this value is not an optimum figure, it is simply a compromise between maximum output current and core size.

A second consideration on primary inductance is the transition from continuous mode to discontinuous mode. At light output loads, the flyback pulse across the primary will drop toward zero before the end of switch off-time. The LT1070 interprets this as a drop in output voltage and raises the duty-cycle accordingly to compensate, which results in an abnormally high output voltage. To avoid this situation, the output should have a minimum load equal to:

$$I_{out}(min) = \frac{(V_{pri} \cdot V_{in})^2}{(V_{pri} + V_{in})^2 (2V_{out})(f)(L_{pri})} \qquad (6\text{-}23)$$

with $V_{pri} = 18 \ V$, $V_{in} = 5 \ V$, $V_{out} = 15 \ V$, $L_{pri} = 200 \ \mu H$:

$$I_{out}(min) = \frac{(18 \cdot 5)^2}{(18 + 5)^2 (2 \cdot 15)(40 \times 10^3)(200 \times 10^{-6})} = 64 \ mA$$

This current may be shared equally on each output at 32 mA per output. If a lighter minimum load is desired, the primary inductance must be increased. But this also increases leakage inductance, so some care must be used. Leakage inductance (which is that portion of the primary not coupled to the secondary) creates a flyback spike following switch opening. The height of this spike must be clamped with a snubber ($R4$, $C3$, $D2$) to avoid overvoltage on the switch. The width of the leakage inductance spike is equal to:

$$t_L = \frac{I_{pri}(L_L)}{V_M - V_{pri} - V_{in}} \qquad (6\text{-}24)$$

$L_L$ = leakage inductance
$I_{pri}$ = peak primary current
$V_M$ = peak switch voltage

This spike width is important because it must be less than 1.5 $\mu s$ wide. The LT1070 has internal blanking for approximately 1.5 $\mu s$ following switch turn-off. This blanking time ensures that the flyback error amplifier will not interpret the leakage inductance spike as the actual flyback voltage to be regulated. To avoid poor regulation, the spike must be less than the blanking time.

If transformer $T1$ is trifilar wound for minimum leakage inductance, $L_L$ may have a typical value of 1.5% of $L_{pri}$. Assuming $L_{pri} = 200$ $\mu$H, $L_L$ would be 3 $\mu$H. To calculate $t_L$, a value needs to be assigned to $V_M$. In this case, with $V_{in} = 5$ V, a conservative value for maximum switch voltage would be $V_M = 50$ V. If a maximum primary current of 5 A is assumed for maximum output current, the spike width is:

$$t_L = \frac{(5)(3 \times 10^{-6})}{50 - 18 - 5} = 0.56 \ \mu s$$

This is well within the maximum value of 1.5 $\mu$s. Note, however, that the pulse width grows rapidly as the sum of $V_{pri} + V_{in}$ approaches maximum switch voltage. The following formula allows one to calculate the maximum ratio of leakage inductance to primary inductance in a given situation.

$$\frac{L_L}{L_p} \text{(max)} = \frac{t_L \cdot (V_M - V_p - V_{in})(\Delta I)(f)\left(1 + \dfrac{V_{in}}{V_p}\right)}{I_{pri}(V_{in})} \qquad (6\text{-}25)$$

With a fairly large $V_{in}$ (36 V), and using a less conservative value of 60 V for $V_M$, with $t_L = 1.5$ $\mu$s, $V_p = 18$ V, $\Delta I = 0.5$ A, and $I_{pri} = 5$ A:

$$\frac{L_L}{L_p} \text{(max)} = \frac{(1.5 \times 10^{-6})(60 - 18 - 36)(0.5)(40 \times 10^3)\left(1 + \dfrac{36}{18}\right)}{(5)(36)}$$

$$= 0.003 = 0.3\%$$

This low ratio of leakage inductance to primary inductance would be nearly impossible to wind, so some compromises must be made. If maximum output current is not required, $I_{pri}$ will be less than 5 A (see equation 6-28). Ripple current ($\Delta I$) can also be increased. Finally, an LT1070HV (high voltage) part can be used, with a switch rating of 75 V. Substituting $I_{pri} = 2.5$ A, $\Delta I = 1$ A, $V_M = 70$ V into the above calculation yields $L_L/L_{pri} = 3\%$, which is easily achievable.

Maximum output power with an isolated flyback converter is less than an ordinary flyback converter because the transformer turns ratio is fixed by the output voltage. This fixes the duty-cycle at:

$$DC = \frac{V_{pri}}{V_{pri} + V_{in}} \qquad (6\text{-}26)$$

and maximum power is limited to:

$$P_{out}(\text{max}) = \left(\frac{V_{pri}}{V_{pri} + V_{in}}\right)\left[V_{in}\left(I_p - \frac{\Delta I}{2}\right) - I_p^2 R\right](0.8) \qquad (6\text{-}27)$$

$R$ = LT1070 switch on-resistance
$I_p$ = maximum switch current
0.8 = loss factor to account for losses in addition to $R$

With $V_{pri}$ at a nominal 18 V, $V_{in}$ = 5 V, $I_p$ = 5 A, $\Delta I$ = 0.5 A, the duty cycle is 78% and the maximum output power is:

$$P_{out}(\text{max}) = \left(\frac{18}{18 + 5}\right)\left[5\left(5 - \frac{0.5}{2}\right) - (5)^2 0.2\right](0.8) = 11.74 \text{ W}$$

An analysis of the power formula shows that at low $V_{in}$ maximum output power is proportional to $V_{in}$, and at high $V_{in}$, maximum power approaches 50 W.

Peak primary current for loads less than the maximum is found from:

$$I_{pri} = \frac{(V_{out})(I_{out})(V_{pri} + V_{in})}{0.8(V_{pri})(V_{in})} + \frac{\Delta I}{2} + \frac{(I_{pri})^2 R}{V_{in}} \qquad (6\text{-}28)$$

This formula is actually a quadratic, but rather than solve it explicitly, a much simpler technique, for the range of $I_{pri}$ involved, is to calculate the first two terms on the right, then use this value of $I_{pri}$ to calculate the last term. For the circuit of Figure 6-61 with $I_{out}$ = 0.25 A on each output, $V_{pri}$ = 18 V, $V_{in}$ = 5 V, $\Delta I$ = 0.5 A, and $R$ = 0.2 Ω:

$$I_{pri} = \frac{(15)(0.5)(18 + 5)}{(0.8)(18)(5)} + \frac{0.5}{2} + \frac{(2.64)^2(0.2)}{5} = 2.92 \text{ A}$$

The transformer must be sized so that the core does not saturate with 2.92 A in the primary winding. Note that there is plenty of margin on the 5 A maximum switch current. A smaller core could be used if $\Delta I$ were increased to 1 A, cutting primary inductance in half.

Flyback regulators do not utilize the inductance of the transformer as a filter, so all filtering must be performed by output capacitors C1 and C4. They should be low ESR types to minimize output ripple. In general, output ripple is limited by the ESR of the capacitor, not the actual capacitance. Output ripple in peak-to-peak volts is given by:

$$V_{p-p} = \frac{I_{pri}}{2 * N} \text{(ESR)} \qquad (6\text{-}29)$$

With $I_{pri} = 2.92$ A, $N = 0.872$, and assigning an ESR of $0.1$ $\Omega$, output ripple is

$$V_{p-p} = 167 \text{ mV}_{p-p} \text{ @ full load}$$

Had the output ripple formula been based on the actual output capacitance, rather than its ESR, the result would have been approximately 10 mV, showing that ESR effects do dominate. The 0.1 $\Omega$ value chosen for ESR is probably higher than typical for a good 500 $\mu$F capacitor, but less than guaranteed maximum. Note that one reason for high output ripple in this circuit is that the converter is operating at a rather high duty-cycle of 78% because of the low input voltage. This leaves only 22% of the time for the secondary to be delivering current to the load. As a consequence, secondary peak currents, and therefore output ripple, are high.

If low output ripple is required, an output filter may be a better choice than simply using large output capacitors.

---

*This factor is 2 because of dual outputs.

# 7

# Magnetic Amplifiers

## INTRODUCTION

In the present state of development, a magnetic amplifier has many advantages over other types of amplifying devices. Magnetic amplifiers have extremely good reliability, long life, ruggedness, no warm-up time, high efficiency, few maintenance problems, and capability of high-temperature operation—all these virtues in combination! Magnetic amplifiers are considered in this chapter only as they apply to DC/DC converters or DC/AC inverters. The primary use of magnetic amplifiers in power supply design has been as voltage regulators.

## MAGNETIC AMPLIFIER DESIGN PRINCIPLES

The design of a magnetic amplifier should begin with the preliminary consideration of employing the most ideal components, such as a toroidal core configuration and rectifiers with the least possible reverse currents for the self-saturation amplifiers. Under such conditions, achieving the best correlation between theoretical and experimental amplifiers can be expected. If extreme high performance is not a necessity, the use of U-laminations and lower-cost rectifiers possessing a limited reverse leakage is generally advantageous. Such amplifiers are critically dependent on the air gaps in the cores that are inherently associated with geometries of the U-lamination and C-core types. This allows wider tolerances for the magnetic properties of the cores and facilitates production control. However, the presence of these air gaps in the reactors also makes it more difficult to predict an exact amplifier design. Consequently, once the theoretical amplifier limits are known, based upon a toroidal amplifier, the designer can more accurately extrapolate any designs performed with gap-type core configurations.

The design method, therefore, is limited to a consideration of the most sensitive magnetic amplifiers, which are those employing toroidal cores in the self-saturated circuits.

In practice, today, amplifiers other than those of the self-saturation type are rarely used except in cases where special applications are required. For most purposes, the ideal magnetic amplifier design would yield maximum power gain from a saturable reactor of a given core volume, subject to cost, weight, and space limitations. This means that for a given set of specifications the designer should attempt to meet the required circuit conditions with the smallest, least expensive core, using the maximum core window area, and operating at optimum efficiency consistent with an allowable temperature rise. In many instances the specifications indicate requirements that are divergent, and for such amplifiers it is possible only to achieve a suitable compromise among the various parameters.

Since the power dissipated in the saturable reactor itself is wasted power (reactor iron and winding copper losses), the loss should be kept to a minimum. For the better materials, having low coercivity flux current loops, the core losses are negligibly small; hence, the primary consideration is to reduce the load winding resistance to a reasonable minimum for the best practical performance. For the feedback circuits the forward resistance of the rectifiers being used influences the magnitude of the load winding resistance.

The majority of power magnetics in present-day power supply design use toroidal cores; therefore, the step-by-step procedure of core selection discussed here applies to toroidal cores. However, the equations are basic magnetic relationships which are applicable to all core geometries.

## Fundamental Design Equations

A basic magnetic amplifier is shown in Figure 7-1 along with the $B$-$H$ characteristic of each core; the fundamental design equations are presented below. The $B$-$H$ curve is assumed to be ideally square. The applied voltage is a square wave of amplitude $E_{in}$ and frequency $f$ for the following equations, which apply to this configuration:

$$N_1 = \frac{E_{in} \times 10^8}{4 A_c B_m f} \qquad \text{For square waveforms}$$

$$N_1 = \frac{E_{in} \times 10^8}{4.44 A_c B_m f} \qquad \text{For sinusoidal waveforms} \qquad (7\text{-}1)$$

where: $A_c$ = the iron area,
$B_m$ = the saturation flux density,
$E_{in}$ = the supply (input) voltage,
$N_1$ = the primary turns.

The average output current is

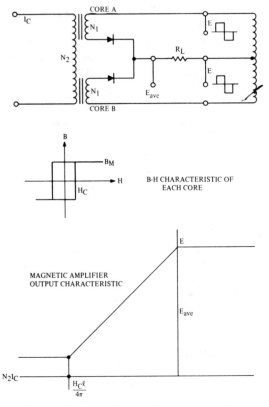

Figure 7-1  Magnetic amplifier characteristics.

$$I_{\text{ave}} = \frac{E_{\text{in}}}{R_L} \frac{(T - t_1)}{T} + \frac{2 H_c \ell}{.4 \pi N_1} \left(\frac{t_1}{T}\right) \qquad (7\text{-}2)$$

where: $\ell$ = the iron path length,

$H_c$ = the coercive force,

$t_1$ = the time at which the core saturates during a half-cycle interval,

$T = 1/f$.

The average output voltage is

$$E_{\text{ave}} = R_L I_{\text{ave}} \qquad (7\text{-}3)$$

The average control current is given by

$$N_2 I_C = \frac{H_c \ell}{.4\pi} \left(\frac{t_1}{T}\right) \tag{7-4}$$

where

$$0 \le I_C \le \frac{H_c}{.4\pi}$$

or

$$I_{ave} = \frac{E_{in}}{R_L} \left(1 - \frac{.4\pi}{H_c \ell} N_2 I_C\right) + \frac{2N_2}{N_1} I_C \tag{7-5}$$

The output voltage is

$$E_{ave} = E_{in} \left(1 - \frac{0.4\pi}{H_c \ell} N_2 I_C\right) + \frac{2N_2}{N_1} I_C R_L \tag{7-6}$$

The output characteristic is also shown in Figure 7-1.

Adding another control winding and feeding back the output current regeneratively, can cause the magnetic amplifier curve to have an almost infinite gain characteristic as shown in Figure 7-2.

The addition of a bias winding and application of the proper amount of current results in the control characteristic being shifted to the right as shown in Figure 7-3. In addition, by using a constant-current diode and a Zener diode, the curve may be shifted to that of Figure 7-4. The dotted line represents the error that is generated over the entire range of outputs. The maximum error over the tem-

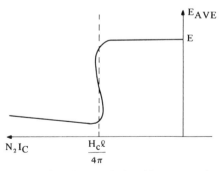

Figure 7-2  Magnetic amplifier characteristic with regenerative current feedback.

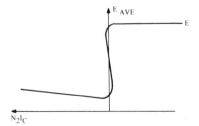

Figure 7-3   Magnetic amplifier characteristic with regenerative feedback plus bias.

perature range is +0.02 A turns over the entire range of outputs. At low levels, the error tends to increase. If $I_B$ represents any additional current through the bias winding, $N_B$, in the same sense as the input current, then we may describe the final output as:

$$I_{in} + N_B I_B - I_C N_2 = \Delta\varepsilon \tag{7-7}$$

where $\Delta\varepsilon$ is the ampere turn error.

Solving for $I_C$ yields

$$I_C = +\frac{I_{in} + N_B I_B - \Delta\varepsilon}{N_2} \tag{7-8}$$

But

$$I_C = I_E\left[1 - \frac{1}{(\beta_1 + 1)(\beta_2 + 1)}\right] \tag{7-9}$$

where: $\beta_1$   is the current gain of $Q_1$,
          $\beta_2$   is the current gain of $Q_2$.

Also

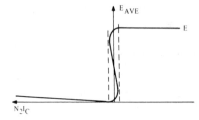

Figure 7-4   Same as Figure 7-3 but shifted down to axis.

$$E_{\text{out}} = R_E I_E$$

$$= R_E \frac{I_{\text{in}} + N_B I_B - \Delta\varepsilon}{N_2 \{1 - [1/(\beta_1 + 1)(\beta_2 + 1)]\}} \tag{7-10}$$

$$E_{\text{out}} = \frac{(R_E I_{\text{in}}/N_2) + (R_E N_B I_B/N_2) - (\Delta\varepsilon R_E/N_2)}{1 - [1/(\beta_1 + 1)(\beta_2 + 1)]} \tag{7-11}$$

## Selecting Toroidal Cores

Listed below is a step-by-step procedure used to select toroidal cores for magnetic circuitry.

(1) Analyze the circuit to determine:
   (a) the operating frequency ($f$),
   (b) voltage the core must support ($E_{\text{in}}$),
   (c) current flowing in the gate or power winding. This can be found if the total power output is known.
(2) Solve equations (7-12) or (7-13) for the maxwell turns, $N\phi$ (the product of total core flux capacity and turns), required to support the applied voltage ($E_{\text{in}}$) at operating frequency ($f$).
   For a sinusoidal flux condition in which the flux excursion extends from just under positive saturation to just under negative saturation, the excitation voltage for a single reactor is given by the relation

$$E_{\text{in}} = 4.44 f B_m k A_c N \times 10^{-8} \tag{7-12}$$

where: $E_{\text{in}}$ = rms magnitude of the supply voltage (volts),
   $f$ = frequency of the supply voltage (hertz),
   $B_m$ = maximum flux density (gauss),
   $k$ = stacking factor of the core material,
   $A_c$ = cross-sectional area of the core (square centimeters),
   $N$ = load turns of reactor.

For a DC voltage or peak voltage condition this equation becomes

$$E_{\text{in}} = 4 f B_m k A_c N \times 10^{-8} \tag{7-13}$$

(3) Select the wire size of the gate or power winding from the current requirement, using standard wire tables, and record the wire area ($A_w$).
(4) Assume that the available portion of the window area ($K$) for the gate winding equals 0.3 (for power converters or inverters use a $K$ of 0.1) and solve the formula:

$$W_a\phi_t = \frac{N_1\phi_t A_w}{K} \qquad (7\text{-}14)$$

where: $W_a\phi_t$ = product of window area and total core flux capacity,
$\quad\quad K$ = ratio of window area used for power winding to the total window area,
$\quad\quad A_w$ = wire area in circular mils.

(5) Select the material type and thickness to be used from the application requirements and the frequency.
(6) Determine the core $W_a A_c$ from the formula:

$$W_a A_c = \frac{W_a\phi_t}{2B_m} \qquad (7\text{-}15)$$

where: $W_a A_c$ = product of window area and effective core cross-sectional area.
$\quad\quad B_m$ = nominal flux density of core material selected.

Next, select the appropriate core size from core tables found in various core design manuals (such as *Magnetics Incorporated Design Manual featuring Tape Wound Cores—TWC 300 and Ferroxcube Bulletin 330*) by selecting the core with the $W_a A_c$ nearest the value calculated above. The core design manuals present applicable design equations, applications, typical *B-H* curves for various core materials, curves depicting temperature variation of magnetic material core sizes and selection tables, wire and winding data, toroidal winding information, etc.

(7) Determine the correct number of turns. Find the core flux capacity $\phi_t = 2B_m A_c$. When the core flux and the operating frequency are known, the turns per volt can be found. Multiply the turns-per-volt by the applied voltage to find the correct turns, $N$, to be wound on the core selected in step (5).
(8) To calculate magnetizing current, note that mean length $(\ell)$ of the core selected in step (5). Determine the magnetizing force $(H)$ of the material from the curves presented in the core design manuals. Solve for magnetizing current $(I_m)$ from the formula:

$$I_m = \frac{0.794 \times H \times \ell}{N} \qquad (7\text{-}15)$$

The window area of the toroid available for winding turns is influenced by two important considerations:

- The minimum diameter to which the coil winder may safely wind.
- The total space that the copper itself will occupy.

The minimum diameter is the diameter which just allows the final turn to be wound with sufficient clearance all along the inner circumference. For the total space, it can be shown theoretically that the ratio of wire area to area required for the turns can never be greater than 0.91 for circular cross-section wire placed upon a flat form. Practically, this constant is a function of the type of feed employed on the coil winder, the winding sector employed, the thickness of insulation on the wire, and the minimum diameter to which the coil is to be wound; consequently, it is always less than this theoretical figure.

The best approximation for $K$ was found to be 0.75 for double Formvar insulated wire (step [5]).

The function relating the coil winding resistance to the maximum number of turns capable of being wound on a particular size core is given by:

$$R_c = \frac{\rho \ell_T N^2}{KW_a} \tag{7-17}$$

This follows from elimination of wire size from the equations

$$R_c = \frac{\rho \ell_T N}{a}$$

and

$$K = \frac{aN}{W_a}$$

where: $W_a$ = available window area,
$\ell_T$ = mean length of a turn of winding.

These formulas can be used to calculate the winding resistance.

## Winding Data

The estimate of the greatest number of turns that can be machine-wound on a toroidal core is made by taking several factors into consideration. These factors are based on the core diameter and height, the type of machine and shuttle used, and the size of wire used. These variables can be expressed in a formula which takes all of the above points into account. This expression is:

$$\text{Maximum Number of Turns} = \frac{\text{Effective Window Area}}{\text{Area of 1 Turn of Wire and Insulation}}$$

(7-18)

### Magnetic Leakage and Core Construction

When current flows through a coil wound on an iron core, most of the flux produced passes through the core, linking all of the windings. Although the permeability of the materials used for cores is high, not all of the magnetic field is confined to the core. Some of the flux flows through the air surrounding the core, as shown in Figure 7-5, or through the insulating material on which the core is wound, so that the two illustrated windings are not linked by the complete flux. This is referred to as magnetic leakage. Furthermore, as core magnetic saturation is approached, the magnetic leakage increases. The leakage must be considered in the design and construction of a saturable reactor core since it reduces the efficiency of the device and increases the time delay in its response. In some cases, the problems caused by the leakage are approached mathematically by assuming the coil to be subdivided into two sections, the main portion producing a flux confined to the core, and the second portion generating the leakage flux. The mathematical results obtained from this basic assumption are an excellent approximation to the truth.

Several practical methods are used to minimize the amount of leakage produced in a reactor. In a single-core device, both coils may be wound on the same limb of the core. Although this construction does reduce the leakage, it necessitates a high degree of insulation between the coils to prevent voltage breakdown and arcing across the adjacent portions of the coils. In a two-core arrangement, the windings are also placed on the same limb of the core to minimize leakage, Figure 7-6(a), and then the two cores may be overlapped, Figure 7-6(b), to reduce the space.

In general, and regardless of whether the reactors under discussion have one, two, three, or four cores, the cores of reactors designed for single-phase applications may be subdivided according to their construction as follows:

- Toroids or ring cores
  Stacked cores
  Spiral cores

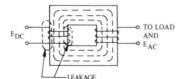

Figure 7-5   Magnetic leakage in a saturable reactor.

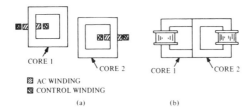

CORE 1

CORE 2

CORE 1

CORE 2

▨ AC WINDING
▩ CONTROL WINDING

(a)                                   (b)

Figure 7-6    Reducing leakage in a two-core reactor.

■ Rectangular cores
Stacked cores
Spiral cores

The stacked toroidal core comprises ring-type laminations of core material arranged in a pile to the desired height of the core, as in Figure 7-7(a). The coil windings are wound toroidally over the core. The spiral toroidal core construction is used to overcome the difficulties present in handling thin laminations. In this core, a strip of core material is wound in the form of a spiral (see Figure 7-7(b)). Again the coil windings are applied by using toroidal winding machines. Because of its method of construction, this core is sometimes referred to as a tape-wound core.

The stacked rectangular core consists of rectangular laminations, which may be arranged in UI fashion. To understand the derivation of the "UI" terminology, consider the uncut piece of core material illustrated in Figure 7-8(a). If a bar of the metal is stamped out along the dotted lines of Figure 7-8(b), the two UI-shaped pieces of material shown in Figure 7-8(c) result. If these two pieces are arranged as in Figure 7-8(d) and are joined along their junction, a rectangular lamination is produced. Rectangular laminations are preferred because they permit the coils to be wound on the straight arbor of a standard high-speed winding machine. In order to maintain a uniform flux density throughout all regions of

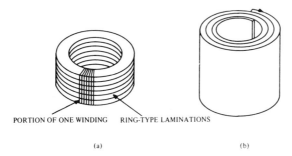

PORTION OF ONE WINDING    RING-TYPE LAMINATIONS

(a)                                   (b)

Figure 7-7    (a) Construction of a stacked toroidal core. (b) Construction of a special toroidal core.

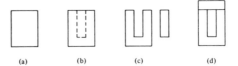

Figure 7-8   Procedural steps in the construction of a UI rectangular lamination.

the core, the rectangular UI laminations are alternated throughout the core. Thus, for the first lamination, the I is placed in a given position (at the top of the U, as in Figure 7-8(d)). The next lamination is placed with the I portion in the opposite direction (Figure 7-8(d) reversed, with the I section on the bottom). In this manner, the laminations are alternated throughout the height of the core.

The stacked rectangular core may also be constructed in EI fashion, as illustrated in (a) through (d), Figure 7-9. The core material is stamped to produce E and I pieces, which are then joined together to form a three-legged rectangular lamination. These laminations, too, are alternated to produce a constant flux density throughout the core.

The spiral rectangular core is constructed in much the same manner as the spiral toroidal core, except that the strip of material is wound in rectangular, rather than spiral, form. The spiral configuration is preferred if high, consistent frequency performance of the magnetic amplifier is essential. This, in part, is because the tape-wound core can be readily fabricated in very thin gauges of fixed thickness. The stacked core, on the other hand, must be made from carefully punched laminations which cannot be assembled after annealing for fear of straining the core material.

Three-phase reactors can be produced in several ways. The most common method is by placing three spiral toroidal cores, each having one load winding, on top of each other. The control winding is wound around the three cores and the result is a three-phase assembly. In another type of construction, three rectangular cores, each with its own load winding, are placed in a Y relationship. The control winding is placed over each of the load windings, and again, the final assembly is a three-phase reactor.

## Multiple Control Windings

The saturable reactor does not necessarily have only one control winding; in fact, it may have as many as four or six. The control action produced in a reactor

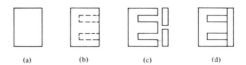

Figure 7-9   Procedural steps in the construction of an EI rectangular lamination.

with multiple DC windings is determined by the resultant of the magnetic fields produced by each winding. Thus, the addition of several control windings produces a reactor of great versatility. Control signals from several points in a system may be introduced independently, and remain isolated from each other. The sum of the magnetizing forces produced by the control signal then acts to initiate the required control function.

As an example of the versatility achieved in a device by the use of multiple control windings, consider a reactor with three such windings. One of the windings may be selected to be responsive to current change in one section of a system; the second winding may be adjusted to be sensitive to a voltage change in another part of the system; and the third may respond to a frequency change in a different portion of the system. The net result is that the reactor can now respond to changes of different types in many different sections of the given system, and thus can provide control action that would be difficult with a different type of circuit.

One control winding of a multiple-winding reactor frequently is used to set the state of the core to a convenient operating condition. This winding, the bias winding, is usually fed with DC produced by rectifying an output of the main AC supply. Another control winding may be used to inject feedback into the reactor.

## Limitations on Magnetic Amplifier Performance

There is, of course, a limit to the number of load turns that can be placed on an amplifier core. In the power frequency range, the maximum number of turns and minimum core area are set by the resistance of the load winding and the resultant internal impedance of the amplifier which limits the output power. It is generally accepted that the major factor limiting the amplifier figure of merit is, for a given core, the load-winding resistance (sometimes spoken of in terms of the available load-winding area). As supply frequency increases, fewer load turns are required to handle a given power and the limitation on performance of load-winding area becomes less severe. The output power can be handled on a smaller core which requires less control power and produces shorter response times. Therefore, there is a possibility of achieving increased amplifier performance at high frequencies.

Because of increased core losses, the differential permeability of magnetic materials (except possibly ferrites) decreases rather drastically with increasing frequency. Above the audio frequency range, the differential permeability of even a $\frac{1}{8}$-mil thickness magnetic tape is considerably lower than the differential permeabilities which can be obtained in the power frequency range. In contrast to the behavior of load-winding area, the limitation imposed on amplifier performance by the characteristics of core materials becomes more severe at high frequencies. High-frequency magnetic amplifier operation produces power gains per unit response time which are much higher than those attainable in the power frequency range. This increased performance is possible primarily because the beneficial effects which arise from the less severe limitations of load-winding area are greater than the detrimental effects which increased frequency has on the magnetic characteristics of the amplifier core materials.

### Frequency

If the minimum area has been selected to meet the volume requirements for input power and response time, the supply voltage for the core is a function of the number of gate winding turns and the frequency. If a large amount of the available winding area is to be used for control copper, the supply voltage and power output can be increased only by increasing the frequency. This is the real advantage to be gained by the higher carrier frequencies. Of course, the upper limit on the voltage for many applications is imposed by the rectifier ratings. An upper limit on frequency also exists because of winding capacitance and exciting currents. This upper limit on frequency, however, is well above the audio-frequency range. If a transistor power supply is used, a reasonable frequency for efficient operation is 20 kHz to 50 kHz.

An important frequency effect is that, as frequency increases, eddy currents in the core decrease the effective AC permeability. The approximation equation is $\mu_d = C/A + \sqrt{f}$ where $C/A$ is the DC permeability. Over a frequency range from approximately 3 to 10 kHz, $\mu_d$ may be approximated by $\mu_d = D/\sqrt{f}$.

In addition to decreasing $\mu_d$, high frequencies also accentuate capacitive effects, and increase exciting current and the saturated impedance of the core. These effects, which are difficult to evaluate quantitatively, prevent accurate prediction of amplifier characteristics by means of the design equations developed in this chapter. Despite the lack of accuracy, however, the equations show that it is advantageous to decrease input power and time constant by decreasing the core size, and that the figure of merit can be increased by operating the amplifier at a high frequency.

### MAGNETIC AMPLIFIER APPLICATION TO TRANSISTORIZED POWER SUPPLIES

Increased magnetic amplifier performance can be achieved by using higher frequencies. The primary advantage of higher-frequency operation for full-wave magnetic amplifiers is that, for a given power output level, the core size can be greatly reduced. This reduction in core size reduces the input power and time constant, thus increasing the figure of merit.

The design considerations for high-frequency amplifiers are in many respects different from those of 60 Hz or 400 Hz amplifiers. The core of a magnetic amplifier when used for high-frequency applications must possess the following characteristics:

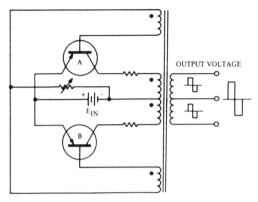

Figure 7-10 Transistor power supply.

- The core must be easily saturable.
- Eddy-current effects must be at a minimum.
- Hysteresis losses must be low.
- The physical mass of the core must be small.
- The effective retentivity of the core must be low.

The circuit diagram for a typical power supply, already discussed in chapters 2 and 3, is shown in Figure 7-10.

Magnetic amplifiers can be used either to produce chopped DC at the output or to preregulate the input. If a nonreversing square wave is applied to a square loop core, it blocks the first forward biasing until it saturates (assuming $t \leq N_2/E_{in}$). On the second forward biasing, it will have had no reset and will conduct much sooner (see Figure 7-11). The ratio $t_1/t$ will be approximately the squareness (Figure 7-12).

The magnetic-amplifier/diode combination produces nicely controlled and pulsed DC (from an AC source) if some means of reset is available. Reset either

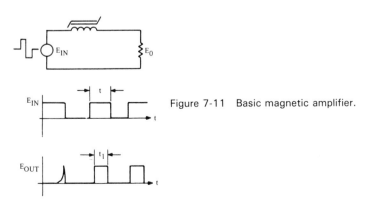

Figure 7-11 Basic magnetic amplifier.

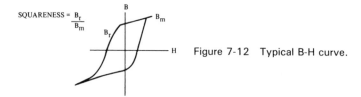

Figure 7-12   Typical B-H curve.

can be continuous or can be applied from another AC source during the time the primary (gate) is nonconducting. In this latter case reset is applied out of phase in both time and polarity (Figure 7-13). In either case, the reset current reverses the magnetism in the core requiring that the flux change further to saturate in the forward direction.

As the reset current increases from zero, the gate conduction time decreases. If the reset current is equal to the magnetizing current ($H_c$), the gate conduction will be zero and only magnetizing current will flow in the gate. (The gate load must be high enough to allow magnetizing current to flow.) A further increase in reset current beyond $H_c$ results in an increasing output voltage. The gain of this second rise ($\Delta\varepsilon_o t_o / I_{reset}$) is much lower than the gain with $I_{reset} H_c$ (see Figure 7-14). The reset current (if flowing in the same direction as the gate current) pushes the conduction time beyond the squareness ratio toward 100% conduction.

Most of the magnetic amplifiers in present use are full-wave types rather than half-wave types (Figure 7-15). The DC common-current reset style will now be discussed. Each core has a winding of its own which is a gate. Around both cores is a common winding which is the reset. If the cores are perfectly matched, there is no voltage induced into the reset winding. The usual tolerance on matched cores is 5%; however, some second harmonics are induced. These can be quenched fairly well by shunting the reset winding with a resistor. The current, which flows in the reset winding because of the shunt resistor, flows in a direction to set the core rather than reset it, thus resulting in a low induced second harmonic. The reset winding should have fewer turns than the gate for

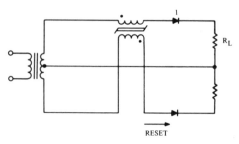

Figure 7-13   Magnetic amplifier with reset.

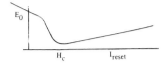

Figure 7-14   Reset current characteristic.

best operation and for low second harmonics. If there are substantially more reset turns than gate turns, the control loop develops negative resistance. For increasing reset current, the output falls until point A (Figure 7-16). Then it switches to point *B* and the output rises with more reset. As the reset is decreased, the output follows the curve back to point *C* and then switches to *D*. A similar effect can be generated by running the output current back through the core as another reset.

The common-current reset style magnetic amplifier is fairly simple to build, requires few parts, and works fairly well with reasonably matched cores. Because the reset does not work independently of the gate conduction, a step change in reset current takes several cycles before it causes the output to come to a new level.

The number of turns required for a gate winding is given by $N \gg E_{in} t 10^8 / \phi$. The time response decreases as the reset turns are increased in excess of the gate windings. The cores can be roughly matched by applying a signal to the reset winding and measuring the output at each gate. If the voltages are the same, the cores are fairly well matched for reset current. If $N_1 = N_2 = N_3$, then $E_1 = \frac{1}{2} E_3$. The cores appear as series loads to the common winding.

The other reset scheme (Figure 7-17) to be considered can be looked at as either independent voltage or current reset. As connected, it is current reset. Each core is set and reset on opposite half-cycles. As such there is very little interaction between the two processes. The time response in this scheme is as

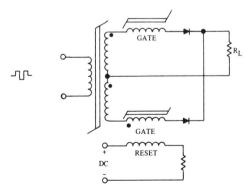

Figure 7-15   Full-wave magnetic amplifier with common current reset.

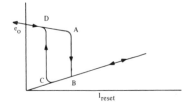

Figure 7-16 Reset current control loop for Figure 7-15.

fast as can be obtained from magnetic amplifiers. The output changes one cycle after a change in reset current. In practice, the rate at which reset current can change should be slowed down or a half-frequency oscillation results as each side corrects for what the other side did during the previous half-cycle.

If, instead of a resistive reset load, a Zener diode or equivalent circuit is used (Figure 7-18) the output will be approximately equal to $E_z$ (discounting diode and IR drops). Waveforms which can be expected in these two circuits are shown in Figure 7-19.

Until a forward-biased gate saturates, $I_{mag}R_L$ appears at the output. After the drive switches, the energy stored in the saturated inductance of the core which was on keeps current flowing for a short period. Since the cores do not saturate instantaneously, the rise time of the output is not too fast. The rise time will be faster if an LC filter is used.

## INPUT VOLTAGE REGULATION

Instead of, or in addition to, regulating the DC output voltage, a power supply input voltage may be regulated to obtain the desired regulated output voltage. Some power supply designs use both input and output regulation. There are as

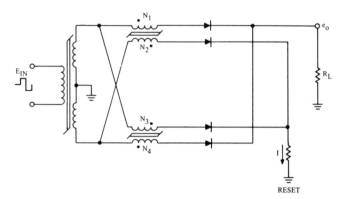

Figure 7-17 Full-wave magnetic amplifier with independent voltage or current reset.

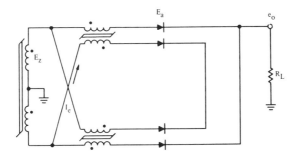

Figure 7-18   Circuit of Figure 7-17 with Zener diode load.

many types of input-regulating circuits as there are output-regulating circuits; a typical circuit is discussed here.

Figure 7-20 depicts a boost preregulator which utilizes a Zener diode as the regulating element. This can be mechanized, but not with a single Zener. A slight negative temperature compensation is required for temperature regulation. A string of CD-3136 (nominal 5.1 V) Zeners has the proper temperature compensation. Unfortunately, the low voltage Zeners have a rather soft knee; thus, the line regulation suffers. By adding $CR_6$, $CR_7$, and $R_2$, some line voltage feedback is obtained which will modify the Zener characteristics enough to give very good line regulation. The average voltage at the cathodes of $CR_6$ and $CR_7$ is inversely proportional to line voltage. Thus more current flows at low line, which decreases the amount of reset current and raises the output. With enough feedback through $R_2$ the output can be made to go down for increased line voltage.

For an even simpler regulating scheme, the Zener stack can be replaced by a resistor or by an integrated circuit sharp-knee Zener diode such as the LM103. Proper selection of the Zener replacement and $R_2$ will give good regulation against line changes.

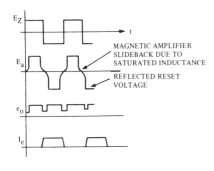

Figure 7-19   Waveforms of Figure 7-17 and 7-18.

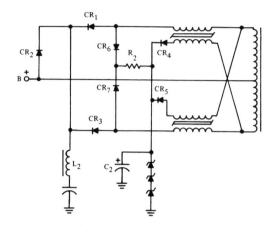

Figure 7-20   Magnetic amplifier/Zener diode boost regulator.

Even better regulation is possible by sensing the voltage across $C_3$ with a differential amplifier and using it instead of the Zener to correct the reset current.

Pulse-width circuits of the type discussed earlier are most often used as input voltage preregulators.

## THE MAGNETIC AMPLIFIER OUTPUT REGULATOR

Frequently, magnetic amplifiers are used in the output of a DC/DC converter (before the filter) to provide the required voltage regulation. The simplest magnetic amplifier is a square hysteresis loop magnetic core with two windings: one load winding and one control winding. When no control current is applied, the impedance to alternating current in the load winding is very high. A small control current, however, saturates the core, reducing this impedance to near zero. Figure 7-21 shows a typical magnetic amplifier, such as might be used between the output transformer and the filter of a static inverter.

If no DC signal is applied and the input AC waveform is a square wave, the output AC waveform will also be a square wave. $T_1$ and $T_2$ will alternately be saturated by the input square wave. If a DC signal is applied in the proper direction, this saturation may be delayed, the delay being proportional to the amplitude of the direct current. This results in the output shown in Figure 7-22(b) or (c). The output waveform becomes a square wave with variable dwell time. If this wave is filtered, a sine wave of variable amplitude is produced.

$CR_2$ is a commutating diode and provides a path for the output current when the magnetic amplifiers are blocking. The filter inductor will pull current through the commutating diode. $L = R_{max}/W$ represents the minimum value for the filter inductor. Due to the inherent inductance of the magnetic amplifiers it is possible to work a lower load than $L = R/W$ would indicate.

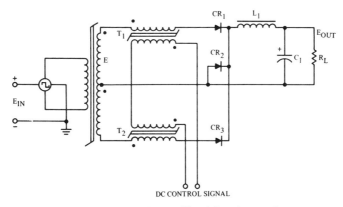

Figure 7-21 Magnetic amplifier AC series regulator.

The magnetic amplifier voltage regulator is simple in nature, provides some degree of automatic regulation, but is quite heavy physically—its main disadvantage. It is best utilized in the driver stage rather than in or following the power stage.

A voltage-controlled magnetic amplifier requires more component parts than a common-current controlled magnetic amplifier but offers better regulation. Almost any transformer will work as an AC source. It is not necessary to have a magnetic amplifier winding referenced to the input. (While the input-reference winding does not detract from the isolation, it does help couple noise from input to output.)

The output is proportional to the Zener voltage and $N_1/N_2$ (of Figure 7-23). A different drive is used for the gate and reset windings in order that a near zero temperature-compensated Zener diode can be used for the reset and have

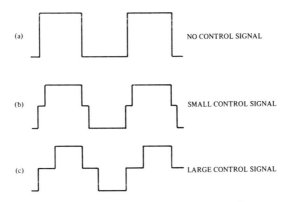

Figure 7-22 Magnetic amplifier output waveforms.

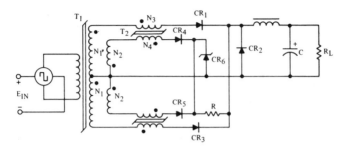

Figure 7-23   A voltage-controlled magnetic amplifier.

an output voltage other than 6 or 8 V. The output varies less than 2 to 3% from input voltage and temperature variations. $N_1$ should provide 125% of the output voltage at minimum $E_{in}$. $N_3$ must be able to support this drive level for a full half-period. (Drive voltage times half-period is a constant.) Nominal input voltage should be used to select $N_4$ and $N_2$.

In Figure 7-23 better regulation can be obtained by adding transistors and by sensing the actual output voltage to control the reset current or effective reset voltage. A differential amplifier works very nicely for this control.

Figure 7-24 depicts the schematic diagram of a magnetic amplifier-oscillator combination boost regulator. The circuit operates in the following manner. When power is applied, current flows through $CR_1$, $CR_2$, and $CR_3$ to start the oscillator. Assume that $Q_1$ starts to conduct making all $T_1$ dots positive: $T_{3A}$ is then forward-biased and blocks current; $T_{3B}$-$CR_3$ is reverse-biased; current continues to flow through $CR_2$; and $T_{3A}$ saturates. Now the input current flows through a voltage rise in $T_1$, causing $CR_2$ to be back-biased. $E_{in} + E(N_3) - E(T_1)$ is applied to the choke input until the oscillator switches. The cycle is then repeated with $T_{3B}$ forward-biased. $L_2$ and $C_3$ filter this pulsing DC to a clean DC and apply it to the oscillator. The voltage at $C_3$ is approximately the voltage across $CR_6$. Thus, constant-amplitude square waves are fed to the output and are easily rectified back to DC. The only losses in boosting the input voltage are the diode drops and $I^2R$ losses. Figure 7-25 illustrates typical waveforms of the magnetic amplifier regulator.

The pulse-width modulated circuits are difficult to use where the load resistance is large because of the large output filter chokes required. In the boost system, a pulsing voltage must still be filtered, but it never goes to zero. A large boost filter choke keeps the slope of the boost current down, reduces its maximum amplitude, and reduces the input ripple. A small filter choke reduces the current in $C_3$ when the magnetic amplifier switches, but increases the current when the transistor switches. It also reduces the switching transients that are generated when the magnetic amplifiers saturate. If the boost filter is too small, the magnetic amplifier comes out of conduction at light loads, and the waveforms

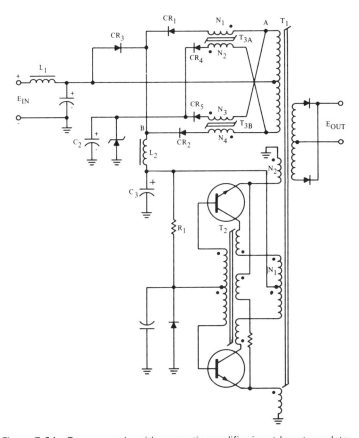

Figure 7-24 Power supply with magnetic amplifier input boost regulator.

in the boost loop change (Figure 7-26), but the output voltage remains essentially unchanged. While a small boost choke does have some disadvantages, it improves the regulator's frequency response by reducing the time delay through the boost filter. All things considered, it is usually best to use the largest choke possible, consistent with physical limitations and permissible IR losses.

## Output Voltage Regulation Example[1]

Figure 7-27 shows the block diagram of a multiple-output, forward converter switching power supply in which a square-loop core provides a controllable

[1]"Output Regulation Using Mag Amp Control for Switched-Mode Power Supplies," Applications Engineering Staff, Magnetics Division, Spang Co., *Power Conversion International*, April 1985. Used with permission.

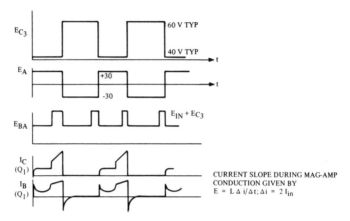

Figure 7-25   Waveforms of Figure 7-24.

delay at the leading edge of the pulses at the transformer secondary. The use of saturable reactor (magnetic amplifier) output regulators, which have been used in power converters with frequencies up to 1 MHz, provide advantages in high output applications. Figure 7-28 is a schematic diagram of the output regulation technique along with typical waveforms.

In the pulse-width modulator (PWM) of Figure 7-28, the primary pulse width is controlled by sensing the 5 V output, comparing it to a reference, and using the error signal to adjust the pulse duration. If there were no saturable core (*SC*) in the circuit, the 15 V output would be "semi-regulated," since the primary control loop would provide line regulation. But the output would vary with load and temperature.

To produce 15 V DC at the output, the average value of the rectified waveform

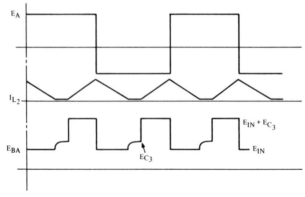

Figure 7-26   Expanded waveforms of Figures 7-24 and 7-25 showing effect of boost filter.

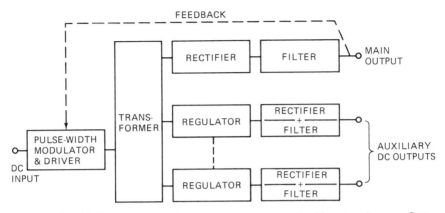

Figure 7-27   Multiple output switched mode power supply. (Courtesy Intertec Communications, Inc.)

applied at the input of inductor $L$ must be 15 V. Given the pulse height of 50 V and a repetition period of 10 $\mu$s, the required width of the positive pulse at $e_2$ must be:

$$PW = (15 \text{ V}/50 \text{ V}) \cdot 10 \ \mu s = 3 \ \mu s$$

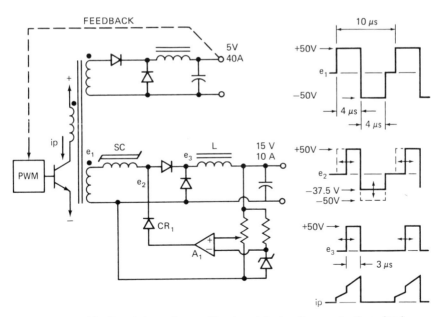

Figure 7-28   Regulation scheme. (Courtesy Intertec Communications, Inc.)

Because the input pulse ($e_1$) is 4 $\mu$s wide, the saturable core must delay the leading edge by 1 $\mu$s. Since the amplitude of the pulse is 50 V, the core must "withstand" 50 V · 1 $\mu$s, or 50 volt-microseconds. To accomplish this, the core is reset by this amount during each alternate half-cycle. The waveform at $e_2$ (Figure 7-28) illustrates this. As the input to the core swings negative, diode $CR1$ conducts and allows error-amplifier $A1$ to "clamp" the output side of the core at $-37.5$ V for a duration of 4 $\mu$s, providing a reset of:

$$A = 12.5 \text{ V} \cdot 4 \text{ } \mu s = 50 \text{ V} \cdot \mu s$$

(where $A$ = withstand)

As the output load varies, the error-amplifier changes this value to ensure that the output is regulated at 15 V DC in spite of changes in the rectifier voltage drops.

The waveform of the primary current, $i_p$, shows the increase in current when the core saturates and begins to deliver current to the output inductor. This provides an added bonus: the primary switching transistor has already turned on and saturated, and thus the 15 V output does not contribute to turn-on switching losses in the transistor.

The design of the saturable reactor requires three steps:

1. Determine $A$, the withstand volt-seconds, to delay the leading edge of the pulse and achieve the required output voltage. Two choices exist: (1) Does the output need to be capable of independent "shutdown" (for short-circuit protection or turn-off from an external logic signal), or (2) simply regulated at a fixed value?

$$\text{Withstand} = \text{Excluded Pulse Area} = A$$
$$A = V \cdot t \tag{7-19}$$

Where $V$ = pulse amplitude, and $t$ = delay at leading edge.

**Case 1: Shutdown.** The required withstand is simply the area under the entire positive input pulse. In the circuit of Figure 7-28, it would be 50 V · 4 $\mu$s = 200V · $\mu$s.

**Case 2: Regulation Only.** Assuming that the output inductor has been designed for continuous conduction, the reactor must only reduce the input pulse width enough to furnish the required average value (equal to the DC output voltage) at the input of the filter inductor.

In both cases, one must allow a safety margin to accommodate load transients, a result of the choice of turns on the secondary winding of the transformer which feeds the regulator. This must precede the calculation of the volt-seconds which the reactor must support. For example, one might design for a control range of $\pm$ 20% to allow the pulse width to increase or decrease by 20% when the load current steps up or down. To allow the pulse width to increase, the input pulse width must be 20% greater than the nominal pulse at the output of the reactor. Depending on the operating frequency and core used, one must allow an additional margin due to the rise-time of current in the core after it saturates. This is typically on the order of one microsecond and implies that the secondary voltage be at least 20% higher than it would be to produce the desired output voltage if the saturable reactor were not present. To allow the pulse width to decrease, the reactor must withstand additional volt-seconds to reduce the pulse width 20% below the nominal value.

In the circuit of Figure 7-28, a "regulation only" design would require a withstand of $A = 50$ V $\cdot$ $\mu$s $+ 20\%$, or 60 V $\cdot$ $\mu$s.

2. Choose the core. There are two widely used methods of determining the size of the required core. Each results in a minimum area product, $W_a A_c$, to provide the necessary withstand and accommodate the wire size (which determines the temperature rise). One method begins with the desired temperature rise and power to be handled (withstood), the core geometry, and the fill factor. The other requires an initial choice of the wire size, which must be estimated based on intuition about the ultimate temperature rise. Although the latter is admittedly pragmatic, it is popular because of its simplicity.

In the latter method, the steps are as follows:

A. Pick the wire size, based on the current. A reasonable value is 500 circular mils per amp of current (rms) for a temperature rise of 30° to 40°C in core sizes 5 to 1 in. o.d. This yields $A_w$, the cross sectional area of one conductor.

B. Choose a core material, to determine the saturation flux density, $B_o$. In this application, Square Permalloy 80 is a good choice, since it has low coercive force and a very square $BH$ loop; its $B_o$ is approximately 7000 gauss. In fact it has been used in power converter applications operating at 1 MHz.

C. Choose the fill factor, $K$, in accordance with the Core Selection information of Magnetics Tape Wound Core Catalog TWC-300R, for example. (In this section, Magnetics, Inc. Transformer Core and Bobbin Design Data Books will be used exclusively and are so referenced.) Values of .1 to .3 are appropriate, with the lower values for power applications.

D. Calculate *WaAc* as follows:

$$W_a A_c = \frac{A_w \times A \times 10^8}{2 \times B_o \times K}, \text{ in circ. mils} \times \text{cm}^2 \qquad (7\text{-}20)$$

E. Select a core from the selection table of Catalog TWC-300R with at least this area product. Tape thicknesses of .0005 and .001 in. are recommended for frequencies up to 100 kHz, with the thinner tapes found in the Magnetics bobbin core catalog BC-303 preferred at higher frequencies.

In the circuit of Figure 7-28, the current during conduction of the core is 10 A, and the duty ratio is 15/50, or .3. Thus the rms current is $(10^2 \times .3)^{1/2}$, or 5.5 A. An appropriate wire size is 16 gauge, since its cross-sectional area, $A_w$, is 2581 c.m. Again, using the "regulation only" case, $W_a A_c$ is as follows:

$$W_a A_c = \frac{2581 \times 60 \times 10^6 \times 10^8}{2 \times 7000 \times .1} = .011 \times 10^6 \text{ c.m.} \times \text{cm}^2$$

Note that a fill factor of .1 has been used, since the wire size is relatively large.

Since the converter frequency is 100 kHz, a tape thickness of .0005 in. was chosen. In consulting Catalog TWC-300R, the $W_a A_c$ figures must be altered by a factor of approximately .013/.022 (the typical ratio of the cross-sectional areas of cores with .0005 in. and .002 in. tape thickness). The most convenient way to do this is to alter the value of the desired $W_a A_c$, and then find the appropriate core in the table. Using this approach, the listed value must be at least .011 × (.022/.013) × $10^6$, or .019 × $10^6$. Two logical candidates are the 5_374 and 5_063 cores, whose $W_a A_c$ (× $10^6$) values are .028 and .026, respectively.

For the purpose of this example, the 5_063 core was chosen. Its effective core cross-sectional area, $A_c$, is .050 cm$^2$; its mean length of magnetic path, $\ell$, is 5.98 cm.

3. Determine the number of turns. The number of turns is determined by the withstand, $A$, to produce the desired output of the regulator:

$$N = \frac{A \times 10^8}{2 \times B_o \times A_c} \text{ turns} \qquad (7\text{-}21)$$

where

$A$ = withstand, in volt-seconds

$B_o$ = saturation flux density in gauss

$A_c$ = core cross sectional area in cm$^2$

The control circuit can now be designed. In doing so, it is helpful to estimate the current required to reset the core, and thus calculate the average control current based on the duty ratio of the resetting (negative portion) of the input pulse. The current is related to the magnetizing force as follows:

$$I_m = \frac{.975 \times H \times \ell}{N} \text{ amps} \tag{7-22}$$

where

$H$ = magnetizing force in Oersteds

$\ell$ = magnetic path length in cm.

$H$ is not simply the DC coercive force, but rather the value corresponding to the flux swing and frequency, as given in Magnetics Catalog TWC-300R.

Again, using the circuit of Figure 7-28 and the chosen core, the required number of turns is:

$$N = \frac{60 \times 10^6 \times 10^8}{2 \times 7000 \times .050} = 8.57 \text{ turns (round off to 9 turns)}$$

Completing the example, the magnetizing current is calculated as follows: Since the regulator must swing across the entire $BH$ loop during transients, the curves shown in Catalog TWC-300R will give a worst-case estimate of the magnetizing force. At a frequency of 100 kHz, the 1/2 mil curve has the value of $H = .215$ Oersteds. Thus, the magnetizing current will have a maximum value of:

$$I_m = \frac{.975 \times .215 \times 5.98}{9} = .11 \text{ A}$$

An alternative control circuit is given in Figure 7-29. It has two notable features:

1. The reset current is derived from the output, providing a ''pre-load''—a means of preventing the magnetizing current of the reactor from raising the output voltage at zero load.

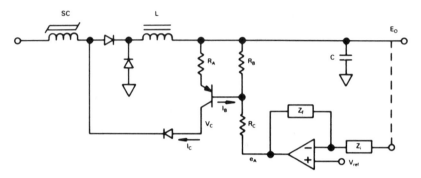

Figure 7-29 Alternative control circuit. (Courtesy Intertec Communications, Inc.)

2. The core is reset from a current source, rather than a voltage source. This minimizes the phase shift of the control transfer function. In this circuit, $R_A$ degenerates the transconductance of the transistor, making the transfer function more independent of the transistor. $R_B$ and $R_C$ simply shift the level of the amplifier's output, which is unnecessary if the amplifier is powered from a voltage higher than the output.

The compensation networks, $Z_f$ and $Z_i$, can be designed using techniques for conventional buck-derived regulators. Note, however, that this circuit actually has two feedback loops: one through the error amplifier, and one directly from the output through $R_A$ and the transistor.

It is sometimes useful to be able to translate the voltage required to reset the core, change its level, or trade voltage for current. In these cases, a second winding can be placed on the core, with a larger or smaller number of turns than the power handling winding, and with its end opposite the control transistor being returned to a convenient bias voltage. For example, a control winding with less turns will exhibit less voltage swing, but will require more control current than the main winding.

The regulation technique just discussed is applicable to other power supply topologies, as well. A power supply with full-wave outputs is shown schematically in Figure 7-30.

## A MAGNETIC AMPLIFIER/SWITCHING POWER SUPPLY DESIGN EXAMPLE[2]

This section describes the practical design of a magnetic amplifier controlled switching regulator. Operating at 40 KHz, the switching regulator produces five regulated output voltages with a total output load of 130 W.

---

[2]A Magnetic-Amplifier-Controlled 40 KHz Switching Regulator by R. Dulskis, J. Estey, and A. Pressman, *Solid State Power Conversion*, September/October 1977. Used with permission.

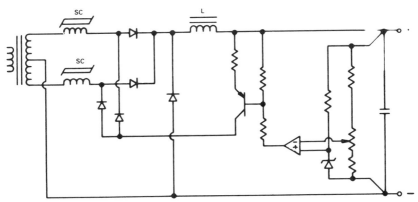

Figure 7-30  Full wave saturable core regulator. (Courtesy Intertec Communications, Inc.)

Prime power for the supply described herein is a 30 V battery, but the technique is just as easily applicable to direct off line AC switching regulators with 150 V or 300 V supplied from the AC line rectifier.

Higher output power than that realized with the present 30 V input is easily attainable. With either the higher DC input, or with additional paralleled output power transistors and the 30 V prime input voltage, the same magnetic amplifier (a cylinder of about 0.5 in. in diameter and 0.6-in. height) can easily control outputs of 1 kilowatt or much higher powers.

In the supply, 75% of the output power is at the 5 V level, where the 1 V drop of conventional rectifier diodes alone limits efficiency to a maximum of 83%. With Schottky diodes or with a less fraction of the output power at the 5 V level, efficiencies of 85% to 90% can be achieved.

The magnetic amplifier control method offers the following significant advantages over conventional switching regulator control means:

1. The magnetic amplifier serves both as the low level pulse width modulator and as the error voltage amplifier. This eliminates the need for the usual auxiliary DC supply voltage required to supply conventional control circuits. All the time race problems involved in the time sequence where auxiliary and main DC supply voltages turn on and off are avoided. Failures in the main power switch transistors caused by partial or complete failures in the auxiliary DC source are avoided.

Slow start circuitry, which is usually required in conventional pulse width modulators to avoid a disastrously high duty cycle at initial turn on is unnecessary as the mode of operation of the magnetic amplifier ensures that turn on occurs at minimum duty cycle. Slow start circuitry itself has subtle failure modes. Unless precautions are taken in the detailed circuit design, normal turn on without failures is possible, but turn off followed by fast return turn on can still cause failures.

The ability to survive turn off followed by fast return on may require a half dozen to a dozen components, and associated time race problems must be carefully examined. Since slow start circuitry is unnecessary with magnetic amplifier control, all its associated problems are avoided.

2. Magnetic amplifier control results in minimized parts count and hence greater reliability and higher output power per cubic inch.

3. Control and output windings of the magnetic amplifier are DC isolated and can be operated at different and varying DC voltage levels. The magnetic amplifier with its control and output winding isolated offers the simplest solution to the problem of output and input grounds being at different or varying DC voltage levels.

4. Temperature insensitivity of the magnetic amplifier eases the problems of thermal design. Where voltage settings and stability to mV (millivolt) rather than microvolt accuracy is adequate, the magnetic amplifier is superior to the conventional semiconductor operational amplifier as an error amplifier. The magnetic amplifier, using a tape wound moly-Permalloy core, operates with no significant variations in properties over a far larger temperature range than do semiconductors.

Figure 7-31 shows the circuit of the magnetic amplifier controlled switching regulator. It is a half cycle width modulated DC/DC converter in which the

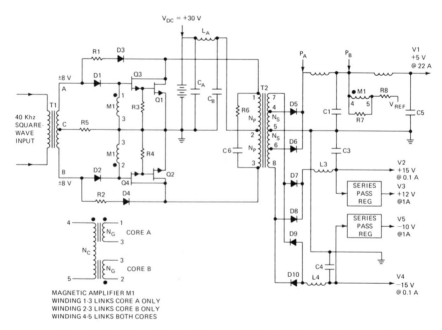

Figure 7-31  Magnetic amplifier controlled switching regulator circuit.

power transistors are Darlington pairs $Q_3$, $Q_1$, and $Q_4$, $Q_2$. The converter transistors on each side are not on for a full half cycle as in conventional DC/DC converters. Rather their on time in each half cycle is controlled by the negative feedback through the magnetic amplifier in such a manner as to yield a constant DC output voltage at point $P_B$ which is independent of line and load change.

The voltage at the cathodes of rectifier diodes $D5$, $D6$ (point $P_A$) consists of the set of half cycle width modulated power pulses shown in Figure 7-32. The base of the power pulses is one diode drop (rectifier diodes $D_5$, $D_6$) or about one volt below ground; their peak amplitude, as shown in Figure 7-32 is determined by the DC supply voltage, the transformer turns ratio, the Darlington pair on voltage (about one volt), and the output rectifier forward drop (about one volt).

The on time, $T_{on}$, of these output power pulses is equal to the transistor on time. It starts with, and terminates before the completion of, the input square wave half cycle ($T/2$). The DC or average voltage at the rectifier cathodes is then (assuming to a close approximation the base of the power pulse is at ground rather than at $-1$ V), equal to

$$V_{DC}(AV) = \left[ (V_{DC} - 1) \left( \frac{N_s}{N_p} \right) - 1 \right] \left[ \frac{T_{on}}{T/2} \right] \qquad (7\text{-}23)$$

as shown in Figure 7-32. Thus by controlling the ratio $T_{on}/(T/2)$ through the control winding of the magnetic amplifier, the average or DC output voltage is kept constant under conditions of varying DC input voltage or output current.

Figure 7-31 shows feedback elements $R_1$-$D_3$ and $R_2$-$D_4$. These comprise a Baker clamp which prevents Darlington drivers $Q3$, $Q4$ from saturating, and speeds up turn off time.

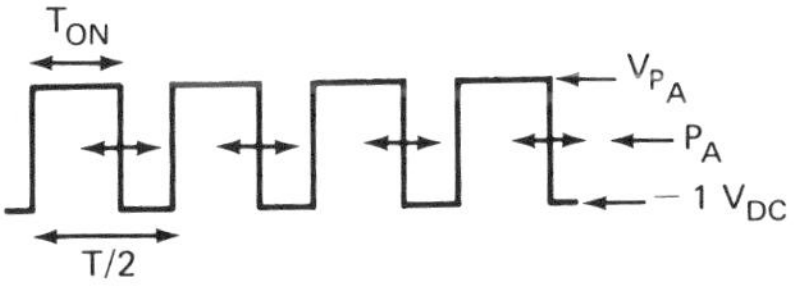

$$V_{P_A} = \left\{ V_{DC} - \left[ V_{BE}(Q1) + V_{CE}(Q3) \right] \right\} \left( \frac{N_S}{N_P} \right) - V_{D5}$$

$$\cong (V_{DC} - 1) \left( \frac{N_S}{N_P} \right) - 1$$

AVERAGE DC VOLTAGE AT RECTIFIER CATHODES ($P_A$)

$$\cong \left[ (V_{DC} - 1) \left( \frac{N_S}{N_P} \right) - 1 \right] \left[ \frac{T_{ON}}{T/2} \right]$$

Figure 7-32   Output rectifier cathode waveform.

The *L1-C1* network filters out the AC components of the waveforms of Figure 7-32 and yields, at $P_B$, a clean DC output whose ripple frequency is twice the input square wave frequency and whose amplitude can be made arbitrarily small by proper selection of $L_1$-$C_1$.

The voltage waveform at the rectifier cathodes and the operation of the $L_1$-$C_1$ circuit after that point is identical to the more familiar stepdown switching regulator. The function of the usual "free wheeling" diode is performed here by the rectifier diodes themselves. A "free wheeling" diode may be added from $P_A$ to ground to minimize dissipation in the rectifier diodes.

The novelty in this scheme lies in the magnetic amplifier $M_1$, which combines, in one small passive core and wire combination, the features of a voltage comparator and low level pulse width modulator without requiring the use of an auxiliary DC supply voltage.

The magnetic amplifier consists of two square hysteresis loop (Square Permalloy 80) toroids placed vertically above one another. As shown in Figure 7-31 a "gate" winding (1-3) links core $A$ only, a second gate winding (2-3) links core $B$ only, and a control winding (4-5) links both cores.

The "gate" windings are placed from the base of the Darlington drivers to input ground. The secondary of the input transformer supplies, at points A, B of Figure 7-31, out of phase 40 kHz, $\pm 8$ V square waves. These square waves drive the gate windings through diodes $D_1$, $D_2$. The common gate winding terminal, pin 3, connects through voltage dropping resistor $R_5$ to the center tap of transformer $T_1$.

Significant waveforms of the circuit are shown in Figure 7-33.

At the start of each positive half cycle, roughly 8 V is applied via diodes $D_1$ or $D_2$ to the series combination of $R_5$ and one of the gate windings $M1$-1-3 or $M1$-2-3. If the gate winding has a high impedance corresponding to the core state of being on the vertical part of the hysteresis loop of Figure 7-34, the voltage across it is clamped to the sum of the base emitter voltage of a Darlington driver and that of its output stage, or about $+1.6$ V.

Now for the duration of time that the core of a gate winding is on the high impedance vertical part of its hysteresis loop, its Darlington driver and output stage are energized, and so is the $T_2$ primary and secondary on the same side. Later in the positive half cycle when the core of the gate winding reaches the top of its hysteresis loop, its impedance drops to zero. All of the input transformer secondary voltage is now dropped across $R_5$, base voltage and current to the Darlington driver drop to zero, and voltage across the $T_2$ primary and secondary on the same side drop to zero. The result at the output of the secondary rectifiers is the sequence of power pulses shown in Figure 7-32. These pulses are at the high level of about:

$$(V_{DC} - 1) \left(\frac{N_s}{N_p}\right) - 1$$

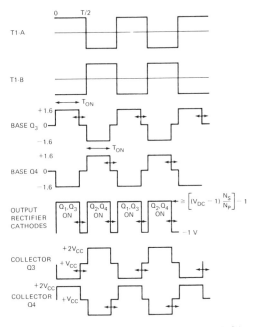

Figure 7-33  Circuit waveforms of Figure 7-31.

for a time equal to $T_{on}$, and at $-1$ V for $(T/2 - T_{on})$. As stated before in eqn 7-23, this corresponds to an average voltage of about:

$$\left[ (V_{DC} - 1) \left( \frac{N_s}{N_p} \right) - 1 \right] \left[ \frac{T_{on}}{T/2} \right]$$

and voltage regulation is achieved by controlling the ratio $T_{on}/(T/2)$.

The on time control is achieved by controlling the time at which the core, moving vertically up its hysteresis loop in Figure 7-34, reaches saturation at the magnetic-field intensity $B_s$. This time is determined by the initial field intensity, $B_0$, to which a core has been reset at the start of the positive half cycle. The starting point in field intensity, $B_0$, is controlled by the DC current in the magamp control winding (winding 4-5 in Figure 7-31). By bridging the winding between the output voltage and a standard reference voltage, the current is made proportional to the difference in regulator output voltage and the constant reference voltage.

The polarity of the control winding is chosen such that if the DC output voltage rises due to an increase in DC input voltage or a decrease in output current, the initial flux, $B_0$, to which a gate winding has been reset at the start of a positive half cycle, moves up and closer to $B_s$, the point at which core

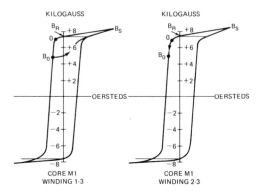

Figure 7-34   Operating points in flux density.

saturation occurs. This decreases the time required for core saturation, the transistor on time, and the ratio $T_{on}/(T/2)$. And from Equation 7-23, decreasing this ratio brings the output voltage back down to its original value.

The relation between $B_0$, the initial starting point in flux density, and the transistor on time is calculated from the fundamental magnetic relation as shown in Figure 7-35. It can be seen from the transistor on time expression of the equation in Figure 7-35, that decreasing the magnitude $(B_s - B_0)$ decreases the gate winding "firing time" or transistor on time.

The operating points in flux density of the two cores of the magnetic amplifier are shown in Figure 7-34. At the start of one half cycle when, say, $T1$-$4$ is positive, $T1$-$6$ is negative, core $A$ had been reset to $B_0$ on its hysteresis loop, and core $B$ had, on the previous half cycle, been forced up to saturation at $B_s$. As a constant 1.6 V ($V_{BE}$ of $Q_1$ plus that of $Q_3$) is forced across core $A$, it moves linearly up its hysteresis loop towards $B_s$. After a time $T_{on}$, (given in the equation

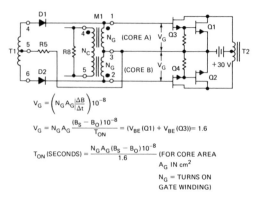

$$V_G = \left(N_G A_G \frac{|\Delta B|}{\Delta t}\right) 10^{-8}$$

$$V_G = N_G A_G \frac{(B_s - B_0) 10^{-8}}{T_{ON}} = (V_{BE}(Q1) + V_{BE}(Q3)) = 1.6$$

$$T_{ON} \text{ (SECONDS)} = \frac{N_G A_G (B_s - B_0) 10^{-8}}{1.6} \quad \text{(FOR CORE AREA}$$

$$A_G \text{ IN cm}^2$$

$$N_G = \text{TURNS ON}$$

GATE WINDING)

Figure 7-35   Transistor on time relation.

in Figure 7-35), core $A$ saturates, its impedance drops to zero and base current to $Q_3$ ceases, as discussed below.

But during $T_{on}$, as core $A$ moves from $B_0$ to $B_s$, core $B$ is forced from $B_s$ back down toward $B_0$. This is done by the voltage induced in the control winding, $N_C$, by transformer action resulting from the 1.6 V impressed on primary winding 1-3. Since winding $N_C$ links both cores, the voltage induced in it by transformer coupling in core A is impressed as a source voltage across an $N_C$ turn control winding that links cores $A$ and $B$ in the same direction, the voltage forced across winding 2-3 in core $B$ is equal and opposite to that applied across winding 1-3 of core $A$ by the transformer winding $T1$-4-5. Since the volt-second areas applied to cores $A$ and $B$ are equal and opposite, and since $\Delta B = E \Delta T / NA$, the values of B through which they are driven are equal and opposite. Thus as core $A$ moves up from $B_0$ to $B_s$, core $B$ is driven down from $B_s$ to $B_0$. This procedure reverses on the next half cycle.

The two cores thus always move through equal and opposite flux swings. The termination of a transistor on time occurs when the core, moving from $B_0$ toward $B_s$ reaches $B_s$. The value of $B_0$, the initial flux starting point, is forced by the negative feedback through the magamp control winding to be such that the value of $T_{on}$ of the equation in Figure 7-32, yields the desired output voltage in Equation 7-23.

The magnetic amplifier contains two identical tape-wound magnetic toroids. The magnetic tape is $\frac{1}{4}$ mil thick Square Permalloy 80, wound on bobbins of 0.290 in. OD, 0.160 in. ID and 0.175 in. high. The cores are #80523 supplied by Magnetics, Inc., and have a total flux capacity of 100 Maxwells.

The gate windings (winding 1-3 which links core $A$ only and 2-3 which links core $B$ only) consist of 40 turns of #35 wire each. The control winding 4-5, which links both cores, consists of 250 turns of #37 wire.

The power transformer uses a ferrite core with an air gap to prevent core saturation resulting from unequal transistor currents on alternate half cycles. Such unbalanced currents are a common problem in DC/DC converters and result from unequal widths on alternate half cycles, or unequal current gains or $V_{CE}$(sat) in opposite halves of the converter. This causes the core to "walk" up or down its hysteresis loop to the point where core saturation occurs at the end of a half cycle. In extreme cases, this can bring the transistor out of saturation near the end of the half cycle and cause failure.

This supply uses a Ferroxcube core type 3622-PA1600-3B7, which has a 3.5 mil built-in air gap. With a 12 turn primary at a peak input voltage of 34 V, its peak flux excursion is only 420 gauss. At this peak flux density and 40 kHz operating frequency, core losses are only 2 mW per $cm^3$. Total core loss for its 10.7 $cm^3$ volume is only 22 mW. $I^2R$ losses are under 5%.

The inductor $L_1$ carries 18 amperes of DC and must have an "air gap" to avoid saturation. Rather than using a gapped ferrite core, it is found more

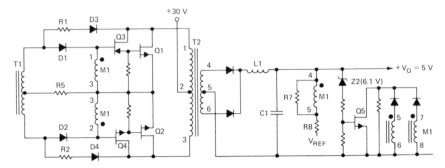

Figure 7-36  Turn-on overshoot protection via $Q_5$.

economical in volume to use a moly-Permalloy powder core. This type of core consists of a slurry of moly-Permalloy powder particles with the magnetic particles insulated from one another with a ceramic binder which forms a uniformly distributed air gap. The core used is the Arnold Engineering type 189043-2 with a 10 turn winding.

Figure 7-36 shows the means used to prevent turn-on overshoot. The relatively slow response time of the magamp control winding makes it difficult to prevent output overshoot at turn-on. Windings 6-8 (linking core $A$ only) and 7-8 (linking core $B$ only) transistor $Q_5$ and Zener $Z2$ are added to solve the overshoot problem. If overshoot occurs, normally nonconducting transistor $Q_5$ is turned on via Zener $Z2$. This puts shorts across secondary windings 6-8 and 7-8 and hence, by transformer action, also across gate windings 1-3 and 2-3. This terminates the transistor on time and immediately reduces output voltage without requiring a current change in the slowly responding control winding.

Figure 7-37 shows a superior way of feeding the magamp gate windings. In the original circuit of Figure 7-31, narrow positive going "ringing spikes" occur at a control gate at the instant the opposite gate winding turns off. These can

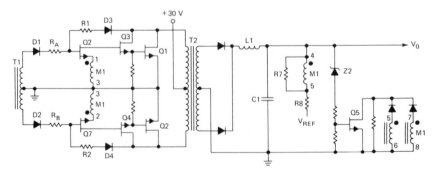

Figure 7-37  De-spiking by addition of $Q_6$, $Q_7$.

cause false extraneous turn-ons prior to the normal turn-on. By placing the gate windings in the emitters of $Q_6$, $Q_7$, these spikes are blocked from getting into the bases of Darlington drivers $Q_3$, $Q_4$ by decreasing the stored energy in the transistors from saturation.

Additional line regulated output voltages become available by simply adding additional secondaries to the power output transformer. Since the feedback loop adjusts the $T_{ON}(T/2)$ ratio to keep the master output voltage independent of DC input, the same duty cycle control keeps the other rectified secondary voltages relatively independent of line voltage.

With a 30 V supply voltage and $\pm 10\%$ line changes, the output voltage change was measured at $\pm 20$ mV and the actual efficiency was found to be 70%.

# 8

## Electromagnetic Compatibility

### INTRODUCTION

Electromagnetic energy covers a vast range of frequency and wavelength. The low-frequency end of the spectrum is occupied by the power and radio frequencies. These are followed by infrared, visible light, ultra-violet light, X-rays, gamma rays, and cosmic rays.

Electromagnetic interference (EMI) is defined as any man-made or natural electromagnetic disturbance or signal which causes an unacceptable response or malfunctioning of any electrical or electronic device or system. This interference to equipment may involve any or all of three separate conditions:

- Interference Generation. EMI is generated by a varying electric or magnetic field. Almost any equipment carrying an electric current is a possible source of interference. The more abrupt the variation in energy flow, the broader will be the frequency range of the interference generated. Electric motors, switches, relays, and radio and radar transmitters are major sources of such interference.
- Interference Transmission. The energy contained in an interference signal can be transmitted from a source to a susceptible unit by means of conductive coupling, a common inductive or capacitive impedance, free space radiation, or by any combination of these means.
- Interference Susceptibility. Receptors or susceptible equipment can be affected by EMI energy penetrating their enclosures, through enclosure openings, or by conduction along the enclosure input and output conductors. Once introduced within an enclosure, undesired energy can be coupled from section to section of the equipment by the same modes as are instrumental in its transmission from one equipment to another.

## Design Objective for EMI Suppression Components

The basic design objective for EMI suppression components is to provide the degree of EMI suppression required in each case, with a minimum increase in weight, bulk, circuit complexity, and cost and a minimum decrease in circuit performance. To achieve maximum effectiveness of the control of EMI with a minimum of equipment, provision for adequate EMI suppression components must be included in the initial formulation of the equipment design. Then, as the design progresses, these EMI suppression components must be included as an integral part of both the electrical circuit design and the packaging design. This is especially true for most EMI suppression components; for, to a large degree, their effectiveness in the control of EMI is directly determined by the manner in which they are packaged in relation to the rest of the basic equipment and its enclosure.

The purpose of suppression components, basically, is to attenuate, by blocking or bypassing, interference present on a line and prevent it from reaching receptors. This can be done by introducing a high impedance into the path of the interfering currents, by shunting them to ground through a low impedance, or a combination of the two. Ideally, a suppression unit should attenuate the interfering current and not affect the functionally required current flowing in the line. High series impedance is usually provided by inductance in the line and low shunt impedance by capacity to ground. One of the most common applications of suppressors is in power lines where it is desired to pass power currents from DC to approximately 2000 Hz, and suppress higher-frequency interference currents. Since the series impedance (inductive reactance) of an inductor increases with frequency and the shunt impedance (capacitive reactance) or a capacitor decreases with frequency, these components are well adapted to this use.

## CONTROLLING EMI

The obvious way of suppressing conducted electromagnetic interference (EMI) (or radio frequency interference (RFI)) is to place filters on input and output leads. This is usually not enough, however, since noise is induced into every wire in the supply. The best method of avoiding conduction of noise going out of the regulator is to filter every lead coming out of the unit, although even this is not always a cure, because the filters shunt radio frequency (rf) noise to ground (either circuit ground or chassis ground). In either case, the rf noise must be brought to system ground. Since the system ground wire contains inductance enough to produce a substantial voltage drop at radio frequencies, system ground can be contaminated by EMI. Even though the regulator contains millivolts of EMI at the output, there may be volts of EMI at the load—due to a badly routed ground.

There are times when the only way to eliminate EMI is to relocate the regulator in the system rack. In other instances, the problem can be brought under control by shielding, filtering, and rerouting system grounds, in which case the regulator should be mounted as closely as possible to the point where circuit ground joins the cabinet or chassis ground.

In addition to conducted EMI, there exists the problem of radiated EMI. This can usually be reduced by enclosing the regulator in a metal box.

There are design techniques which reduce the EMI problem. For example, a linear Class B amplifier used to replace the chopper stage will chop the DC input sinusoidally (without harmonics). Regulation can be accomplished by varying signal input voltage. However, since the efficiency of Class B amplifiers when swinging between positive supply and ground is only 70%, a tradeoff exists. Another approach is to use a switched mode stage with a tuned load, i.e., a Class D amplifier. The voltage waveform may have rapid rise and fall times, but the current is sinusoidal (a high derivative of current as a function of time ($di/dt$) is what produces EMI). An efficiency of 80% is feasible here.

## NOISE SOURCES IN SWITCHING POWER SUPPLIES: GENERAL DISCUSSION

Compact, highly efficient power supplies are made with DC/DC converters and IC switching voltage regulators. However, the switching operation generates broadband RF noise in the kilohertz and megahertz ranges, and also poses other EMC (Electromagnetic Compatibility) problems.

The primary source of this noise is the sudden charging and discharging of parasitic capacitances within the power supply; transformer capacitances, semiconductor-to-heat-sink space, and wiring-to-chassis capacitances predominate. And although most of the noise-producing loops within the supply should ideally be constructed such that noise currents would circulate only within them, "real world" circuits and loop impedances are just not that tidy. Irregular loop impedances and capacitances-to-case cause both normal-mode and common-mode noise to flow not only within the larger (non-noise producing) loops of the supply, but through the entire load and source system.

Figure 8-1 shows typical rf generating sources in a power supply.[1]

Six of the major EMC design considerations of switching power supplies are:

1. DC isolation
2. Circuit grounding

---

[1]"Minimizing Noise in Switching Regulators," by D. R. Gordon, *Solid State Power Conversion*, November/December 1975.

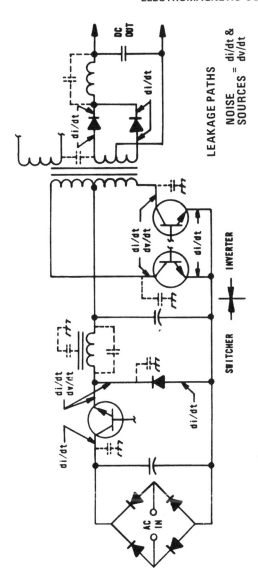

Figure 8-1 Noise generators and leakage paths in a typical switched-mode converter.

3. Susceptibility to audio-frequency and RF noise conducted to the supply on power lines
4. Interference generated in the supply and conducted to other parts of the system on power lines
5. Radiated interference and susceptibility
6. Turn-on and turn-off transients

These problems can all be solved. In fact, several of them can be solved simultaneously by one well-placed and properly chosen component. However, the time to make tradeoffs and integrate the solutions is at the concept design and breadboard stage of development. EMC requirements can be met at this time with the least sacrifice of efficiency and performance, unlike making "fixes" after a supply fails EMI tests.

From an efficiency standpoint, DC/DC converters usually generate, transform, and rectify square wave voltages in the low kilohertz frequency range. This process generates interference that manifests itself in several forms, the most obvious being a long series of harmonic components of the square wave that typically extend well into the low megahertz region. The energy content of these various harmonics decreases with frequency at a rate that is inversely proportional to some integral order of the frequency. The rapidity of this decrease of harmonic intensity with frequency is essentially dependent on the amount of rounding off that can be effected in the transition of the wave or step function at its inception and termination. The drop-off rate with increasing frequency of the harmonic content of a true rectangular wave is only 20 dB/decade, while for the less acute transition of a trapezoidal wave it is 40 dB/decade; for a rounded cosine-squared function it is 60 dB/decade and for a fully rounded pulse with the least abrupt transition feasible it is about 80 dB/decade. Thus, the mere rounding off of the corners of the basic square wave can result in a reduction in the level of interference generated at VHF frequencies ranging from an 80 dB minimum to a probable maximum of 240 dB.

Such reductions in the interference levels generated at VHF frequencies, with even greater reduction at the higher frequencies, not only justify but also impose a mandatory requirement that the designs of such power supplies incorporate the necessary and simple networks required to properly smooth and round off the corners of what would otherwise be essentially square waves. However, square waves with steep slopes are required to minimize transistor power dissipation and increase efficiency. Thus, a design tradeoff or compromise must be reached: minimize interference and maximize efficiency.

The abrupt transitions of the switching function for such inverters and converters also tend to shock-excite "ringing" or oscillations in the parasitic capacitances and inductances of the primary windings of these units. The ringing occurs at low frequencies where coupling to adjacent circuitry is enhanced. This "ringing" can be damped out by the use of small RC networks across these windings. A Faraday shield between the primary and secondary windings will prevent capacitive coupling of all types and frequencies of interference into the power supply's output.

A third manifestation of interference associated with DC/DC converters is in the form of magnetic coupling resulting from the switching, transforming, or rectifying of relatively large currents. A fourth source of potential interference is the generation of fast rise-time current spikes resulting from the sudden reverse biasing of diodes used in the transformer rectifier circuitry. Each of these four sources is capable of generating objectional interference that can appear on external transmitter leads as conducted interference or that couple into the RF section and cause unwanted modulation of the RF carrier. In addition to basic ripple and high-frequency ringing, power supplies produce interference due to regulating characteristics. Many power supplies, when regulating, will produce a low-frequency variation termed "motorboating."

If maximum efficiency is the sole controlling design parameter for a DC/DC converter, very little can be done to prevent the prolific generation of the harmonics inherently present in a square wave. However, additional circuit elements must then be used to confine and absorb the unwanted harmonics. If a sacrifice in efficiency is acceptable, pulse-shaping techniques are used to limit the generation of the higher-frequency harmonics of the switching or chopping frequency. A slight rounding off of the corners of the square wave by the use of small RF inductances with low flux leakage characteristics and small R-C networks can effect major reductions in the level of higher-order harmonics generated by the converter without any substantial reduction in the overall efficiency of the converter. In any case, symmetry may be used to advantage in canceling the effects of certain harmonics in the output transformer of push-pull circuits. The cancellation of selected harmonics will not fully solve the EMI problem, however, but it can result in filtering advantages by shifting the filter cutoff frequency to a higher value. This would allow the use of somewhat smaller filter components and reduce the size and weight of the filter. In push-pull circuits, matched transistors are used to provide symmetrical switching with equal conduction angles. This tends to equalize the harmonics generated, thus resulting in more complete cancellation. Electrostatic shielding of the transformer is used to minimize capacitive coupling between the primary and secondary windings.

The use of a fixed rather than a variable chopper frequency facilitates filtering, in that tuned circuits, bridge-T filters, and other narrow-band methods may be used to attenuate one or two offending frequencies. A low-pass filter is recommended for use on power input leads.

The other three major sources of interference in DC/DC converters may be eliminated by proper design to prevent their generation. Oscillations may be damped out by inserting proper resistances, if the resonant circuit can be localized. If not, small RF capacitors may be used to bypass the high-frequency components to the transmitter case. Magnetic coupling can be reduced by the use of magnetic shielding, using thin "Netic" or "Co-netic" foil or small "cans" made of similar magnetic material. Magnetic coupling, due to wiring to and from the DC/DC converter, is minimized by the use of twisted leads. Coupling can be further minimized and controlled by the use of conductive shielding on the twisted pair. The effectiveness of such shielding increases with frequency and can afford substantial additional reduction of magnetic coupling at the higher frequencies involved in most cases. These techniques do not eliminate conducted interference on leads containing the interference, but prevent coupling to other leads which might not be filtered. Other methods of minimizing magnetic coupling are: (1) the use of toroidal transformers and inductors to minimize any magnetic cross-coupling and achieve low leakage fields, (2) quadrature placement of such toroids in reference to the adjacent toroids can further minimize any such magnetic cross coupling, and (3) 90° orientation of high-level and low-level twisted pairs where they must cross.

The prevention of fast rise-time current spikes in the rectifier diodes of converters can be accomplished by inserting small resistors in series with the diodes, if the resulting loss in efficiency can be tolerated. Reactive elements can also be used for this purpose, but they often create additional problems due to resonances, so resistive elements are preferred. The current spikes generated are a function of the diode recovery time and may be minimized by proper choice of diodes. The electrostatic shield of the transformer minimizes coupling of this high-frequency interference to the primary circuits.

The solid-state switching devices, whether diodes, SCR's, or transistors, should be chosen to obtain units with the minimum switching time that are available in the types required. This minimizes generation of initial switching voltage spikes which not only generate interference but also endanger the operation of the unit.

The operating frequencies of all such switching devices in the power supplies must be controlled to a reasonable degree of accuracy so that interference suppression filters which are frequency-sensitive will be constant and dependable in operation.

## Transients

Inductive kicks and the general effects of suddenly collapsed magnetic fields produce transients ranging from sub-audio to megahertz regions depending on the effective inductance and capacitance in the path of the current produced by the collapsing field. The resistance in the path controls the damping, which in turn controls the envelope of the oscillatory discharge.

It is usually assumed that loop inductance is unchanged during the time under consideration. Hence, the current through an inductive loop cannot change instantaneously in the theoretical case or even nearly instantaneously in the practical case. In power supplies, complex inductances in one or more loops are switched. Each switching is accompanied by a rapid current surge with a high voltage surge. Voltage surges of hundreds of volts actually occur and can be measured. It is often assumed that the voltage across a capacitor cannot change instantaneously or even nearly so. This is true only when the total capacitance connected to a node (junction of two or more circuit elements) remains unchanged. When capacitors are switched in or out of circuits, voltage changes occur across them and across other elements attached to the node. It is the total charge at a node that cannot be changed instantaneously.

## Electromagnetic Interference of Transformers

In AC subsystems iron-core transformers are often used. These devices exhibit nonlinear characteristics between the voltage and current or between the flux and magnetizing force. Hence, a transformer produces numerous harmonics in both higher and lower frequency regions. In addition, because the permeability of the iron core is not constant, the inductance varies and the input impedance of the transformer varies in a nonlinear fashion. This nonlinearity results in additional generation of harmonics in the power distribution subsystem. In order to minimize the generation of harmonics and interference, high-permeability material must be used as a core for power transformers. Less harmonic generation exists when using mu metal and other high-permeability magnetic materials rather than silicon steel.

Transformer cores for aerospace systems are often tape-wound; but cores made from grain-oriented silicon steel, such as "Hipersil" are also used. The high permeability of this material results in low exciting voltamperes and low core loss. Such cores yield better efficiencies, smaller size, and reduced weight. These characteristics make the cores applicable for use in high-power and high-frequency applications. However, because the cores operate at very high flux densities, the high inrush currents are of particular concern. The maximum inrush current occurs when the circuit is closed at the instant the voltage is zero.

## Electromagnetic Interference of Integrated Circuits (ICs)

The practical application of IC's at the system and subsystem level is presently an area of rapid progress. However, in the design of IC's and their application, certain inherent characteristics of these circuits must be taken into account to meet system and subsystem electromagnetic interference requirements. Reduced size (increased density) will affect the degree of compatibility due to common-mode conductive coupling and due to induction resulting from both magnetic and electric fields.

The use of miniature discrete components results in greater spurious capacitances due to greater packing density. These deteriorate compatibility due to electric-field interactions.

Semiconductor integrated circuit construction involves numerous parasitic substances and coupling paths. When a multiplicity of elements is interconnected on a single substrate, the equivalent circuit due to coupling among them through the substrate becomes very complicated. There is greater interaction among all circuits on a common substrate. Coupling by means of various elements can result in a serious electromagnetic compatibility problem—namely, deterioration of compatibility. The designer should place RF interference sources on different substrates from receivers of such energy.

Susceptibility to RF interference from magnetic fields is reduced by ICs while electric-field susceptibility is increased. The IC circuit designer, because of this, must use lower-impedance circuits than he would ordinarily use with conventional circuits.

The relative inflexibility in modification of integrated circuits requires that the electromagnetic compatibility aspects of them be considered in the early stages of design.

## DC/DC CONVERTER EMI SPECIFICS

The self-oscillating, saturating transformer converter in Figure 8-2 owes its continuing popularity to the availability of fast-switching power transistors and good core materials. The circuit operates as follows. A DC input is chopped by the oscillator and generates complementary square-waves that are transformed and rectified to produce a DC output. Efficiency generally runs 80% or better.

Alternate saturation and cutoff of push-pull transistors $Q_1$ and $Q_2$ produces squarewave voltages at their collectors (see Figure 8-3). Alternate conduction of the two diodes rectifies the transformed output. The feedback windings of saturable transformer $T_1$ provide positive feedback to drive $Q_1$ and $Q_2$. The voltage across the saturated, or "on" transistor is small—about 0.2 V—so the voltage across that half of the primary winding nearly equals the supply voltage.

The primary's voltage remains constant for a half-cycle, during which time the magnetizing flux in the core steadily increases. When the core becomes saturated, the device tries to maintain the same rate of change of flux. Magnetizing current rises rapidly to the peak limit, the flux-rate change can no longer

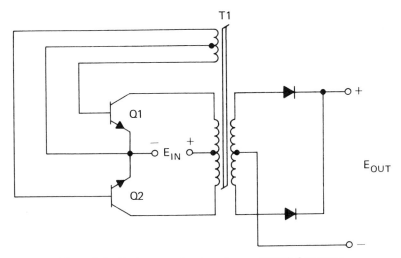

Figure 8-2   Basic saturating transformer DC/DC Converter.

be maintained, and the rate decays to zero. The voltages across the windings collapse, removing base drive from the transistor that was on.

Magnetizing current drops toward zero, and total flux falls as the transistor turns off. The negative flux-change rate reverses winding polarity, completing cutoff of the first transistor. Now the second transistor turns on and quickly saturates, due to regeneration. Then, transformer saturation in the reverse direction cuts off the second transistor and turns the first back on to complete the cycle. Switch frequency is determined by the time required for the transformer flux to saturate in each direction.

Note in Figure 8-3 the current spikes in each half-cycle at the end of each transistor's conduction. These current peaks are caused by core saturation, and thus are relatively independent of load. Peak collector current, $i_c$, is

$$i_c = \beta i_b \qquad (8\text{-}1)$$

where beta is transistor current gain and $i_b$ is its base current.

The peak collector current, coming from the saturation of the transformer rises sharply just before transistor cutoff. The ''collector spikes'' are reflected on the power input.

Although short, the spikes occur while collector voltage is rising and result in instantaneous peak-power dissipations many times steady-state dissipation. Besides reducing converter efficiency, the peaks may overheat the transistor junctions, degrading reliability. It can also be deduced from Eq. (8-1) that if the transistors are not fairly well matched, the spiking and power dissipation

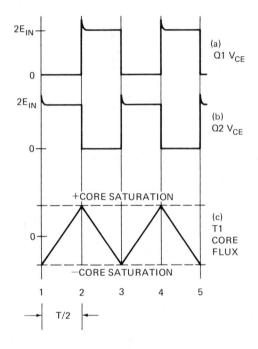

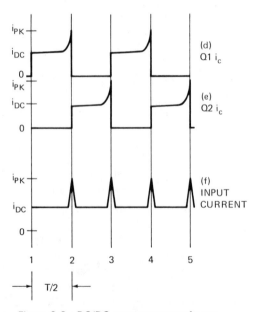

Figure 8-3   DC/DC converter waveforms.

may become excessive and the output waveforms asymmetrical. The first may destroy the transistors and the asymmetry affect system operation.

Moreover, the spikes create EMI conducted by the power lines. Lowering the load current only worsens the interference. That is, spike height is fixed so the distance between $i_{dc}$ and $i_{pk}$ in Figure 8-3f increases as $i_{dc}$ decreases.

Interference and power losses can be decreased with the two-transformer design in Figure 8-4. Transformer $T_1$ does not saturate, which almost eliminates collector spiking as indicated in Figure 8-5. The transistors are driven and the oscillation frequency determined by a separate, small, saturable transformer, $T_2$. Transformer design is less critical for $T_1$ and the total circuit may be more efficient.

Another converter uses a driven, squarewave oscillator stage to govern switching. However, to ensure sufficient base-drive current to the transistors under worst-case operating conditions, the transistors are over-driven. This improves efficiency by keeping $V_{CE(sat)}$ losses low when a transistor is on, but makes spiking unavoidable.

The spiking is due to the transistor storage time. One transistor may be driven on before the other turns off, resulting in a spike that lasts through the storage time. Storage time is directly proportional to the amount of overdrive. Power losses caused by spiking increase with frequency, making it difficult to achieve high efficiency.

Another way to build a converter with a nonsaturable output transformer is to insert a saturating inductor into the switching-transistor base circuit as shown in (Figure 8-6). Square-loop material is used for inductor, $L_1$, and linear magnetic material for the output transformer, $T_1$.

The design of $T_1$ and $L_1$ is such that $L_1$ saturates before $T_1$ does. When $L_1$

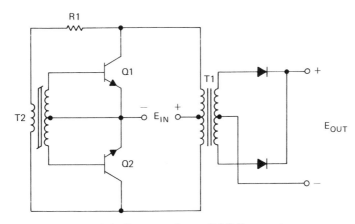

Figure 8-4   Two-transformer DC/DC converter.

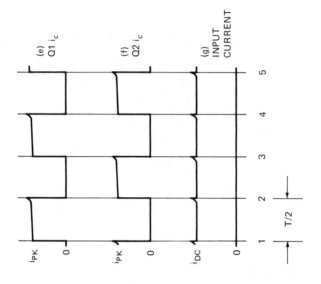

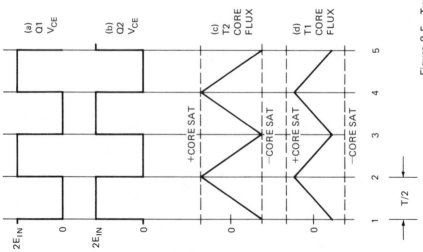

Figure 8-5  Two-transformer DC/DC converter waveforms.

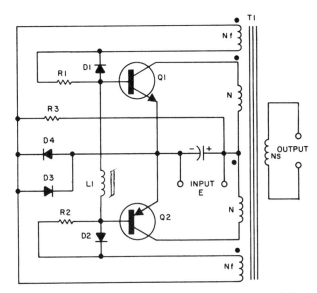

Figure 8-6    Another way to eliminate spikes is by inserting a saturable inductor, $L_1$, into the base circuit.

saturates, it turns off the conducting transistor, simultaneously turning on the nonconducting transistor.

Diodes $D_1$ and $D_2$ insure fast turn-off times for $Q_1$ and $Q_2$, and diodes $D_3$ and $D_4$ insure proper operation of $L_1$. Converter frequency can be set at any desired value by adding or subtracting turns on the $L_1$ winding.

Another circuit that reduces spiking is called a pseudosaturating converter. In this circuit (which is shown in Figure 8-7) inductor $L_1$ and Zener diode $CR_1$ prevent the output transformer from going too deeply into saturation—that is, when the transformer core (a high-permeability material) begins to saturate, the current through $L_1$ increases rapidly, thus raising the voltage level at the emitters of $Q_1$ and $Q_2$. Because of base-emitter capacitances, the bases of $Q_1$ and $Q_2$ will follow the emitter until the base voltages are clamped by $CR_1$. The transistors will then turn off and limit the current into $T_1$.

Note that $L_1$ is now in the emitter circuit rather than, as in Figure 8-6, the base circuit.

A unique two-transformer converter circuit using two cores, one within another is shown in Figure 8-8. The feedback winding is placed on the inner core, while the primary and secondary windings are on both cores. This converter operates as follows:

If it is assumed that $Q_1$ is on, a positive voltage will be applied from $B$ to $A$, and the induced voltage at $E$ will—before saturation of the outer core—collapse

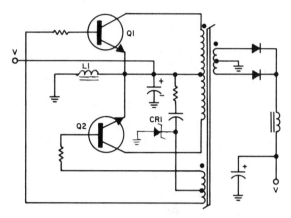

Figure 8-7   Spikes are reduced when the output transformer is permitted to saturate only partly by combined action of $L_1$ and the Zener diode $CR_1$.

as the inner core saturates. As voltage $E$ decreases, $Q_1$ turns off, and the flux in the inner core begins to change in the other direction, turning on $Q_2$. The latter transistor will remain on until the inner core saturates in this direction.

The inner core saturates before the outer because its mean circumference is smaller, while the magnetizing ampere turns are the same. Thus the switching is accomplished without excess magnetizing current and associated losses, thereby raising over-all converter efficiency and greatly reducing electromagnetic interference.

### Spike Reduction Techniques

Careful selection of core material and good transformer design will alleviate spiking. Spike risetime depends on the shape of the "knee" of the core material's hysteresis loop. If the squareness ratio $(B_r/B_m)$ is low, imperfect saturation and undesirable flux changes will increase the spike duration. Spike falltime depends on transistor switching speed and circuit reactances.

Loosely coupled, asymmetrical transformer windings can cause output spiking, too. It is good practice to spiral half the collector winding over the full periphery of the core, and bifilar wind the remaining turns between the first spiral's turns. The feedback windings should be applied the same way. If the center-tapped collector windings must have many layers, try winding collector and feedback windings simultaneously. At least, do the feedback winding in one 360-degree sweep and place it directly next to the collector winding. If possible, use a large-area core to minimize the number of secondary-winding turns. This will reduce output "ringing" (damped oscillations) caused by capacitance. Use progressive or sector windings if a large number of turns cannot be avoided.

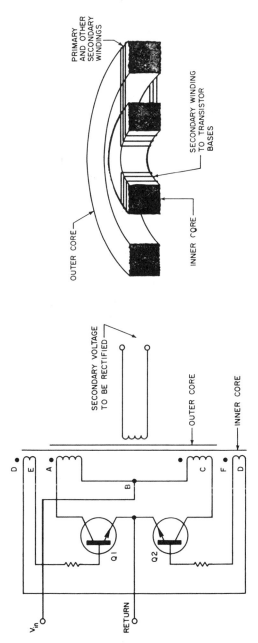

Figure 8-8   A unique two-core transformer reduces spikes by switching the transistors before the outer core saturates. The technique improves conversion efficiency and reduces electromagnetic interference.

Feedthrough capacitors on the DC/DC converter outputs are used to prevent internally generated switching spikes from leaving the power supply; that is, containment of EMI within the unit that is generating it. Faraday screens also are used around transformers to contain the switching spikes.

Devices that suppress heavy transients on the power lines can be placed where they will help suppress collector spikes. Several appropriate locations are shown in Figures 3-35 and 3-36.

These networks protect the transistors from catastrophic failure when externally generated spikes are large enough to destroy the transistor collector junctions. Such spikes can come from load transients, line transients and supply turn-on transients. The networks do suppress some of the generated interference; however, they cannot do the total EMI filtering job.

## AC Susceptibility

Input power line noise at frequencies higher than the switching frequency will not pass through the input filters and the converter. But its output will tend to follow low-frequency noise conducted on the power lines and source-voltage fluctuations. If filters built into the converter do a good job of filtering switching noise, they will usually do a good job of suppressing input noise at the switching frequency. If not, undesirable effects such as beat frequencies and slow drifting of the output will occur.

In precision applications a pre-regulator is advisable because of the susceptibility of the converter to audio and other low-frequency noise, and to changes in source voltage. IC switching regulators make excellent pre-regulators.

## EMI AND THE SWITCHING REGULATOR

The switching voltage regulator is preferred because it is several times more efficient than a conventional linear regulator when large variations of input voltage must be regulated. The switching regulator needs less cooling hardware and permits smaller battery packs. But its efficiency stems from its squarewave operation, which unfortunately generates switching noise. The result is a design tradeoff: The noise spectrum is controlled and suppressed, but efficiency and weight are sacrificed. The problem for the EMC engineer is to participate in the tradeoff and help hold down the sacrifice.

### Basic Switching Regulators

The switching regulator is a functional unit which may be used alone in a power system having no requirement for primary to secondary power return isolation. It is frequently used as a preregulator ahead of a DC/DC converter where it absorbs large input line voltage changes efficiently and presents a nearly constant input voltage to the converter. If a precision output voltage is required from the converter, a conventional series type of postregulator is used. However, in some cases, the switching regulator will also be used as a postregulator, after a DC/DC converter or an AC/DC converter. The concepts presented here are applicable to all cases provided that proper consideration is given to source and load characteristics.

The on-off switching period of a regulator may be fixed by some external frequency source or may be variable and free-running. There are distinct advantages to the EMC engineer when the frequency is fixed because the filter design range will then be well defined. In the free-running regulator, the frequency will vary over a range which must be predicted in order to provide proper filtering.

Switching regulators are further classified by their input-output relationships. They may be buck, boost, or buck-boost in nature. A buck regulator is used when the input voltage is always higher than the output voltage. The boost regulator is just the opposite because the input voltage is lower than the output voltage. The buck-boost regulator is in between the two; it provides a regulated output voltage when the input is either higher or lower than the output. A buck pre-regulator is frequently used ahead of a DC/DC converter in low voltage power supplies.

The basic buck regulator is shown in Figure 8-9. This regulator is typical of either an externally synchronized fixed frequency regulator or a free-running variable frequency regulator. The external frequency source is used when the regulator is fixed frequency. In the fixed frequency regulator, one switch transition is always performed by the external source and the sensed voltage causes the other transition when the output reaches a predetermined threshold voltage level. The sampling period is fixed by the external frequency and the duty cycle is free to vary. In the free-running limit cycle regulator, both switching transi-

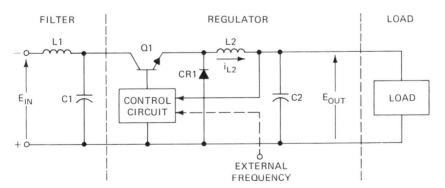

Figure 8-9    The buck switching regulator.

tions are controlled by the sensed voltage. The regulator cycles between an upper and a lower threshold of output voltage. Thus, both the duty cycle and the frequency are free to vary. Frequency is a complex function of $L_2$, $C_2$, the threshold range, $E_{in}$, the load current and amplifier hysteresis. At light loads, $L_2$ may not store enough energy to maintain conduction during the entire off time of $Q_1$. The result may be large transients and low regulator frequency. Therefore, interference filters must have a very wide frequency range or the load must be kept above the critical minimum. The control circuits of the two types of regulators will vary but both will include the necessary power switch drivers, reference voltage and operational amplifier. An input filter, $L_1C_1$, is shown because transistor $Q_1$ draws current in pulses. The capacitor supplies the on-off current demand and the inductor provides a smooth flow of current from the supply.

Figure 8-10 is a practical implementation of the buck switching regulator shown in Figure 8-9 using the LM105 IC voltage regulator.

The basic buck regulator components are $L_1$, $CR_1$, $C_1$ plus the IC regulator. The internal switch transistor of the IC is used to drive $Q_1$. The reference is internal and the output level is set by a voltage divider composed of $R_1$ and $R_2$. Saturation of the booster transistor is assured without excessive overdrive by $R_3$. $R_4$ and $C_3$ provide amplifier compensation. Since an IC regulator is an operational amplifier with a large amount of feedback, frequency compensation is required to prevent oscillations.

Positive feedback is improved by $R_4$, which works into the IC's input impedance. $C_2$ makes ripple appear on the feedback terminal (pin 6), which minimizes ripple, while $C_3$ prevents the shunt capacitance of $R_4$ from coupling input spikes into the IC.

A voltage regulator has compensation problems in addition to those encountered in an operational amplifier. For example, the compensation method must provide a high degree of rejection to input voltage transients. It must also be stable with reactive loads which are far heavier than those normally encountered with operational amplifiers and it must minimize the overshoot caused by large load and line transients. These factors have been considered in the design of modern IC regulator chips. The trend is for more fool-proof, high current circuits with a minimum of external components, whether they are active or passive. A useful high power IC regulator should include everything within one package, including the power control element, or pass transistor.

As an alternative to letting the IC regulator run free, with the frequency variation problems mentioned previously, it can be driven at a fairly stable frequency by the converter, when both are used.

Both $Q_1$ and $CR_1$ must be capable of switching fast to minimize losses. Losses arise primarily during transitions between saturation and cutoff when semiconductor devices are resistive. However, fast switching increases the amount of

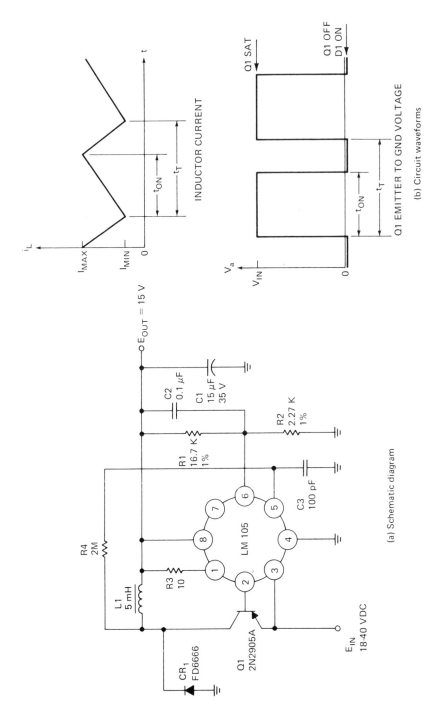

(a) Schematic diagram

(b) Circuit waveforms

Figure 8-10   IC buck switching regulator. (a) Schematic diagram, (b) circuit waveforms.

noise that must be controlled, so this parameter must be traded off against EMI requirements. Also, it is necessary to return the catch diode (and any bypass capacitance on the unregulated input) to ground separately from other parts of the circuit. The diode carries large current transients and large voltage transients can develop across even a short length of wire.

The core of $L_1$ should have soft saturation characteristics. Cores that saturate abruptly produce excessive peak currents in the switching transistor if the output current goes high enough to run the core close to saturation. A material that has a gradual reduction in permeability with excessive current, yet retains excellent high-frequency characteristics, is powdered molybdenum-permalloy.

Don't overlook the filter capacitor ripple rating. The ripple current through the regulator capacitors will be quite high. Thus, the capacitors should have both low dissipation factors and voltage ratings that are higher than the DC circuit voltages.

The boost and buck-boost regulators are shown in Figures 8-11 and 8-12. There are basic similarities with all three classes of regulator: Current is switched on and off by $Q_1$; output voltage is sensed by the control circuits; fast recovery diodes are used; power switch driver and voltage reference circuits are required; and an external frequency may be used to convert the circuit from free-running to fixed frequency. The choice of which of the three regulator circuits to use will be based on functional reasons and not on EMI characteristics. From an EMI viewpoint, the buck regulator has a large ripple current appearing on $C_1$ (this will be discussed later) while the boost regulator has a large ripple current appearing on $C_2$. These ripple currents are due to the nature of each regulator's normal operating mode. The buck-boost regulator has "large ripple currents" appearing on both $C_1$ and $C_2$ but of lower magnitude than those of the buck regulator or the boost regulator. These ripple currents must be controlled. In order to limit the size of this section, only the widely used buck regulator will be examined in detail, but most of the information will be applicable to all three types.

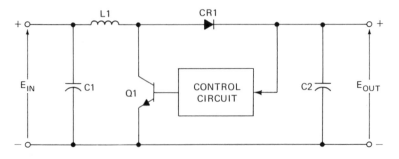

Figure 8-11   The boost switching regulator.

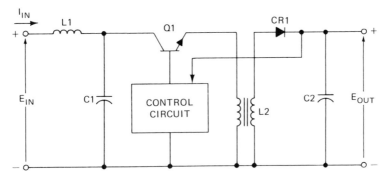

Figure 8-12   The buck-boost switching regulator.

## Regulator Waveforms

The typical waveforms for the free-running buck regulator of Figure 8-9 are shown in Figure 8-13. In figure 8-13A, the voltage at the emitter of $Q_1$ switches alternately from $E_{in}$ to zero. The output DC voltage, $E_{out}$, is the average value of a cycle and is

$$E_{out} = \frac{T_{on}}{T} E_{in} \qquad (8\text{-}2)$$

where $T_{on}$ is the switch on time (switch closed), $T$ is the time for one full regulator cycle.

This equation shows that the duty cycle of the regulator switch will vary with the input line voltage but not with load current. In this discussion, the transistor $V_{ce(sat)}$ and the resistance of inductors, capacitors and source are neglected to simplify the analysis. This is done to allow a discussion of the primary factors in regulator operation in a simplified form. In Figure 8-13B, the current through $Q_1$ will flow in pulses during the $T_{on}$ period. This pulsing current is supplied by the input filter $L_1$, $C_1$. The average output current, $I_{out}$, is related to the average input current, $I_{in}$, by

$$I_{in} = \frac{T_{on}}{T} I_{out} \qquad (8\text{-}3)$$

Therefore, for a given input voltage, the average input current will be directly proportional to the load current since the ratio $T_{on}/T$ is a function of the input and output voltages.

Figure 8-14 shows the waveforms of $i_{L_2}$, the inductor current and $e_0$, the output ripple voltage. The figure is shown with the average output voltage and

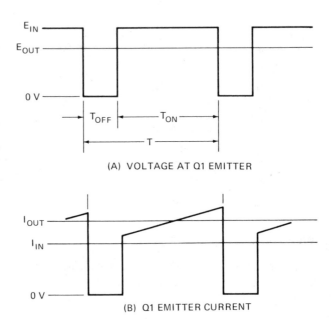

(A) VOLTAGE AT Q1 EMITTER

(B) Q1 EMITTER CURRENT

Figure 8-13   Regulator waveforms of figure 8-9.

current as the zero reference line and the alternating waveforms of $i_{L_2}$ and $e_0$ superimposed to show the timing relationships. The sensed voltage thresholds are $+E_{ss}$ and $-E_{ss}$. In the free-running regulator, a switching cycle is triggered whenever $e_0$ goes higher or lower than the $E_{ss}$ lines. The effect of transistor delays, $t_{on}$ and $t_{off}$, are also shown because, at higher regulator frequencies, these will become significant. The current $i_{L_2}$ shows an almost straight line sawtooth waveform since the current flow is caused by two fixed voltages.

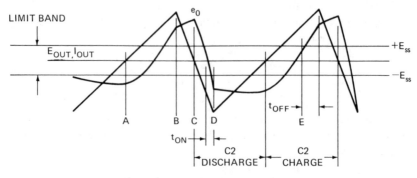

Figure 8-14   Output ripple waveforms.

$$+\Delta i_{L_2} L_2 = (E_{in} - E_{out}) T_{on} \tag{8-4}$$

$$+\Delta i_{L_2} L_2 = (-E_{out})(T - T_{on}) \tag{8-5}$$

When $i_{L_2}$ exceeds its average value, $I_{out}$, at point $A$, charge begins to flow into capacitor $C_2$ and the output ripple starts to rise. At point $B$, the positive peak current is reached but charge still flows into $C_2$ and $e_0$ continues to rise until point $C$ where current crosses below the average line. At that point, the maximum ripple voltage is reached, the ripple cycle reverses and capacitor $C_2$ is discharged below the average output voltage level until a negative peak of ripple is reached. Switch transistor $Q_1$ is driven on at point $D$, after delay $t_{on}$, when the ripple voltage crosses below the $-E_{ss}$ line. The switch remains on until point $E$ is reached where the positive ripple cycle crosses $+E_{ss}$ and, after delay $t_{off}$, the switch opens.

## Regulator Frequency

The switching regulator frequency is usually an independent variable and for functional purposes should be set as high as possible consistent with efficiency. This permits smaller sizes of $L$ and $C$ and increases the response time of the regulator to changes in input voltage or load current. Figure 8-14 shows how the rise and fall of $e_0$ controls the switching of the regulator. This voltage is produced by the integration of the current $i_{L_2}$

$$e_0 = \frac{Q}{C_2} = \frac{1}{C_2} \int i_{L_2} \, dt \tag{8-6}$$

where $i_{L_2}$ is governed by (8-4) and (8-5). The current is in the form of ascending ramps so the voltage will be

$$e_0 = \frac{1}{C} \int at \, dt = \frac{1}{2C_2} at^2 \tag{8-7}$$

where: a = the slope of the current ramp. Therefore, the output ripple waveform is made up of a sequence of curves having the form of equation 8-7. Increasing the size of $L_2$ or $C_2$ will slow this rise and fall time and will decrease the operating frequency. Reducing the $E_{ss}$ sensed voltage band will increase the frequency because the switching thresholds are reached sooner by $e_0$. Reducing the input voltage will reduce the frequency and increase the duty cycle because it takes longer to charge $C_2$ to the $+E_{ss}$ threshold.

Transistor switching time also affects frequency. This is shown in Figure 8-14 as $t_{on}$ and $t_{off}$, both of which tend to increase the length of a regulator

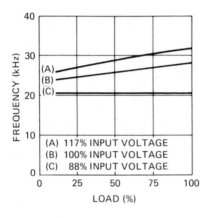

Figure 8-15   Regulator frequency as a function of load.

cycle. At high frequencies, this will be a significant effect. Load current will make no direct change in frequency but there are secondary effects caused by IR drops, $L_2$ saturation and changes in transistor switching time. Figure 8-15 shows a typical regulator frequency variation with load for three input voltages. Only a small change in frequency is seen. Ideally, the frequency should not vary with the load. However, if the inductor ($L_1$) saturates, the frequency will vary somewhat with the load. In Figure 8-16, frequency varies as a function of $E_{in} - E_{out}$ showing how frequency decreases as input voltage is reduced.

### Effect of Diode Recovery Time

Diode $CR_1$ of Figure 8-9 conducts when $Q_1$ is off and blocks when $Q_1$ is on. This commutating function is not ideal. Figure 8-17 shows how an undesired current spike is generated by the diode. This spike occurs at the end of a diode conduction cycle when reverse voltage is just applied by $Q_1$. A short pulse of

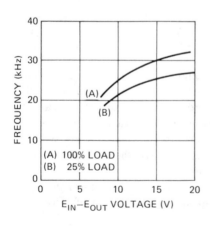

Figure 8-16   Regulator frequency as a function of $E_{in}$-$E_{out}$ voltage.

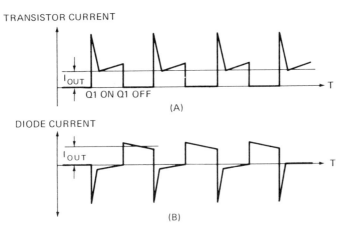

TRANSISTOR CURRENT

DIODE CURRENT

.Figure 8-17   Switching transistor and commutating diode current.

reverse current through the diode is required to sweep out minority carriers and
to establish the reverse biased junction. The transistor must supply this current
in addition to the commutated inductor current. Therefore, there is a corresponding transistor current spike. This effect should be limited by the use of a fast
recovery diode.

### Input Filter Criteria

The input filter should be given much more consideration than appears from
Figure 8-9. Because of the size and weight of large value capacitors and magnetic cores, the filter is one of the largest contributors to total weight and package
size. The filter will also affect the regulator operation. Therefore, the design
criteria should be included in regulator tradeoffs. The filter must supply the
current demands of the switching regulator during each cycle. These are shown
in Figure 8-17A, with amplitude equal to the output current and duty cycle
determined by the voltage ratio as defined in equations 8-2 and 8-3. The filter
output capacitor must be a type having low equivalent series resistance (ESR)
at the operating frequency so as to reduce $i^2R$ losses during the current pulse.
It must also be safely rated for the AC ripple current it is required to conduct.
The filter must have a controlled $Q*$ so that sharp resonant peaks are not produced when audio frequency (AF) test signals are applied to the power input or
if an AF load is applied to the current output. Two methods of filter damping
are described in the section on Conducted Susceptibility and shown in Figure
8-29. The filter must protect the regulator and its load from power line transients

---

*Quality factor, or ratio of resonance frequency to the bandwidth between those frequencies on
opposite sides of the resonance frequency (known as half-power points) where the response of the
resonant structure differs by 3 dB from that at resonance.

while, at the same time, not producing undesired turn-on or turn-off transients on the power system. Proper placement and sizing of filter capacitors will accomplish this objective. In addition to the above criteria, the filter must have sufficient attenuation to reduce the switching regulator harmonics to a level well below the equipment EMI requirement for conducted emission. As a minimum a two-section filter will generally be required to meet these requirements. The different sections can be optimized to perform specific tasks.

### EMI Reduction

The conducted emission on the input power leads will be reduced to the desired levels by the input filter. A regulator current pulse, shown in Figure 8-18 with a diode recovery spike superimposed, is a good choice for predicting the filter requirement. The narrowband interference spectrum resulting from a typical 1 amp pulse with a 1 amp spike is shown. This illustrates how much effect the narrow spike can have on the high end of the spectrum. At the low end, the amplitude is controlled by the basic regulator pulse. The large duty cycle and high frequency of the regulator produces a spectrum in which the harmonic lines are spaced far apart. As a result, they should be treated as narrowband emissions until a frequency is finally reached where several lines are included in one receiver bandwidth.

The information needed to design the input filter attenuation may be derived from the regulator voltage, current, frequency and by using equations 8-2 through 8-7. Diode recovery spike characteristics may be taken from manufacturer's data sheets. Once the extremes of operation have been estimated, conventional pulse spectrum analysis techniques may be used to determine the filtering required.

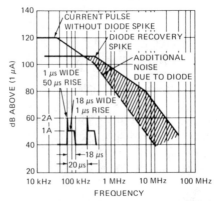

Figure 8-18   Narrowband conducted current spectrum for 50 kHz switching regulator showing effect of current spike.

## Special Considerations

In any of the switching regulators there is a possibility that load current will be reduced to such a low value that the conditions of equations 8-4 and 8-5 cannot be supported during the $T - T_{on}$ period. When this happens, inductor current suddenly stops, the commutating diode no longer conducts and a large ringing is observed on the output of $Q_1$. The regulator frequency will drop to a very low value as $Q_1$ conducts in short pulses. This condition should be avoided by proper proportioning of $L_2$ and $C_2$ and by not operating at such a low load current if possible.

Another condition to be avoided is a possible instability of the regulator at high load currents and certain $L_1$ and $C_2$ combinations. This condition occurs because of the constant power output characteristic of a voltage regulator and the fact that it appears as a negative resistance load to the input filter.

The buck regulator must also be designed such that the peak negative swing of input voltage is not less than $E_{out}$ (plus $IR$ drops) or the regulator will fail to operate during that portion of the cycle. The worst case will occur at the lowest line voltage and at resonance of the input filter.

## SUPPRESSING ELECTROMAGNETIC INTERFERENCE WITH POWER SUPPLY FILTERS

To build noise suppression in the circuitry, filter networks must be placed in the power supply circuit. They affect the switching time constants, and are thus a critical factor in power supply efficiency.

Both input and output filters are required to reduce EMI to acceptable levels in switching power supplies.

For a switching regulator, the input filter must supply the current demands of the regulator during each cycle, with amplitude equal to output current, and duty cycle determined by the voltage ratio equation

$$E_{out} = \frac{E_{in} \times t_{on}}{t_{on} + t_{off}} \tag{8-8}$$

To prevent performance degradation in the regulator, low frequency variations in the load current, after processing by the output filter and the duty-cycle-controlled power switch, must not interact with the input filter. For example, consequent ripple voltage corresponding to load-variation frequency across the input filter capacitor should never exceed the input voltage that can be safely applied to the regulator. At the other extreme, it must never drop below the minimum input voltage level at which output regulation must be maintained.

Several guidelines are offered for filtering switching power supplies. These are:

1. Decouple rf noise at the source to chassis or the rf ground using capacitors having good high frequency characteristics (i.e., low inductance and low ESR). The capacitor selected should provide attenuation at least one octave below the frequency where attenuation is desired.
2. Keep the leads on the capacitors as short as possible. For a rule-of-thumb, a 0.01 $\mu$F capacitor should not have a total lead length of greater than 0.5 in. Notice that if it is possible to decouple the noise with long leads on the capacitor, the capacitor is not at the noise source.
3. Leads going into clean areas, or out of the power supply should be decoupled by a capacitor or by a feed-through EMI filter at the shield to prevent radiation on the leads after they are filtered.
4. Use few capacitors to chassis, and as low a value of capacitance as possible to minimize the AC leakage paths.

An input filter is required to attenuate all signals from at least 10 kHz to 50 MHz in order to meet the requirements of MIL Standard 461.

### Input Filters

Several kinds of filtering are effective. An input capacitor is usually sufficient to protect the converter from itself. To get the noise level down enough to keep from interfering with other equipment including some bench power supplies, an input inductor is required. For 10 kHz operation, for example, a 0.1 mH and 47 $\mu$F LC combination is appropriate. Thus as a minimum, a simple LC filter as shown in Figure 8-19 is used to attenuate the ripple current that the regulator would normally reflect back into the battery. Inductor $L_1$ has the primary task of attenuating the ripple current created by the regulator to a level acceptable to the system. Because the addition of the inductor makes the battery a high impedance, as seen by the converter, capacitor $C_1$ is added. It converts the input back into a low impedance, capable of both delivering and absorbing power, and it protects the regulator from noise generated by other equipment in the system.

However, an inductor that works at 10 kHz, without saturating on the peak currents flowing into the bridge, will not necessarily attenuate the 1 MHz to 50 MHz frequencies.

The interwinding capacitance of the transformer creates a generator between

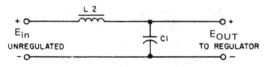

Figure 8-19 A simple LC input filter.

the input and output returns. This generator produces current spikes in the loop between input and output returns every time the transformer switches. The interwinding capacitance can be lowered but never completely eliminated. In order to tolerate the problem, the oscillator must be isolated from the input and output lines so that it can float on its own noise instead of forcing the noise onto the lines. One way to let the oscillator float is to insert inductors into both input lines. To further reduce the noise coupled through the chokes, chokes should be wound with only a single layer (this reduces distributed capacitance). Another way to help the oscillator float is to place a filter transformer in series with the lines. This allows the input to stay constant while the whole oscillator moves (see Figure 8-20). Similar techniques can be applied to the output return lines to keep the switching noise in the converter.

Tantalum capacitors are usually required to provide the energy storage for the pulse currents and for the dead time while transistors are switching or are off. While tantalum is the only material that provides high energy storage in a small volume, its effective series resistance and inductance get rather high above 10 kHz. Thus, it does not filter all the noise that it should. Ceramic capacitors are used to shunt the tantalum capacitors to improve their frequency characteristics. Solid tantalum capacitors have better temperature and frequency characteristics than wet tantalum capacitors. Other than being slightly larger for a given $CV$ product, solid tantalum capacitors have one serious drawback—they are prone to exhibit a phenomenon known as current flicker. Solid tantalum capacitors will break down at unpredictable intervals and give a good imitation of a short circuit. If they are used in series redundancy or have 3 ohms/V series resistance, they will recover. The former case quadruples the parts required and the latter case is completely unusable as a filter element. In places where a converter has current limiting or there is enough series impedance (3 ohms/V minimum), solid tantalum is usable. Wet tantalum capacitors also suffer from current flicker, but the liquid electrolyte can absorb enough heat to heal the breakdown. For this

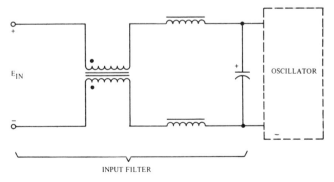

Figure 8-20  A series transformer input filter.

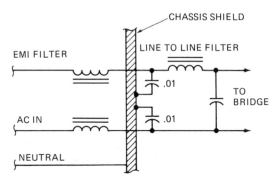

Figure 8-21 Input filter combines low frequency and RF sections.

reason, and for the better $CV$ product, wet tantalum finds wide use in converter applications.

A very effective input filter (shown[2] in Figure 8-21) is made up of two parts: a low-frequency line to line filter and an RF filter. The line-to-line filter consists of an inductor and a polycarbonate or mylar capacitor. The corner frequency of this filter should be about 8 kHz for a 200 W power supply to insure that the current components above 20 kHz will not exceed the specified limit. However, this corner is directly dependent on the input power, and can be determined using the Fourier Series for the input line currents.

The EMI filter works on the premise that high frequency currents must be bypassed to chassis while being blocked from the input line. In this case, a dissipative core material, such as powdered iron, is used while high frequency capacitors bypass the RF to chassis. It is important not to make capacitors larger than necessary to filter the RF, since they do constitute an AC leakage path.

In Figure 8-22, which is a modification of Figure 8-23, capacitors $C_A$ and $C_B$ are low-impedance capacitors, center-tapped to ground. $L_1$ and $L_2$ are high-impedance inductors which are wound oppositely (bifilar) on the same core to reduce power supply size. $L_1$ and $L_2$ show their full inductance to AC currents—thus attenuating them; the inductor fluxes cancel each other for normal DC currents. Capacitor $C_9$ acts to prevent load transients from entering the power supply when the filter is on the regulator output. Or, if the device is used as an input filter, $C_9$ is positioned to prevent transients from entering the power supply.

The filters of Figures 8-21 and 8-22 can be used on both the input and output leads of a power supply.

An approach to rounding off the square-wave corners at the source of the square wave—at the switch—with small RC networks is shown in Figure 8-23. The RC networks suppress higher-order harmonics. The inductors are wound on toroids that are 0.30 or 0.34 in. in diameter and usually have a value of 50

---

[2]Ibid.

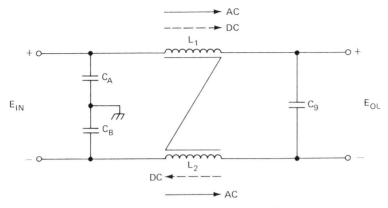

Figure 8-22   Commonly used input filter.

to 100 mH. Lower values may be used if only the high-frequency noise is of concern.

Noise suppression begins at about 50 kHz; the noise is reduced by, typically, 80 dB. Switching spikes are suppressed by the network across the switch. Such networks as these are now commercially available as small components with $R$ and $C$ values of, typically, 15 to 25 ohms and 100 pF.

The two inductor toroids can be potted together if shielding is placed between them. Output, input and switch leads must be well separated, of course, to avoid coupling effects.

### Output Filter[3]

Meeting the output conducted interference requirements at the low frequencies (that is, the converter frequency) is much more difficult. This is because the

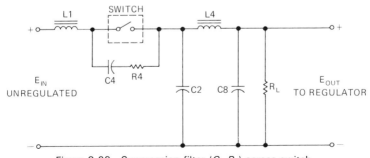

Figure 8-23   Suppression filter ($C_4$-$R_4$) across switch.

---

[3]Ibid.

conducted current, as specified in MIL-STD-461, is an absolute limit, while the actual ripple current at the converter frequency is proportional to the DC load current, if the ripple voltage is held constant. For instance, for a 5 V output (assuming a triangular ripple):

Spec. limit at 20 kHz = 10 mA RMS (50 A load with 20 mV, P/P, at 20 kHz, ripple = 120 mA RMS; 5 A load with 20 mV, P/P, at 20 kHz ripple = 12 mA RMS).

High frequency spikes are not necesarily proportional to load current, and in fact, it is not unusual to see the same spikes occurring on a 50 A, 5 V output as on a 5 A, 5 V output. Most users frown on outputs which have a high impedance at high frequencies, even though the distance between the power supply and the loads, because of line inductance, often negates the advantage of having a low output impedance.

The output of the converter must be an LC filter to ensure that the converter gap is filtered. The width of this gap and the DC current determine the size and type of inductors. The main filter capacitors are usually solid tantalum for minimum ESR. It is important to note that if the inductor is laminated, a capacitance may exist in parallel with the inductor, which can bypass the inductor at high frequencies (Figure 8-21).

Once the basic filter design is complete, the primary problem is to remove the very high frequencies that remain on the bus. This can be done quite adequately using the technique shown in Figure 8-24. The object, like that of the input filter, is to ensure that the high frequency spikes on the output lines are absolutely common mode and are coupled to the chassis or RF ground.

Again, the actual lengths of lead on the bypass capacitors should not exceed $\frac{1}{4}$ in. and high frequency capacitors must be used. Furthermore, the outputs, once filtered, must not be exposed to the high energy RF fields within the power supply.

The design of power supply filters can be simplified by considering the amount of ripple voltage that can be tolerated at the output of DC/DC converters.

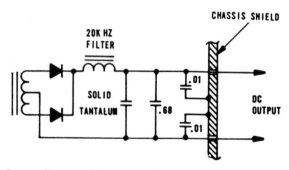

Figure 8-24   Output filter provides both rf bypass and smooths the rectified PWM waveforms.

## Filter Application

A switching-regulator circuit that employs some of the techniques discussed is shown in Figure 8-25. Commercial versions of the filter of Figure 8-22 are used. They are "pi" and "T" section filters, made by stringing ferrite beads on a center conductor. High-dielectric ceramic material is fused around the beads, and an outer metal shell with feedthrough connections completes the construction.

Additional elements are added if noise must be more strongly suppressed. Capacitors $C_4$ and $C_5$ are added to reduce high-frequency switching transients, by slowing the switching speeds of the output transistors. They also protect the semiconductor devices from transients. Capacitor $C_4$ protects the emitter-base junction of $Q_2$ while capacitor $C_5$ protects both the IC and the collector-base junction of $Q_1$ by controlling the risetimes of the switching voltages.

In addition capacitors $C_6$ and $C_7$ suppress transients on the unregulated input, $C_8$ minimizes the input impedance seen by the regulator, and $C_9$ improves the filtering of the switching noise in the regulated output. The other capacitors have the same function as before, except that $C_1$ is much larger.

This circuit oscillates in the range of 20 kHz to 40 kHz. Resistors $R_1$ and $R_2$ determine the output voltage level, and $R_6$ insures enough positive feedback.

Inductor $L_1$ has the same function as in Figure 8-10, but the lower $E_{out}$ and the added inductive resistance provided by $R_4$ allow $L_1$ in this case to be only 550 $\mu$H. Inductors $L_2$ and $L_3$ are high-frequency input and output filters.

To avoid the possibility that problems in the power supply might reverse the input-voltage polarity, or that heavy negative-going transients might be generated on the unregulated input, diodes are placed on the collector of $Q_2$.

Although $I_{max}$ is only 500 mA, the higher dissipation of this slower-switching circuit, and the fact that the regulator has to operate at relatively high temperatures, make it necessary to use two external transistors. Pass transistor $Q_2$ is an *npn* type for the same reason that $Q_1$ is a *pnp*—so cascade connections on the IC booster output can be used. Finally, $R_5$ is added to terminate $Q_2$ and for low off leakage.

Losses in this circuit are 30 to 40%, depending on $E_{in}$. Most of the extra loss builds up in the pass transistors. The efficiency can be raised to over 75% by eliminating $C_4$ and $C_5$.

Over the input voltage range of 22 to 31 V, the switching rate of this circuit varies from about 20 to 40 kHz. At temperatures from $-30°F$ to $150°F$, the total variation in $E_{out}$ over the $E_{in}$ range is $\pm1.1\%$. The efficiency is 67% at room temperature, with $E_{in} = 22$ V, and 61% with $E_{in} = 31$ V. Maximum power output is 2.3 W.

At the same voltage drops, a comparable linear regulator would have efficiencies ranging from only 15 to 24%.

To determine this regulator's noise suppression characteristics, four tests were performed. First, positive and negative 50 V, 10 $\mu$s transients were injected on

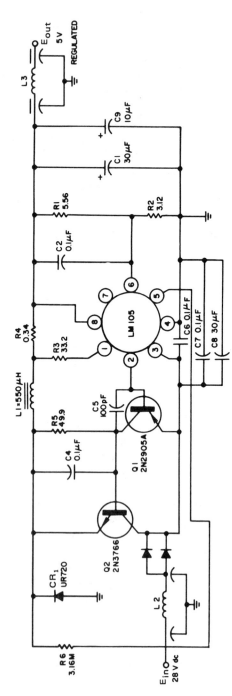

Figure 8-25  In the noise-suppressed regulator, capacitors $C_4$ and $C_5$ slow the switching speeds of the switching transistors, $C_6$ and $C_7$ suppress transients on the unregulated output; $C_8$ lowers the impedance seen by the regulator, and $C_9$ reduces switch noise in the output. Efficiency is 60 to 70%.

top of the DC input voltage to the regulator. Noise measured after the input filter (at $Q_2$) was 1.5 V peak-to-peak, and noise at the regulator output was 100 mV peak-to-peak.

Next, a 50 V DC square wave input with a 10 ms period was injected at the input of the regulator. A positive shift of the output voltage of 0.5 V DC was observed. The frequency of the switching regulator decreased to 10 kHz, indicating that the input filter wasn't effective for blocking this type of waveform. Some of the 10 kHz noise appeared on the output as a 400 mV peak-to-peak modulated signal.

In the next test, the output was driven with a 60 mA-to-80 mA load. Time-domain ripple measurements were made on the input. This ripple appeared as a half sine wave (rectified) with the same period as that of the switching regulator. With a $E_{in}$ of 22 V DC, there was 4.5 mA peak-to-peak of input ripple. At the same time the output ripple remained at 100 mV peak-to-peak.

Finally, audio noise was injected at the regulator power input as follows: 4 V peak-to-peak over a frequency range of 6 Hz to 100 kHz, and at least 1 V peak-to-peak over the range of 100 to 250 kHz. The response of the input filter to these signals was essentially flat up to 300 Hz, where it began to roll off at 20 dB/decade. The 3 dB point was at 1 kHz. The regulated output had a flat response of −37 dB to 30 kHz, then decreased to −30 dB to 60 kHz (indicating a resonant peak), then went back to −37 dB above 60 kHz (dB are with respect to the input).

## EMC REQUIREMENTS

### Thermal Design[4]

Unfortunately, the primary noise generators that couple RF noise into the chassis are components that get hot. Since transistors, power rectifiers, and to a lesser extent transformers and inductors will be damaged or destroyed if not properly cooled, they are normally fastened to the chassis.

The very act of this mounting provides RF leakage paths that allow RF ground currents to flow in the chassis. These currents cause RF noise to appear at all inputs and outputs. The packaging designer is frequently asked to locate these bulky, hot, noise generators in such a way that they won't burn up, won't radiate noise to sensitive leads, and can be replaced in 10 minutes, preferably without a soldering iron.

The capacitance between a TO-3 transistor and a mica-insulated heatsink is typically on the order of 100 pF . . . given a single part on a grounded heatsink, a 200 V input and an almost perfectly rectangular 20 kHz switching waveform, over 1 mA at 1 MHz could flow through this path alone.

The solution, of course, is to restrict this noise path to the primary circuit,

---

[4]Ibid.

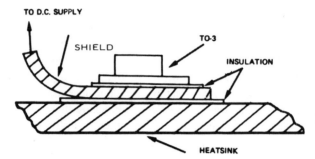

Figure 8-26   Capacitance-coupling to the heatsink may be reduced through utilization of an intermediate shield connected back to the primary circuit.

and not let it into the case. One technique would be to connect the heat sink to the emitter. Another possibility would be the introduction of a shield between the transistor and the heat sink (as shown in Figure 8-26), with the shield connected to the DC supply line.

Thus when planning the power supply package, remember to:

1. Keep high-current $di/dt$ and $dv/dt$ lines as short as possible to reduce the effective area of the noise transmitter.
2. Keep the input and output leads as far away as practical from the electrostatic and electromagnetic noise generators.
3. Keep the switching current paths simple to keep from creating ground loops and thereby minimizing the introduction of additional noise spikes.
4. Provide shielding between the noise generating source and the sensitive input and output leads. This shielding may be aluminum, mu metal, steel, or copper clad board (for electrostatic shielding).
5. Minimize capacitive coupling to the chassis.

## DC Isolation

Transformer primary-to-secondary isolation should be used if the system requires a high degree of isolation between input and output return lines. Multiple outputs may also have to be isolated from one another. Otherwise, volume, weight and cost can be reduced by using alternatives such as autotransformers, buck or boost types of switching regulators, series regulators, or some combination of these.

The non-isolated designs may allow a loop to exist through the power supply's load-return circuit and back to the input. Noise may be picked up by sensitive circuits in the system. A decision based on a system analysis, should be made whether to prevent noise pickup by supply design or a combination of supply and system design.

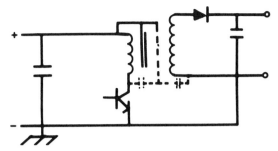

Figure 8-27  Unwanted noise may flow into the output through interwinding capacitance unless a shield is provided to retain the energy within the primary circuit.

## Interwinding Capacitance[5]

A significant path for unwanted noise currents is the interwinding capacitance of the output transformer, which couples the switching waveform harmonics directly into the output.

The solution, for low-voltage equipment, is an interwinding shield connected in such a manner as to return capacitive current to the supply line (see Figure 8-27). Noise currents may also flow through the winding-to-core capacitance, therefore the core (if possible) should be connected to the positive supply line, and the primary located adjacent to the core, thereby retaining these currents within the primary circuit loop.

## Circuit Grounding

Supply ground connections also vary with system usage. If the return leads are grounded at the supply, noise conducted from the returns to the ground plane is not very important. But if the ground point is some distance away from the supply, it may be essential to filter the input or output returns.

Explicit test specifications will ensure properly located filters. For example, the test setup in Figure 8-28 calls out the location of grounds of the source and load returns to the reference ground plane. These points must be designated. Both the DC/DC converter and switching regulator generate broadband conducted interference voltages, creating noise currents that could circulate from the output to input returns through the ground plane.

## Conducted Susceptibility

The supply is generally expected to prevent power source noise from disturbing the load, and source transients from damaging the supply itself or the load. Until

---

[5]Ibid.

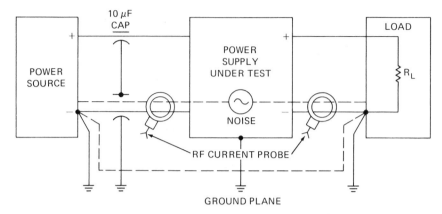

Figure 8-28   Conducted interference test setup showing grounds and noise current loops.

susceptiblity levels are defined and specified, the supply concept design is not complete. Also, success criteria should be well defined so that supplies can be tested on a pass/fail basis.

Conducted-interference filters increase susceptibility thresholds but are not necessarily compatible with all susceptibility requirements. Input filtering to remove lowest-frequency audio noise is seldom practical, since such filters may be too bulky. The power-line input filters may actually contribute to audio noise. Check them for sharp, resonant peaks in the AF range. L-C components with very high $Q$ may double or quadruple the applied test-signal levels. Rather than increase the regulation range to handle this, it is better to provide some clamping of the filter circuits. Two schemes for filter damping are shown in Figure 8-29.

Give serious consideration to the combination of lowest input voltage and maximum low-frequency susceptibility test level. Initial specs should spell out if the supply should or should not be tested at line-voltage extremes. This is important because regulators are designed to operate at the lowest input-voltage

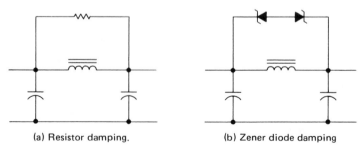

(a) Resistor damping.                    (b) Zener diode damping

Figure 8-29   Filter damping to prevent audio resonances.

swing. Specifying only the degree of regulation actually required will minimize power supply cost, weight and size.

In switching regulator design, a tradeoff must also be made between high-frequency rolloff and input filter capability. The regulator's response bandwidth—its ability to correct rapidly for load and line variations—improves as frequency rises, but the error amplifier's frequency is limited by external capacitance and by the IC compensation used to prevent undesirable oscillation. A check should be made to insure that when the response is rolling off, the filter attenuation stays high enough to prevent signals in the susceptibility range from passing through the regulator to the converter and out to the loads.

Finally, make sure that the transient spike susceptibility test signals used will be reduced to a safe level by suppression devices. Spikes must be well below the voltage breakdown thresholds of semiconductor devices and capacitors.

High frequencies allow smaller, lighter transformers, inductors and capacitors to be used, but high frequencies require fast transitions which produce increased EMI. Another tradeoff in switching stage design is switching speed versus efficiency, which may be stated as conducted interference versus power loss.

Power loss occurs mainly in the time when the pass transistor makes a transition between cutoff and saturation. At high frequency, the number of transition times per second is large. Consequently efficiency drops unless the transition time is very short, which sharpens the current spikes and increases broadband EMI.

The losses in $Q_1$ of Figure 8-10 will be (at 50% duty cycle)

$$P_L = \frac{2t_{tr}}{t_T} P_{tr} + i_{DC} \frac{V_{CE(sat)}}{2} \tag{8-19}$$

where:

$P_L$ = average power loss, watts
$t_{tr}$ = time of switching transition, sec
$t_T$ = time of one regulator cycle, sec
$P_{tr}$ = average power dissipation during each transition, watts
$i_{DC}$ = DC output current
$V_{CE(sat)}$ = collector to emitter voltage across transistor when on and saturated

Assume, for example, that the regulator cycle is 20 $\mu$s (50 kHz), current = 1 A, $V_{CE(sat)} = 0.2$ V, output voltage = 25 VDC, and average power dissipation = 10 W. Then, if transition time $t_{tr}$ is varied from 5 $\mu$s to 0.5 $\mu$s:

1. At $t_{tr} = 5$ $\mu$s,     $P_L = 5.1$ W
2. At $t_{tr} = 0.5$ $\mu$s,     $P_L = 0.6$ W

Without changing switching frequency, regulator efficiency has been increased from 83 to 97.7% by reducing switching time.

Look, though, at the broadband interference spectra of the two examples, in Figure 8-30. Obviously it will be necessary to filter all of the supply inputs and outputs (and use an RFI-tight package). If both a switching regulator and a converter are used, they will need an RFI-tight compartment and feedthrough filters on the connecting pins.

It is generally not desirable to place slowdown filters in the power switching transistor circuits because the increase in transition power loss will not be acceptable. But it is desirable to limit unnecessary overdrive current spikes in the driver circuits. Only as much drive should be used as is necessary for reliable operation under all conditions. The driver transistors will usually be much faster switches and therefore will generate even higher interference frequencies.

### Radiated Interference

The circuits generate noise at frequencies whose detectable harmonics may reach 30 to 50 MHz. These components could easily leak out of supply packages that are not RF-tight. Good design practice, including proper bonding of package case and cover, proper filtering and lead isolation will control radiated noise. The same techniques control susceptibility to radiated noise.

### Turn-On and Turn-Off Transients

Turn-on spikes are generated by the sudden charging current taken by filter capacitors in the supply. They can be reduced by using inductive filters or smaller filter capacitors, usually at the expense of efficient interference filtering. When power is first applied, large filter capacitors may pull 10 times as much, or more, current than steady-state input current. Turn-off spikes will be produced when the energy stored in filter inductors discharges into the opened switch. A damping network across the switch will absorb this energy. In some cases diodes may be required to discharge large filter inductors safely.

Figure 8-31 is a typical transient-measurement circuit. The power source is a low-impedance regulated supply. Point $X$, $X'$ represents a local distribution point where the supply lines branch to feed other equipment in the system. An oscilloscope is connected at $X$, $X'$ to measure the transient produced when relay $K$ is closed or opened.

Current flows suddenly into the input filter capacitors when $K$ closes. This initial surge must be limited by some input inductance. Then the regulator, the capacitors in the DC/DC Converter, or both, begin to draw charging current. Finally, after the supply circuitry starts oscillating, the output filter capacitors

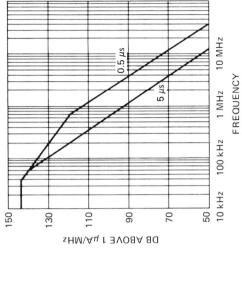

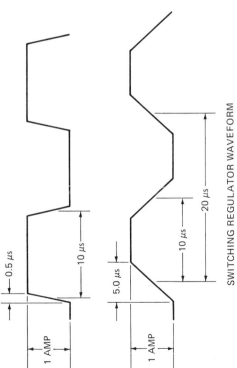

SWITCHING REGULATOR WAVEFORM

Figure 8-30 Broadband conducted-interference spectrum of switching-regulator input current pulse.

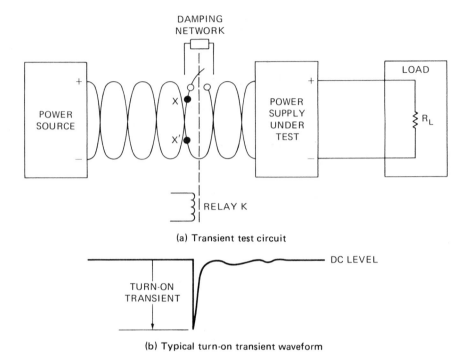

(a) Transient test circuit

(b) Typical turn-on transient waveform

Figure 8-31   Turn-on and turn-off transient test.

(and possibly load capacitors) will charge. All of these produce a large and sustained current spike on the input power line.

When the switch is opened, the input filter and line inductance will produce an oscillatory spike unless a damping network is placed across the switch contact. Since line impedance is a factor, the length of line between the source and the switch or supply input should be determined from the system design and specified for the test setup.

## External Measures

Switching regulators often generate noise at frequencies whose detectable harmonics may reach 30 to 50 MHz. These noise components could easily leak out of packages which are not rf-tight. Good design practice, besides filtering, includes proper bonding of package case and cover, completely surrounding the regulator in a shielding enclosure (Figure 8-32), and lead isolation to control radiated noise. (Both electric and magnetic field radiation are reduced by the enclosure.)

When using a shield be careful to construct it such that there are no slots (joints, etc.) perpendicular to the induced current in the sensitive area, otherwise

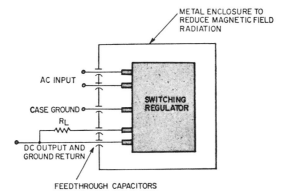

Figure 8-32   Shielding a switching regulator with a metal enclosure and using feed-through capacitors on all wires going into and out of the regulator is an effective means of reducing radiated interference.

the induced shield current times the slot impedance may generate a local field right where you don't want it (Figure 8-33). If you're not sure that all joints are adequately short-circuited, use your scope to compare the voltage developed across them to other points of similar spacing on the shield.

In addition, feedthrough capacitors used on all wires (input, output, and ground) entering and leaving the enclosure prevent internally generated switching spikes from leaving the regulator. They can also put prime power voltages on the shield if it is not securely grounded. Faraday screens are also used around magnetic devices to contain switching spikes.

Adding a large high-frequency capacitor between DC output and AC input (Figure 8-34) reduces externally circulating ground currents significantly. A series resistor is added to dampen oscillation, since the capacitor is part of a large tank circuit whose inductances are created by wiring and magnetic devices.

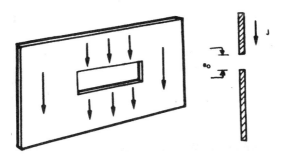

Figure 8-33   When constructing a shield, be sure to avoid discontinuities adjacent to the "sensitive" circuitry because induced shield current ($J$) can develop RF voltages ($e_0$) across slots and joints, providing "leaks".

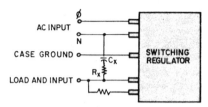

Figure 8-34  Noise can be shorted in a switching regulator by placing a high frequency capacitor, $C_x$, between AC and DC input and output lines to dampen tank circuit oscillations.

Capacitors should be tied to case (system) ground for all input and output wires. Shorting one side of the DC output directly to the system ground is an effective way of keeping noise from entering the external world.

A common mode choke can be utilized to reduce circulating ground noise currents that normally exist in a system—as is often the case when circuits, subassemblies, or racks of equipment are interconnected. The choke is a bifilar-wound broadband transformer which allows equal and opposite current to flow through its windings while suppressing unequal and opposite currents, such as those due to ground noise. Because of the bifilar winding, no net flux is generated in the choke when its two currents are balanced; therefore, balanced signals encounter no inductance when passing through it. For unbalanced currents, however, the choke acts as an inductance and effectively breaks up the ground path current. Figure 8-35 illustrates one way the common mode choke may be used. In summary, it balances the current of a cable regardless of whether the imbalance is due to faulty circuit design, ambient environment, or conductance from a connecting circuit.

## EMC DESIGN EXAMPLE

The supply in Figure 8-36 is a practical example of integrating most of these tradeoffs. To provide DC isolation from output to input, it uses a converter similar to that shown in Figure 8-2. The preregulator is similar to that of Figure 8-10, but based on an IC amplifier (LM101) that was selected as a standard component for a military system using the supply. Operation is practically iden-

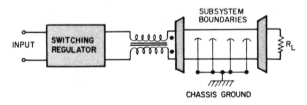

Figure 8-35  Common mode choke used for critical circuit protection.

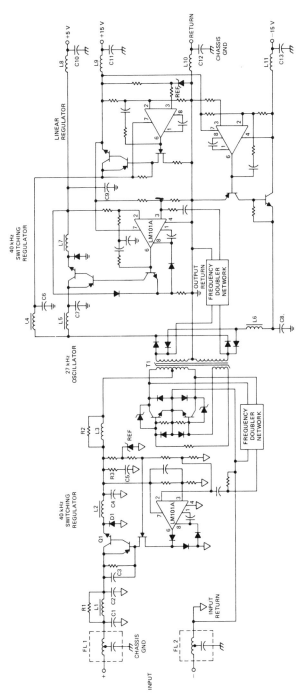

Figure 8-36   Example power supply.

tical to that of an IC regulator, except that the output and booster transistors are a complementary pair and a Zener diode is the reference.

The switching regulator is not allowed to run free. It is driven at twice the converter frequency. Source and load circuits are grounded similar to Figure 8-19. Therefore, both input and output leads have balanced RF filters. Startup components are not shown, to simplify the schematic. This rather complex design was needed to meet the following specifications:

| | |
|---|---|
| Input current | 0.8–1.1 amps, full load |
| Input voltage | 24–33 volts DC |
| Outputs | +15 V, 0.3 amps |
| | −15 V, 0.3 amps |
| | +5 V, 0.5 amps |
| Output regulation | ±15 V, +1% |
| | +5 V, ±2% |
| Output ripple | ±15 V, 50 mv p-p |
| | +5 V, 40 mv p-p |
| Electrical isolation | greater than 10 megohms |
| EMI requirement | MIL STD 826, Class Au, on input power |
| Converter frequency | approximately 22 KHz |
| Regulator frequency | approximately 44 KHz |

The high efficiency of each stage, around 80%, gave the total supply an efficiency over 50%. A conventional supply with the same power and voltage range would have been much less efficient and considerably more bulky.

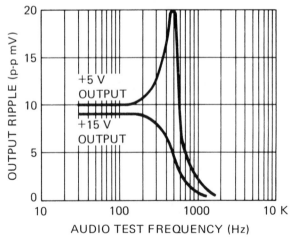

Figure 8-37   AF susceptibility test response of power supply with 1.5 V rms applied to DC input.

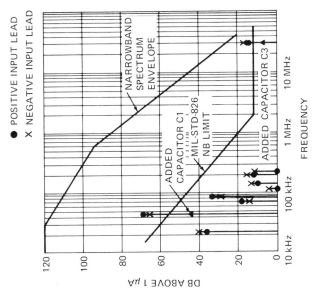

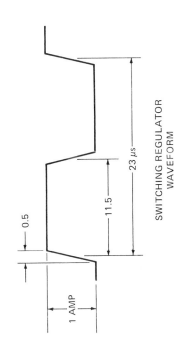

Figure 8-38   NB conducted interference on power input leads of example power supply.

This design posed quite an EMC challenge because both circuits on the primary side switch at rather high frequencies with heavy current loads. Since the pre-regulator chops the incoming line current at the higher frequency, it generates more conducted interference on the input side than the converter. The network around $L_1$ is the low frequency filter (here $FL_1$ and $FL_2$ are feedthrough filters for the RF-tight package).

The RF filtering capability of $L_1$ had to be traded off somewhat to reduce susceptibility to audio noise and AF resonances of the input filters. The resonances are damped by $R_1$ of 4 ohms. Audio susceptibility tests showed an ample attenuation of audio tests signals (Figure 8-37) and attenuation of RF susceptibility test signals was also good. During AF susceptibility tests, peak negative swings were applied at line voltages of 24 VDC (the low line voltage).

Figure 8-38 shows conducted interference on the power input of the breadboard and final designs. Capacitor $C_1$ was added to improve filtering at 44 KHz. Then $C_3$ was added to reduce the high harmonics. $C_3$ slowed down the transitions of $Q_2$, which proved to be switching much faster than required. (Note that slowdown capacitors are not used in the secondary regulators.)

$L_2$ has the same function as the inductor in Figure 8-10. $L_2$ with $C_4$ and $L_3$ with $R_2$ also filter the current spikes arising in the converter. Rise-times and peak values of the converter current spikes (similar to Figure 8-3f) conducted on the converter input line are first limited by $L_3 R_2$ then filtered by $L_2 C_4$.

The filtering on the secondary side is much less stringent since the primary circuits bear the brunt of suppressing input susceptibility noise. Input inductors, decoupling capacitors, and fairly simple output filtering do the job. Of course, the packaging of the entire supply was made RF tight.

A maximum turn-off or turn-on transient of 12 V superimposed on the input DC voltage was specified. This spec was met by limiting the filter capacitor sizes. The addition of $C_1$ brought the turn-on transient to the $-12V$ peak limit. Turn-off transients produced by opening the circuit to the inductive input filters, were controlled by a small RC network placed across the contacts of the off/on switch contact. Capacitive input filters could not be used because any larger capacitance would produce a large turn-on spike.

Integrated circuits, high-power switching transistors, and good core materials make high efficiency DC/DC converters and switching voltage regulators practical. However, there are close relationships among the circuit functions and interference generation and susceptibility characteristics. Therefore, it is necessary to begin planning in the concept design phase the circuit techniques used to solve EMI problems. This means that personnel responsible for system electromagnetic compatibility should understand supply design requirements enough to help the supply designer make the initial tradeoffs needed to develop a concept that satisfies both performance and EMI specifications.

# 9

## Converter and Inverter Design Considerations and Examples

### INTRODUCTION

This chapter provides a summary of DC/DC converter design equations as an introduction. Then, discussions of specific converter and inverter designs are presented by means of practical examples encountered by the author. Listed first in each section are the electrical performance parameters (specifications) which each design is to attain. Following this are discussions of the problems that were encountered in meeting the specification requirements, how these problems were solved, and test data that were taken to verify the designs' conformance with the specification requirements. Thus, this chapter is concerned with the problems which may be encountered by a designer and offers practical solutions to these problems.

### SUMMARY OF TWO-TRANSFORMER DC/DC CONVERTER DESIGN EQUATIONS

The design of a high-speed transistorized converter is based on the available supply voltage range, the required output voltage and power, and the range of ambient temperature over which the converter must operate. Moreover, the converter's specifications usually provide additional preliminary design information, such as efficiency, size, weight, operating frequency, and stability. The following discussion pertains to a two-transformer DC/DC converter; however, the principles are the same as for a one-transformer inverter except for the extra transformer. Mention is made of this difference.

The following are used in the design of a converter circuit. Refer to Figure 9-1 for the basic schematic of a typical two-transformer DC/DC converter.

(1) The power delivered to the output transformer $T_2$ is computed as follows:

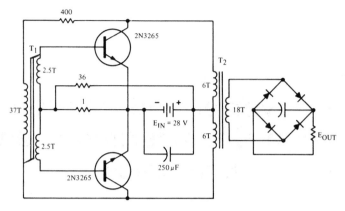

Figure 9-1    Schematic diagram of two-transformer DC/DC converter.

$$P'_{\text{out}} = \frac{P_{\text{out}}}{\eta_2} \tag{9-1}$$

where $P_{\text{out}}$ is the required output power of the converter and $\eta_2$ is the transformer efficiency. A value of 90% to 95% is usually assumed.

(2) An estimate of the transistor collector current for a square wave ($I'_C$) can then be obtained from the ratio of $P'_{\text{out}}$ to the supply voltage, $E_{\text{in}}$, that is,

$$I'_C = \frac{P'_{\text{out}}}{E_{\text{in}}} \tag{9-2}$$

(3) Determine from the manufacturer's data sheet the transistor saturation voltage, $V_{CE}(\text{sat})$, that corresponds to the collector current, $I'_C$, and case temperature, $T_c$. The transistor collector current should now be recomputed as follows:

$$I''_C = \frac{P'_{\text{out}}}{E_{\text{in}} - V_{CE}(\text{sat})} \tag{9-3}$$

(4) From the manufacturer's data sheet determine the base-to-emitter voltage, $V_{BE}$, required for the collector-to-emitter saturation voltage, $V_{CE}(\text{sat})$, as given in step (3), at the collector current, $I''_C$, and the case temperature, $T_c$. Also find the common-emitter forward-transfer ratio, $H_{FE}$, at this collector current and case temperature. A value of $H'_{FE}$ that is low enough to insure saturation (usually $H'_{FE}$ is about half of $H_{FE}$) is then used, together with the value found for $V_{BE}$, in the following equation to estimate the base-circuit input power, $P_{\text{in}}$:

$$P_{in} = V_{BE}\left(\frac{I_C''}{H_{FE}'}\right) + \left(\frac{I_C''}{H_{FE}'}\right)^2 R_B \qquad (9\text{-}4)$$

The base stabilizing resistance, $R_B$, is small and is usually chosen so that the voltage dropped across it will be about half of $V_{BE}$.

(5) The input power to base-drive transformer, $T_1$, can be approximated on the basis of the base circuit input power, $P_{in}$, and the transformer efficiency, $\eta_1$, as follows:

$$P_{in}' = \frac{P_{in}}{\eta_1} \qquad (9\text{-}5)$$

(6) The collector current can now be approximated on the basis of the total power developed in the converter circuit:

$$I_C = \frac{P_{out}' + P_{in}'}{E_{in} - V_{CE}(sat)} \qquad (9\text{-}6)$$

If the collector current given by equation (9-6) is significantly higher than that given by equation (9-3), steps (4), (5), and (6) should be repeated with this higher value of collector current substituted for $I_C''$.

(7) The turns ratio of output transformer $T_2$ may now be computed using the specified load impedance, $(Z_L)$, and the reflected impedance, $(Z_L')$, determined by

$$Z_L' = \frac{E_{in} - V_{CE}(sat)}{I_C} \qquad (9\text{-}7)$$

Thus, the turns ratio for $T_2$ is determined from the following relationship:

$$N_2 = \frac{Z_L}{Z_L'} \qquad (9\text{-}8)$$

(8) The value of the feedback resistor, $R_{FB}$, is usually chosen so as to drop about half of the available voltage. Thus, $E_{pri} = (E_{in} - V_{CE}(sat))$ with a primary current as follows:

$$I_{pri} = \frac{P_{in}'}{E_{pri}} \qquad (9\text{-}9)$$

where $P_{in}'$ is the value determined from equation (9-5).

(9)  The turns ratio for transformer $T_1^*$ is given as follows:

$$N_1 = \frac{V_{BE} + I_B R_B}{E_{\text{pri}}} \tag{9-10}$$

where

$$I_B = \frac{I_C}{H'_{FE}}$$

The general design procedure outlined above presupposes that the transistors to be used are selected on the basis of the operating requirements specified for the converter and that the transformers used in the circuit are the best types for high-speed converter applications. Also, because of the variety of arrangements possible, no provisions are included in the general procedure for the design of a bias-starting network.

## Transistor Requirements

The selection of a transistor for use in a high-speed circuit is dictated by the following conditions:

- In the high-speed converter, the peak value of the collector-to-emitter voltage of each transistor is equal to twice the supply voltage plus the amplitude of the voltage spikes generated by transient elements. Therefore, the collector-to-emitter breakdown voltage, $V_{CEO}$, of the transistors should be slightly greater than twice the supply voltage (usually an additional 20% is sufficient).
- The transistors must be able to handle the currents that are necessary to produce the required output power at the given supply voltage, and their saturation voltage at these currents must be low enough so that the desired efficiency can be obtained. Most transistors can be safely expected to switch eight times their Class A power rating. Hence, a transistor with a 2 W Class A rating can be used in a converter to switch 16 W; and two transistors will switch a total of 32 W. Thus, assuming an efficiency of 80%, 25.6 W are available at the output.
- The frequency cutoff characteristics of the transistor must be high compared to the actual switching frequency. If the transistor cannot switch rapidly between the states of saturation and cutoff, excessive junction heating results. Therefore, the frequency cutoff characteristics of the transistor should be from five to ten times the frequency of oscillation. When the transistor characteristic is ten times the frequency of the multivibrator, the output

waveform would be more nearly a square wave than if the frequency were only five times that of the multivibrator.

■ The junction-to-case thermal resistance of the transistors must be low enough so that for the given ambient temperature and the available heat-sink and cooling apparatus, the manufacturer's maximum ratings are not exceeded.

■ To maximize the efficiency of a power converter under load, the transistor should switch the maximum voltage possible. Because of junction heating there is a maximum collector current which can be switched and is independent of the supply voltage. The maximum collector current, dissipation and the heat-sink thermal resistance of the transistors can be approximated on the basis of the following limiting conditions.

The maximum collector current is approximately

$$I_C = \frac{P_{out}/\eta_2}{E_{in} - V_{CE}(\text{sat})} \tag{9-11}$$

where $E_{in}$ is the supply voltage, $V_{CE}(\text{sat})$ is the transistor collector-to-emitter saturation voltage (for a specific $I_C$), $P_{out}$ is the required power output and $\eta_2$ is the desired efficiency of the output transformer (usually 90% to 95%).

The transistor dissipation can be approximated as follows (because the base dissipation is very small, it is neglected here):

$$P_D = \frac{T_1}{T}\left[V_{CE}(\text{sat})I_C + 2I_{CEX}E_{in} + \left(\frac{t_{on} + t_f}{T}\right)\left(\frac{E_{in}I_C}{3}\right)\right] \tag{9-12}$$

where $E_{in}$ is the supply voltage, $V_{CE}(\text{sat})$ is the transistor saturation voltage (for a specific $I_C$); $I_C$ is the collector current, as given by equation (9-11); $I_{CEX}$ is the collector current with the base reverse-biased (for $V_{CE} = 2E_{in}$); $t_{on}$ is the transistor on time at $I_C$ given by equation (9-11) and $H'_{FE}$ given in step (4) of the general procedure; $t_f$ is the transistor fall time at $I_C$ given by equation [9-11] and $H'_{FE}$ given in general procedure; and $T_1 = \frac{1}{2}[T - (t_{on} + t_f)]$.

Equation (9-12) is used as a guide for the first stages of design; the exact dissipation is determined experimentally. The transistor saturated switching characteristics must be fast enough so that the transient dissipation terms do not become excessive.

The required heat-sink thermal resistance may be approximated by

$$\theta_{CA} = \frac{\Delta T}{P_D} - \theta_{jc} \tag{9-13}$$

where $\Delta T$ is the permissible junction temperature rise ($\Delta T = T_j(\text{max}) - T_A$); $P_D$ is the transistor dissipation; and $\theta_{CA}$ is the case-to-air thermal resistance, including mounting, interface, any insulation material, and heat sink.

The estimate of the required heat-sink thermal resistance together with the manufacturer's maximum rating curve of safe operating region will complete the determination of transistor requirements.

### Transformer Design

High flux density material is required to achieve maximum miniaturization and high efficiency. The efficiency is high because the core losses are small compared to the output power, which may be many watts. High efficiencies can be attained at high audio frequencies because as the frequency is increased, the core size is decreased; hence the increased core losses are offset by the reduction in core volume as the frequency increases. Transformer design is discussed in detail in Chapter 4.

Referring to Figure 9-1, the output power transformer is designed to satisfy the following familiar equation from Chapter 4:

$$N_1 = \frac{E_{\text{in}} \times 10^8}{4fA_cB_m} \tag{9-14}$$

In the design of the output transformer for high-power, high-frequency converters, excessive primary turns should be avoided (1) for minimum power dissipation, (2) to assure that the transformer can be made in view of the large wire sizes and the relatively small cores which are usually employed, and (3) to assure that a low value of leakage inductance is maintained. Good balance and close coupling between primaries is normally achieved by the use of bifilar windings. Flux density for the output transformer is determined by the usual compromise—the wire size selected on the basis of a 50% duty cycle must be large enough so that power dissipation will be low. If the wire size is inadequate, dissipation will be appreciable and high transformer-core temperature will result.

The design of the saturable base-drive transformer is not as straightforward as that of the output transformer, because when the transformer saturates, a sharp drop in the applied primary voltage must be produced. Thus, the magnetizing current must increase considerably from a small value to one that is comparable with primary current, as given by equation (9-9). The following equation must be used in addition to equation (9-14) to arrive at the number of primary turns because of the saturation requirement:

$$H_s = \frac{1.26N_pI_m}{\ell} \tag{9-15}$$

where $N_p$ is the number of primary turns, $I_m$ is the value of magnetizing current at saturation in amperes (chosen to be comparable with $I'_{pri}$—a value of $\frac{1}{2}I'_{pri}$ is usually acceptable), $\ell$ is the length of the magnetic path in centimeters, and $H_s$ is the value of the magnetizing field strength at saturation in oersteds (a value of five to ten times the coercive force is usually used). Equations (9-14) and (9-15) must both be satisfied for the proper design of base-drive transformer $T_1$.

The value of feedback resistor ($R_{FB}$) for a given primary current, $I_{pri}$, is computed such as to drop one-half the collector-to-collector voltage of the two transistors. The other half of the voltage is applied to the primary of transformer $T_1$. The optimum value of the feedback resistor is then determined experimentally. A decrease in the value of $R_{FB}$ will cause the magnetizing current to increase, thus increasing the voltage across the primary. As may be inferred from equation (9-14), the operating frequency will then increase.

An increase in $R_{FB}$ causes a greater voltage drop across it, and less voltage is then available to the primary of transformer $T_1$. However, if the value of $R_{FB}$ is increased excessively, the frequency will increase, because sufficient base drive will not be available to saturate the transistor for the proper period, and the saturation of the base-drive transformer will not be complete. Thus, $R_{FB}$ can be used to control frequency only over a limited range.

## Converter Starting

To ensure converter starting under all conditions, a starting bias may be applied so that the circuit has a loop gain greater than unity. Two methods are possible. The bias can be applied only during starting or a permanent bias arrangement can be used.

Two practical starting circuits are shown in Figure 9-2.

- The converter in Figure 9-2(a) uses a resistive voltage-divider network to supply the necessary starting bias. The required value of resistor $R_1$ can be found from

$$R_1 = \frac{E_{in}}{I_C} = \frac{E_{in}}{2I_B + [(V_{BE} + I_B R_B)/R_2]} \qquad (9\text{-}16)$$

The denominator of the fraction in equation (9-16) is equal to the desired value of starting current. Assuming a value of starting current $I_s$, this relationship can be used to determine the value of $R_2$. A compromise of reliable starting current and minimum bleeder current must be reached by trial and error.

- The diode starting circuit of Figure 9-2(b), in which the bases of the two inverter transistors are supplied by a resistance, $R_1$, is determined as follows:

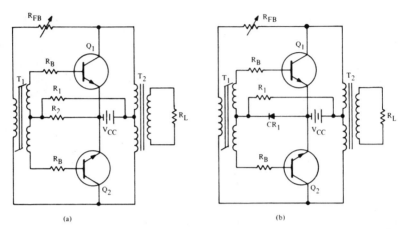

(a)                                              (b)

Figure 9-2   Two practical starting circuits. Converter on the left uses a resistive voltage-divider network to supply the necessary bias. Converter on the right uses a diode-type starting circuit.

$$R_1 = \frac{E_{\text{in}}}{2I_B + I_d} \qquad (9\text{-}17)$$

As the converter begins to oscillate, the base current flows through the base-emitter diode and in the forward direction through the starting diode. Usually, additional drive is needed to compensate for the diode voltage drop. Low-voltage silicon diodes, which must be capable of carrying the base current continuously, generally are used. Both of these starting circuits were discussed in detail in Figures 2-66 and 2-67.

### Second-Breakdown Effects

High-speed, high-power converters require transistors with high power-handling capabilities and very fast saturated-switching speeds. Second breakdown is a factor that must also be considered in the design of these circuits. A comprehensive discussion of second breakdown is presented in Chapter 5.

### MULTIOUTPUT DC/DC CONVERTER DESIGN CONSIDERATIONS

A DC/DC converter was required which would yield four tightly regulated (1%) DC output voltages from an unregulated DC battery voltage. The significant design parameters are:

Input voltage:       22 VDC to 30 VDC
Output voltages:    ±5 VDC ± 1% at 0.25 W
                    ±10 VDC ± 1% at 0.50 W

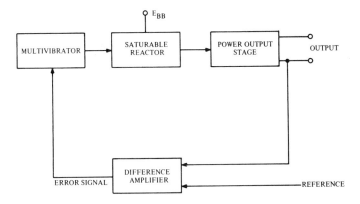

Figure 9-3 Complete regulator block diagram.

| | |
|---|---|
| Efficiency: | 85% minimum |
| Temperature range: | $-30°F$ to $+160°F$ |
| Output ripple: | $\pm 10$ mV peak-to-peak |

Figure 9-3 depicts a block diagram of the converter. A circuit description of each block of this diagram is presented in the following sections, ultimately leading to a schematic diagram of the converter along with breadboard test results.

## SR (Saturable Reactor) Timing Circuits

The SR circuit is shown in Figure 9-4. The circuit operation proceeds as follows. Assume that the multivibrator has just switched and that the output polarities are such as to turn $Q_1$ on. Capacitor $C_1$ provides a turn-on current spike to $Q_1$.

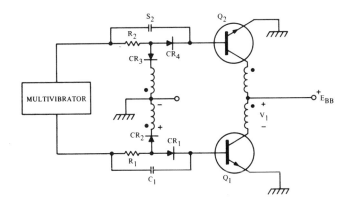

Figure 9-4 SR timing circuit.

As the core voltage $(V_1)$ builds up, a voltage is induced in the feedback winding that back-biases diode $CR_2$. $Q_1$ conducts and $CR_2$ remains back-biased until the core saturates, at which time the core voltage drops and the base drive current is shunted to ground through $CR_2$. Thus, $Q_1$ is turned off and the core voltage drops to zero and remains there until the multivibrator again changes state, turning $Q_2$ on. The cycle of events is then repeated with a core voltage of the opposite polarity. The principal limitation of this configuration is in the turnoff transient. When the core saturates, energy is stored in the saturated inductance that must be returned to the circuit. This generally leads to overshoot and ringing during turnoff. Aside from the possibility of coupling these transients to the output stage, there is a distinct danger of premature switching of the multivibrator (and loss of regulation) due to these transients. A circuit in which the appropriate switching transistor conducts for the full half-period of the multivibrator is shown in Figure 9-5.

This circuit is designed so that most of the drive current from the multivibrator flows into the base, while the core is unsaturated. The collector current for the on transistor consists of the core exciting current, plus the reflected load current (base drive for the power output stage). When the core is unsaturated, the appropriate transistor is in the saturated region and $R_E$ is chosen so that the drop across this resistor is small relative to the supply voltage. When the core saturates, the collector current will rise, but the base current will drop since the drop across $R_E$ increases. With the transistor in the active region, the base current will be approximately

$$i_b = \frac{i_1 R_3}{R_3 + \beta R_E} \tag{9-18}$$

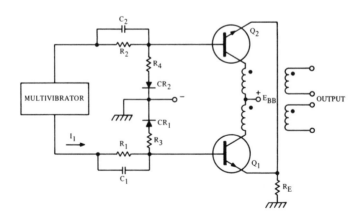

Figure 9-5 Modified SR timing circuit.

The collector current will rise to a value

$$i_c = \beta i_b = \frac{\beta R_3 i_1}{R_3 + \beta R_E}$$  (9-19)

Although the core voltage drops to zero, this collector current will continue to flow until the multivibrator switches. There are sufficient degrees of design freedom so that the rise of collector current following saturation of the core will be very small. In this circuit, the energy stored in the saturated inductance is returned to the circuit only when the multivibrator switches and actually improves the switching speed. Thus, a large spike of collector current at the instant of core saturation, followed by a potentially dangerous turnoff transient, has been replaced by a well-controlled low-amplitude current step. The turnoff transient of the SR transistor has been deferred until the multivibrator has switched, at which time the effects are beneficial. The price that has been paid is a small net increase in dissipation since collector current in the SR transistor flows for the full half-cycle.

## Power Output Stage

A push-pull power output stage is shown in Figure 9-6. The transistors are operated as switches whose on and off times are controlled by the SR timing circuit. The output transformer is designed to prevent saturation under any conditions. The rectified voltage from winding $N_k(v_k)$ thus consists of a series of

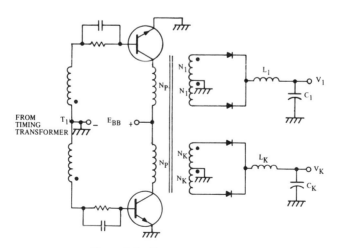

Figure 9-6   Power output stage.

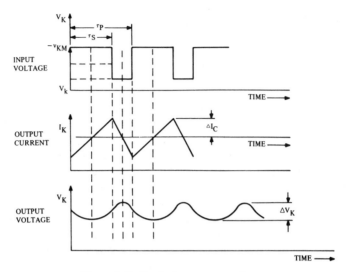

Figure 9-7   Typical output waveforms.

pulses of duration $\tau_s$ with a leading edge-to-leading edge spacing of $\tau_p$. The pulse amplitude is

$$\frac{N_k}{N_p} [E_{BB} - V_{CE}(\text{sat})] \tag{9-20}$$

where $V_{CE}(\text{sat})$ is the saturated drop for the output transistor.

Typical waveforms for the input voltage ($v_k$), inductor current ($i_k$), load current ($I_K$), and output voltage ($V_k$) are shown in Figure 9-7. Note that when the core voltage drops to zero, both output rectifiers become forward-biased and the load current divides equally between the two paths. This follows from the requirement that there can be no net core mmf with zero voltage across the core. Thus, the voltage $V_k$ drops to the forward-biased diode potential $V_d$ when the core voltage drops to zero.

In order to fix the filter requirements, consider the simple L-section when the peak-to-peak output ripple is small (say less than $0.1\ V_k$). If we denote the peak value of $V_k$ as $V_{km}$ and assume that $V_k$ is essentially constant, the inductor current has a sawtooth variation about the average value ($I_k$) with $\Delta i_k$ given simply by

$$\Delta i_k = \frac{V_{km} - V_k}{2L} \tau_s = \frac{V_k}{2L} (\tau_p - \tau_s) \tag{9-21}$$

Since the rectifiers constrain the current to be unidirectional, the value of $\Delta i_k$

must be less than the average load current in order that regulation be maintained. Solving equation (9-21) gives the anticipated value for the average voltage.

$$V_k = V_{km} \frac{T_s}{T_p} \tag{9-22}$$

The peak-to-peak ripple may now be found by integrating the alternating component of the current waveform

$$\Delta V_k = \frac{1}{C} \int_0^{T_s/2} \Delta i_k \frac{t}{T_s/2} \, dt + \frac{1}{C} \int_0^{(T_p - T_s)/2} \Delta i_k \left(1 - \frac{t}{T_p - T_s/2}\right) dt \tag{9-23}$$

Integrating:

$$\Delta V_k \simeq \frac{\Delta i_k}{LC} T_p = \frac{V_{km} - V_k}{8LC} T_p T_s = \frac{V_k T_p (T_p - T_s)}{8LC} \tag{9-24}$$

### Difference Amplifier

A very simple difference amplifier is shown in Figure 9-8. One of the output voltages ($V_k$) is fed back and compared to the reference diode voltage $V_{b_1}$. The collector current ($I_C$) of transistor $Q_1$ is used to control the period of the timing multivibrator. If the collector of $Q_1$ is connected to a point whose potential is such that $Q_1$ remains in the active region and if the load impedance for $Q_1$ is much lower than the output impedance for $Q_1$, then the current $I_C$ is essentially independent of the load characteristics. The equation describing the static behavior of this current is:

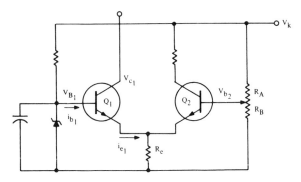

Figure 9-8   Difference amplifier.

$$V_E = V_{b_1} - V_{be_1} = V_{b_2} - V_{be_2} = (i_{e_1} + i_{e_2})R_e$$

$$i_{b_2} = \frac{V_k - V_{b_2}}{R_A} - \frac{V_{b_2}}{R_B} \tag{9-25}$$

where $V_{be_1}$ and $V_{be_2}$ are the forward-biased base-to-emitter drops.

Using the supplemental transistor gain equation:

$$i_c = \beta i_b \tag{9-26}$$

the solution for $i_c$ is

$$i_c = \frac{\beta_1}{\beta_1 + 1} \{V_{b_1}[G_E + (\beta_2 + 1)(G_A + G_B)] - V_{be_1}G_e$$

$$+ (V_{be_2} - V_{be_1})(\beta_2 + 1)(G_A + G_B) - (\beta_2 + 1)V_kG_A\} \tag{9-27}$$

where the symbol $G$ is used for the reciprocal of resistance. If the terms involving base-to-emitter drops are neglected:

$$i_c \simeq \frac{\beta_1}{\beta_1 + 1} \{V_{b_1}[G_E + (\beta_2 + 1)(G_A + G_B)] - V_kG_A(\beta_2 + 1)\} \tag{9-28}$$

### Timing Multivibrator

A modified free-running multivibrator is shown in Figure 9-9. The period for the conventional free-running multivibrator would be determined by $R_1$ and $C$.

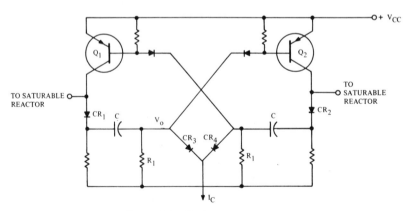

Figure 9-9    Timing multivibrator.

In this circuit, this period is altered by the difference amplifier current $I_C$. For example, when $Q_1$ is on and $Q_2$ is off, diode $CR_3$ is forward-biased and the current $I_C$ speeds up the discharge of the timing capacitor, thereby shortening the period. Diodes $CR_1$ and $CR_2$ decouple switching transients in the SR circuit from the timing circuit.

To write the timing equation, consider the instant that $Q_1$ comes on. The value of the voltage $V_o$ prior to switching is $V_{cc} - V_{cb} - V_d$ where $V_d$ and $V_{cb}$ are forward-biased diode- and collector-to-base potentials respectively. When $Q_1$ comes on, the collector potential of $Q_1$ is $V_{cc} - V_{ec}$, where $V_{ec}$ is the saturated emitter-to-collector drop. Thus, the value of $V_o$ immediately following turn-on of $Q_1$ is

$$V_o = 2V_{cc} - 2V_d - V_{ec} - V_{cb} \tag{9-29}$$

If the base of $Q_2$ were open, $Q_1$ would remain on indefinitely and the final value of $V_o$ would be

$$V_\infty = -I_C R_1 \tag{9-30}$$

Now, the solution for $V_o$ involves a single time constant $(R_1 C)$, so we may write

$$V_o(t) = V_\infty + (V_o - V_\infty)e^{-t/R_1 C} \tag{9-31}$$

The multivibrator switches at the time $\tau_p$ when $Q_2$ returns to the active region, that is

$$V_o(\tau_p) = V_{cc} - V_{cb} - V_d \tag{9-32}$$

Solving for $\tau_p$:

$$\tau_p = R_1 C \ln\left(2 + \frac{V_{cb} - V_{cc} - I_C R_1}{V_{cc} - V_d - V_{cb} + I_C R_1}\right) \tag{9-33}$$

Ideally, the multivibrator period should vary only with $I_C$ and that variation should be a linear function of $I_C$. In order to reduce the effects of supply voltage variations, it would be desirable to run the multivibrator from one of the output voltages. Obviously, some starting scheme would be necessary if this were to be done. Such a starting circuit is shown in Figure 9-10. When $E_{BB}$ is connected, $V_k$ is initially at zero. If the multivibrator is represented as some effective DC resistance $R_m$, then the starting voltage is

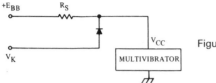

Figure 9-10   Starting circuit.

$$V_S = \frac{R_m}{R_s + R_m} E_{BB} \tag{9-34}$$

As the output voltage, $V_K$, rises, the starting diode becomes forward-biased, clamping the multivibrator voltage at a value

$$V_{cc} = V_K - V_d \tag{9-35}$$

The necessary condition for the steady-state multivibrator supply to be regulated is that $V_{cc}$ be greater than $V_S$ for the maximum value of $E_{BB}$.

With regard to the variation of $\tau_p$ with $I_C$, equation (9-33) obviously is not linear over a wide range of $I_C$. However, due to the open loop regulation provided by the saturable reactor, the variation in $\tau_p$ over the full range of input voltage is quite small. If the design center value of the control current $I_C$ is denoted as $I_o$ and the corresponding multivibrator period $\tau_p$ as $\tau_{po}$, the first two terms of a Taylor series expansion of equation (9-33) about the point $I_o$ are

$$\tau_p = \tau_{po}\left[1 - k_1\left(\frac{I_c - I_o}{I_o}\right)\right] \tag{9-36}$$

At $I_C = I_o$,

$$\tau_{po} = R_1 C \ln\left[\frac{2V_{cc} - 2V_d - V_{eb} - V_{cc} + I_o R_1}{V_{cc} - V_d - V_{cb} + I_o R_1}\right] \tag{9-37}$$

Similarly, equating the first derivatives evaluated at $I_C = I_o$:

$$k_1 = \frac{I_o R_1^2 C}{\tau_{po}} \times \frac{V_{cc} - V_d - V_{ec}}{(2V_{cc} - 2V_d - V_{eb} - V_{cc}) + I_o R_1(V_{cc} - V_d - V_{cb} + I_o R_1)} \tag{9-38}$$

## Closed Loop Operation

The regulated output voltage as given by equation (9-22) is

$$V_k = V_{km} \frac{\tau_s}{\tau_p} \qquad (9\text{-}22)$$

$\tau_p$ is related to $I_C$, which in turn is related to $V_k$. Then

$$I_C R_1 = K_2 V_{b_1} - K_3 V_k \qquad (9\text{-}39)$$

Substituting for $I_C$ in equation (9-35), we have

$$\tau_p = \tau_{po}(1 + K_1) - \frac{K_2 V_{b_1}}{I_o R_1} + \frac{K_3 V_k}{I_o R_1} \qquad (9\text{-}40)$$

The resulting schematic diagram for this multioutput DC/DC converter is shown in Figure 9-11. Various critical circuit waveforms are shown in Figures 9-12 through 9-15. These waveforms verify the design's compliance with (a) the specification requirements, (b) push-pull power amplifier design principles, and (c) maximum transistor ratings.

## FREQUENCY-CONTROLLED RESONANT DC/DC CONVERTER[1]

Figure 9-16 depicts the schematic diagram of a power supply using a Jensen two-core variable-frequency oscillator operating at 20 kHz in conjunction with a resonant output control circuit. The 60 Hz 117 V AC line power is converted to a high-frequency power input to the bridge-converter which supplies a saturable-core constant voltage transformer, diode rectifier, and filter elements, for powering the load.

The oscillator produces a variable-frequency signal proportional to the voltage error. The frequency signal is applied to a bridge-power converter through a current-drive buffer to produce a square-wave input to a saturable-core power transformer whose secondary output is rectified and filtered to supply the load power demand.

The input to the primary of the saturable-core transformer includes a series ferroresonant circuit consisting of an inductor and a capacitor. At resonance, the voltage across the circuit capacitance reaches maximum, and by design, the increasing voltage is sufficient to drive the transformer core into saturation.

---

[1]Work sponsored by NASA under Contract NAS 7-100.

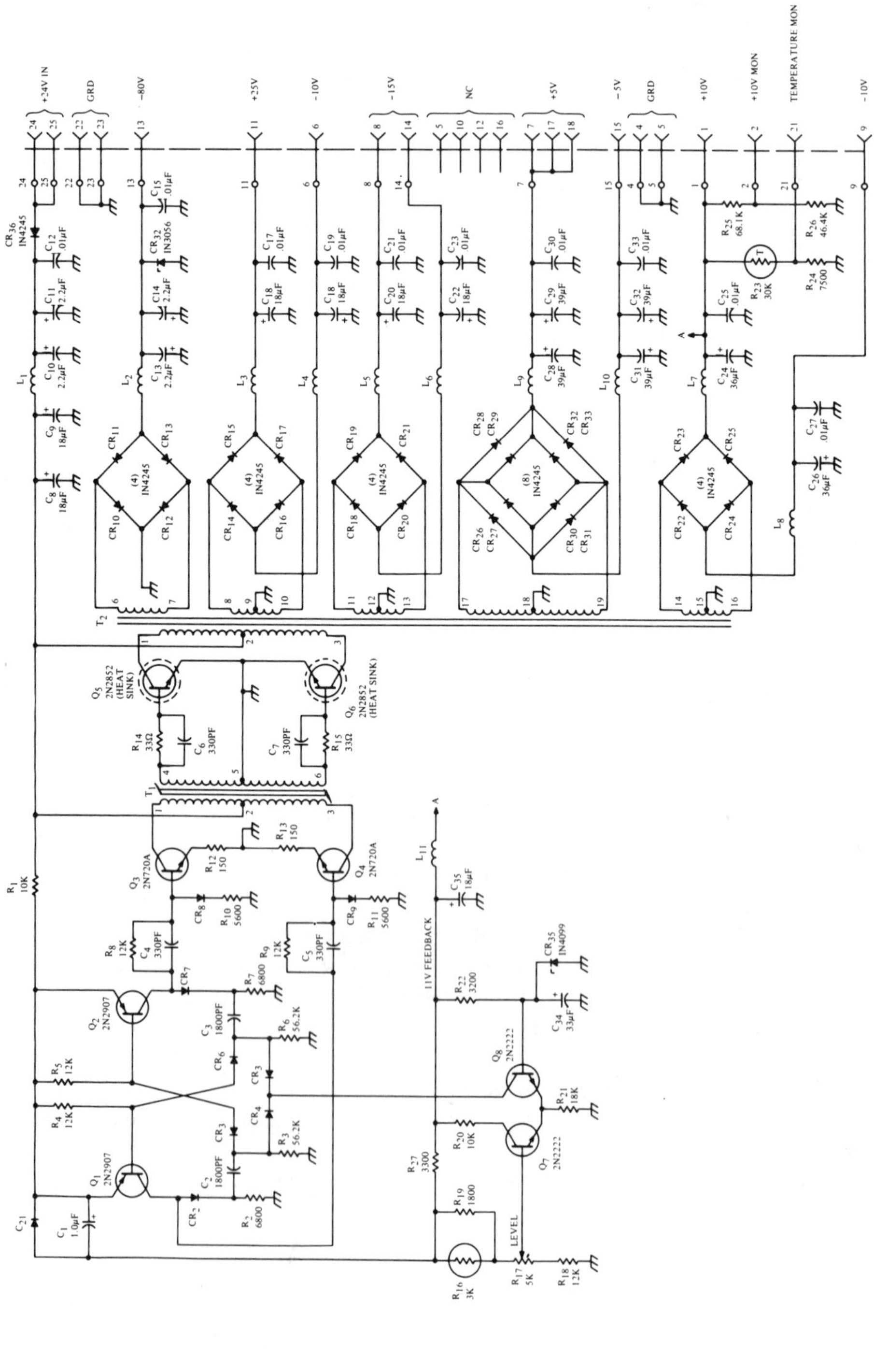

Figure 9-11 Multioutput DC/DC converter schematic diagram.

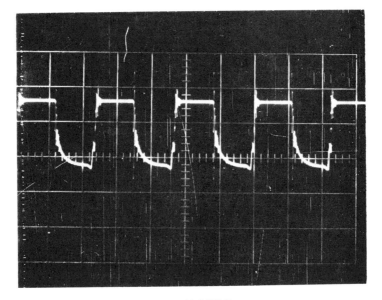

$E_{\text{in}} = 20.0$ VDC

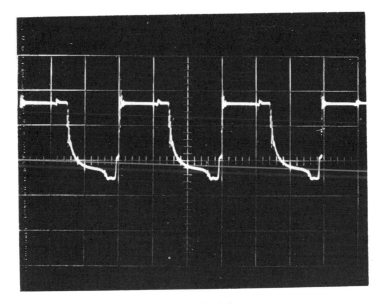

$E_{\text{in}} = 25.5$ VDC

Figure 9-12   $Q_3$ collector voltage waveforms. (Vertical = 5 V/cm; horizontal = 10 $\mu$sec/cm.)

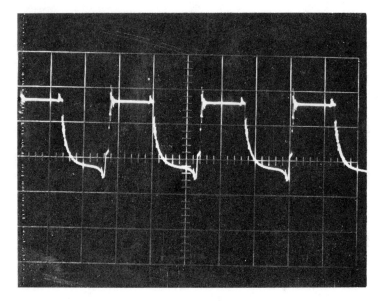

$E_{in} = 22.0$ VDC

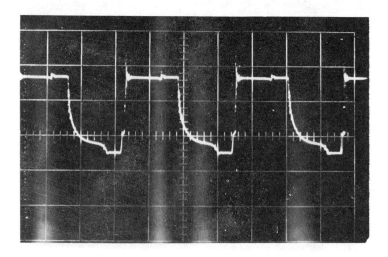

$E_{in} = 29.0$ VDC

Figure 9-12 (*Continued*)

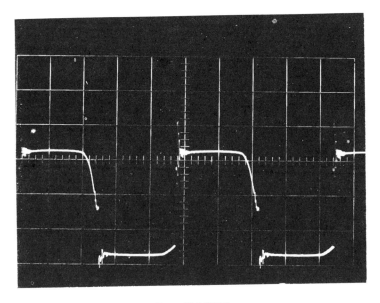

$E_{in} = 20.0$ VDC

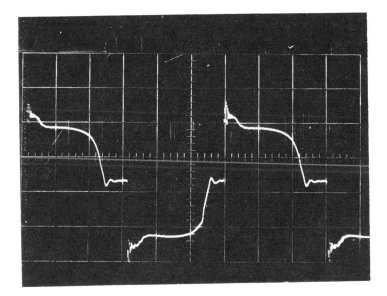

$E_{in} = 25.5$ VDC

Figure 9-13   $Q_5$ collector voltage waveforms (Vertical = 5 V/cm; horizontal = 10 $\mu$sec/cm.)

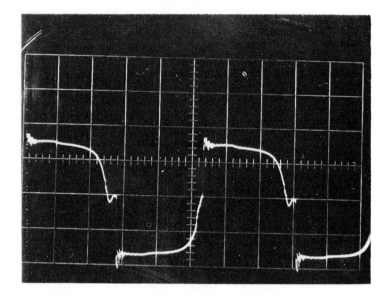

$E_{in} = 22.0$ VDC

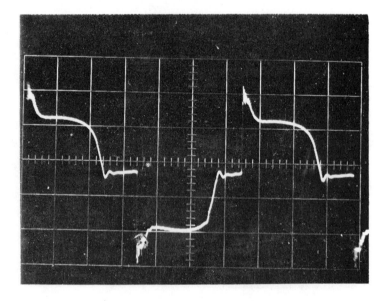

$E_{in} = 29.0$ VDC

Figure 9-13 (*Continued*)

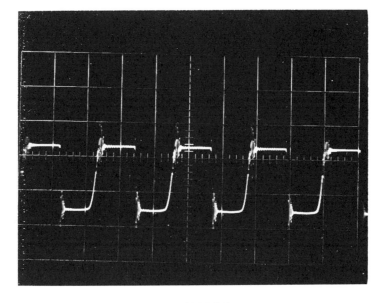

$E_{in} = 20.0$ VDC

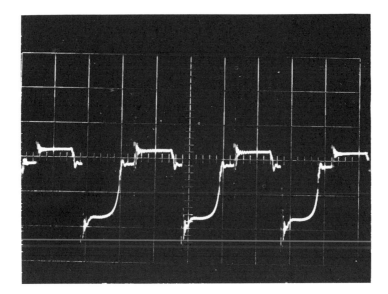

$E_{in} = 25.5$ VDC

Figure 9-14  $Q_7$ base voltage waveforms (Vertical = 2 V/cm; horizontal = 20 $\mu$sec/cm.)

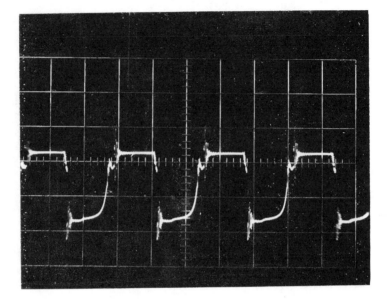

$E_{in} = 22.0$ VDC

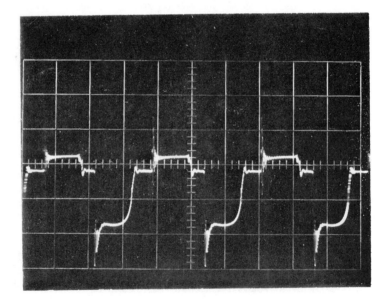

$E_{in} = 2.90$ VDC

Figure 9-14   (*Continued*)

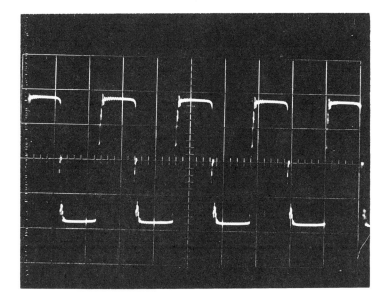

$E_{\text{in}} = 20.0$ VDC

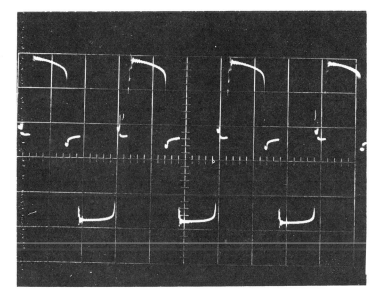

$E_{\text{in}} = 25.5$ VDC

Figure 9-15  $Q_7$ collector voltage waveforms. (Vertical = 10 V/cm; horizontal = 20 $\mu$sec/cm.)

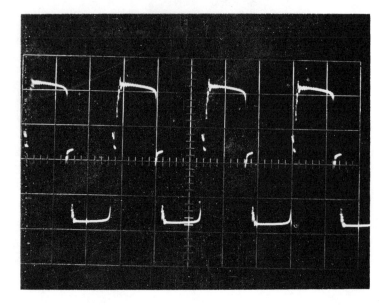

$E_{\text{in}} = 22.0$ VDC

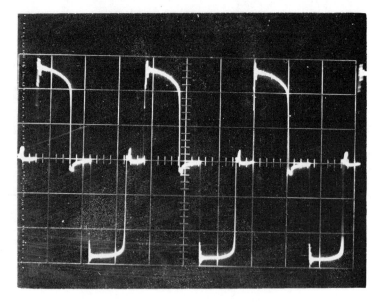

$E_{\text{in}} = 29.0$ VDC

Figure 9-15   (*Continued*)

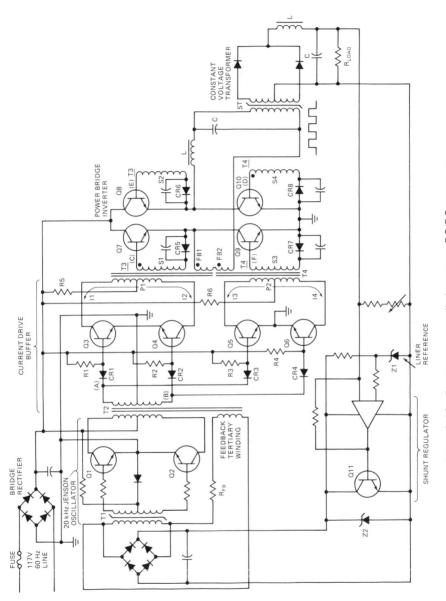

Figure 9-16  Frequency controlled resonant DC/DC converter.

When the core saturates, the transformer operates in the constant volt/second area of the characteristic curve resulting from plotting supply voltage versus time. In that region, the voltage appearing across the transformer secondary is hard-limited to a value which is selected to meet the design requirements, since by definition there can be no further increase in flux lines to support a higher voltage.

In the saturation mode, the transformer voltage varies proportionally to its drive frequency, regulating the circuit output. The error signal obtained by monitoring the load voltage is fed to one input of an op amp. The other input is a reference Zener diode. Any difference between these inputs is amplified and applied to a shunt regulator to vary the resistance across a bridge rectifier. Changes in the bridge impedance at the input to transformer $T_1$ adjust the oscillator frequency to compensate for changes in the load voltage. The ferroresonant circuit is inherently short circuit proof.

Using Figure 9-16, a detailed discussion of circuit characteristics, operation, and tradeoffs is now presented.

Load $R$ is powered by the rectified and filtered output of constant-voltage, saturable-core transformer $ST$. Power for the transformer is obtained from the switching inverter designated "Powerbridge" which includes transformers $T3$ and $T4$. As mentioned, the input power is chopped to convert the input frequency of the power supply system to a much higher frequency, on the order of 20 kHz. This provides a reduction in the size of the inductive components with a corresponding weight reduction, along with noise-free operation of such components at a frequency above the audible range. Conversion to the higher frequency is accomplished by the Jensen oscillator circuit which operates at a nominal frequency of 20 kHz.

Proper driving of the constant voltage output transformer $ST$ requires the bridge converter. However, it is not possible to drive the bridge converter directly from the Jensen oscillator; thus, a current drive buffer is used to interface the oscillator and the inverter.

Operation of the current-drive buffer is as follows. Transistors $Q3$, $Q4$, $Q5$, and $Q6$ are turned on and driven into saturation by biases applied through resistors $R1$, $R2$, $R3$ and $R4$, causing currents $I1$, $I2$, $I3$ and $I4$ to flow from the power source through current limiting resistors $R5$ and $R6$, through the transistors to the ground side of the power source. However, currents $I1$, $I2$, and currents $I3$ and $I4$, each flow in opposite directions in primaries $P1$ and $P2$ of transformers $T3$ and $T4$. Thus, the flux produced by the opposite current flowing in each primary winding $P1$ and $P2$ cancel out. Drive for transformers $T3$ and $T4$ produces a switching operation of power transistors $Q7$, $Q8$, $Q9$ and $Q10$, but only when one of the pairs of opposing currents $I1$ and $I3$ or $I2$ and $I4$ is turned off. Turn-off of one of the pairs of opposing currents is accomplished by the alternating output on the secondary of transformer $T2$ in conjunction with rectifiers $CR1$ and $CR2$ as follows.

When current flow in the secondary of transformer $T2$ is such that a negative potential appears at point $A$, rectifier $CR1$ will become forward-biased, and this will short out the bias voltage on the base of transistor $Q3$ provided by resistor $R1$. The shorting path is through the diode and the upper half of the secondary winding to ground. This will cut off current flow in transistor $Q3$. At the very same instant, point $B$ will be at a positive potential, while rectifier $CR2$ will be back-biased, allowing current flow in transistor $Q4$ because of the bias applied to its base through resistor $R2$. On the next half-cycle, points $A$ and $B$ will change polarity, and current flow will begin in transistor $Q3$ and transistor $Q4$ will turn off. Similar operation occurs with respect to transistors $Q5$ and $Q6$ and transformer $T4$ for switching the bridge converter.

The dual transformer arrangement provides some unique features and advantages in this current drive buffer. One advantage can be seen by referring to the schematic representation of the conventional single-transformer drive arrangement for a converter which is shown in Figure 9-17. Transistors $Q_a$, $Q_b$, $Q_c$ and $Q_d$ are the converter transistors, and the current path through one pair of transistors on one half-cycle is indicated by the arrows. If $Q_a$ has a low $V_{BE}$ of 0.75, and $Q_d$ has a higher $V_{BE}$ of 0.95, the current through $Q_d$ will be limited by the clamping action of the base-emitter junction of transistor $Q_a$, despite $Q_d$'s much higher $V_{BE}$. Thus, it will not be possible to drive transistor $Q_d$ into hard saturation in that event. This problem normally can only be overcome by very careful matching of the $V_{BE}$'s of the power transistors.

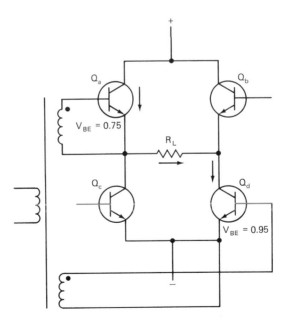

Figure 9-17  Bridge converter section.

Another advantage, as mentioned in Chapter 2, is that the transistors used in the bridge converter can have a much lower high-voltage rating, which reduces costs. In push-pull converter circuits driven by a single transformer, the voltage appearing across the junctions of each of the output transistors is twice the supply voltage, because when current flows in the active half of the transformer winding, a corresponding voltage is induced in the other half of the winding. This doubled voltage is applied to the junction of the cut-off transistor, and hence higher voltage transistors must be used.

There is a further advantage which requires understanding of the operation of the bridge converter circuit. Transformers $T3$ and $T4$ provide the high-frequency switching of converter transistors $Q7$, $Q8$, $Q9$ and $Q10$. Each transformer has a primary winding $P1$, $P2$ and two pairs of secondary windings $S1$, $S2$, and $S3$, $S4$. In addition, each transformer has a regenerative, or feedback, winding $FB1$, $FB2$. On one half-cycle, points $C$ and $D$ of secondaries $S1$ and $S3$ will be at a positive potential, while points $E$ and $F$ will be at a negative potential. This will turn on transistors $Q7$ and $Q10$, and turn off transistors $Q8$ and $Q9$. Current from the bridge rectifier will then flow through transistor $Q7$, windings $FB1$ and $FB2$ (which are series connected), the primary of transformer $ST$, inductor $L$, and transistor $Q10$ to ground. On the next half-cycle, current will flow through transistor $Q8$, inductor $L$, the primary of transformer $ST$, winding $FB2$ and $FB1$, and transistor $Q9$ to ground.

If, for any reason, both transistors $Q7$ and $Q9$ turn on for any period of time, then the full supply voltage from the line-bridge rectifier would be shorted. Since the circuit is regenerative and there is no current limiting provision, considerable damage could result. However, there is little possibility that this could happen momentarily during operation without turn-on of both transistors, because transistor turn-on is three times as fast as turn-off time.

Figure 9-18 shows the turn on/turn off relationship of one transistor pair, plotted against the square-wave drive input. Assuming that transistor $Q3$ is turning off, it will take three units of time for collector current to fall to zero, and only one unit of time for collector current to rise to maximum in transistor $Q4$. In a conventional arrangement, this would hold both output transistors on at the same time, which could be disastrous. However, it should be remembered that when currents $I1$ and $I2$ flow at the same time in the primary of transformer $T3$ that the fluxes set up by each counterflowing current cancel, and there is no output from transformer $T3$. This provides a notch or dead band in the drive to the output transistor which ensures that one set of power transistors will turn off fully before the other set of transistors can turn on.

Some other features of this power supply design are of interest. For example, the use of $FB1$ and $FB2$ windings on transformers $T3$ and $T4$ assists in providing proportional drive to match load demand. When secondary $S1$ begins to drive transistor $Q7$, the current flow in $FB1$ acts regeneratively to induce a greater

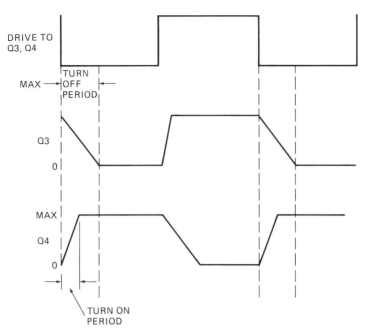

Figure 9-18   Bridge converter section transistor waveforms.

drive in secondary $S1$, quickly driving transistor $Q7$ to saturation. When additional demand is made by the load, increased current flow occurs in the primary of the constant voltage transformer $ST$. Since this current also flows through winding $FB1$, there is an action which increases current flow through transistor $Q7$ or $Q10$, (subject to the limiting effect of resistors $R5$ and $R6$), which is proportional to load demand.

If the shunt regulator, which controls the frequency of the Jensen oscillator fails, there can be no runaway condition because Zener diode $Z2$ clamps the bridge-resistance effectively across the secondary of transformer $T1$. Since this limits maximum voltage, the shunt regulator only controls the DC level below the maximum limited by bypass conduction in Zener $Z2$.

Diodes $CR5$ to $CR8$, in conjunction with the capacitors speed up the turn-off time of output transistors $Q7$, $Q8$, $Q9$, and $Q10$.

## SINGLE-PHASE INVERTER DESIGN

An interesting and informative miniaturized power supply design is now presented. This design is somewhat unique in that the usual, highly efficient push-pull inverter (oscillator) was not used. Instead, an inefficient Wien bridge oscillator was designed, primarily because it produces a sine-wave output without switching transistor operation. Thus, generated EMI is at a minimum.

Many design problems were encountered and had to be overcome. They are presented in depth so that the designer may gain some insight into the causes for these problems and the resulting tradeoffs required to solve them.

Computer-Aided Design (CAD) was used jointly with breadboard tests to compare the inverter's performance with the specification requirements and to verify (or dispute) its design adequacy.

The significant design parameters are listed below.

| | |
|---|---|
| Input voltage: | 20.0 VDC to 30.0 VDC |
| Output load: | 8 W, 10 VAR lagging |
| Output voltage: | 111.5 V rms to 118.5 V rms |
| Frequency: | 400 ± 4 Hz |
| Harmonic distortion: | 5% maximum |
| Temperature extremes: | −30°F to +160°F |
| Weight: | 1.30 lb maximum potted |
| Maximum power dissipation: | 32 W at maximum load and maximum $E_{in}$ |

A block diagram of the inverter is shown in Figure 9-19. From this diagram it is seen that the inverter consists of the following major circuits:

- Wien bridge oscillator with Zener diode regulation
- Pre-amplifier

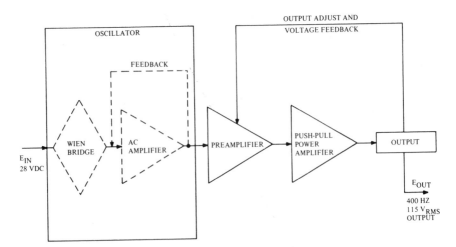

Figure 9-19 Inverter block diagram.

- Push-pull Class AB power amplifier
- Interconnecting feedback circuits

The resultant schematic diagram is shown in Figure 9-20.

### Circuit Description

The operation of the inverter is as follows.

From a 20.0 to 30.0 VDC input, the oscillator generates an AC signal which is regulated and stabilized to compensate for supply voltage variations. The signal is coupled to the preamplifier where the power is raised to a level compatible with the power amplifier input. The push-pull power amplifier smoothes the AC signal to a sine wave and raises the power to the output level. The output circuit provides a 115 Vrms output voltage. Feedback paths from the two amplifiers assure output voltage stability and a constant output voltage level independent of load.

### Oscillator

The Wien bridge (R-C) oscillator provides the inverter with a highly regulated and stabilized AC signal. By means of a frequency-selective Wien bridge, transistors $Q_1$ and $Q_2$, and a negative feedback network, the oscillator generates a 400 Hz signal. Resistors $R_2$, $R_3$, $R_6$, $R_7$, and capacitors $C_2$ and $C_3$ are the Wien bridge components which determine the frequency of oscillation. Resistors $R_8$ and $R_9$ provide the negative feedback which reduces harmonic distortion in the Wien bridge output signal.

The amplifier open loop gain is determined by transistors $Q_1$ and $Q_2$ and exceeds 3000; the closed loop gain of approximately three is determined by resistors $R_8$ and $R_9$, which act as voltage dividers. Filtering to eliminate 50 V spikes in the 28 VDC supply voltage occurs at resistor $R_1$ and capacitor $C_4$. Supply voltage is limited to approximately 11.2 V by Zener diodes $CR_4$ and $CR_5$. Diodes $CR_2$ and $CR_3$ control the amplitude of the 400 Hz AC signal generated by the Wien bridge. $R_7$ is selected during test to set the frequency at 400 ($\pm$4.0) Hz. Distortion in the AC sine wave at the oscillator output is less than 2%.

Modulation in the oscillator output, which can be caused by periodic oscillations or transients in the supply voltage, is minimized by capacitors $C_4$ and $C_5$. Due to the stabilizing effect of the capacitors, a 3.4 V peak-to-peak modulation in the DC supply of anywhere from 30 Hz to 15 kHz will result in modulating the oscillator output envelope by less than 0.25%. Even less output modulation occurs when the DC supply contains 1.5 kHz to 150 kHz modulations. The oscillator output envelope modulation in this frequency range is less than 0.1%.

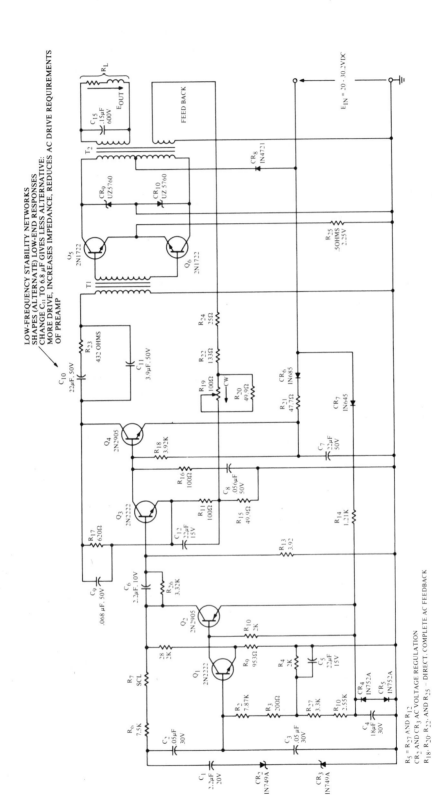

Figure 9-20 Inverter schematic diagram.

The oscillator output contains a very low frequency shift at extreme temperature conditions due to temperature-sensitive resistor $R_3$. This resistor compensates for high or low temperatures as follows:

| Temperature | Frequency Shift |
|---|---|
| +71°C | 0.63% high |
| +25°C | 0 |
| −35°C | 0.84% low |

To overcome any tendency toward oscillation ceasing under cold temperature, the following precautions were taken; (1) overdrive the oscillator, and (2) temperature compensation of the 1N748A diodes (1N749A have a negative temperature coefficient which results in raising the output "clipping level" as the temperature decreases).

## Preamplifier

A closed loop voltage gain of approximately 13 in the preamplifier is determined by resistor $R_{15}$.

Transistors $Q_3$ and $Q_4$ provide the increase in power necessary to drive the push-pull amplifier. The preamplifier, power amplifier, and output circuit form a forward loop which returns to the preamplifier as negative voltage feedback. This negative feedback, comprised of $R_{19}$, $R_{20}$, and $R_{24}$, maintains a constant inverter output voltage with supply voltage variations.

A negative feedback factor of 10 to 15 (ratio of open loop to closed loop gain) reduces the output voltage variation to less than 5% as the load is varied from maximum to minimum. The feedback factor reduces distortion to less than 5% maximum.

The negative feedback circuit contains potentiometer $R_{19}$, which can be used to set the output voltage by adjusting the power amplifier gain. This output voltage adjustment may be necessary in any inverter assembly to compensate for tolerance variations from unit to unit. The value of $R_{19}$ was selected such that the output voltage can rise no higher than 125 Vrms in the event of a potentiometer open circuit.

Temperature-sensitive resistor $R_{24}$ is also included in the feedback circuit discussed above. $R_{24}$ compensates for temperature variations in the preamplifier. By the methods discussed above, the negative feedback loop provides stabilization of the preamplifier and power amplifier output voltage amplitude over the entire temperature and voltage excursions.

Figure 9-21 represents the schematic of the inverter preamplifier.

## Power Amplifier

Figure 9-22 shows the inverter's Class AB power amplifier.

In Class AB operation only a small bias current flows. The amount of bias is adjusted so that the input impedance of the transistor is, under steady-state

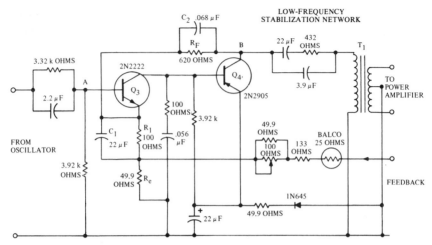

Figure 9-21  Preamplifier.

conditions, approximately equal to the reflected source impedance. The crossover distortion is decreased as the reflected source impedance exceeds the transistor input impedance.

In a push-pull arrangement the distortion is less due to the nonlinearity of each transistor. The push-pull arrangement cancels out in the transformer not only the DC component of the two collector currents but also all even-order harmonics.

Class AB operation minimizes crossover distortion by operating each transistor with a finite steady-state collector current.

In Figure 9-22 the emitter resistor is small (0.5 ohm) and not bypassed with a capacitor in order to compensate somewhat for the nonlinearities of the input resistance and the current transfer characteristic.

The power amplifier, the output circuit, and the preamplifier provide a total

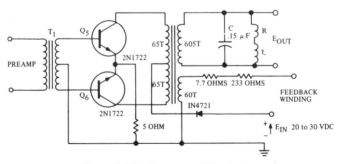

Figure 9-22  Power amplifier design.

voltage gain of approximately 600. The AC signal level is amplified by power transistors $Q_5$ and $Q_6$. The voltage feedback loop discussed previously causes a reduction in output to a value of between 67 and 100 ohms (due to a negative feedback factor of 2); the open loop output impedance varies between 800 and 1200 ohms.

Loads producing conditions other than unity power factor are partially compensated by tuning capacitor $C_{11}$. The capacitor value is a compromise to correct the power factor for all load conditions from approximately 0.625 lagging to 0.8 lagging.

During breadboard testing at the temperature extremes of $-30°F$ and $+160°F$ (for heaviest load and lowest power supply voltage), considerable distortion of the AC output voltage waveform was noticed. Based on these tests and analysis of the entire inverter circuitry the following was established:

(1) A minimum beta of 65 for the 2N1722 is adequate for all loads and voltages specified; (2) biasing the oscillator with a temperature-sensitive network provides the proper operation for the oscillator, namely, increasing output voltage at low temperature; and (3) increasing $C_{11}$ reduces the output voltage requirements of $Q_4$. (However, increasing $C_{11}$ might lead to instability, that is, low-frequency motorboating, if the inverter were operated in an open-circuit fashion, namely, with a capacitive load, leading power factor. Under the load conditions specified, the worst case is 3 W at 3.7 VA reactive.)

### Summary of Breadboard Tests and Solution of Design Problems

Breadboard tests indicated that at low temperature, low voltage, and full load, the output voltage exceeds its specification limit. The compound causes are listed below:

- Oscillator output voltage varies as the input voltage is varied from 20 VDC to 30 VDC.
- The oscillator load increases as the temperature is increased.
- The AC coupling network from the preamplifier to the inter-stage transformer requires a greater output voltage from the preamplifier.

DC-coupling the oscillator to the preamplifier and stabilizing the preamplifier bias by stabilizing the oscillator bias would provide a greater output from the oscillator and would prevent the oscillator voltage from varying as a function of supply voltage.

Since the 1N749A diodes have a negative temperature coefficient, the oscillator output increases as the temperature is decreased.

By stabilizing the oscillator bias, the preamplifier bias remains sufficiently constant to enable the inverter to operate from 20 to 30 VDC input with less than a 1 Vrms change in the output.

Altering the coupling network, that is, increasing the 2.2 $\mu$F ($C_{11}$) capacitor to 6.8 $\mu$F requires less drive from the preamplifier and thereby further increases the safety margin.

$C_{11}$ was compromised to a value of 3.9 $\mu$F. This eased the output requirements of the preamplifier by reducing the impedance by a factor of 2 and increasing the open loop gain by 15%.

Eliminating $R_3$ provides greater frequency stability (less change as a function of temperature) but causes a greater output voltage variation as a function of temperature.

Increasing the Balco resistors from 200 ohms to 400 ohms in the oscillator circuit provided better frequency stability and corrected the voltage variation under temperature.

Tables 9-1 and 9-2 show the effect of varying $R_7$ with the 200 ohm Balco

**Table 9-1   Effect of Variation of 200-Ohm Balco ($R_7$ = 445 ohms) on Inverter Performance.**

| Temp. | Supply (VDC) | Load$^a$ | Oscillator Base, $Q_3$ (DC) | (AC) | Preamp. DC | Output (Vrms) | Frequency (Hz) |
|-------|------|------|-------|------|-------|------|-------|
| 165°F | 20 | 3 | 3.289 | 2.33 | 14.80 | 113 | 403.5 |
|       | 20 | 6 | 3.299 | 2.33 | 14.79 | 113 | 403.5 |
|       | 25 | 3 | 3.311 | 2.33 | 14.89 | 113 | 403.5 |
|       | 25 | 6 | 3.319 | 2.33 | 14.91 | 113 | 403.5 |
|       | 30 | 3 | 3.339 | 2.33 | 15.09 | 113 | 403.5 |
|       | 30 | 6 | 3.340 | 2.33 | 15.09 | 113 | 403.5 |
| −30°F | 20 | 3 | 3.357 | 2.47 | 13.99 | 117 | 399.5 |
|       | 20 | 6 | 3.359 | 2.47 | 13.99 | 117 | 399.5 |
|       | 25 | 3 | 3.381 | 2.48 | 14.19 | 118 | 399.5 |
|       | 25 | 6 | 3.392 | 2.49 | 14.21 | 118 | 399.5 |
|       | 30 | 3 | 3.407 | 2.50 | 14.33 | 118 | 399.5 |
|       | 30 | 6 | 3.424 | 2.50 | 14.36 | 119 | 399.5 |
| 75°F  | 20 | 3 | 3.302 | 2.42 | 14.39 | 116 | 401.5 |
|       | 20 | 6 | 3.325 | 2.42 | 14.30 | 116 | 401.5 |
|       | 25 | 3 | 3.342 | 2.42 | 14.59 | 118 | 401.5 |
|       | 25 | 6 | 3.359 | 2.42 | 14.61 | 118 | 401.5 |
|       | 30 | 3 | 3.365 | 2.43 | 14.72 | 118 | 401.5 |
|       | 30 | 6 | 3.381 | 2.43 | 14.75 | 119 | 401.5 |

$^a$ Load is defined as

| Load No. | Load | | |
|------|------|------|------|
| 1 | 1730 ohms, .86 H | 4 | 712 ohms, .41 H |
| 2 | 1295 ohms, .64 H | 5 | 1058 ohms, .32 H |
| 3 | 1100 ohms, .44 H | 6 | 647 ohms, .32 H |

Table 9-2   Effect of Variations of 200-Ohm Balco ($R_7 = 181$ Ohms) on Inverter Performance.

| Temp. | Supply (VDC) | Load | Oscillator Base, $Q_3$ (DC) | (AC) | Preamp. | Output | Dist. | Frequency |
|---|---|---|---|---|---|---|---|---|
| 80°F | 25 | 6 | 6.159 | 2.45 | 14.79 | 110 | 2.3% | 402 |
|  | 20 | 6 | 6.071 | 2.47 | 14.59 | 110 | 2.3% | 402 |
| −30°F | 20 | 3 | 6.069 | 2.55 | 13.89 | 110 | less | 400 |
|  | 20 | 6 | 6.069 | 2.58 | 13.84 | 112 | than | 400 |
|  | 25 | 3 | 6.110 | 2.58 | 13.99 | 112 | 2% | 400 |
|  | 25 | 6 | 6.153 | 2.58 | 14.09 | 111 |  | 400 |
|  | 30 | 3 | 6.156 | 2.58 | 14.19 | 111 |  | 400 |
|  | 30 | 6 | 6.193 | 2.58 | 14.24 | 112 |  | 400 |
| +165°F | 20 | 3 | 6.031 | 2.35 | 14.79 | 107 | less | 404 |
|  | 20 | 6 | 6.032 | 2.35 | 14.79 | 107 | than | 404 |
|  | 25 | 3 | 6.049 | 2.35 | 14.99 | 107 | 2.5% | 404 |
|  | 25 | 6 | 6.080 | 2.35 | 14.99 | 107 |  | 404 |
|  | 30 | 3 | 6.075 | 2.35 | 15.09 | 107 |  | 404 |
|  | 30 | 6 | 6.109 | 2.35 | 15.15 | 107 |  | 404 |

($R_3$) upon the inverter performance requirements. Figure 9-23 shows a graph of the variation of inverter performance with temperature and $R_7$. The variation of frequency with temperature, $R_3$, and $R_7$ is shown in the graph of Figure 9-24. The data of Table 9-1 shows the oscillator voltage, output voltage, and frequency of oscillation variation with temperature, input voltage, and $R_{FB}$ (where $R_{FB} = R_{24} + (R_{22}) \times [(R_{19}) (R_{20})]/(R_{19} + R_{20} + R_{15})$). All these tests were performed to reach an optimum bias and temperature stability of the Wien bridge oscillator under all conditions of input voltage, loads, and temperature extremes. Table 9-3 lists the inverters' performance values with the final circuit configuration (shown in Figure 9-20).

Figures 9-25 and 9-26 show (1) the distortion of the AC output voltage at low temperature due to the low DC gain of power transistors $Q_5$ and $Q_6$ and (2) the effect on the AC output voltage as the DC gain is increased.

It is seen that a minimum DC gain ($H_{FE}$) for $Q_5$ and $Q_6$ must be specified to keep the distortion of the output within the specification limits. Thus a minimum beta of 65 was specified together with a matching requirement for both transistors. The transistor gains must be matched to 0.1 (10%).

Tests were run by changing the 200 ohm Balco resistor to a 550 ohm Balco resistor and changing the 7.87 kohm resistor to 7.5 kohm; that is, keeping the total of $R_2$ and $R_3$ constant. With these changes the specification requirements were met.

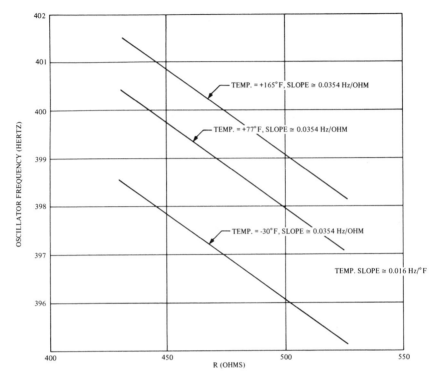

Figure 9-23   Variation of frequency with temperature and $R_5$ and $R_7$.

Test data indicate that the 550 ohm Balco overcompensated the circuit, that is, the output voltage now decreases at low temperature and increases at high temperature. Prior to this modification, the output voltage increased at low temperature and decreased at high temperature.

By proper selection (matching) of the AC regulating diodes (1N749A) and increasing the Balco resistor from 200 ohm to 300 ohms, the inverter performance is enhanced.

The excessive distortion at low temperature (20 V and heaviest load) was still another problem. The distortion under these conditions was between 8% and 10%. The oscilloscope photographs shown in Figure 9-25 and 9-26 show the cause of the distortion—low gain ($H_{FE}$) of transistors $Q_5$ and $Q_6$ under low line, low temperature, and heavy load.

The subsequent photographs (Figures 9-27 through 9-34) show the oscilloscope waveforms of various portions of the inverter (in its final configuration) which were taken during the breadboard testing to verify the design adequacy and to ensure proper component derating. The observance of the waveforms of

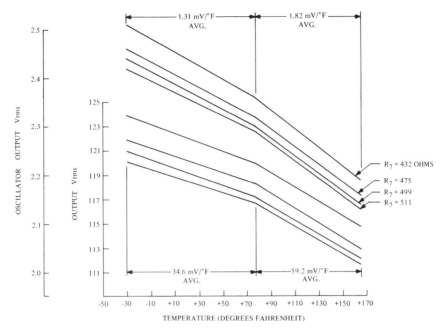

Figure 9-24   Variation of inverter performance with temperature and $R_7$. (*Note:* Data taken only at temperatures of $-30°F$, $+77°F$, and $+165°F$, and *assumed* to be linear throughout the range.)

various portions of a circuit by the designer are an inherent part of the design procedure.

## A FLYBACK HIGH VOLTAGE REGULATED CONVERTER[2]

Though the flyback method is not new, it is not generally known that its theoretical maximum efficiency in regulated supplies surpasses that of both the pulse-modulated square-wave generator and the class C oscillator.

The flyback converter in Figure 9-35 incorporates a number of power saving steps for a 20 V unregulated input, 5 kHz drive frequency and 250 $\mu$A, 2.4 kV load. The input voltage, however, can range from 10 to 50 V. Performance of the optimized converter is given in Figure 9-36 and the individual power losses are listed in Table 9-4.

Regulation of the converter is excellent: The 2.4 kV output voltage changes

---

[2]"Slash High-Voltage Power Supply Drain" by Joseph C. Geck, Jr. Reprinted with permission from *Electronic Design*, 19, September 13, 1974, copyright Hayden Publishing Co., Inc. (1974).

## Table 9-3   Summary of Single Phase Inverter's Electrical Performance

| $E_{osc}$ (Vrms) | $E_{out}$ (Vrms) Load 3 | $E_{out}$ (Vrms) Load 6 | $E_{supply}$ (VDC) | $R_{FB}$ (ohms) | Temperature | Frequency |
|---|---|---|---|---|---|---|
| 2.438 | 107 | 108.5 | 20 | 909 | B. Plate: −17°F | 404.9 |
| 2.445 | 107.5 | 109 | 25 | 909 | Oven: −27°F | 404.9 |
| 2.447 | 108 | 109 | 30 | 909 | | 404.9 |
| 2.572 | 114.2 | 114.8 | 30 | 409 | B. Plate: −17°F | 408.3 |
| 2.570 | 114 | 114.3 | 25 | 409 | Oven: −27°F | 408.3 |
| 2.565 | 113.7 | 113 | 20 | 409 | | 408.3 |
| 2.543 | 117.5 | 118.1 | 30 | 409 | B. Plate: +88°F | 409.6 |
| 2.540 | 117.5 | 118 | 25 | 409 | Oven: +98°F | 409.6 |
| 2.536 | 116.7 | 116.7 | 20 | 409 | | 409.6 |
| 2.479 | 114.4 | 114.1 | 30 | 909 | B. Plate: +88°F | 406.2 |
| 2.469 | 114 | 114.4 | 25 | 909 | Oven: +98°F | 406.2 |
| 2.464 | 113.8 | 114 | 20 | 909 | | 406.2 |

| $E_{osc}$ (Vrms) Load 3 | $E_{osc}$ (Vrms) Load 6 | $E_{out}$ (Vrms) Load 3 | $E_{out}$ (Vrms) Load 6 | $E_{supply}$ (VDC) | $R_{FB}$ (ohms) | Temperature | Preamp Out Load 3 | Preamp Out Load 6 | Frequency |
|---|---|---|---|---|---|---|---|---|---|
| 2.423 | 2.409 | 114.5 | 114.5 | 20 | 409 | B. Plate: +158°F | 1.04 | 1.26 | 409.8 |
| 2.429 | 2.418 | 115 | 115 | 25 | 409 | Oven: +170–175°F | 1.03 | 1.25 | 409.8 |
| 2.430 | 2.420 | 115 | 115 | 30 | 409 | | 1.03 | 1.25 | 409.8 |
| 2.359 | 2.346 | 111.5 | 111.5 | 20 | 909 | | 1.02 | 1.24 | 406.6 |
| 2.364 | 2.352 | 112 | 112 | 25 | 909 | | 1.02 | 1.23 | 406.6 |
| 2.368 | 2.354 | 112 | 112 | 30 | 909 | | 1.02 | 1.23 | 406.6 |

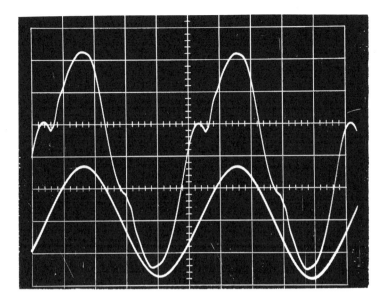

Full Load, $E_{in} = 20$ VDC, $-30°$F    (2) Lower: 2 V/cm (oscillator output)
Vertical:    Horizontal: 0.5 msec/cm
(1) Upper: 50 V/cm (AC output)

Figure 9-25 Photograph showing distortion of AC output. Voltage at low temperature due to low DC gain of transistors $Q_5$ and $Q_6$.

less than 10 V as the unregulated input varies from 10 to 50 V. And, for load changes from no load to 390 $\mu$A, the output varies less than 20 V.

Transformer design parameters of the optimized supply include an air gap, $\Delta \ell_e$, of 0.02 in. and a primary and secondary of 74 and 125 turns, respectively. The U-I core is an Indiana General F740-1, with two 0.005-in. spacers in each leg to set the $\Delta \ell_e$.

In the circuit of Figure 9-35, the $\mu$A776 acts as an error amplifier, the LM322 as a duty-cycle controller, and the combination of $Q_1$, $CR_1$ and $Q_2$ forms a driver circuit for the power transistor, $Q_3$. When a clock signal is applied, the signal is differentiated by the $C_1 - R_1$ combination, and the positive spike triggers the LM322 to the on state. The spike turns on $Q_3$ and energy begins to be stored in transformer $T_1$.

Once the LM322 turns on, an exponentially increasing voltage appears across $C_2$. The slope of the voltage is determined by the unregulated input voltage and by $R_2$ and $C_2$. When the exponential voltage exceeds the voltage at pin 7 of the LM322, $Q_3$ shuts off (via the drive circuit), and $C_2$ discharges and stays at ground potential until another clock pulse arrives.

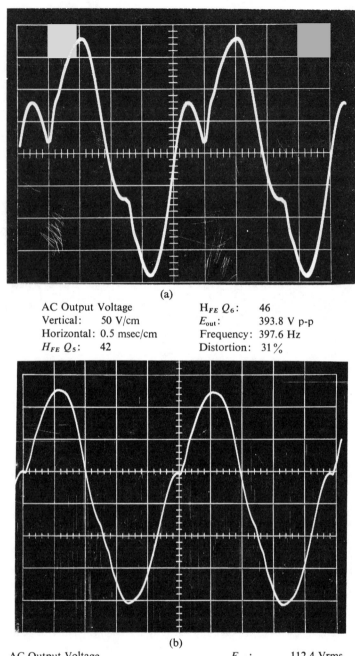

(a)

| AC Output Voltage | | $H_{FE}$ $Q_6$: | 46 |
|---|---|---|---|
| Vertical: | 50 V/cm | $E_{out}$: | 393.8 V p-p |
| Horizontal: | 0.5 msec/cm | Frequency: | 397.6 Hz |
| $H_{FE}$ $Q_5$: | 42 | Distortion: | 31% |

(b)

AC Output Voltage

Heavy load, $E_{in}$ = 20 VDC, −30°F.

| Vertical: | 50 V/cm | $E_{out}$: | 112.4 Vrms |
|---|---|---|---|
| Horizontal: | 0.5 msec/cm | | at $20V_{Ein}$ |
| $H_{FE}$ $Q_5$: | 76 | | 119.6 V |
| $H_{FE}$ $Q_6$: | 74.6 | | at 25 $V_{Ein}$ |

Figure 9-26   Effect of $H_{FE}Q_5$ and $H_{FE}Q_6$ on AC output waveform. Power amplifier limiting due to operating point drift with temperature and low $H_{FE}$; phase shift due to p-f correction capacitor.

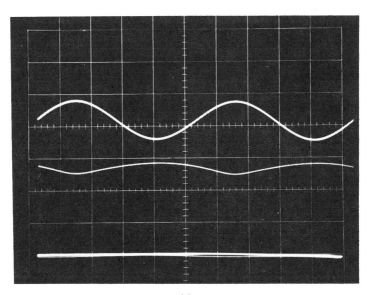

(a)

| | | Scope Scale: | |
|---|---|---|---|
| $Q_1$ Emitter-Collector DC Voltage: | 8.33 | | |
| $R_{10}$ DC Voltage: | 0.641 | $Q_1\ V_{CE}$ Upper: | 2 VDC/cm |
| Collector Current: | 0.34 mA | $V_{R_{10}}$ Lower: | 0.2 VDC/cm (across $2k$) |
| Ambient temperature | | Horizontal: | 0.5 msec/cm |

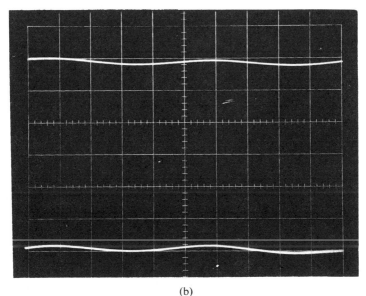

(b)

$Q_1$ Emitter-Base DC Voltage:  0.586
Scope scale:
Vertical:  100 mV/cm
Horizontal:  0.5 msec/cm
Ambient temperature

Figure 9-27  (a) $Q_1$ emitter-collector voltage; (b) $Q_1$ emitter-base voltage.

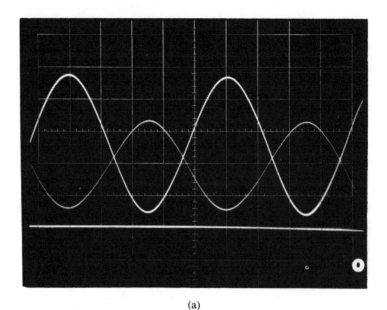

(a)

| $Q_2$ Emitter-Collector DC Voltage: 5.12 | | Scope Scale: | |
|---|---|---|---|
| Scope Scale: | | $Q_2$ $V_{CE}$ Upper: | 2 VDC/cm |
| $R_8$ DC Voltage: | 3.83 | $V_{R8}$ ($I_C$) Lower: | 2 VDC/cm (across 2k) |
| Collector Current: | 3.3 mA | $H$: | 0.5 msec/cm |
| Ambient temperature | | | |

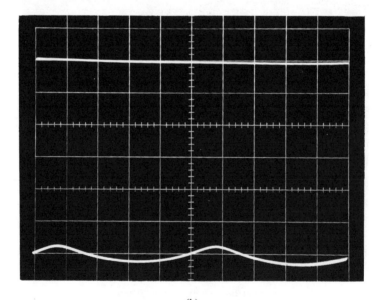

(b)

$Q_2$ Emitter-Base DC Voltage: 0.638
Scope Scale:
$Q_2$ $V_{BE}$:                      100 mVDC/cm
$H$:                              0.5 msec/cm
Ambient temperature

Figure 9-28   (a) $Q_2$ emitter-collector voltage; (b) $Q_2$ emitter-base voltage.

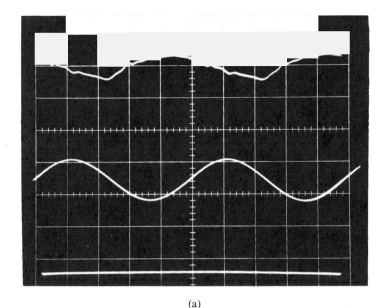

(a)

| $Q_3$ Collector-Emitter DC Voltage: | 15.05 | Scope Scale: | |
|---|---|---|---|
| $R_{18}$ DC Voltage: | 0.662 mA | $Q_3$ $V_{CE}$ Lower: | 5 VDC/cm |
| Collector Current: | 0.18 mA | $V_{R18}$ Upper: | 100 mVDC/cm |
| Ambient temperature | | $H$: | 0.5 msec/cm |

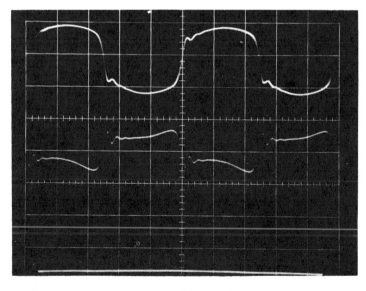

(b)

| $Q_4$ Collector-Emitter DC Voltage: | 13.93 | Scope Scale: | |
|---|---|---|---|
| $R_{17}$ DC Voltage: | 11.65 | $Q_4$ $V_{CE}$ Upper: | 2 VDC/cm |
| Ambient temperature | | $I_C$ Lower: | 5 mA DC/cm |
| | | $H$: | 0.5 msec/cm |

Figure 9-29    (a) $Q_3$ emitter-collector voltage and current; (b) $Q_4$ emitter-collector voltage and current.

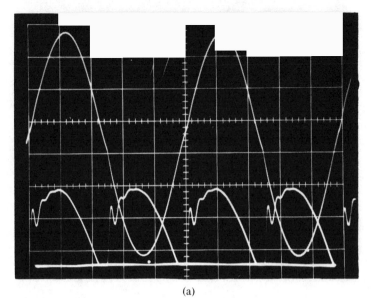

(a)

| $Q_5$ Collector-Emitter DC Voltage: | 19.3 | Scope Scale: | |
|---|---|---|---|
| $R_{25}$ DC Voltage: | 0.148 | $Q_5\ V_{CE}$ Upper: | 5 VDC/cm |
| Collector Current: | 0.48 A peak | $V_{R25}$ Lower: | 100 mVDC/cm |
| | | | (across 0.50 ohms) |
| | | $H$: | 5 msec/cm |
| | | | (Sync. on $V_{CE}$) |

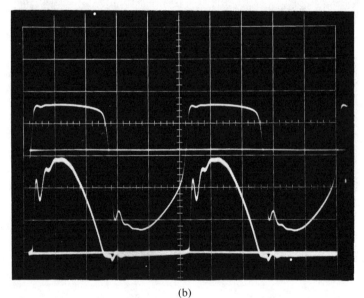

(b)

| $Q_5$ Emitter-Base DC Voltage: | 0.171 | Scope Scale: | |
|---|---|---|---|
| Scope Scale: | | $Q_5\ V_{BE}$: | 0.5 VDC/cm (upper) |
| $R_1$ ohm DC Voltage: | 0.002 | $H$: | 0.5 msec/cm |
| Base Current: | 5.8 mA peak | Lower: | 2 mA/cm |

Figure 9-30  (a) $Q_5$ emitter-collector voltage and current; (b) $Q_5$ emitter-base voltage and current.

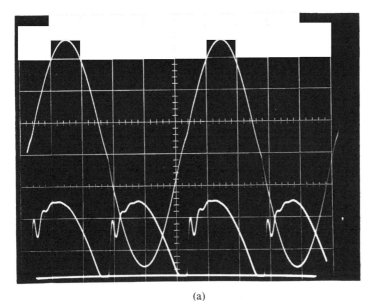

(a)

| | | Scope Scale: |
|---|---|---|
| $Q_6$ Collector-Emitter DC Voltage: | 19.3 | |
| $R_{25}$ DC Voltage: | 0.148 | $Q_6$ $V_{CE}$ Upper: 5 VDC/cm |
| Collector Current: | 0.48 A peak | $V_{R35}$: 100 mVDC/cm (across ½ ohm or 200 mADC/cm |
| | | $H$: 0.5 msec/cm |

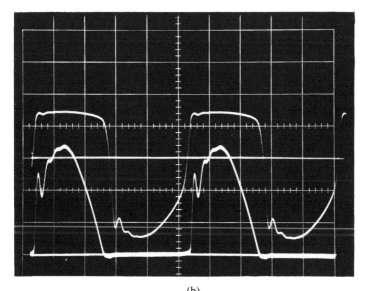

(b)

| | | Scope Scale: |
|---|---|---|
| $Q_6$ Emitter-Base DC Voltage: | 0.174 | $V_{BE}$ Upper: 0.5 VDC/cm |
| Scope Scale: | | $I_B$ Lower: 2 mADC/cm |
| $R_1$ ohm DC Voltage: | 0.002 | $H$: 0.5 msec/cm |
| Base Current: | 6.9 mA peak | |

Figure 9-31 (a) $Q_6$ emitter-collector voltage and current; (b) $Q_6$ emitter-base voltage and current.

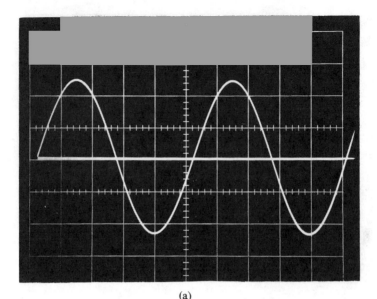

(a)
Anode-to-Cathode DC Voltage: 0
Scope Scale:
Vertical:                 1 VDC/cm
Horizontal:           0.5 msec/cm

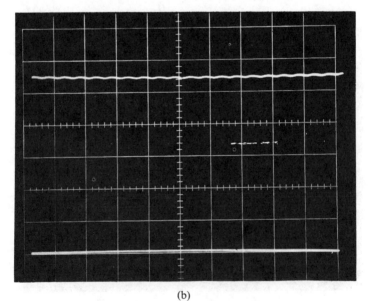

(b)
Anode-to-Cathode DC Voltage: 11.09
Scope Scale:
Vertical:                 2 VDC/cm
Horizontal:           5 msec/cm

Figure 9-32    (a) Diodes $CR_2$ and $CR_3$ voltage waveshape; (b) diodes $CR_4$ and $CR_5$ voltage waveshape.

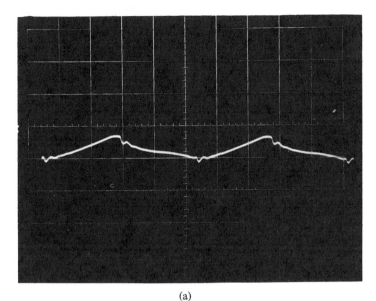

(a)

Scope Scale:
Vertical:    10 mVAC/cm
Horizontal: 0.5 msec/cm

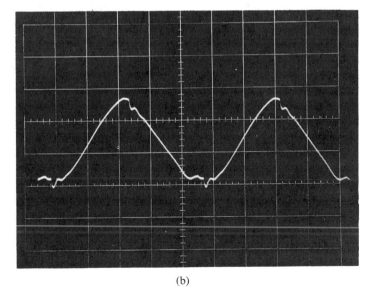

(b)

Scope Scale:
Vertical:    10 mVAC
Horizontal: 0.5 msec/cm

Figure 9-33  Diode $CR_6$ test results (ambient room temperature). (a) Diode $CR_6$ ripple for load 1; (b) diode $CR_6$ ripple for load 6.

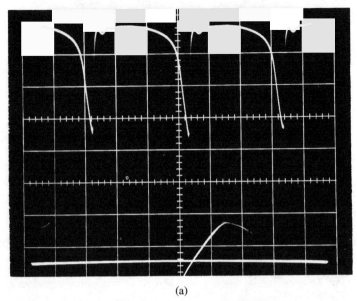

(a)

Scale Factor:                                         Scope Scale:
Anode-to-Cathode DC Voltage: 0.697        Vertical:       100 mVDC/cm
                                                 Horizontal: 0.5 msec/cm

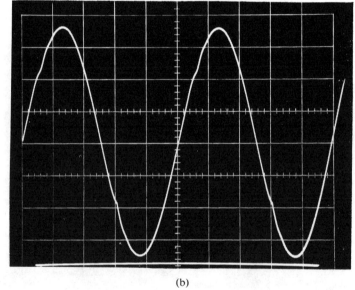

(b)

Scale Factor:
Anode-to-Cathode DC Voltage: 19.5
Scope Scale:
Vertical:                              5 VDC/cm
Horizontal:                       0.5 msec/cm

Figure 9-34   Diodes $CR_8$ to $CR_{10}$ test results. (a) Diode $CR_8$ voltage waveshape; (b) diode $CR_9$ or $CR_{10}$ voltage waveshape.

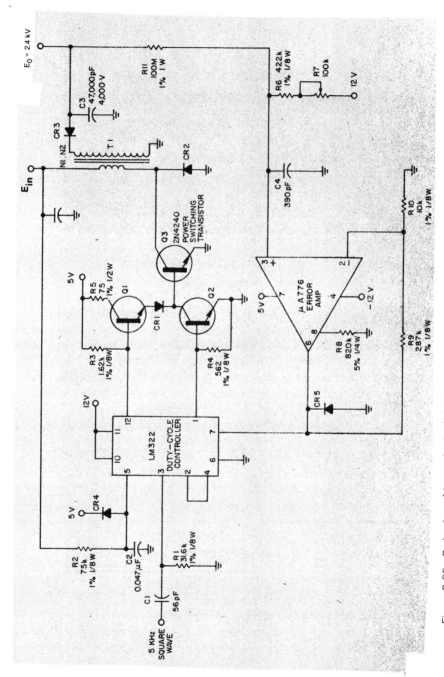

Figure 9-35  Reduction of losses in the flyback converter results in more efficiency. (*Courtesy Electronic Design*).

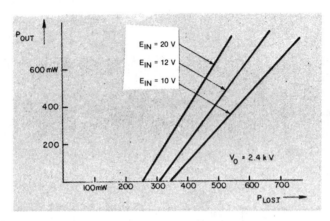

Figure 9-36    Variations of power dissipation and power loss for various input levels at $V_0 = 2.4$ kV. (*Courtesy Electronic Design*).

When $Q_3$ turns off, the energy stored in $T_1$ is pumped into the load circuit, which consists of $CR_3$, $C_3$ and the load resistor. Here, $C_3$ acts as a filter. As the collector voltage falls, $CR_2$ begins to conduct and prevent the collector from going negative. In this way, energy is returned to the unregulated power supply.

Resistors $R_{11}$, $R_6$ and $R_7$ sense the output voltage, divide it down and compare it with the $+12$ V reference. Note that the regulation of the supply is only as good as that of the reference. The $\mu$A776 amplifies the error and applies the amplifier signal to the LM322, which then controls the on time (pulse width) of $Q_3$.

The action of $R_2$ and $C_2$ is interesting: The unregulated input voltage is fed forward to control the slope of the voltage across $C_2$. Thus, the regulation is improved and the conditions for stability improved, since the loop gain depends less on the input voltage.

It is worthwhile to note that a number of possible sources of power loss were neglected in the optimization of this design. These include leakage inductance, switching and dielectric losses.

While leakage inductance affects the waveshape, tests show that efficiency does not depend on leakage; so it can be ignored in a loss analysis.

Switching losses in the power transistor, $Q_3$, are also neglected. Though these can be substantial, they appear to depend more on the transistor type than on the circuit parameters. Consequently the transistor should be selected for minimum switching loss—which implies a fairly expensive device.

Depending on the transformer, dielectric losses may or may not be negligible. With a given transformer construction, the dielectric loss theoretically depends only on the transformer output voltage and the frequencies. Tests of the flyback circuit show little dielectric loss, if any.

### Table 9-4 Measured Flyback Converter Performance (*Courtesy of Electronic Design*)

| | |
|---|---|
| No-load case | $E_0$ = 2.4 kV |
| | Power out = 0 |
| | Power in for $E_i$ = 10 V = 255 mW |
| | for V drive = 5 V = 70 mW |
| | Power to sense resistor = 58 mW |
| | Transistor drop = 10 mW |
| | Diode drop = 24 mW |
| | R loss in transformer = 19 mW |
| | Switching loss in transistor = 74 mW |
| | Core loss & dielectric loss = 70 mW |
| Full-load case | $E_0$ = 2.4 kV |
| | Power out = 750 mW |
| | Power In for $E_i$ = 10 V = 1.37 W |
| | for V drive = 5 V = 120 mW |
| | Power to sense resistor = 58 mW |
| | Transistor drop = 84 mW |
| | Diode drop = 24 mW |
| | R loss in transformer = 91 mW |
| | Switching loss in transistor = 165 mW |
| | Core loss & dielectric loss = 200 mW |

One surprising result is the apparent lack of dependence of the losses on the number of primary or secondary turns.

Figure 9-37 shows $P_{TL} - P_{DR}$ (this is the difference between the total power loss and the base drive loss) plotted againt primary turns for the supply of Figure 9-35 with a 2.5 kV, 315 $\mu$A load. The number of secondary turns is fixed at 1250, and all other parameters are held constant. Base-current drive is increased, however, to keep the power transistor in saturation.

In Figure 9-38, $P_{TL}$ is plotted against secondary turns, with the primary turns fixed at $N_P$ = 115. Again, all other parameters are kept constant. Note that the losses increase dramatically below about 1050 turns.

The large increase is explained as follows: Because the turns ratio controls the peak of the flyback-voltage spike at the transistor collector, the magnitude of the flyback pulse increases as the turns ratio is decreased—and as the control circuit keeps the output voltage constant. As the pulse increases, the transistor turns off with an ever increasing voltage gradient. Consequently switching losses rise rapidly.

Therefore the assumption–of constant power loss with secondary turns must

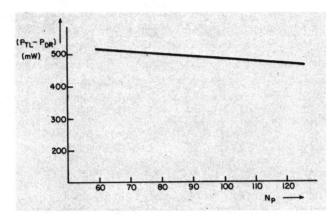

Figure 9-37 The total power loss (minus base-drive loss) is almost independent of primary turns. (*Courtesy Electronic Design*).

be made with care. (The advantage of this assumption is that the turns ratio can be chosen to allow selection of a less-expensive transistor of lower voltage. Lower voltage means faster turn-off time and greater efficiency. But the trade-off is a large number of secondary turns, leading to increased cost and size.)

Both the secondary stray capacitance and the frequency have about the same effect on power loss as shown in Figure 9-39. The losses at 6.6 kHz with $C_l = 32$ pF are about the same as those with $C_l = 11$ pF and $f = 20$ kHz. If size and cost are not constraints, higher efficiency can be attained by lowering $f$ or $C_l$—or both.

Power loss is reduced drastically when $E_0$ drops. With a voltage doubler, the

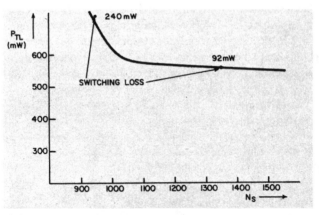

Figure 9-38 Total power loss is independent of the secondary above 1050 turns, where switching loss is low. (*Courtesy Electronic Design*).

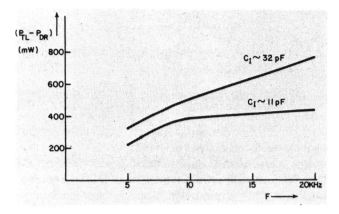

Figure 9-39  Both operating frequency and stray capacitance in the secondary affect power loss equally. (*Courtesy Electronic Design*).

effective $E_0$ (the voltage that charges $C_I$) can be reduced. To see this, the following equations with $E_0$ as the output voltage and $E_0'$ as the peak voltage at the output of the transformer are given.

$$E_T = \frac{E_0^2}{fR_L} + \frac{1}{2} C_I(E_0')^2 + \text{[Energy lost per cycle]},$$

$$E_{MT} = \frac{E_0^2}{fR_{L(\text{min})}} + \frac{1}{2} C_I(E_0')^2$$

$$+ \text{[Energy loss per cycle in the maximum-load case]}. \quad (9\text{-}41)$$

Figure 9-40 shows the output at transformer $T_1$, first with a peak-to-peak-rectifier assembly, then with a peak-to-peak doubler. Since most of the energy is pumped during the flyback pulse, the peak-to-peak assembly gives about the same efficiency as that of a single-diode-and-capacitor peak rectifier. But the peak-to-peak doubler should yield better efficiency, since it reduces $E_0'$ without degrading the output voltage.

Notice that the power input, and thus the power losses, are the same for both the 5 kHz and the 10 kHz peak-to-peak doubler in Figure 9-40. This is consistent with the fact that the $C_I$ term also exhibits a linear frequency dependence on the power loss. The trade-off here is size and output-voltage ripple. Ripple—for the peak-to-peak assembly at 10 kHz—is 1 V; for the peak-to-peak doubler, the ripple is 2.5 V.

Other constraints come into play in the design of a practical high-voltage supply. The most important constraint—the need to pump the required energy

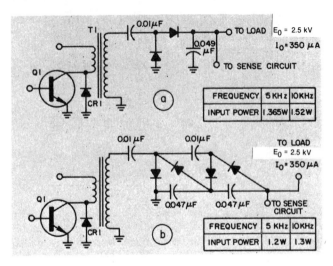

**Figure 9-40**  Power can be saved by use of a peak-to-peak rectifier (a) or voltage doubler (b) after the transformer. (*Courtesy Electronic Design*).

into the core in a limited amount of time—translates into a maximum required inductance:

$$\text{Maximum total energy per cycle} = E_{MT} = \tfrac{1}{2} L I_p^2 \qquad (9\text{-}42)$$

where

$$I_p = \frac{E_i t_{on}}{L} = \text{peak primary current.}$$

Now the maximum $t_{on}$ is a function of the frequency, since a certain time is needed to pump energy into the load and to return most of the energy (given by $\tfrac{1}{2} C_I E_0^2$) to the input supply. Thus $t_{on}$ can be stated as

$$t_{on} = Af,$$

where $A$ is a fraction that can be calculated, and

$$L_{max} = \frac{1}{2} \frac{A^2 E_i^2 f^2}{E_{MT}}. \qquad (9\text{-}43)$$

From Eq. 9-43, the total power lost is

$$P_{TL} = K_1 \frac{\ell_e}{\Delta\ell_e} f E_T + \frac{K_2}{E_i} \left(\frac{\Delta\ell_e}{A_e}\right)^{1/2} E_T^{3/2} + K_2' \frac{f R'}{E_i} \left(\frac{\Delta\ell_e}{A_e}\right)^{1/2} E_T^{3/2}$$

$$+ \frac{V_{\text{drop}}}{E_i - V_{\text{drop}}} f E_T + K_3 \frac{A V_s f^2}{N_p} \left(\frac{\Delta\ell_e}{A_e}\right)^{1/2} E_T^{3/2} \tag{9-44}$$

where $K_3$ is a constant or proportionality. Note that $N_p$, $\ell_e$, and $A_e$ are related by:

$$L_{\max} \; \alpha N_p^2 \frac{A_e}{\Delta\ell_e}.$$

From Eq. 9-44, $K_2$ and $K_3$ can be calculated exactly for any situation, and $K_2'$ can be closely approximated. Unfortunately most manufacturers don't give curves of core loss with rectangular excitation, so $K_1$ is hard to determine.

After the basic system parameters have been identified and the obvious efficiency measures implemented, a major power savings can be obtained: Just optimize $N_p$ and $\Delta\ell_e$. Since $E_{MT}$, $E_i$ and $f$ are determined, $L_{\max}$ is also determined. Thus Eq. 9-44 becomes

$$P_{TL} = K_1 \frac{\ell_e}{\Delta\ell_e} f E_T + K_2 \frac{f\rho}{E_i} \left(\frac{\Delta\ell_e}{A_e}\right)^{1/2} E_T^{3/2} + K_2' \frac{f R'}{E_i} \left(\frac{\Delta\ell_e}{A_e}\right)^{1/2} E_T^{3/2}$$

$$+ \frac{V_{\text{drop}}}{E_i - V_{\text{drop}}} f E_T + K_3 \frac{A V_s f^2}{L_{\max}^{1/2}} E_T^{1/2}. \tag{9-45}$$

Equation 9-54 shows that $\Delta\ell_e$, the transformer air gap, is the only parameter that doesn't have a monotonic relationship with $P_{TL}$. Therefore a value exists for $\Delta\ell_e$ that will minimize the power loss. When this value is found, it's coupled with $L = L_{\max}$ to get the final transformer.

The optimization procedure is carried out for several loads and voltage settings in Figure 9-41. Since $E_i$ and $E_T$ cause a different weighting of the various loss terms, the gap should be optimized for desired load and input voltage.

## LOW VOLTAGE HIGH POWER SUPPLY DESIGN[3]

Traditional power supply designs usually take the transformer-rectifier-low-voltage-regulator route. While this method is adequate for low power ($<200$ W)

[3]"Switch Your High Power-Supply Design to an Off-Line Regulation Technique" by Peter N. Wood. Reprinted with permission from *Electronic Design*, 7, April, 1, 1975, copyright Hayden Publishing Co. Inc. (1975).

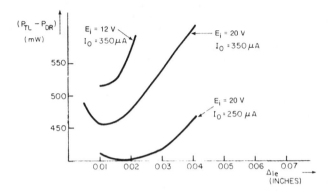

**Figure 9-41**  An optimum value for the transformer air gap can be found to reduce power loss even further. (*Courtesy Electronic Design*).

designs with single-frequency inputs, higher-power applications suffer from high cost and low efficiency—two serious drawbacks of the traditional method.

But these problems can be avoided by using transistor choppers to regulate at high frequency (20 kHz), to design with all TTL to simplify the circuit, and to use Schottky diodes in the output rectifier to hold down switching and forward-drop losses thus increasing efficiency.

The power switching circuit is a half bridge configuration that has several advantages over the usual push-pull approach to off-line regulation: The output transformer primary gets 100% use—as opposed to 50% in push-pull circuits—and the output transistors have just half as much voltage stress. This means that the unregulated bus voltage can be double that of a comparable push-pull circuit but with the same current and voltage stress at the transistors. Note that a doubling of the bus voltage also doubles output power.

The single-phase AC input can be 95 to 130 V, or 190 to 260 V, at frequencies from 50 to 400 Hz (or higher). Or the input can be 190 to 260 V three-phase over the same frequency range. Output is 5 V at 200 A, with a ripple and regulation envelope of 50 mV. Overload protection is to 250 A.

Taking the place of a front-end transformer is an off-line rectifier that delivers unregulated, 300 V DC bus power, which is first chopped to 20 kHz by transistor switches, then transformed down to the secondary voltage required to feed the output rectifiers (Figure 9-42). To achieve regulation, the conduction time of the high voltage switching transistors is controlled at the 20 kHz rate.

Smooth DC is generated at the output terminals by integration of the pulse train after rectification. A DC regulator senses the output voltage and current and supplies a feedback error signal to the pulse-width control circuit.

To accommodate both 115 and 230 V inputs, a voltage doubler is used in conjunction with a conventional bridge rectifier. Self-heating thermistors are

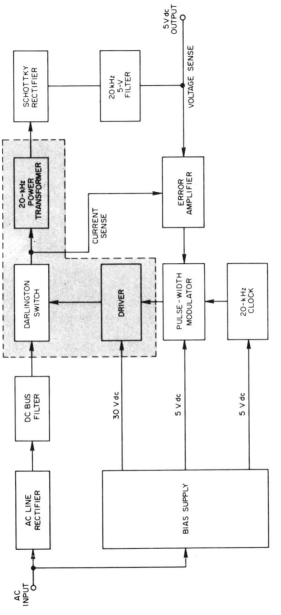

Figure 9-42 High power converter block diagram. (*Courtesy Electronic Design*)

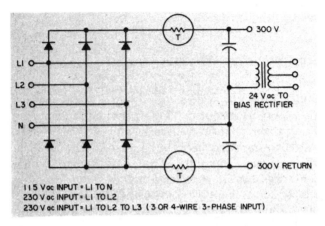

**Figure 9-43** A three-phase bridge rectifies the input, which can be 115 or 230 V, single or three-phase. The thermistors limit line-inrush current. (*Courtesy Electronic Design*).

needed to limit the turn-on line in-rush current caused by the low impedance of the filter capacitors (Figure 9-43). The rectifier shown is a three-phase bridge with a peak-reverse voltage of 660 V minimum and a current capability of 8 A.

Because of the fast response of the switching regulator, input-line variations—including line ripple—are, in effect, filtered by the pulse width modulation action. It is, however, still necessary to keep the DC bus within the voltage regulation range of the control circuit, thus requiring the use of bus filter capacitors. In actual operation from 50 Hz 95 V lines (worst case), the DC bus will have about 20 V peak-to-peak ripple at full load with 2500 $\mu$F filter capacitors.

The peak line currents at the input terminals depend on the line impedance, as does the rectified bus voltage. The surge-limiting thermistors prevent excessive peak line currents at the expense of about a 5 V drop on the DC rectified bus under full-load conditions.

Darlington transistors, connected as a bridge, switch the unregulated bus at the 20 kHz rate (Figure 9-44). Drive signals are applied at the bases of the four Darlingtons so that transistors $Q_1$ and $Q_4$ conduct simultaneously. When $Q_1$ and $Q_4$ turn off, $Q_2$ and $Q_3$ then conduct simultaneously. Thus when the DC bus is at 300 V, a 600 V pk-pk waveform appears across the primary of the power transformer (neglecting the small voltage drop in the series capacitor at the primary winding and the saturated drops of the switches).

In Figure 9-44, a 2 $\mu$F capacitor across the DC bus at the bridge location reduces voltage transients traced to the DC bus impedance. Such transients can be expected—especially when the bridge is located more than a few inches from the main filter capacitors. The capacitor is usually polycarbonate but must have low-loss characteristics to adequately handle the 20 kHz ripple currents.

Another 2 $\mu$F polycarbonate capacitor in series with the transformer primary

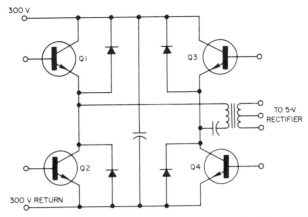

Figure 9-44 Darlington transistors in a bridge arrangement switch the 300 V bus. (*Courtesy Electronic Design*).

blocks DC and thereby prevents possible saturation of the core by small imbalances in the volt-second product applied at each half cycle. There are always some differences in switching times and transistor saturation voltages that cause the core flux excursions to "walk" to one of the saturation regions of the B-H curve. Ferrite cores without air gaps are especially sensitive to such flux imbalance.

The operating frequency precludes the use of steel cores because of the high hysteresis and magnetizing losses at that frequency. Tape-wound toroids with 0.001 in., or thinner, nickel-steel tape can be used, but these are expensive and costly to wind. A better choice is a ferrite material in an E-E, U-U or U-I configuration, though the saturation flux density is somewhat lower than for the nickel steels. Peak flux densities of 2000 G can be used safely in some ferrites. This allows design of compact transformers with low copper and core losses.

With high current windings, the ease of winding must be considered. Usually, the secondaries are split into medium current sections—about 50 A per section—an arrangement that also leads to improved current sharing by the output rectifier anode. Thus current sharing is optimized, and there are no secondary terminals to cause power losses.

The primary winding is layer wound in the usual way. This lends itself to the multiple-coil, stick winding techniques used in large quantity production.

Rectification takes place in Schottky power diodes developed for high-current, high-temperature use. These provide a lower forward drop than p-n diodes and are especially suited to high frequencies because of the absence of recovery current. The ease with which Schottkys can be paralleled makes them ideal for a 5 V, 200 A supply. Individual Schottky diodes can handle 50 A with a maximum forward drop of 0.6 V at a case temperature of 125°C.

Because of the regulation method, the rectified output is a pulse train that must be transformed into the smooth DC required at the output. Conventional L-C filter techniques are used to do this. But like the power transformer, the inductors must have low losses. Consequently powdered permalloy toroids are used and are wound with just two turns of the flexible output bus.

The inductance value of $L_1$ is determined by the allowable peak currents in the Darlington switching transistors, rather than the more conventional approach of ripple attenuation. This can be determined by breadboard measurement allowing a 10 A peak collector current rating for the Darlingtons—a conservative derating for acceptable reliability.

Filter capacitors are low inductance, stacked-foil electrolytics to minimize both the ESR (equivalent series resistance) and the AC impedance at 20 kHz. Stacked foil exhibits inductance values as low as 2 nH and provides superior ripple-current ratings. Small series inductors ($T_5$) are formed between the capacitor sections of the filter by passage of the DC bus bars through powdered permalloy toroids. With $C_{11}$—a low-impedance capacitor located directly across the output—$T_5$ forms a second section of the output filter. This technique further reduces the common-mode ripple voltage at the output.

Two functions are performed by the error amplifier circuit: voltage regulation and current limiting. A 723 IC voltage regulator generates the error signal needed for regulation as shown in Figure 9-45. By rectification of the output voltage of a current transformer in series with the power transformer primary, a voltage proportional to the primary current can be generated. This signal is also applied to the 723 regulator, and in such a way that the error signal becomes current-dependent above a predetermined load current. Thus, a regulation envelope exhibits constant voltage up to the overload point, and then constant current down to zero voltage as shown in the output characteristic VI curve of Figure 9-46.

During start-up, the DC bus voltage is delayed to ensure that all bias voltages are established and drive signals are applied to the Darlington power stage before power is switched. This delay is performed by the inrush limiting thermistors in the input filter. At the end of this delay time the DC bus falls within the regulation limits and normal regulation is established.

To form the pulse width modulator, a free-running TTL multivibrator (MC4024) generates a 20 kHz square wave, which triggers a monostable multivibrator (9601) on the leading and trailing edges. The output of the multivibrator is a pulse train of 20 $\mu s$ width and a 40 kHz repetition rate.

By use of the 723's error signal to vary the monostable's R-C time constant, the pulse width can be varied proportionately from 0 to 20 $\mu s$. The resultant one-shot pulse is applied to one input of two NAND gates and the 20 kHz square wave—and its inverse—are applied to the other inputs as shown in Figure 9-47a.

The gate outputs are therefore pulse width regulated signals, one in phase and the other 180° out of phase with the 20 kHz clock signal. An inverter

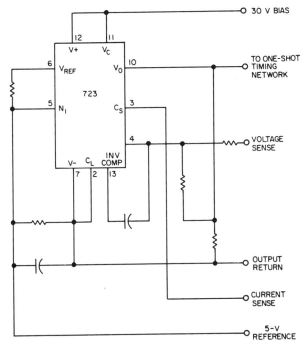

Figure 9-45  Both output voltage and current are sampled and generate an error signal in a 723 regulator. The error signal modulates the pulse width of a one-shot multivibrator to provide the voltage regulation. (*Courtesy Electronic Design*).

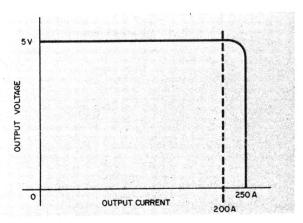

Figure 9-46  Regulation provides a constant-voltage output up to the overload point (250 A) and then goes into a constant-current mode to 0 V. (*Courtesy Electronic Design*).

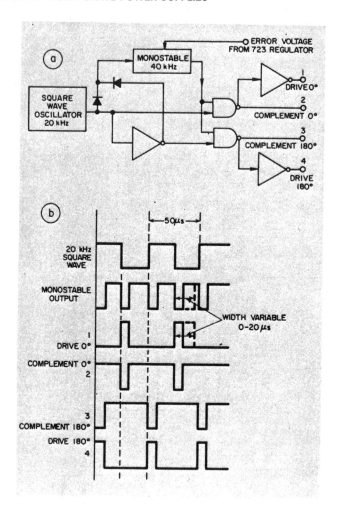

Figure 9-47 To derive the four signals to drive the Darlington bridge, the one-shot output is "mixed" with the 20 kHz clock in two NAND gates (a). The resulting waveforms consist of two out-of-phase signals and their complements (b). (*Courtesy Electronic Design*).

connected to each of the NAND-gate outputs provides drive signals that are also 180° out of phase with each other, but are complements of the NAND-gate outputs (Figure 9-47b).

Thus four pulse width modulated signals—from the NAND gates and the inverters—are generated and constitute the base drives for the power Darlington switches. As shown in Figure 9-47, the single driver transformer has two center-tapped primaries. One primary connects with a pair of push-pull driver transis-

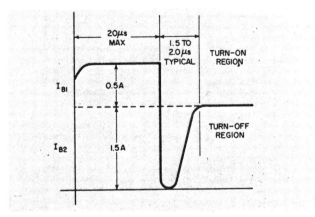

Figure 9-48 Critical to proper operation is the Darlington base-drive current, which must have fast transitions. (*Courtesy Electronic Design*).

tors; the other short circuits the core flux during those periods when zero drive is required at all the Darlington switches. This short-circuiting technique ensures that noise pulses or other spurious signals cannot falsely trigger the power switches during periods of no conduction.

A factor in the successful operation of high-voltage switching regulators lies in the base-drive waveforms supplied to the power switches. Drive signals must have fast rise and fall times plus a low source impedance in the turn-off ($I_{B2}$) direction to minimize storage and fall times (Figure 9-48).

High-voltage Darlingtons are used typically at a forced gain of 20, so that acceptable saturation losses can be obtained without excessive storage time during turn-off. However, it is necessary to drive 1 to 2 A in the reverse direction to ensure minimum turn-off times.

$I_{B2}$ is supplied by a clamp circuit (Figure 9-49). During the forward-drive condition of the Darlington, voltage is dropped across $R_1$, limiting $I_{B1}$. At the same time, the drop across this resistor charges $C_1$ to approximately 4 V DC. The clamp transistor, $Q_2$, is reverse biased by the forward voltage of $CR_1$ and therefore does not conduct. When the winding voltage drops to zero, because of the shorting action of the driver circuit, $Q_2$ is forward driven to saturation. The collector current of $Q_2$ supplies $I_{B2}$—the Darlington reverse current.

A small amount of DC bias is required to operate the TTL and driver circuits. To provide this, a 115-to-24 V, 0.25 A transformer is driven from line to neutral of the AC input. The secondary is rectified full-wave and filtered to give about 30 V at 1 A. Since the 5 V logic load is substantially constant, only a dropping resistor and a 5 V Zener diode are required for the TTL bias (Figure 9-50).

In switching regulators, pulse-width regulation may be performed with any of the following methods:

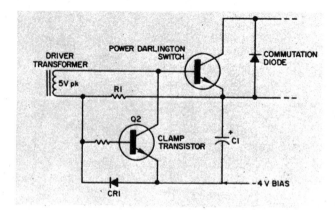

Figure 9-49   Turn-off current, $I_{B2}$, is provided by a transistor clamp arrangement at the base of the Darlington. (*Courtesy Electronic Design*).

- Constant frequency, variable ON time.
- Constant ON time, variable OFF time.
- Constant OFF time, variable ON time.
- Variable ON time, variable OFF time—or random-switching—regulator.

All these methods follow the basic equation

$$E_{\text{out}} = E_{\text{in}} \frac{t_{\text{on}}}{t_{\text{on}} + t_{\text{off}}}. \tag{9-46}$$

Of the four methods, the constant frequency approach used here is preferable because of the predictable performance of magnetic components and the simpler RFI suppression method available at constant carrier frequency. By varying the on time, the regulator modulates the output power in response to the internally generated error signal.

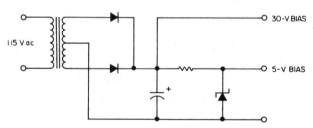

Figure 9-50   Bias power for the TTL and driver circuitry is provided by a simple auxiliary supply. (*Courtesy Electronic Design*).

For the power stage and driver, the bridge connection of four Darlingtons can provide 1000 W. If less power is required ($<500$ W), only two transistors are needed. In this case, the transistors are connected in a half-bridge circuit, and, the transformer primary AC voltage is then half as much as for the full bridge. With either circuit, the same driver and TTL logic are required, but the half-bridge uses only two of the driver secondary windings.

Figure 9-51 shows the power stage with the necessary turn-off circuits and commutating diodes that provide paths for inductive currents in the power transformer (Figure 9-48 illustrates the typical base-current waveform, generated by the turn-off clamp circuits of Figure 9-51). Base current, $I_{B2}$, is limited by the impedance of the clamp transistor at saturation, the ESR of the bias capacitor and the impedance at turn-off of the Darlington. Values of 1.5 to 2.0 A are typical of $I_{B2}$.

For the power transformer, a low-loss ferrite core material is chosen with an E-E configuration to minimize winding costs. Stackpole 24B Ceramag material has a saturation density of 5.1 kG at 9.0 Oe. The core loss is approximately 3 W/lb at 20 kHz and 1 kG. With a duty cycle of 80% (corresponding to a low AC input condition), the secondary voltage must be 6 V to be dropped in the Schottky diodes, and another 0.4 V in the choke and output busses, to give 5 V at the output.

Calculations show that to develop and support 6 V for 25 $\mu$s would require approximately 1600 G with one turn or 750 G with two turns at 3 V per turn. The transformer is designed therefore at a flux density of 1500 G. The complete supply schematic is given in Figure 9-52 along with a parts list. Table 9-5 depicts the transformer winding data.

## LOW POWER DC/DC CONVERTER[4]

Figure 9-53 shows the schematic diagram of a simple low power DC/DC converter suitable for patient worn biomedical instruments. This practical implementation converts +5 V to $\pm6$ V and dissipates less than 300 mW. When supplying $\pm30$ mA, its efficiency is typically 74%. Line regulation over +4.5 to +5.5 V is $\pm1\%$; load regulation from 44 to 135% of the 30 mA load is better than 1%.

In this circuit $IC_1$ (operating as a 100 kHz oscillator), $Q1$, and $Q2$ combine to form a push-pull DC/DC converter. $T_1$ is constructed of No. 32 magnet wire wound quadrifilar on a Ferroxcube 266T12513BT core. The bridge-rectifier uses Schottky diodes because their lower forward-voltage drop improves converter efficiency. The 7663 voltage regulator IC senses only the positive output, and

[4]Used with permission from "DC/DC Converter uses low power" by Warren Jochem, *EDN*, July 26, 1984.

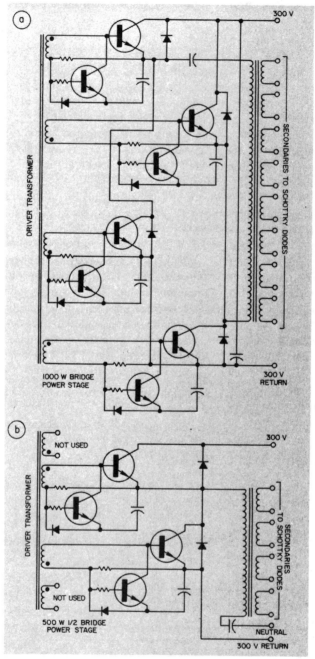

Figure 9-51   Power stage for a 1000 W output (a). Commutating diodes and other circuitry shunt the output-transformer inductive currents to ensure proper turn-off. If only 500 W are needed, a half-bridge connection can be used (b). (*Courtesy Electronic Design*).

as a result, positive and negative loads on the converter are well balanced. A $\pm 20\%$ load imbalance results in a voltage imbalance of less than 0.2%.

The output of the 7663 voltage regulator supplies voltage to oscillator $IC_1$. As the output voltage drops, the supply voltage to $IC_1$ increases, which increases the drive to $Q1$ and $Q2$. This acts to stabilize the output voltage. For best efficiency and regulation, $R$ should be selected such that with a $+5$ V input and $\pm 6$ V output $IC_1$ oscillates at 100 kHz. Because the design uses a linear regulator, the output noise is low: $< 20$ mV p-p. The pi filter network shown reduces reflected input ripple to 30 mV p-p.

The VN10KM MOSFETs were chosen for small size and low cost. Other devices with lower on-resistance, though, could improve the converter's efficiency. The circuit can be configured for other output voltages by changing the transformer's turns ratio and the $R_2/R_3$ divider ratio.

## HIGH-EFFICIENCY Cuk CONVERTER[5]

Figure 9-54 shows the schematic diagram of a high efficiency DC/DC converter that provides output voltages of $\pm 12$ V from a 14 to 40 V DC unregulated source. The converter consists of a precision reference source and error-amplifier ($U_1$), which pulse-width-modulates an astable oscillator ($U_2$) to produce the gate drive for power MOSFET switch ($Q_1$) in a multiple-output Cuk power stage. The control and reference circuits were designed for minimum power consumption.

Use of a 555 CMOS timer as a controlled pulse-width modulator, along with the LM10 low-power op amp and voltage reference, resulted in a total control circuit drain of just under 1 mA—one of the major factors contributing to the high overall efficiency of the converter.

The 555 timer generates the basic switching waveform. A diode across pins 6 and 7 allows separate control of the charge and discharge of timing capacitor $C_6$, making possible the generation of a less than 50% duty-cycle output directly. Resistor $R_7$ controls the pulse width and therefore the switch on-time, while $R_8$ controls the pulse spacing. With no control voltage applied, the pulse width is 10 $\mu$s. It can be controlled by a voltage at pin 5, but can only be decreased because diode $D_2$ blocks an increase in control voltage that would be required to increase the pulse width. Fixing the maximum pulse width does two things. First, it provides a rough limit on output power, and second it allows the circuit to start independent of the control. When power is first applied to the supply, $U_2$ starts oscillating at maximum pulse width. As the output voltage increases to 12 V, which takes about 7 ms, the control loop starts to decrease the pulse width to maintain regulation. Overshoot at turn-on is no more than 5%. Besides

---

[5]This work was performed by the NASA Lewis Research Center.

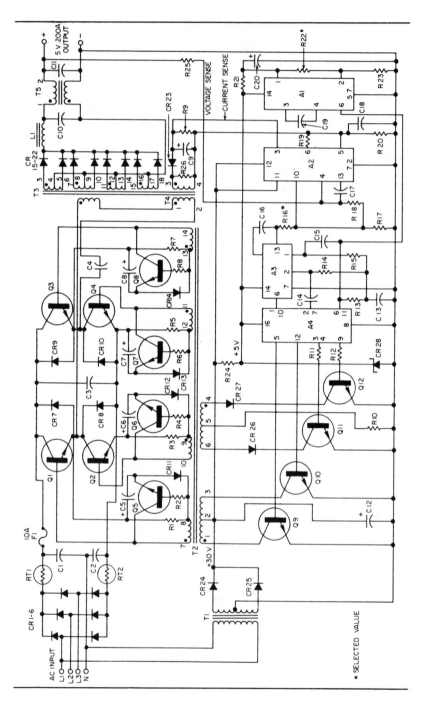

# Parts list

## Semiconductors

| | | |
|---|---|---|
| $Q_{1,2,3,4}$ | TRW SVT6001 | |
| $Q_{5,6,7,8,9,10}$ | TRW SVT60-5 | |
| $Q_{11,12}$ | 2N2222 | |
| $CR_{1,2,3,4,5,6}$ | 1N1204 | |
| $CR_{7,8,9,10}$ | TRW SVD400-12 | |
| $CR_{11,12,13,14,23}$ | TRW DSR3051 | |
| $CR_{16,16,17,18}$ | TRW SD-51 | |
| $CR_{19,20,21,22}$ | TRW SD-51 | |
| $CR_{24,25}$ | 1N4002 | |
| $CR_{26,27}$ | TRW DSR3051 | |
| $CR_{28}$ | TRW LVA-51A | |
| $A_1$ | MOTOROLA MC4024 | Voltage-Controlled multivibrator |
| $A_2$ | NATIONAL LM723C | Voltage regulator |
| $A_3$ | NATIONAL DM8601 | Monostable multivibrator |
| $A_4$ | NATIONAL DM8090 | Dual NAND, Quad Inverter |

## Resistors

| | | |
|---|---|---|
| $RT_{1,2}$ | 5Ω | Themistor, 5DA5R0 (Rodan Industries) |
| $R_{1,3,5,7}$ | 10Ω, 5W | |
| $R_{2,4,6,8}$ | 51Ω, 1/2W | |
| $R_{9,14,16,25}$ | 1K, 1/2W | |
| $R_{10}$ | 10Ω, 1/2W | |
| $R_{11,12}$ | 270Ω, 1/2W | |
| $R_{13,23}$ | 10K, 1/2W | |
| $R_{16}$ | 100K, 1/2W(selected) | |
| $R_{17}$ | 1.5K, 1/2W | |
| $R_{18}$ | 470K, 1/2W | |
| $R_{19}$ | 2K, 1/2W | |
| $R_{20}$ | 5.1K, 1/2W | |
| $R_{21}$ | 100Ω, 1/2W | |
| $R_{22}$ | 6.8K, 1/2W(selected) | |
| $R_{24}$ | 270Ω, 2W | |
| $R_{26}$ | 39Ω, 1/2W | |

## Capacitors

| | | |
|---|---|---|
| $C_{1,2}$ | 2500μF, 200V | Sprague type 36D |
| $C_{3,4}$ | 2μF, 400V | |
| $C_{5,6,7,8,13,20}$ | 22μF, 6V | |
| $C_9$ | 56μF, 6V | |
| $C_{10}$ | 47000μF, 7.5V | Sprague type 432-D |
| $C_{11}$ | 10μF, 50V | |
| $C_{12}$ | 1100μF, 50V | Sprague type 39D |
| $C_{14,15,16}$ | 0.001μF, 50V | |
| $C_{17}$ | 100pF, 300V | Elmenco type DM |
| $C_{18}$ | 0.1μF, 50V | |
| $C_{19}$ | 0.01μF, 50V | |

## Transformers

| | | |
|---|---|---|
| $T_1$ | FT-14 | TRW/UTC Transformers |
| $T_2$ | Special Driver | TRW/UTC Transformers |
| $T_3$ | Special Power | TRW/UTC Transformers |
| $T_4$ | Special Current | TRW/UTC Transformers |
| $L_1$ | Special Filter | TRW/UTC Transformers |
| $T_5$ | Special Balun | TRW/UTC Transformers |

## Fuses

| | | |
|---|---|---|
| $F_1$ | 10A fast blow | Bussman type 3AG |

Figure 9-52   Low voltage high power supply detailed schematic diagram. (*Courtesy Electronic Design*).

## Table 9-5   Transformer Winding Data

---

DRIVER TRANSFORMER ($T_2$)
  Core:
    TDK, part number H6A-P18/14-52H, pot core.
  Windings:
    (1-2)(2-3)        60T Bifilar, #32 single sodereze.
    (4-5)(5-6)        30T Bifilar, #36 single sodereze.
    (7-8)(9-10)       10T Quad,  #26 single sodereze.
    (11-12)(13-14)
  Terminations:
    Self leads to 14 pin header mounted to core.

POWER  TRANSFORMER ($T_3$)
  Core:
    Stackpole, part number 50-631, ceramag 24B, 0244.
  Windings:
    (1-2) 32T, #14 Teflon covered flex.
    (3-4) thru (17-18), 1T, #8 PCV covered solid core.
  Terminations:
    Self leads, length as required.

CURRENT SENSE TRANSFORMER ($T_4$)
  Core:
    Magnetics Inc., part number 52056-1D.
  Windings:
    (1-2)  Bar primary formed by passing $T_3$ primary lead thru core.
    (3-4)  34T, #30HF
  Terminations:
    Self leads, length as required.

BALUN TRANSFORMER ($T_5$)
  Formed by passing output bus bars through Magnetics, Inc., core part number 55436-A2,
  1 turn each bus bar.

FILTER INDUCTOR ($L_1$)
  Formed by winding 2 turns of positive bus bar through 2 Magnetics, Inc., cores part number 55436-A2.

---

limiting in-rush current during turn-on, this control design prevents the initial error signal from forcing excess pulse widths that would saturate the input transformer and cause destructive overcurrents. When driving the design load, the on- and off-times are about 4 and 38 $\mu$s, respectively.

Power for $U_2$ and voltage reference for the error amplifier are provided by one half of the LM10. Its internal 200 mV reference signal is amplified to 10 V for these functions. The other half of the LM10 is used as the error-amplifier, where it compares a portion of the +12 V output to the 10 V reference and drives the modulation input of $U_2$.

One limitation of the control as shown is that the minimum pulse width

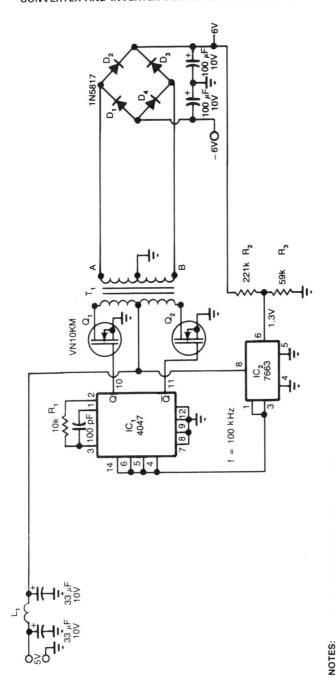

Figure 9-53  Schematic diagram of high-efficiency low-power DC/DC converter. (Reprinted with permission from "DC/DC Converter Uses Low Power," *EDN*, July 26, 1984 © Copyright Cahners Publishing Company, a division of Reed Publishing, USA).

**NOTES:**
$L_1$ = 3 μH: 13T #26, MICROMETALS T25-26 CORE
$T_1$: FERROXCUBE 266CT121513BT CORE
PRIMARY = 26T #32 (TWO SEPARATE 13T WINDINGS)
SECONDARY = 40T # (TWO SEPARATE 20T WINDINGS)

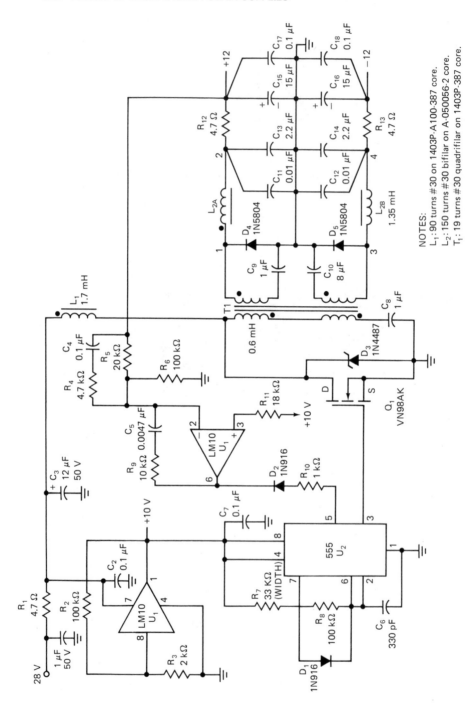

NOTES:
$L_1$: 90 turns #30 on 1403P-A100-387 core.
$L_2$: 150 turns #30 bifilar on A-050056-2 core.
$T_1$: 19 turns #30 quadrifilar on 1403P-387 core.

Figure 9-54   Multiple output high efficiency Cuk converter.

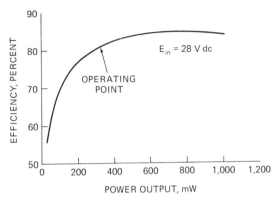

Figure 9-55 Efficiency vs. Power Output for the ± 12 V DC/DC converter of Figure 9-54.

possible is 1 μs. This is not narrow enough to maintain output voltage regulation when the load is removed completely. A minimum load of 80 mW is required.

Since the maximum output power is 1 W, it results in a usable range of power output of greater than 12 to 1. If desired, the uncontrolled duty-cycle could be set lower, restricting the maximum power output and shifting the operating range downward. The remainder of the circuit in Figure 9-54 is the Cuk power converter, with its input and output filters.

Limiting maximum pulse width provides a rough current limit and improves turn-on characteristics. Here operating frequency is 25 kHz and was chosen low to minimize losses. The circuit, though, has been operated at above 100 kHz with somewhat reduced efficiency.

When operated at 28 V DC input, the circuit has an efficiency of 80% or greater from the design power level of 300 mW up to 1 Watt, as shown in Figure 9-55. Efficiency increases at lower input voltages. The circuit can deliver 300 mW over an input voltage range of 13 to 40 V DC, making it ideal for battery-powered equipment.

The high efficiency at low power level and the ability to provide an output either larger or smaller than the input voltage makes this converter well adapted for use with CMOS logic circuits.

## MULTIOUTPUT MEMORY SYSTEM PWM POWER SUPPLY

The high-efficiency, low weight, and small size of switching power supplies makes them particularly useful in computer, instrumentation, and industrial applications. This section presents the detailed design of a pulse width modulated (PWM) power supply designed for use in a portable memory terminal.

The combined battery discharge and memory loads necessitated use of a PWM approach. The supply provides power for driving a microprocessor, memory, and digital read-out device.

Pulse width modulation (PMW) techniques have a real advantage in power conversion applications, especially in systems where wide input voltage variations and large load fluctuation exists.

In conventional regulators, as the input voltage is increased, the operating frequency also increases. This occurs because a fixed number of volt-seconds is required to saturate the core. Thus frequency varies inversely with the input voltage. If however, the operating frequency is held constant by introducing a period of off time after the saturation of each flux reversal, the average value of rectified voltage is independent of input voltage. The required constant frequency variable duration waveform is developed with the PWM circuit of Figure 9-56.

Most DC converters achieve maximum efficiency at full load only. This is undesirable with varying loads, since full load capability is utilized only during a small percentage of the duty cycle.

Declining efficiency is caused by not so much the relatively minor fixed losses in the control circuits, but by the losses in driving the power stage. Since the power transistors must be driven to full-load capability, the drive circuit dissipation can be justified only when the converter is delivering maximum power. Thus a base-drive circuit for the switching transistors, that would automatically adjust the base current to the load was required.

Since $E_{AV}$ is a function of $T$, $T$ can be automatically adjusted to provide tighter regulation. The output is sampled and compared to a temperature compensated Zener diode in a differential amplifier. Any error signal is used to vary the duty cycle of the multivibrator, adjusting $T$ to bring the output voltage back to the desired level.

Even with unmatched transistors in the differential amplifier, it is feasible to achieve $\pm 1\%$ regulation. Voltage regulation is accomplished by transformer $T_1$. $T_2$ does not saturate. The input voltage is applied at the primary center tap of $T_1$, and tapped for the primaries of $T_2$, as shown. When $T_1$ primary current is flowing through $Q_1$ to drive $Q_3$, the collector current of $Q_3$ is reflected as a load on transformer $T_1$, and the $T_1$ primary current automatically adjusts itself to this load according to the turns ratio. But $T_1$ primary current is the base drive for $Q_3$.

Hence

$$I_B = N_1 I_C / (N_1 + N_2). \tag{9-47}$$

By setting

$$(N_1 + N_2)/N_1 = \beta_{min}, \text{ then } I_B = I_C/\beta_{min}. \tag{9-48}$$

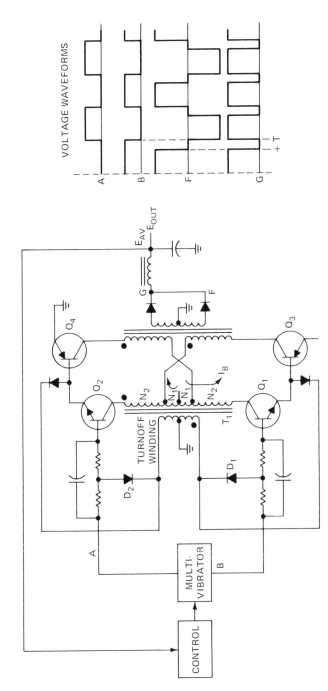

Figure 9-56  PWM converter. Efficiency is increased and short-circuit protection is obtained by driving the output stage with variable width pulses from $Q_1$ and $Q_2$. Only transformer $T_1$ saturates: not $T_2$.

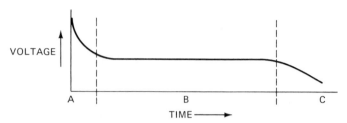

Figure 9-57   Characteristics of rechargeable NiCd batteries.

If the load current increases, the immediate result is that current $I_C$ also increases; primary current in transformer $T_1$ also increases and this in turn causes current $I_B$ to increase. Thus base-drive current regulation is both instantaneous and automatic. As a further benefit from the circuit, no power has to be applied only to be wasted by being dissipated in base-drive resistors, and transistors $Q_3$ and $Q_4$ are operated at their full power gain.

The circuit of Figure 9-56 was modified in order to meet the requirements of the memory system peculiarities. Specifically a power supply was needed that would operate from a battery voltage source that varied from 8 to 15 V DC. Most rechargeable NiCd batteries have the characteristics shown in Figure 9-57.

Once the initial "top off" discharge occurs (A) the battery puts out a rather constant voltage (B) until final discharge (C). The power supply must be able to handle the short times at the extremes of operation (A) and (C) in addition to the nominal operating voltage (B). A strictly step down switching regulator approach was not feasible because of the fact that to obtain one of the desired outputs (12 V) when the battery was at 8 V necessitated a step up transformer circuit.

Separate outputs were required to provide the desired voltages and currents to drive the associated memory, microprocessor and display devices:

1. +5 V @ 2 A
2. +12 V @ 0.200 A
3. −5 V @ 5 mA
4. +5 V @ 0.500 A

The power supply also had to have an efficiency in excess of 70% at full load and it had to be producible (less than $50 in 1000-piece quantities). Thus exotic core materials and state-of-the-art components could not be used. The use of proven parts was a must.

Figure 9-58 shows a block diagram of the converter. Each section will be discussed and then a schematic diagram showing the completed circuit will be presented. The design philosophy used emphasized the use of integrated circuit components to minimize size.

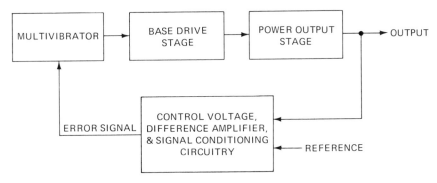

Figure 9-58   PWM power supply block diagram.

## Multivibrator

A multivibrator using two 555 (or one 556) timers was constructed as shown in Figure 9-59. One 555 timer is connected as an astable multivibrator and provides the basic operating frequency for the power supply. The oscillator frequency must be two times the desired converter switching frequency. $R_1$, $R_2$ and $C_1$ are used to establish the operating frequency in accordance with the following equations.

$$f = \frac{1.44}{(R_1 + 2R_2)C_1} \tag{9-49}$$

$$D = \text{duty cycle} = \frac{R_2}{R_1 + 2R_2} = 0.1 \tag{9-50}$$

$$t_{\text{on}} = 0.693(R_1 + R_2)C_1 \tag{9-51}$$

$$t_{\text{off}} = 0.693\,R_2 C_1 \tag{9-52}$$

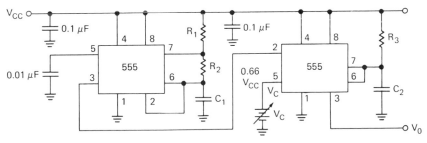

Figure 9-59   Circuit of the pulse width modulator.

In the present example, $f_o \cong 10$ kHz for a converter frequency of approximately 5 kHz.

$$C_1 = 0.01 \ \mu\text{F}$$

$$R_1 = 12 \ \text{k}\Omega$$

$$R_2 = 1.5 \ \text{k}\Omega$$

which gives

$$T_1 \cong 100 \ \mu\text{s}, \ t \cong 10 \ \mu\text{s}$$

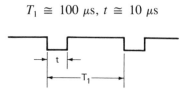

The output of the astable multivibrator is used to provide a trigger of the one-shot multivibrator which sets the pulse width.

The monostable multivibrator is triggered with a continuous pulse train developed by the 555 astable multivibrator via pin 2 and the threshold voltage is modulated by the signal (varying input voltage) applied to the control voltage terminal (pin 5). This results in modulating the pulse width as the control voltage varies.

The data used in designing the monostable multivibrator (one-shot) is as follows:

$$t_D = 1.1 \ R_3 C_2 = 100 \ \mu\text{s (nominal)}$$

$$R_3 = 9.1 \ \text{k ohm}$$

$$C_2 = 0.01 \ \mu\text{F}$$

### Control Voltage

A control signal for modulating the one-shot multivibrator is developed by comparing the rectified output of the transformer with a Zener reference voltage. The transformer output voltage will vary as a function of battery voltage and output load requirements. A dual general purpose op amp is used to provide a low cost differential and summing amplifier as shown in Figure 9-60. As the feedback voltage varies compared to the Zener reference, the control voltage $V_c$ varies the one-shot duty cycle. The circuit parameters are designed to provide

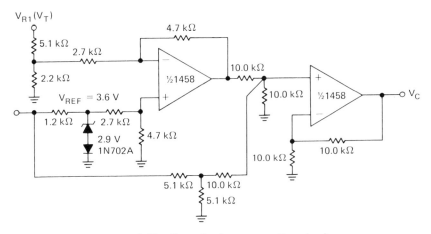

Figure 9-60  Control voltage generation circuit.

a control voltage of one half of the 555 supply voltage when the error voltage is zero.

## Signal Conditioning

The output of the one shot monostable multivibrator is a pulse train signal fed into a signal conditioning network, that causes a dead time (delay) to be built into the waveform and provides the PWM waveform to the base-drive stage. A dead time is necessary to ensure that in the base-drive stage and power stage both transistors are not on simultaneously and thus dissipating power. One transistor must turn off before the other can turn on. Thus higher efficiency is achieved.

Figure 9-61 shows the signal conditioning network. The associated waveforms are shown in Figure 9-62. Use of CMOS circuits was desired to minimize power drain of the battery source.

## Base Drive and Power Stage Circuits

Figure 9-63 shows the schematic diagram of the base-drive and power stage. This somewhat self regulatory circuit design utilized nonsaturating pot cores for both transformers because pot cores are 30%–40% cheaper than toroids, pot cores have less audio noise, pot cores are self-shielding and the efficiency for the power levels handled here is about the same as that of toroidal cores.

In designing the base-drive/power stage circuitry one has to work backwards from the outputs in order to size the power transistors and transformer core. The following steps are used in designing this circuitry:

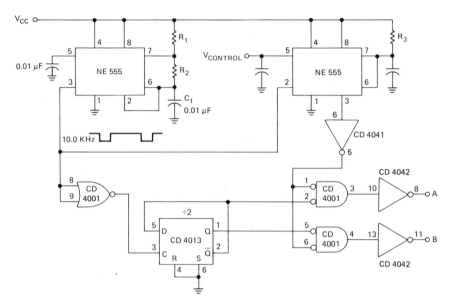

Figure 9-61   Signal conditioning network.

1. Calculate total output voltage based on known currents and voltages at all outputs

$$P_0 = V_1I_1 + V_2I_2 + V_3I_3 + V_4I_4 \qquad (9\text{-}53)$$

2. Compute power delivered to output transformer $T_2$.

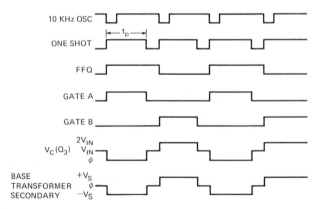

Figure 9-62   Waveforms associated with the signal conditioning network in Figure 9-61.

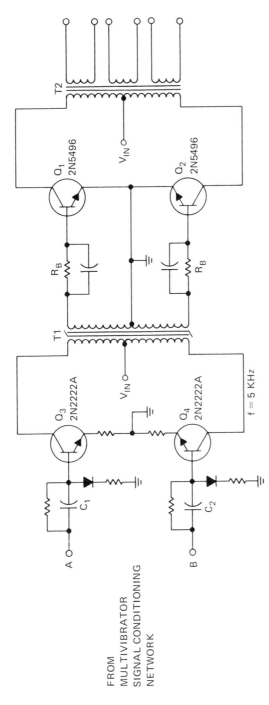

Figure 9-63   Schematic diagram of the base drive and power stage.

$$P^1_{out} = \frac{P_0}{efficiency} \tag{9-54}$$

Assume 90% for the transformer efficiency.

3. Estimate the collector current for a square wave using the following formula.

$$I_C = \frac{P^1_{out}}{E_{in}} \tag{9-55}$$

4. Choose the power transistor based on:
  a) The value of $I_C$ in step (3).
  b) Maximum $BV_{CEO} = 2$ Vin + 20% for spikes due to turn on/off spikes induced in primary winding.
  c) Turn on/off time and storage time must be much less than frequency of operation so that excessive heating does not result.
  d) Low junction-to-case thermal resistance.

5. From transistor manufacturer's data sheets determine $V_{CE(sat)}$ corresponding to $I_C$ (step 3). Then recalculate collector current

$$I^1_C = \frac{P^1_{out}}{E_{in} - V_{CE(sat)}} \tag{9-56}$$

6. Find common emitter forward transfer ratio ($H_{FE}$) at this collector current. A value of $H_{FE}$ that is low enough to ensure saturation ($H^1_{FE} = H_{FE}/2$) is usually used.

7. Calculate base current

$$I_B = \frac{I_C}{H_{FE}} \tag{9-57}$$

8. Determine the value of base emitter voltage ($V_{BE}$) required for $V_{CE(sat)}$ at $I^1_C$.

9. Calculate $R_{in}$ from

$$R_{in} = \frac{V_{BE(sat)}}{I_B} \tag{9-58}$$

10. The base stabilizing resistor, $R_B$ is small and is usually chosen so that the voltage dropped across it is about

$$\frac{V_{BE}}{2}$$

11. Total Base Circuit Input Resistance $R_{in}^1$ can be found from:

$$R_{in}^1 = R_{in} + R_B \tag{9-59}$$

12. The Base circuit input voltage is determined from:

$$E_{in}^1 = V_{BE(sat)} + I_B R_B \tag{9-60}$$

13. Next, calculate the secondary power of $T_1$:

$$P_{SEC} = E_{in}^1 I_B \tag{9-61}$$

14. Calculate the primary power of $T_1$ using the same formula as in step (2):
$P_{pri} = E_{in}^1 I_B$

15. Calculate the base-drive circuit collector current from

$$I_C = \frac{P_{pri}}{E_{in}} \tag{9-62}$$

16. Choose base-drive transistors using the same criteria of step (4).

17. Find the value of $h_{FE}$ at this collector current (step 6).

18. Calculate the base current $I_B = I_C/h_{FE}$. This is the current sampled by the multivibrator signal conditioning circuit.

In the present example, RCA Type 2N5496 power transistors were selected for $Q_1$ and $Q_2$, and the standard 2N2222A general purpose transistor was used for $Q_3$ and $Q_4$. Two important features in the design of the base-drive and power stages are the selection of proper power transistors and careful transformer design.

## Power Transistor Selection

When selecting transistors for use in DC/DC converters, they must be selected based on the following electrical characteristics:

- Voltage rating
- Power rating
- Cutoff frequency
- Efficiency

In the high-speed converter, the peak value of the collector-to-emitter voltage of each transistor will be equal to twice the supply voltage plus the amplitude of the voltage spikes generated by transient elements. Therefore, the collector-to-emitter breakdown voltage, $V_{CEO}$ of the transistors should be slightly greater than twice the supply voltage (usually an additional 20% is sufficient).

One problem encountered in power supply design is that of voltage spikes.

Turn-on spikes are generated by the sudden charging current taken by filter capacitors in the supply. They can be reduced by using inductive filters or smaller filter capacitors, usually at the expense of efficient interference filtering.

When power is first applied, large filter capacitors may pull 10 times as much, or more current than steady-state input current. Turn-off spikes will be produced when the energy stored in filter inductors discharges into the opened switch (transistor). A damping network across the switch (transistor) will absorb this energy. In some cases diodes may be required to discharge large filter inductors safely.

Although short, the spikes occur while collector voltage is rising and result in instantaneous peak-power dissipations many times steady-state dissipation. Besides reducing converter efficiency, the peaks may over-heat the transistor junctions, degrading reliability. It can also be deduced that if the transistors are not fairly well matched, the spiking and power dissipation may become excessive and the output waveforms asymmetrical.

The first may destroy the transistors and the asymmetry can affect system operation. To ensure sufficient base-drive current to the transistors under worst-case operating conditions, the transistors are over-driven. This improves efficiency by keeping $V_{CE(sat)}$ losses low when a transistor is on, but makes spiking unavoidable.

The spiking is caused by the transistor storage time. One transistor may be driven on before the other turns off resulting in a spike that lasts throughout the storage time. Storage time is directly proportional to the amount of overdrive. Power losses caused by spiking increase with frequency, making it difficult to achieve high efficiency.

Storage time varies from transistor to transistor and even within the same device family. When two transistors with unequal storage times are combined, an asymmetrical square wave which has a DC component results. Careful selection of core material and good transformer design will alleviate spiking.

## Transformer Design

One of the first steps in designing a typical converter is selection of the transformer core. In cross-section, it should be just large enough for the applied voltage to induce the necessary level of flux every half cycle. The window area (the area available for building up the windings) must be of a size that will accommodate all the necessary windings.

The core can be of the pot or toroidal variety. At the frequencies used in high-speed converters, iron cores cannot compete successfully with ferrite cores, either in performance or in economy. Even at the low end of the converter frequency range, ferrite cores are more economical because the iron must be in the form of very thin laminations or tape-wound toroids.

Ferrites also have fairly constant losses. Up to about 40 kHz the losses increase as $f^{0.1}$ where $f$ is the operating frequency. At higher frequencies, the losses increase as $f^{0.6}$. These rates are much lower than would be possible with an iron core.

The core of a transformer is selected on the basis of power-handling requirements, frequency, and temperature of operation. Temperature is an important consideration in the selection of a ferrite core because the Curie temperature for many ferrites is low. (Magnetization is zero above the Curie temperature.)

Another important feature in the core material selection is the desired transformer efficiency. The efficiency can be used to obtain an approximation for the magnetic power to be dissipated, $P_m$. The necessary volume of core material can then be estimated on the basis of the value of $P_m$ and the core loss factor for the chosen material at the operating frequency. The core loss factor is determined from the core-material manufacturer's data sheets.

In the design of the transformer, excessive primary turns should be avoided to assure minimum power dissipation, to assure that the transformer can be made in view of the large wire sizes and the relatively small cores which are usually employed, and to assure that a low value of leakage inductance is maintained. Good balance and close coupling between primary windings is normally achieved by the use of bifilar windings.

Flux density for the output transformer is determined by the usual compromise—the wire size selected on the basis of a 50% duty cycle must be large enough so that power dissipation will be low. If the wire size is inadequate, dissipation will be appreciable and a high transformer-core temperature will result.

Manufacturers tabulate core sizes by the product of window area, $W$, and crossectional area, $A_c$. The transformers in the present PWM power supply are both nonsaturating types. The following procedures and formulas listed in Chapter 4 are used in designing a nonsaturating square wave transformer.

1. Choose square core material by looking at manufacturer's flux density curves. The more square the B-H curve, the less the losses.

2. Calculate the secondary-to-primary turns ratio for each output winding as follows:

$$\frac{N_{sec}}{N_{pri}} = \frac{E_{sec}}{E_{pri}} \tag{9-63}$$

3. Choose the operating frequency.

4. Assuming 1 V per turn, select the number of primary turns. For a first approximation use 20 turns for the primary winding.

5. Then, using this information and the fact that $E_{pri} = E_{Batt} = E_{in}$ calculate the available winding area of the core using the following formula:

$$A_C = \frac{E_{in} \times 10^8}{4fN_p B} \tag{9-64}$$

Where $E_{in}$ is the primary input voltage in volts, $f$ is the frequency of operation in hertz, $A_c$ is the transformer core area in cm$^2$, $B$ is the flux density in Gauss and $N_p$ is the number of primary turns.

6. From step (5) select the core size and bobbin size necessary to provide the required $A_c$ plus some tolerance.

7. Calculate the number of turns for each output winding (secondary winding) using $N_{sec} = (E_{sec}/E_{pri})N_{pri}$.

8. Select the wire size of each winding based on the current it has to carry. (For the secondary ($N_{sec}$), use 100% duty factor and for the primary ($N_p$) use a 50% duty factor.)

9. Calculate the total cross-sectional area of the wire. Make sure that depending on which manufacturer's core you have selected, you keep the same units of measure (square inches or square centimeters).

NUMBER
WINDINGS

Total wire area = $\sum$ (Number of turns)(Cross-Sectional Area of wire)$\times$

(mean core length) $\tag{9-65}$

10. If the area calculated in step (9) is less than the available winding area of the bobbin of step (6), the first cut design is satisfactory. If not, choose a large core size and recalculate the wire area.

11. Calculate winding resistance by first calculating mean turn length from $I_m$ = (core box O.D. $-$ core box I.D.) + 2 (core box height) so that winding resistance $R_w$ becomes

$$R_w = [N(L_W/12) \text{ ohms}/1000 \text{ ft}]/1000 \tag{9-66}$$

The iron and copper losses are then calculated to estimate the efficiency of the transformer.

12. Calculate the copper losses in each winding from the winding resistance and current in each winding

$$P_L = I^2 R \tag{9-67}$$

13. Obtain the core from the manufacturer's data sheet. The sum of the copper losses should approximate 20% of the input power to transformer.

This provides a first cut at the transformer design. The procedure followed in all power supply design is to use this as a starting point. Then with the circuit operating and debugged, adjust each winding as necessary to provide the optimum voltage/current output for the particular design.

In the memory circuit discussed here, the base-drive circuit is a low-level push-pull circuit using two 2N2222A transistors in the switching mode to drive a nonsaturating pot core. Two 510 ohm resistors are used to establish the base drive current to the required levels.

The previously developed signal from the astable multivibrator and CMOS signal conditioning network ensures that a definite dead time exists between the turn-off of one of the 2N2222A's and the turn-on of the other—this ensures that both transistors cannot be on simultaneously which would decrease efficiency. An RC snubber is used across the primary of the transformer winding during the programmed off time.

The base circuit transformer uses a pot core for maximum noise reduction, size and efficiency of operation. It is center-tapped on both its primary and secondary windings and performs a stepdown function. The primary current of this transformer is calculated to be 30 mA at the center voltage of 12 V. The secondary current which is essentially the base current of the power drive stage, is 1428 mA and the secondary voltage is 2.2 V.

The power stage consists of two RCA 2N5496 power transistors operated in push-pull configuration and driving a nonsaturating pot core. The 2N5496 power transistors were chosen because of their fast rise and falltimes, short storage time, $BV_{ceo}$, $H_{FE}$ and $I_C$ ratings.

The primary winding sustains a current of 1.5 V at a voltage of 12 V. An RC snubber circuit is connected across the primary winding. As with the base drive circuit, because of the pulse shaping technique employed there is a programmed dead time in which neither transistor is on and thus cross-conduction does not exist. The primary winding is center tapped as are the three secondary windings.

## The Output Circuits

The four output circuits utilize well established and proven regulation techniques employing popular IC regulators. The 5 V, 2 A output uses a LM305 switching regulator to minimize power dissipation. Two other outputs use fixed three-terminal regulators and the −5 V, 5 mA output uses a $\mu$A741 op amp to develop the regulated output voltage.

Figure 9-64 is the schematic diagram for the complete memory power supply. Efficiency of this circuit was 80% at full load, with maximum output ripple of

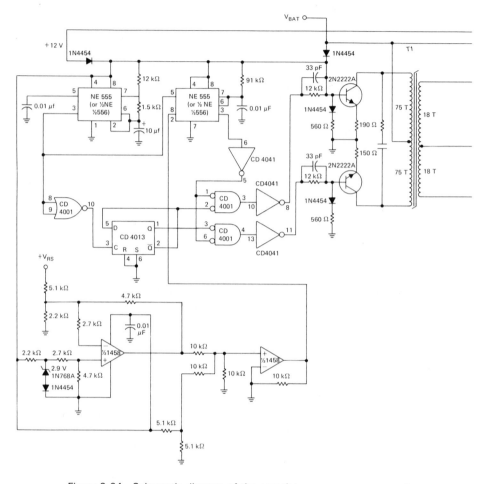

Figure 9-64   Schematic diagram of the complete memory power supply.

less than 100 mV p-p on each output. Additionally over the temperature range
of 0 to 70°C each output varied by less than 3%.

## HIGH SPEED MATRIX PRINTER CURRENT DRIVEN HALF BRIDGE SUPPLY[2]

Table 9-6 summarizes the voltage and current requirements for a high speed
matrix printer whose electromechanical print head and drive motors demand

---

[6]LaBach, Frederick A., *Implementing Switchmode Power Supply Design*, Quality in Electronics
Proceedings, 1983. Used with permission.

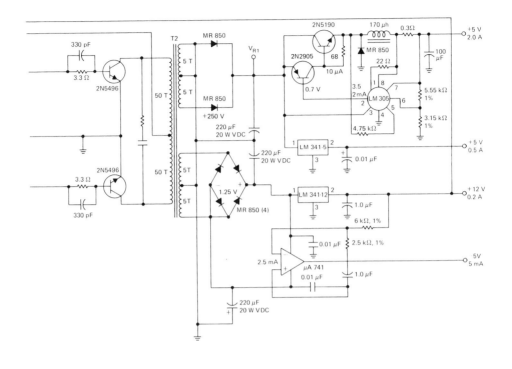

Figure 9-64  *(Continued)*

large pulse currents and high print quality demands close operation on all outputs under all operating conditions.

The general requirements for the supply include efficiency of 70% or better, short circuit protection on all outputs, independent voltage adjustment and overvoltage protection for the +5 V logic, and the ability to operate off any line voltage from 90 to 132 V AC and from 180 to 264 V AC. The power supply must also operate in a 50°C maximum ambient and meet all applicable UL, CSA, VDE, and Telecom (Australia) safety standards.

The complete current-driven half bridge supply that was designed to meet these requirements is shown in Figure 9-65 in schematic form. The design will be discussed by subdividing the complete circuit into its building block constit-

## Table 9-6 Power Supply Output Requirements.

| Output | Usage | Required Regulation | Max p-p Noise And Ripple | Minimum Current | Maximum Current |
|--------|-------|---------------------|--------------------------|-----------------|-----------------|
| +5V | Logic | 2% | 100mV | 1.0A | 5A |
| −12V | Logic | 5% | 50mV | 0 | .10A |
| +12V | Logic | 5% | 50mV | 0 | .10A |
| +12V | Analog | 5% | — | 0.5 | 3.0A[DC] + 2.8A[AC] |
| +30V | Analog | 5% | — | 0 | 1.5A[DC] + 2.0A[AC] |
| +60V | Analog | 5% | — | 0 | 2.0A[DC] + 3.1A[AC] |

Note: Total Power requirement not to exceed 200 W

uent components. Figure 9-66 shows a commonly used line input rectification and filtering circuit. Jumper J1 changes the diode-capacitor network from a full-wave to a voltage doubling rectifier. Thus either a 115 V AC or a 230 V AC input will produce ~320 V DC across the capacitors. Resistors R1 and R2 provide a bias current sufficient to overcome capacitor leakage current and ensure an equal voltage drop across each capacitor. $R_{th}$ is a thermistor with a cold resistance of 10Ω; $R_{th}$ limits the inrush current at power on, protecting the input diodes and capacitors. Components T1, C1, and C2 act as a line filter to meet FCC conducted emission requirements.

Figure 9-67 is a schematic diagram of a front end buck regulator that uses two parallel transistors. The effective case-to-heat sink thermal resistance of the pair is half that of a single transistor, while the pair's drive requirements can be optimized using less control circuitry. This allows the FET pair to run cooler and more efficiently for the same cost as an equivalent single FET. The output of the transistor pair goes to $B^+$ and the FET gates are connected to primary return to allow the buck regulator FETs to be driven directly from the control circuitry without voltage isolation which would not have permitted the wide range of pulse widths needed by this stage.

T4 is an off-the-shelf current sense transformer used to guard against destructive overcurrents in the front end.

Figure 9-68 depicts the half-bridge portion of the supply, which normally operates from ~72 V DC. Q5 and Q6 drive the main power transformer T2 with a combined duty cycle of 98% leaving a 2% dead time to prevent power loss and component stress due to overlapping on times. C23 and R29 form a snubber to protect the voltage sensitive FETs from high voltage spikes. C23 should be a film-foil polypropylene capacitor; a metalized film type could fail due to the large voltage slew across it. CR19 and C21 also protect the FETs by forming a clamp which absorbs the current from the buck regulator during the half-bridge dead time and returning it during the on time through R28. Since C21 must handle high peak currents (up to 6 A) a metalized polypropylene capacitor was chosen to provide long life while conserving board space.

SCR *Q7* and its associated components form an overvoltage crowbar operating at ~95 V. This protects the half-bridge FETs and prevents severe overvoltages on the secondaries in the event of a failure in the secondary rectifiers, feedback or control circuitry, or buck regulator. Circuit impedances are kept low and *C22* is added to ensure against false triggering due to noise spikes.

The front end control circuitry for the power supply is shown in Figure 9-69. The heart of this circuit is a pair of SG3525A pulse width modulators whose output characteristics are ideally suited to driving power FETs while providing the necessary oscillators, amplifiers, and voltage references.

*U3*'s primary purpose is to operate continuously at maximum duty cycle (98%) and drive the half-bridge section. The 3525A's totem-pole outputs have sufficient current capability to directly drive a small signal transformer (*T5*) which then drives the half-bridge FETs. *R26* and *C20* form a snubber to curtail ringing on the secondary of *T5* and prevent glitching of the FETs during switching. The remaining function of *U3* is to provide a +5 V reference used by the buck regulator's current limit and soft-start circuitry; *U3*'s other functions are disabled, and its oscillator section is controlled by sync pulses from *U2*.

*U2* directly drives the buck regulator FETs while also incorporating the feedback amplifier, soft-start control, and master oscillator for the supply. *R23* and *C18* set the oscillator frequency at ~50 kHz; this runs the single-ended buck regulator at 50 kHz while the push-pull attribute of the half-bridge runs the switching power transformer *T2* at 25 kHz. It should be noted here that since the 3525A's two outputs are enabled alternately and not in unison, the buck regulator's parallel FETs are also switched on one at a time and not simultaneously. This does not compromise any of the advantages of the parallel arrangement while allowing the FETs to be driven directly from *U2*'s outputs.

Since the gates of all the front end power FETs can be damaged by voltages exceeding ±20 V, the supply voltage for the output stages of *U2* and *U3* must be limited. This function is performed by *U1*, an inexpensive 12 V regulator. *C13* is necessary to supply the high peak currents drawn by the capacitive FET gates.

*U2*'s +5 V reference is tied directly to the feedback amplifier's noninverting input. *R13-15* and *C10-12* form a standard phase/gain compensation network with values selected to provide good transient response while protecting the supply from oscillation.

*R21*, *R22*, and *C16* are connected to the soft-start input of *U2*. This circuit reduces stress on the supply's components by gradually increasing the buck regulator's pulse width from zero during turn on and also limits the maximum pulse width to ~55%. This is sufficient to operate the supply in brownout conditions while allowing the current limit circuit (which operates by discharging both *C16* and *C12*) to restrict *U2*'s outputs rapidly. This is also the reason that *CR17* is used to limit the voltage across *C12*, the output of the feedback amplifier.

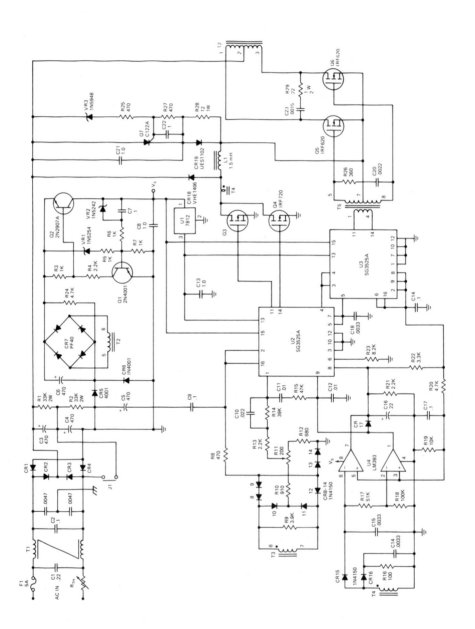

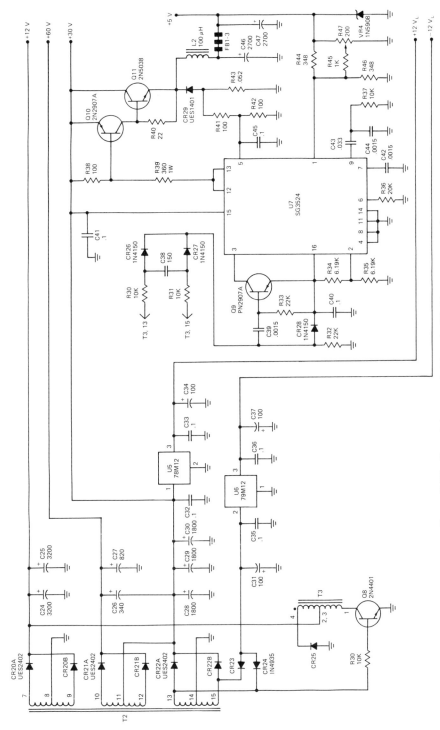

Figure 9-65 Complete current-driven half-bridge power supply.

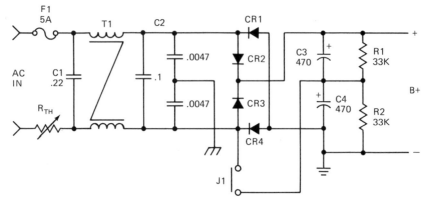

Figure 9-66   Line input circuit.

The current limiting portion of the control circuitry is shown in Figure 9-70. *T4*, mentioned in the buck regulator section, is a single wire running through a toroid with a 100 turn secondary wound on it. It was chosen for its low cost and high noise immunity. *CR16* and *R16* set the V/A relationship at 1 V per A of current flowing through the buck regulator FETs during their on-time. *C14* filters the leading edge spike caused by the reverse recovery current of the buck regulator catch diode, *CR18*. *CR15*, *C15*, *R17*, and *R18* form a peak detector with a fast rise, slow decay characteristic so that the voltage across *C15* accurately represents the current delivered by the buck regulator inductor *L1* at any point in time. If this voltage exceeds a given reference, the top comparator of *U4* acts to discharge the soft-start capacitor *C16*. This requires several cycles to complete and determines the maximum continuous output power of the supply. If the current through *L1* increases faster than *C16* can be discharged (i.e., a short circuit on the secondaries) the voltage across *C15* will increase until the input of the bottom comparator exceeds the given reference. At this point *U4* will immediately discharge *C12*, the output of the feedback amplifier, and shut

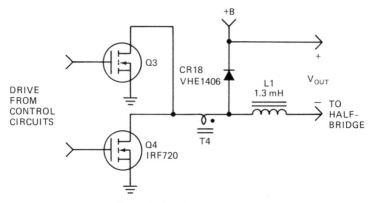

Figure 9-67   Buck regulator.

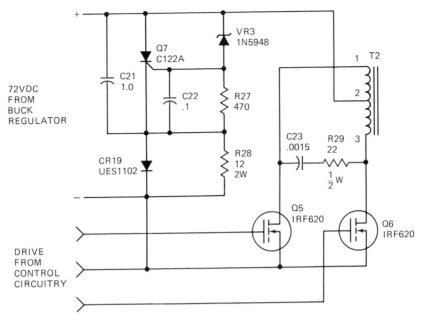

Figure 9-68   Half-bridge.

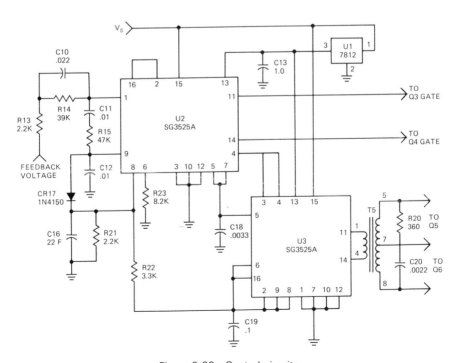

Figure 9-69   Control circuitry.

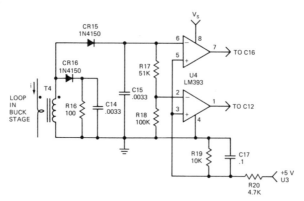

Figure 9-70   Current limit circuit.

off *U2*'s outputs on a cycle by cycle basis. This determines the peak output power of the supply; the ratio of *R17* to *R18* sets this point at about 1.5 times the maximum continuous output. *R19* and *R20* determine the current limit reference so that the maximum continuous output power is 220 W while the peak power is 330 W.

Many power supplies in this wattage range use a small line frequency transformer to provide the necessary low voltage to run their control circuitry; the circuit of Figure 9-71 was developed as a less expensive alternative. *Q1* and *Q2* are initially off when AC power is applied to the printer, and a small current from the B+ voltage flows through *R2* and begins to charge *C5*. When *C5*'s voltage exceeds the Zener voltage of *VR1* ($\sim 30$ V), *Q1* is switched on which switches *Q2* on. Current is then dumped from *C5* through *CR5*, bridge *CR7*, and *Q2* into the control circuitry as the supply voltage $V_s$. *VR2*, a 12 V Zener, holds *Q1* on after *C5* begins discharging and shuts off *Q1* and *Q2* when $V_s$ drops below 12 V. Before *C5* discharges to that point, a 20 V winding on the main switching transformer *T2* begins to feed voltage back into this circuit, adding

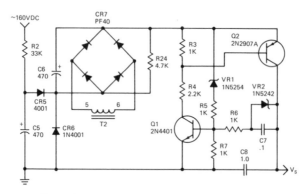

Figure 9-71   Start-up and low voltage supply.

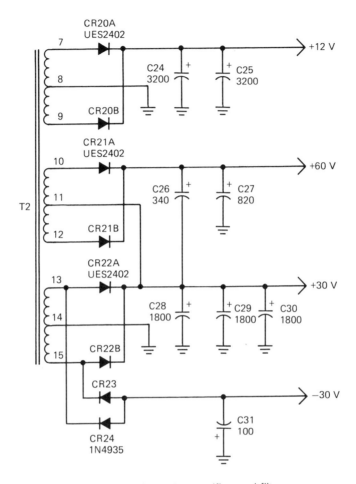

Figure 9-72 Secondary rectifiers and filters.

it to *C5*'s voltage. Eventually *C5* discharges completely and the control circuit's power is supplied entirely from *T2*. In the event of a continued overcurrent or short on the supply's outputs, the voltage across *T2* will drop, *Q1-2* will switch off, and the whole supply will shut down and restart.

Figure 9-72 shows the secondaries of *T2* and its associated rectifiers and capacitors. Since the logic ±12 V and +5 V are derived via post-regulators, only the analog 12 V, 30 V, and 60 V are generated here. To simplify the winding of *T2* and reduce the voltage requirement of *CR21*, the +60 V was derived by adding two 30 V outputs. *C24* through *C30* are high ripple current capacitors capable of handling the input current ripple from *T2* plus the maximum AC current created by the motors and print hammers. *C31* is a small capacitor used only to filter the −30 V for use by the −12 V post-regulator.

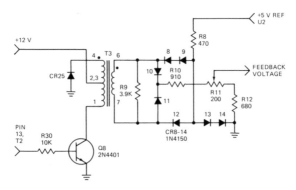

Figure 9-73   Isolation and feedback.

The secondary to primary voltage feedback circuit is shown in Figure 9-73. *Q8*'s drive signal comes from the main switching transformer's 30 V winding. *Q8* then switches the feedback transformer *T3* in unison with the power transformer *T2*; when *Q8* is off, *T3* sources current through *CR25*. *Q8* was selected for its low saturation voltage and linear operation over temperature.

*T3*'s output is a square wave with amplitude equal to the +12 V output voltage. Diodes *CR8* through *CR12* form a bridge which produces a replica of the DC voltage on *T3*'s primary. The extra series diode in the bridge compensates for the voltage drop created by *CR25*, and *CR13-14* along with current sourcing resistor *R8* compensate for the drop created by the bridge. *R9* helps to linearize the output of *T3*, while *R10*, *R11*, and *R12* provide the voltage adjustment and centering for feedback into the primary control circuitry.

Figure 9-74 shows the linear 12 V post-regulators for the logic. *U5* and *U6* are inexpensive three terminal devices with internal thermal and overcurrent shutdown circuits. *C32* through *C37* decouple noise from the devices.

The circuit for the 5 V switching post-regulator appears in Figure 9-75. This circuit features an SG3524 pulse width modulator, similar to the devices used in the front end but with a current limit comparator instead of a soft-start control. An important aspect of this switching regulator is that it is synchronized with

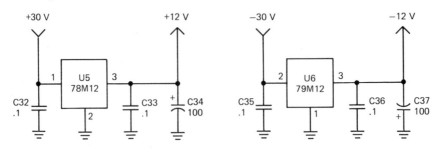

Figure 9-74   ±12V Post-regulators.

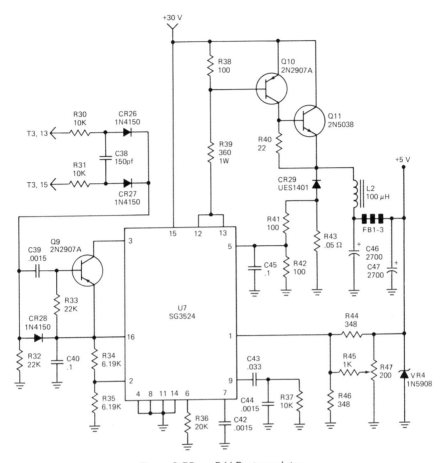

Figure 9-75   +5 V Post-regulator.

the front end oscillator. *R30*, *R31*, and *C38* take the complimentary square waves from *T2*'s 30 V winding and integrate the edges. *CR26* and *CR27* combine these signals while *CR28* clamps them to *U7*'s + 5 V reference. This creates a 50 kHz sync train which is fed through *C39* into *Q9* which drives the sync input of the SG3524. In the event of a loss of sync (i.e., a momentary overcurrent shutting back the primary) *U7* will continue to run the +5 V regulator at a slightly lower frequency until power is lost or sync is restored. *R36* and *C42* set this frequency at 40 kHz.

## SOURCES OF FURTHER INFORMATION

1. Silicon General Application Notes in 1986 Product Catalog.
2. Motorola Semiconductor Products Inc. *Linear/Switchmode Voltage Regulator Handbook* HB206, Rev. 2, 1987.

3. Unitrode Corporation Application Notes in 1984-85 Semiconductor Databook.

4. National Semiconductor Corporation *Linear Applications Databook*, 1986.

5. Linear Technology Corp. *Linear Applications Handbook*, 1986.

6. *MOSPOWER Applications Handbook*, Siliconix, Inc. 1984.

7. RCA Power MOSFETs Databook, 1986.

8. Motorola Semiconductor Products, Inc., Power MOSFET Transistor Data, 1986, DL135, Rev. 1.

9. *High-Frequency Switching Power Supplies, Theory and Design* by G.C. Chryssis, McGraw Hill Book Co., 1984.

10. *DC-DC Switching Regulator Analysis* by Daniel M. Mitchell, McGraw Hill Book Co., 1987.

11. *Transformer and Inductor Design Handbook* by Col. Wm. T. McLyman, Marcel Dekker, Inc., 1978.

12. *Magnetic Core Selection for Transformers and Inductors* by Col. Wm. T. McLyman, Marcel Dekker, Inc., 1982.

13. *Switchmode Power Conversion, Basic Theory and Design* by K. Kit Sum, Marcel Dekker, Inc., 1984.

14. *Modern DC to DC Switchmode Power Converter Circuits* by Rudolf P. Severns and Gordon E. Bloom, Van Nostrand Reinhold, Inc., 1985.

15 *Switching and Linear Power Supply, Power Converter Design* by Abraham I. Pressman, Hayden Book Company, 1977.

16. *Transformers for Electronic Circuits* by Nathan R. Grossner, McGraw Hill Book Co., 2nd Edition, 1983.

17. *Advances in Switched-Mode Power Conversion* by Dr. R.D. Middlebrook and Dr. Slobodan Cuk, Teslaco.

18. *Power Integrated Circuits*, Edited by Paolo Antognetti, McGraw-Hill Book Company, 1986.

19. *Power Electronic Converters*, by Guy Seguier, McGraw-Hill Book Company, 1987.

20. Unitrode Power Supply Design Seminar, SEM-500, 1986.

21. Proceedings of High Frequency Power Conversion Conferences.

22. Proceedings of Power Electronics Conferences.

23. Proceedings of Power Conversion International (PCI) Conferences.

24. Proceedings of the Applied Power Electronics Conferences (APEC).

25. Proceedings of the Power Conversion and Intelligent Motion Conferences.

26. Proceedings of the SATECH Conferences.

# Index